Electric Machinery and Drives

IEEE Press
445 Hoes Lane
Piscataway, NJ 08854

Electric Machinery and Drives

An Electromagnetics Perspective

Nabeel A. O. Demerdash
Marquette University
Milwaukee, WI, USA

JiangBiao He
University of Tennessee
Knoxville, TN, USA

Hao Chen
Zhejiang University
Hangzhou, China

IEEE Press Series on Power and Energy Systems
Ganesh Kumar Venayagamoorthy, Series Editor

IEEE*Press*
WILEY

Published by John Wiley & Sons, Inc., Hoboken, New Jersey.
Published simultaneously in Canada.

Library of Congress Cataloging-in-Publication Data

Names: Nabeel A.O. Demerdash, JiangBiao He, and Hao Chen.
Title: Electric machinery and drives : an electromagnetics perspective /
Nabeel A. O. Demerdash, Marquette University, Milwaukee, WI, USA;
JiangBiao He, University of Tennessee, Knoxville, TN, USA;
Hao Chen, Zhejiang University, Hangzhou, China.
Description: First edition. | Hoboken, New Jersey : Wiley, [2025] |
Includes index.
Identifiers: LCCN 2024008446 (print) | LCCN 2024008447 (ebook) | ISBN
9781119985723 (hardback) | ISBN 9781119985532 | ISBN 9781119985549
(epub)
Subjects: LCSH: Electric machinery.
Classification: LCC TK2000 .C486 2025 (print) | LCC TK2000 (ebook) | DDC
621.31/042–dc23/eng/20240319
LC record available at https://lccn.loc.gov/2024008446
LC ebook record available at https://lccn.loc.gov/2024008447

Cover Design: Wiley
Cover Images: © natatravel/Shutterstock, © Engineer studio/Shutterstock, © Phonlamai Photo/Shutterstock, © Fasttailwind/Shutterstock, © TebNad/Adobe Stock Photos

Set in 9.5/12.5pt STIXTwoText by Straive, Chennai, India

In Memory
of
My Dear Parents Mr. Aly O. Demerdash and Mrs. Aziza M. Demerdash
and

In Gratitude
to
The Support of My Dear Spouse Mrs. Esther A. Demerdash and My Dear Children Dr. Yvonne A. N. A. Demerdash-Stubbs, Dr. Omar N. A. Demerdash, Dr. Nancy N. A. Demerdash, and All My Former Students.

Contents

About the Authors

Nabeel A. O. Demerdash received the BSc EE degree (distinction with first-class Hons.) in electrical engineering from Cairo University, Giza, Egypt, in 1964, and the MS and PhD degrees in electrical engineering from the University of Pittsburgh, Pittsburgh, PA, USA, in 1967 and 1971, respectively.

From 1968 to 1972, he was a development engineer in the Large Rotating Apparatus Development Engineering Department, Westinghouse Electric Corporation, East Pittsburgh. From 1972 to 1983, he was an assistant professor, an associate professor, and then a professor in the Department of Electrical Engineering, Virginia Polytechnic Institute and State University, Blacksburg, VA, USA. From 1983 to 1994, he was a professor in the Department of Electrical and Computer Engineering, Clarkson University, Potsdam, NY, USA. Since 1994, he has been a professor with the Department of Electrical and Computer Engineering, Marquette University, Milwaukee, WI, USA, where he was the department chair from 1994 to 1997. He was the director of the SEMPEED Consortium at Marquette University. He is the author or co-author of over 150 papers published in various IEEE TRANSACTIONS and magazines. His current research interests include electric machines and drives, electromechanical propulsion and actuation, computational electromagnetics in machines and drives, and fault diagnostics and modeling of motor-drive systems.

Prof. Demerdash is an IEEE Life Fellow and the recipient of the 1999 IEEE Nikola Tesla Technical Field Award for pioneering contributions to electric machine and drive system design using coupled finite-element and electrical network models. He has also received two 1994 Working Group Awards and two 1993 Prize Paper Awards from the IEEE Power and Energy Society (PES) and its Electric Machinery Committee, and a 2012 Prize Paper Award from the PES, as well as a 2012 Prize Paper Award from the IEEE Industry Applications Society Electric Machines Committee and the 2015 Marquette University Lawrence G. Haggerty Faculty Award for Research Excellence.

JiangBiao He is an associate professor in electrical engineering at the University of Tennessee, Knoxville, TN, USA. He was an assistant professor and then associate professor in electrical engineering at the University of Kentucky between 2019 and 2024. He previously worked in industry, most recently as a lead engineer at GE Global Research, Niskayuna, New York. He received the PhD degree in electrical engineering from Marquette University, USA. His research interests focus on motor-drive systems and power electronic converters for various emerging applications. He has authored and co-authored over 150 technical articles and 10 patents in this area. Dr. He is an IEEE Senior Member and has served as an editor or associate editor for multiple prestigious IEEE journals. He also served in the organizing committees for numerous IEEE international conferences and has been an active member of multiple IEEE technical committees. He is a recipient of several achievement awards, including the 2018 GE Whitney Technical Excellence Award, 2019 AWS Outstanding Young Member Achievement Award with IEEE Industry Applications Society, and the 2023 Faculty Excellence in Research Award at the University of Kentucky.

Hao Chen received the BSc degree in electrical engineering from the School of Electrical Engineering, Beijing Jiaotong University, China, in 2012, and the PhD degree in control science and engineering from the School of Automation, Beijing Institute of Technology, Beijing, China, in 2019. From 2019 to 2021, he was a Postdoctoral Research Fellow with the School of Electrical and Electronic Engineering, Nanyang Technological University, Singapore. From 2022 to 2023, he was a Researcher with the Department of Electrical Engineering, Chalmers University of Technology, Gothenburg, Sweden. He is currently an Associate Professor with the College of Electrical Engineering, Zhejiang University, Hangzhou, China. His research interests include the design and optimization of electric machines, power electronic drives, and motor control.

Preface

There is a widespread and almost universally accepted realization that climate change and associated global warming, caused by carbon emissions attributed to energy generation from petroleum, coal, natural gas, and other forms of hydrocarbons, are leading to major efforts toward electrification of all modes of transportation on land, sea, and air. This includes automobiles, buses, trucks, trains, and other forms of transportation. This is besides the widespread desire and urge to move toward renewable energy generation from wind turbine generators and solar energy systems of various sorts. This is also in addition to robotic systems that are at the heart of automation of manufacturing processes. All these forms of new energy conversion and generation systems rely heavily on knowledge and deep understanding of the principles of modeling, performance analysis, and design of electric machinery and adjustable-speed drives (ASDs). That is, motors and generators, as well as their associated control systems consisting of power electronic components. These electric motors, generators, and their control systems/drives are the subjects of the various chapters of this textbook, which is intended for a wide range of electrical/mechanical/automotive engineering senior/junior, first- and/or second-year graduate MS and PhD-level students, and engineers working in the related domains. Accordingly, this textbook includes topics and subjects on rotating electric machinery and drives, as well as associated electromechanical energy conversion concepts and devices that are hereby delineated below.

We start with Chapter 1 on electric circuit notations and elementary circuit concepts including AC phasor notations. This is in addition to linear algebra matrix and vector notations, as well as time-domain instantaneous power and AC complex-power, real and reactive, notations. In Chapter 2, we cover fundamentals of power electronic conditioning and processing. This includes the conventional silicon and the emerging wide bandgap semiconductor devices, DC choppers of the buck and boost DC–DC type, as well as three-phase inverters.

Meanwhile, in Chapter 3 we review magnetostatic and magnetic circuit concepts and notations. This is in addition to the review of the fundamentals of electromechanical energy conversion, including magnetomotive forces (MMFs) and consequent concepts of production of electromechanical torque in both AC and DC machines. Accordingly, in Chapter 4 we address the mechanical dynamics of motor-load and generator-prime mover systems. This includes the modeling and approaches to simulation of motor-load starting transients.

In Chapter 5, we review AC and DC winding layouts, induced electromotive forces (EMFs), MMFs, in such windings, and analytical relationships between MMFs and flux density waveforms (distributions) in airgaps of such machines. This is in addition to analytical relationships between such MMFs, flux density waveforms, and resulting electromechanical torques. Meanwhile, we present in Chapter 6 the poly-phase AC induction machine, all the way till covering its scalar torque–speed control and its vector torque-speed control in induction motor drives. Accordingly, we present in Chapter 7 the poly-phase AC synchronous machine, both as a mainline generator of AC power and as a motor in special, more limited applications, all the way till covering its scalar torque–speed control, and its vector torque–speed control in synchronous motor-ASD systems. Furthermore, in Chapter 8, we present the conventional/legacy-type brush-commutator DC machine and the more modern permanent magnet (PM) brushless DC machine. This includes modern power electronic based torque–speed control of brush-commutator legacy DC motors and PM brushless DC motors.

In this textbook, Chapter 9 is dedicated to the more advanced graduate-level state-space modeling of three-phase synchronous machines, including the all-important full effects of rotor saliency, in the natural ABC frame-of-reference, and in the two-reaction (Park's transformation) *DQ*0 frame of reference. This material includes the full details of the algebra and calculus of the *DQ*0 transformation models and the resulting phasor-space vector diagrams. This also includes full details on the modeling and computation of initial conditions for a given load condition, including no-load. Meanwhile, in Chapter 10 we give graduate-level students and engineers full details of the derivation of three-phase induction machines in the natural *ABC*–*abc* frame of reference, as well as three forms of *DQ*0–*dq*0 frame-of-reference transformations and corresponding/resulting three state-space models. This chapter includes further details on the stator and rotor leakage inductances as key parameters in the modeling and performance analysis of induction motors. Also, in this chapter, we revisit modern torque–speed vector control in light of the advanced induction machine models developed here in Chapter 10. In this textbook, Chapter 11 is the subject of single-phase induction motors and other special motors, such as split-phase AC motors, and capacitor start motors, as examples of such special machines. Meanwhile, in the final chapter of this textbook, Chapter 12, we review the major emerging applications in electric machinery and drives, such as electric transportations, robotics, drones, wind turbine powertrains, and other applications. Additionally, prospective innovative technical trends in electric machinery and drives are discussed, such as ultra-compact integrated motor-drive systems, axial-flux machines, superconducting machines, as well as additive manufacturing of electric machinery and drives.

Finally, throughout the chapters of this textbook, we give ample numerical examples with solutions, as well as an adequate amount of problems for use by instructors/professors as homework and/or projects for students, for purposes of assignments throughout a semester or two, whatever the class level is, junior, senior, or graduate level.

Nabeel A. O. Demerdash
Milwaukee, WI, USA
JiangBiao He
Knoxville, TN, USA
Hao Chen
Hangzhou, China

About the Companion Website

This book is accompanied by a companion website:

www.wiley.com/go/demerdash/electricmachineryanddrives

The website includes:

- Animations

1

Electric Circuit Notations and Elementary Concepts

In this introductory chapter, we start by defining the electric circuit notations and elementary circuit concepts, including symbolisms used in representation of time-domain and AC phasor variables, such as voltage and currents, used throughout Chapters 1–12 of this textbook. These include multivariable arrays, the so-called vectors in matrix linear-algebra parlance, as well as concepts of time-domain instantaneous power and complex power in AC phasor formulation of electric circuits.

A basic and fundamental complex-algebra concept to all that will follow is the well-known Euler formulation that yields the following identities:

$$e^{j(\omega t+\phi)} = \cos(\omega t + \phi) + j\sin(\omega t + \phi) \tag{1.1}$$

$$\cos(\omega t + \phi) = \mathrm{Re}[e^{j(\omega t+\phi)}] \tag{1.2}$$

and

$$\sin(\omega t + \phi) = \mathrm{Im}\left[e^{j(\omega t+\phi)}\right] \tag{1.3}$$

With these identities, Eqs. (1.1)–(1.3), we are ready to discuss frequency-domain root mean square (RMS) phasor representation of time-domain AC voltages and currents.

1.1 Frequency-Domain RMS Phasor Representation of Time-Domain AC Voltages and Currents

Here, lower-case variables (symbols) such as v, i, e, and p stand for time-domain voltages, currents, electromotive forces (emfs), and real powers, respectively. Also here, upper-case variables (symbols) such as, V, I, E, and P stand for RMS magnitudes (absolute values) of voltage phasors, current phasors, emf phasors, and average real power in phasor complex power computations, respectively.

Electric Machinery and Drives: An Electromagnetics Perspective, First Edition.
Nabeel A. O. Demerdash, JiangBiao He, and Hao Chen.

Companion website: www.wiley.com/go/demerdash/electricmachineryanddrives

Meanwhile, variables (symbols) such as $\overline{V}$, $\overline{I}$, $\overline{E}$, and $\overline{S}$ stand for RMS phasor complex form voltages, currents, emfs, and complex power (real and reactive) in phasor complex power computations, respectively.

Accordingly, in steady-state AC circuit analysis, one can write the following:

$$v(t) = \sqrt{2}V\cos(\omega t + \phi) \tag{1.4}$$

where V is the RMS AC voltage magnitude (absolute value), $\omega = (2\pi f)$ is the angular frequency in electrical radians per second, f is the AC frequency in Hertz, t is the time in seconds, and ϕ is the phase angle of the voltage signal.

Hence, based on the Euler identities given earlier, Eqs. (1.1)–(1.3), one can rewrite $v(t)$ as follows:

$$v(t) = \sqrt{2}V\left\{\mathrm{Re}\left[e^{j(\omega t+\phi)}\right]\right\} \tag{1.5}$$

or

$$v(t) = \sqrt{2}V\mathrm{Re}\left[e^{j\omega t}\cdot e^{j\phi}\right] \tag{1.6}$$

In AC phasor form, the term $(e^{j\omega t})$ is common to all voltage, current, and other signals. Hence, all the information that is needed is in terms (V) and $(e^{j\phi})$. Hence, in RMS phasor notation, the voltage, $\overline{V}$, can be written as follows in exponential phasor form:

$$\overline{V} = V(e^{j\phi}) \tag{1.7}$$

or in Cartesian coordinate polar and rectangular complex forms as follows:

$$\overline{V} = (V\angle\phi) = (V\cos\phi) + j(V\sin\phi) \tag{1.8}$$

Similarly, for an instantaneous steady-state time-domain, current, $i(t)$, given by the following equation:

$$i(t) = \sqrt{2}I\cos(\omega t + \psi) \tag{1.9}$$

The corresponding Euler formulation gives

$$i(t) = \sqrt{2}I\left\{\mathrm{Re}\left[e^{j(\omega t+\psi)}\right]\right\} \tag{1.10}$$

and for the corresponding RMS phasor notation, the current, $\overline{I}$, can be written as follows:

$$\overline{I} = I(e^{j\psi}) \tag{1.11}$$

or, in Cartesian coordinate polar and rectangular complex forms, the current, $\overline{I}$, can be written as follows:

$$\overline{I} = (I\angle\psi) = (I\cos\psi) + j(I\sin\psi) \tag{1.12}$$

In this text, $\overline{V}^*$ is the conjugate of the phasor $\overline{V}$ and $\overline{I}^*$ is the conjugate of the phasor $\overline{I}$. Therefore,

$$\overline{V}^* = \text{Conj}(\overline{V}) = (V\angle - \phi) = (V\cos\phi) - j(V\sin\phi) \tag{1.13}$$

and

$$\overline{I}^* = \text{Conj}(\overline{I}) = (I\angle - \psi) = (I\cos\psi) - j(I\sin\psi) \tag{1.14}$$

1.2 Time-Domain and RMS Frequency-Domain Power Concepts Using Consumer System Formulation and Notations

Consider the two-terminal "system or device" shown in Figure 1.1. The current i (or $\overline{I}$) is taken to be flowing in a positive orientation when flowing into the terminal designated with a positive voltage polarity, v (or $\overline{V}$). In such a case, the instantaneous input power, $p(t)$, is given by the following:

$$p(t) = vi = v(t)i(t) \tag{1.15}$$

where a positive $p(t)$ means that real power (watts) is being consumed and a negative $p(t)$ means that real power (watts) is being generated.

Meanwhile, in phasor frequency-domain, the complex power, $\overline{S}$, is given by the following:

$$\overline{S} = \overline{V}\cdot\overline{I}^* = P + jQ \tag{1.16}$$

where P is the real power in watts and Q is the reactive power in vars, and once again, P is positive means that watts is being consumed (as in a motoring mode) and P is negative means that watts is being generated (as in a generating mode), and Q is the reactive power in vars, in which Q is positive means that vars is being consumed (as in inductive loads) and Q is negative means that vars is being generated (as in capacitive loads). The reader is urged to examine the complex powers, $\overline{S}_1$, $\overline{S}_2$, $\overline{S}_3$, and $\overline{S}_4$, associated with the current phasors, $\overline{I}_1$, $\overline{I}_2$, $\overline{I}_3$, and $\overline{I}_4$, respectively, relative to the terminal voltage phasor, $\overline{V} = V\angle 0°$, associated with the two-terminal system (or device), shown in Figure 1.1.

In the Consumer System Notation:

$$p(t) = vi = v(t)i(t)$$

Positive $p(t)$ means watts is consumed (load/motoring).
Negative $p(t)$ means watts is generated (generator/source).

Now, for a multiterminal (or port), n, system (or device), the voltages can be written in matrix/array (vector) form in the time-domain, $\underline{V}$, or phasor frequency-domain, $\underline{\overline{V}}$, respectively, as follows:

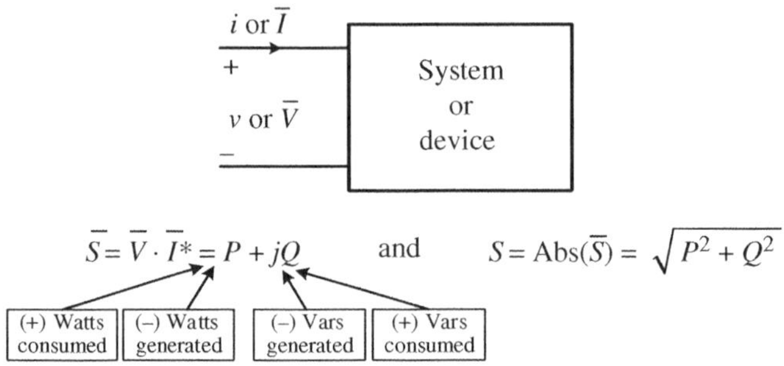

The complex power $\overline{S}$ plane:

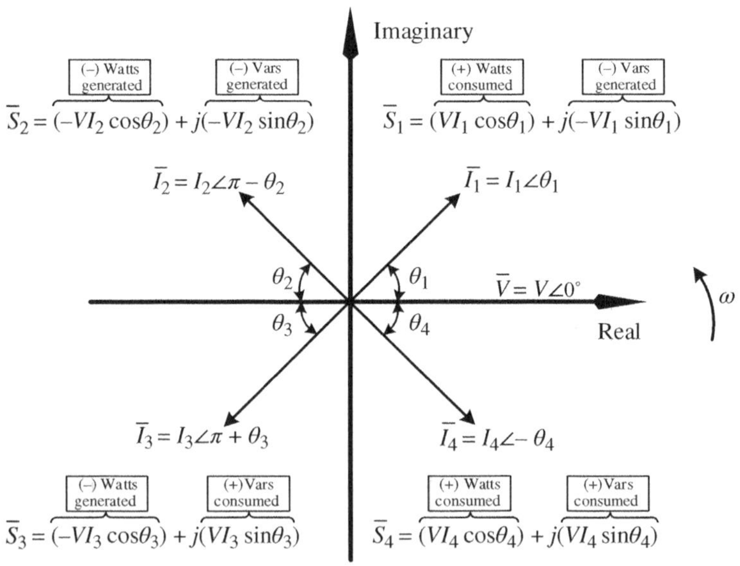

Figure 1.1 A summary graphical and formulation representation of the consumer system in the time-domain and frequency-domain phasor power computation.

$$\underline{V} = \begin{bmatrix} V_1 \\ V_2 \\ V_3 \\ \vdots \\ V_n \end{bmatrix} \tag{1.17}$$

and

$$\overline{\underline{V}} = \begin{bmatrix} \overline{V}_1 \\ \overline{V}_2 \\ \overline{V}_3 \\ \vdots \\ \overline{V}_n \end{bmatrix} \tag{1.18}$$

Similarly, the currents can be written in matrix/array (vector) form in the time-domain, $\underline{I}$, or phasor frequency-domain, $\overline{\underline{I}}$, respectively, as follows:

$$\underline{I} = \begin{bmatrix} i_1 \\ i_2 \\ i_3 \\ \vdots \\ i_n \end{bmatrix} \tag{1.19}$$

and

$$\overline{\underline{I}} = \begin{bmatrix} \overline{I}_1 \\ \overline{I}_2 \\ \overline{I}_3 \\ \vdots \\ \overline{I}_n \end{bmatrix} \tag{1.20}$$

Therefore, the instantaneous time-domain power, $p(t)$, can be written as follows in an n-polyphase device:

$$p(t) = \underline{V}^t \cdot \underline{I} = v_1 i_1 + v_2 i_2 + \cdots + v_n i_n \tag{1.21}$$

where $\underline{V}^t$ is the transpose of $\underline{V}$.

Meanwhile, the frequency-domain phasor computation of the complex power, $\overline{S}$, can be written as follows:

$$\overline{S} = \overline{\underline{V}}^t \cdot (\overline{\underline{I}}^*) \tag{1.22}$$

That is,

$$\overline{S} = \overline{V}_1 \cdot \overline{I}_1^* + \overline{V}_2 \cdot \overline{I}_2^* + \cdots + \overline{V}_n \cdot \overline{I}_n^* \tag{1.23}$$

where $\overline{\underline{V}}^t$ is the transpose of $\overline{\underline{V}}$.

1.3 Elementary Concepts of Complex Real and Reactive Power in Balanced Three-Phase Circuits and Devices Using Consumer System Notations

A balanced three-phase set of current phasors, $\overline{I}_a$, $\overline{I}_b$, and $\overline{I}_c$, is shown in Figure 1.2. We will always assume counterclockwise rotation for such phasor diagrams throughout this textbook, unless it is explicitly stated otherwise. Accordingly, for an "observer" located at the star-point in this diagram, the observer will see an *a, b, c, a, b, c,* … sequence, that is, a positive (+) sequence as designated in this figure. Meanwhile, if the rotation of this phasor diagram is reversed to a clockwise orientation, the sequence would become an *a, c, b, a, c, b,* … one, that is, a negative (−) sequence.

If one would interchange the locations of the $\overline{I}_b$ and $\overline{I}_c$ phasors, as shown in Figure 1.3, and still preserve the counterclockwise rotation and the location of the

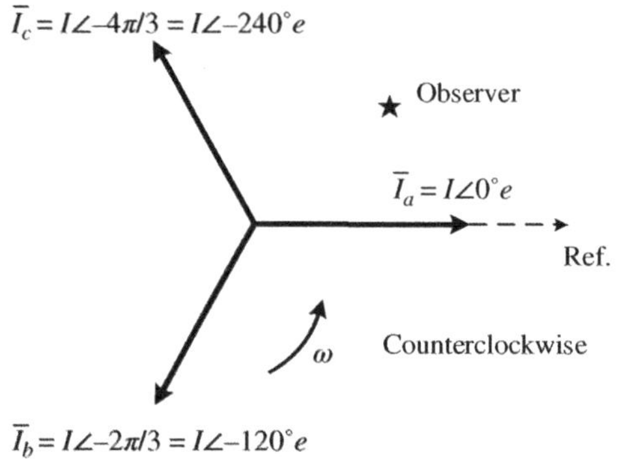

The sequence is *a*, *b*, *c*, *a*, *b*, *c*, …

(+) Positive sequence

If the rotation is reversed to a clockwise one, the sequence would become:

a, *c*, *b*, *a*, *c*, *b*, …

(–) Negative sequence

Figure 1.2 A balanced three-phase set of current phasors, $\overline{I}_a$, $\overline{I}_b$, and $\overline{I}_c$, and the concepts of positive *a*, *b*, *c* and negative *a*, *c*, *b* sequencing.

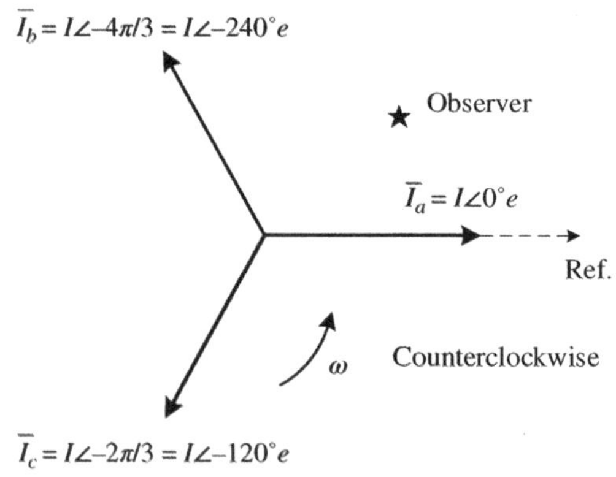

The sequence is *a*, *c*, *b*, *a*, *c*, *b*, …

(–) Negative sequence

Figure 1.3 Another approach to negative *a*, *c*, *b* sequencing in balanced three-phase set of currents, $\overline{I}_a$, $\overline{I}_c$, and $\overline{I}_b$.

star-point "observer," the result will be an *a*, *c*, *b*, *a*, *c*, *b*, … sequence. That is, one will be seeing a negative (–) sequence as designated in Figure 1.3.

Meanwhile, we discuss complex power, $\overline{S}$, in the context of a balanced three-phase Y-connected impedance load with an isolated neutral, *n*. See Figure 1.4 in which the impedance per-phase is $\overline{Z}_Y$, and the phase-currents, $\overline{I}_a$, $\overline{I}_b$, and $\overline{I}_c$, which are also equal to the *a*, *b*, *c* line-currents as depicted in this figure.

Y-Connected load:

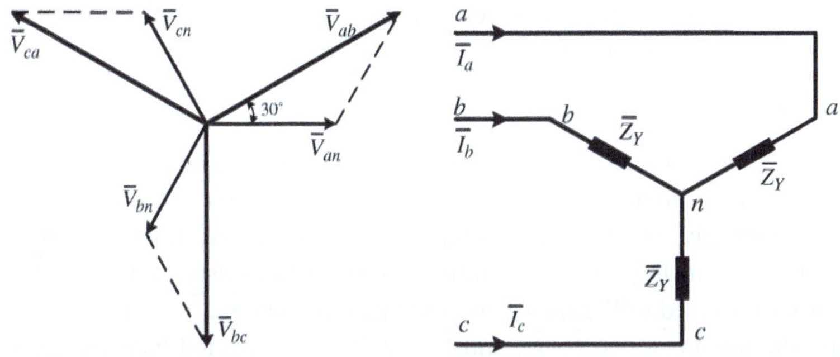

$\overline{V}_{an} = V\angle 0 = V\angle 0°$

$\overline{V}_{bn} = V\angle{-2\pi/3} = V\angle{-120°}$

$\overline{V}_{cn} = V\angle{-4\pi/3} = V\angle{-240°}$

$\overline{Z}_Y = Z_Y\angle\theta$

$\overline{I}_a = \overline{V}_{an}/\overline{Z}_Y = V\angle 0°/\overline{Z}_Y = V/Z_Y = \angle{-\theta}$

$\overline{I}_b = \overline{V}_{bn}/\overline{Z}_Y = V\angle{-120°}/\overline{Z}_Y = V/Z_Y = \angle{-\theta-120°}$

$\overline{I}_c = \overline{V}_{cn}/\overline{Z}_Y = V\angle{-240°}/\overline{Z}_Y = V/Z_Y = \angle{-\theta-240°}$

Figure 1.4 A balanced *Y*-connected impedance load and its associate line-to-neutral and line-to-line voltages and line currents, $\overline{I}_a$, $\overline{I}_b$, and $\overline{I}_c$.

Also, given in Figure 1.4 are the formulations for the phase (line-to-neutral) voltages, $\overline{V}_{an}$, $\overline{V}_{bn}$, and $\overline{V}_{cn}$, as well as phase-currents. The phasor diagram of the voltages depicts the graphical/schematic phasor representation of the line-to-line voltages, $\overline{V}_{ab}$, $\overline{V}_{bc}$, and $\overline{V}_{ca}$. Here, one can write the complex power, $\overline{S}$, as follows:

$$\overline{S} = \overline{V}_{an} \cdot \overline{I}_a^{\,*} + \overline{V}_{bn} \cdot \overline{I}_b^{\,*} + \overline{V}_{cn} \cdot \overline{I}_c^{\,*} = P + jQ \tag{1.24}$$

where P is the real power in watts and Q is the reactive power in vars. Again, here, one can write the following:

$$S = \text{Abs}(\overline{S}) = \sqrt{P^2 + Q^2} \tag{1.25}$$

and in the case of a balanced three-phase load, the power factor, PF, is

$$PF = \cos(\theta) = P/S \tag{1.26}$$

where the angle, θ, is the power factor angle given by

$$\theta = \cos^{-1}(P/S) = \cos^{-1}(PF) \tag{1.27}$$

which is the angle of the impedance per-phase, $\overline{Z}_Y$, where $\overline{Z}_Y = Z_Y \angle \theta$; see Figure 1.4. Furthermore, in this balanced three-phase Y-connected impedance load, one can write the following for $\overline{S}$:

$$\overline{S} = P + jQ = 3\overline{V}_{an}\overline{I}_a^* = 3\overline{V}_{bn}\overline{I}_b^* = 3\overline{V}_{cn}\overline{I}_c^* \tag{1.28}$$

The schematic of a balanced Δ-connected impedance load is shown in Figure 1.5, in which an impedance, $\overline{Z}_\Delta$, interconnects lines a to b, lines b to c, and lines c to a, respectively. The formulations for the line-to-line voltages, $\overline{V}_{ab}$, $\overline{V}_{bc}$, and $\overline{V}_{ca}$, are depicted in Figure 1.5, including the delta (Δ) branch currents, $\overline{I}_{ab}$, $\overline{I}_{bc}$, and $\overline{I}_{ca}$. Also, depicted in Figure 1.5 are the line currents, $\overline{I}_a$, $\overline{I}_b$, and $\overline{I}_c$.

Comparing the line currents, $\overline{I}_a$, $\overline{I}_b$, and $\overline{I}_c$, for the Y-connected balanced load of Figure 1.4 and the Δ-connected balanced load of Figure 1.5, one can conclude

Δ-Connected load:

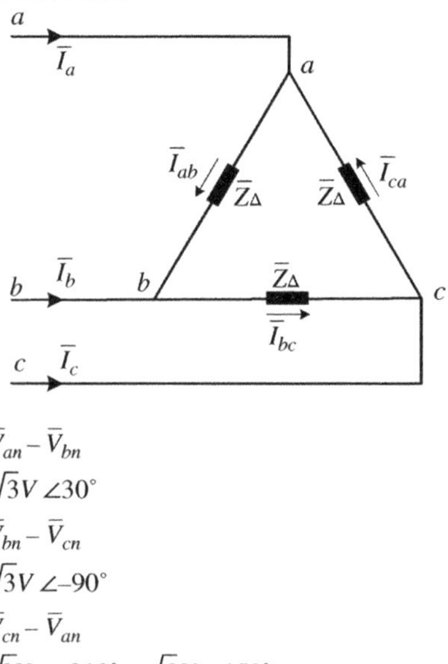

$\overline{V}_{ab} = \overline{V}_{an} - \overline{V}_{bn}$

$\overline{V}_{ab} = \sqrt{3}V \angle 30^\circ$

$\overline{V}_{bc} = \overline{V}_{bn} - \overline{V}_{cn}$

$\overline{V}_{bc} = \sqrt{3}V \angle{-90^\circ}$

$\overline{V}_{ca} = \overline{V}_{cn} - \overline{V}_{an}$

$\overline{V}_{ca} = \sqrt{3}V \angle{-210^\circ} = \sqrt{3}V \angle 150^\circ$

$\overline{I}_{ab} = \overline{V}_{ab}/\overline{Z}_\Delta = \sqrt{3}V \angle 30^\circ/\overline{Z}_\Delta$

$\overline{I}_{bc} = \overline{V}_{bc}/\overline{Z}_\Delta = \sqrt{3}V \angle{-90^\circ}/\overline{Z}_\Delta$

$\overline{I}_{ca} = \overline{V}_{ca}/\overline{Z}_\Delta = \sqrt{3}V \angle 150^\circ/\overline{Z}_\Delta = \sqrt{3}V\angle{-210^\circ}/\overline{Z}_\Delta$

$\overline{I}_a = \overline{I}_{ab} - \overline{I}_{ca} = \sqrt{3}V \angle 0^\circ/\overline{Z}_\Delta$

$\overline{I}_b = \overline{I}_{bc} - \overline{I}_{ab} = \sqrt{3}V \angle{-120^\circ}/\overline{Z}_\Delta$

$\overline{I}_c = \overline{I}_{ca} - \overline{I}_{bc} = \sqrt{3}V \angle{-240^\circ}/\overline{Z}_\Delta$

Figure 1.5 Balanced Δ-connected impedance load and its line-to-line voltages and line currents, $\overline{I}_a$, $\overline{I}_b$, and $\overline{I}_c$.

that for the two loads to draw the same line, a, b, and c, currents, one must be able to write the following for each line current, $\overline{I}_a$, $\overline{I}_b$, and $\overline{I}_c$:

$$\overline{I}_a = (V\angle 0°/\overline{Z}_Y) = (3V\angle 0°/\overline{Z}_\Delta) \tag{1.29}$$

$$\overline{I}_b = (V\angle -120°/\overline{Z}_Y) = (3V\angle -120°/\overline{Z}_\Delta) \tag{1.30}$$

and

$$\overline{I}_c = (V\angle -240°/\overline{Z}_Y) = (3V\angle -240°/\overline{Z}_\Delta) \tag{1.31}$$

Hence, for the two Y-connected and Δ-connected loads to be equivalent at their a, b, and c terminals, we must have

$$\overline{Z}_Y = (\overline{Z}_\Delta/3) \tag{1.32}$$

$$\overline{Z}_\Delta = (3\overline{Z}_Y) \tag{1.33}$$

In summary, Eqs. (1.32) and (1.33) constitute the Δ-to-Y impedance transformation and the Y-to-Δ impedance transformation for balanced three-phase impedance loads, respectively. The complex power, $\overline{S}$, drawn from the source is identical for both impedance loads in Eqs. (1.32) and (1.33).

2

Power Electronics and Converters

A power electronic converter (i.e., power conditioner) is the most critical apparatus for an adjustable speed motor-drive system, since it can control an electric motor to operate at a desired torque and speed through the regulation of the voltage, current, and frequency. Power converters mainly consist of semiconductor power devices and passive devices (resistors, capacitors, inductors, etc.), and the performance of power converters, such as efficiency and power density, is directly affected by the types of semiconductor devices utilized in such power converter circuits. In this chapter, first, we will review the commonly used semiconductor devices, including power diodes, thyristors, insulated-gate bipolar transistors (IGBTs), and the emerging wide bandgap (WBG) transistors, including silicon carbide (SiC) metal-oxide-semiconductor field-effect transistors (MOSFET) and gallium nitride (GaN) high-electron-mobility transistors (HEMT). After that, fundamental topologies and operating principles of DC–DC converters and DC–AC inverters will be introduced.

2.1 Semiconductor Devices

2.1.1 Power Diodes

Similar to a regular diode, a power diode is a two-terminal unidirectional semiconductor device frequently used in power electronic circuits. A power diode has two electrodes, and the electrode connected to the P-zone is called the anode and the electrode connected to the N-zone is called the cathode, as schematically shown in Figure 2.1.

The output characteristic of a power diode is also shown in Figure 2.1. As we can see, a diode operating in the forward direction (i.e., positive voltage applied at the anode terminal) shows an approximately linear increase of the forward current with the increase of the forward bias voltage. The current that increases

Electric Machinery and Drives: An Electromagnetics Perspective, First Edition.
Nabeel A. O. Demerdash, JiangBiao He, and Hao Chen.

Companion website: www.wiley.com/go/demerdash/electricmachineryanddrives

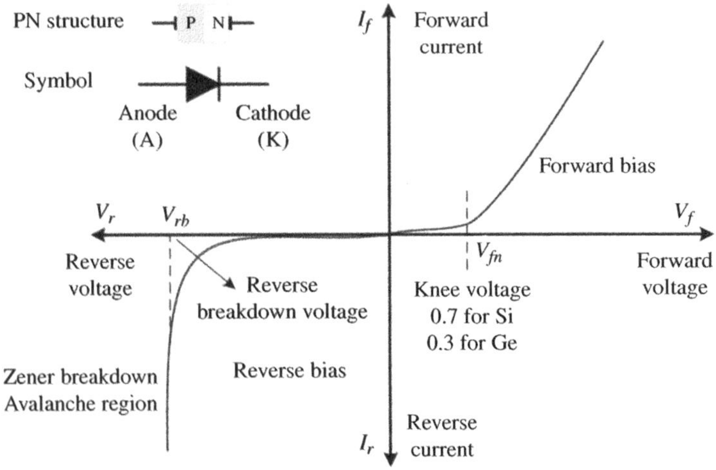

Figure 2.1 Symbol and output characteristic of a power diode.

exponentially with increasing forward voltage begins at 0.2–0.3 V in low-power germanium (Ge) diodes and 0.7 V in diodes made from silicon (Si). During reverse bias direction, if the maximum permissible reverse bias voltage, which depends on the structure and doping, is exceeded, the blocking current will increase dramatically. If the current is not limited with an impedance, an avalanche effect will be triggered, leading to thermal breakthrough and eventually causing irreversible damage of the diode.

From the perspective of applications, the common power diodes can be classified as rectifier diodes, fast recovery diodes, Zener diodes, and avalanche diodes. Rectifier diodes are used for converting AC power into DC power, and the voltage and current ratings of such diodes in the present market range up to multiple kilovolts of blocking voltage and multiple kiloamperes of forward current. Fast recovery diodes are generally used as the freewheeling antiparallel diodes for semiconductor switches to conduct current in the opposite direction (if the switch cannot conduct in the reverse direction), especially in applications with inductive loads. This is to protect the semiconductor switches from transient overvoltage damages by providing a path for reserve current flow. Antiparallel diodes are usually co-packaged with the power switch such as an IGBT or MOSFET. Additionally, Zener diodes and avalanche diodes are usually utilized to stabilize the voltage or current.

2.1.2 Thyristors

A thyristor is a solid-state semiconductor switch consisting of four layers of alternating P-type and N-type materials for high-power applications, as shown in Figure 2.2. The switch has three terminals, that is, anode, cathode, and gate

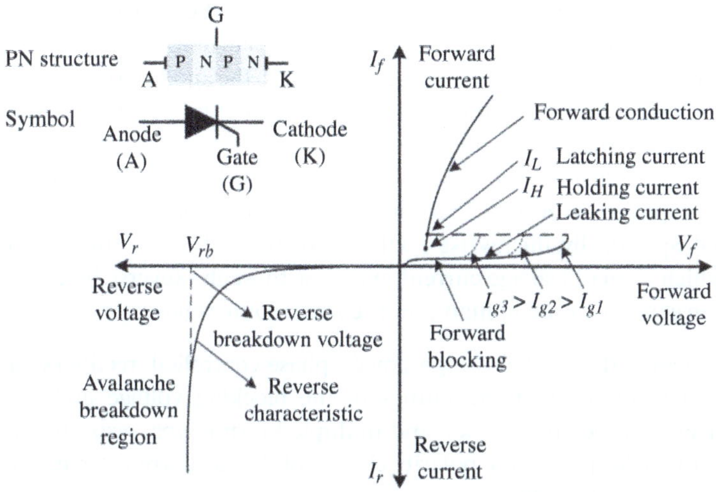

Figure 2.2 Internal structure, symbol, and output characteristics of an SCR thyristor.

control terminal. Thyristors actually represent a family of such semiconductor switches, including silicon-controlled rectifier (SCR), gate turn-off thyristors (GTOs), triode for alternating current (TRIAC), and integrated gate commutated thyristors (IGCTs). As a well-known trade name that General Electric Corporation introduced to describe a specific kind of thyristors in 1950s, SCR has been used intensively in power industries and is sometimes used interchangeably with thyristors, although the two terms are not completely synonymous. SCR is a semicontrolled switch, which can be turned on by a gate current signal, but it cannot be turned off unless the anode current (I_A) is interrupted. The forward and reverse output characteristics of an SCR thyristor are shown in Figure 2.2. As can be seen, during the forward operation mode (i.e., the first quadrant), the SCR starts conducting once the forward voltage (i.e., anode voltage with respect to the cathode) reaches a breakover voltage. Under this condition, the voltage across the SCR suddenly decreases and most of the supply voltage appears across the load. For the reverse biasing mode (i.e., the third quadrant), as the reverse voltage increases gradually, the anode current increases from zero to a very small value of leakage current, until the avalanche breakdown occurs and the SCR starts conducting in the reverse direction.

As mentioned earlier, a thyristor can be turned on to conduct current when the gate receives a positive current pulse, assuming that it is forward biased. Once a thyristor is turned on, it will be latched on until a negative anode current is applied. To sum up, a thyristor has three operation states as follows:

(a) *Forward bias conducting mode*: A forward voltage (from anode to cathode) is applied and a positive current pulse is provided to the gate terminal. Once a thyristor is turned on, it will remain conducting until the forward current drops below a threshold value known as the "holding current," I_H.

(b) *Forward blocking mode*: A positive voltage is applied to the anode terminal and a negative voltage is applied to the cathode terminal, but no positive current pulse is provided to the gate terminal.

(c) *Reverse blocking mode*: A thyristor will be in reverse blocking mode if a negative voltage is applied to the anode terminal and a positive voltage to the cathode terminal. Specifically, the thyristor will behave as two diodes connected in series, in which a small leakage current may flow through. Reverse blocking thyristors are generally used in high-power current source converters.

Thyristors are frequently used for high-power phase-controlled rectifiers or load-commutated inverters, and the ratings of the blocking voltage and the conducting current can be above 10 kV and multiple kiloamperes, respectively. During the control of AC power, the polarity change of the AC current forces the thyristor to turn off automatically.

2.1.3 IGBT

An IGBT is a three-terminal semiconductor switch, which consists of four alternating layers (P–N–P–N) that are controlled by a MOS gate structure, as shown in Figure 2.3. The three terminals of an IGBT are collector, emitter, and gate. Different from a thyristor, an IGBT is a voltage-controlled solid-state switch. Generally, an IGBT can be turned on if the gate is provided with a positive gate voltage pulse (e.g., 15 V), V_{ge}, and can be turned off with zero or negative voltage, while a negative gate voltage turn-off is generally recommended for the sake of electric noise immunity. An IGBT can be turned on and off much faster than a thyristor, with turn-on and turn-off time typically less than 200 ns.

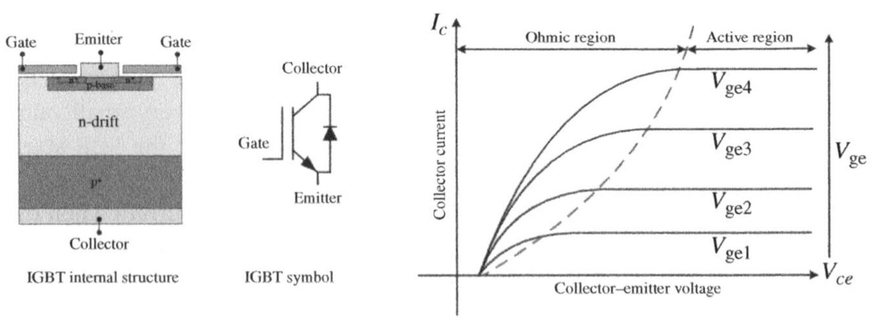

Figure 2.3 Internal structure, symbol, and output characteristic of an IGBT switch.

A typical output characteristic of an IGBT is shown in Figure 2.3. During the ohmic linear region, the IGBT collector current increases linearly with the increase of the collector to emitter voltage, V_{ce}. As V_{ce} increases over a certain value, the switch will operate in the active region, in which the collector current will gradually become saturated. Also, it can be seen that, the lower the gate to emitter voltage, V_{ge}, the lower the value of the saturation current, provided that the V_{ge} is larger than the threshold gate voltage. Such characteristics can be used to rapidly constrain short-circuit current for fast fault protection applications.

IGBTs have been widely used in electric power conversion (conditioning) systems, ranging from a few kilowatts to tens of megawatts scale. In terms of device packaging, discrete IGBTs are mainly used for low-power (up to hundreds of kilowatts scale) applications, and various module packaging, including half-bridge, full-bridge (i.e., H-bridge), and inverter modules (i.e., six-pack), have been developed for high-power applications.

2.1.4 SiC MOSFET

Driven by the requirements of high efficiency and high power density in many applications, such as transportation electrifications and renewable energies, SiC MOSFETs have been experiencing rapid development in both packaging technologies and voltage/current levels. At present, 650 V, 1.2 kV, 1.7 kV and 3.3 kV half-bridge SiC MOSFET modules are already commercially available in the market, and SiC modules with higher voltage ratings are under development and will possibly be released to market soon, including 4.5, 6.5, 10, and 15 kV. Various packages of 1.7 and 3.3 kV half-bridge SiC MOSFET modules are shown in Figure 2.4. Compared to the conventional silicon IGBT counterparts, SiC MOSFET modules can withstand higher voltage, higher temperature, and higher switching frequency, as shown in Figure 2.5. As can be seen, the five vertexes of the pentagons represent the level of electron mobility, electric field, energy bandgap, thermal conductivity, and electron saturation velocity, respectively. Compared to the silicon counterparts, SiC MOSFETs have much higher thermal conductivity and energy bandgap, enabling higher temperature operation capability. Also, the much higher electric field with SiC MOSFETs versus silicon counterparts indicates higher voltage blocking capability and lower on-state conduction losses. The much faster switching speed with SiC MOSFETs enables power converters to switch at much higher switching frequency, leading to better quality of output voltage and current waveforms, as well as higher output fundamental frequencies. Furthermore, higher switching frequency will be helpful for reducing the size and weight of electric machines and other magnetic components, leading to higher power density of motor-drive systems, especially for electrified transportation applications, such as electric vehicles and electric aircraft.

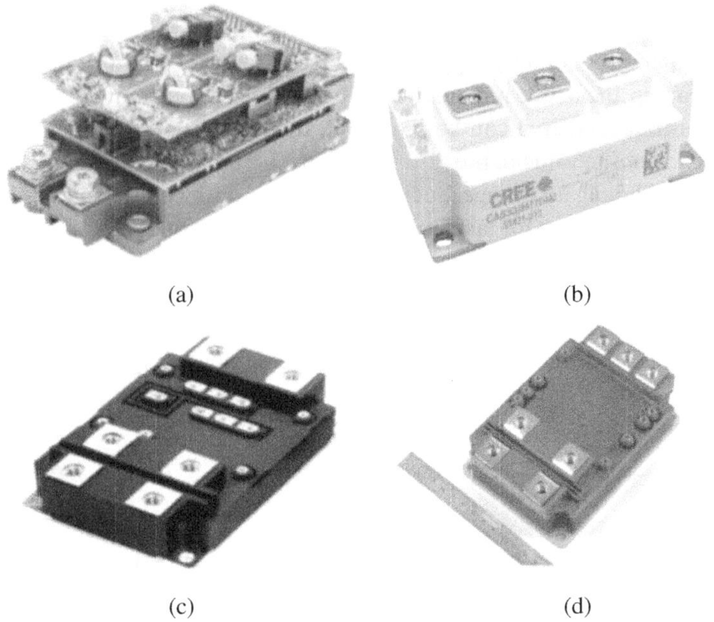

Figure 2.4 Half-bridge SiC MOSFET modules with various packages: (a) GE 1.7 kV/450 A in EconoDual package, (b) Cree/Wolfspeed 1.7 kV/225 A in 62 mm package, (c) Hitachi 3.3 kV/800 A in nHPD2 package and (d) Mitsubishi 3.3 kV/750 A in LV100 package.

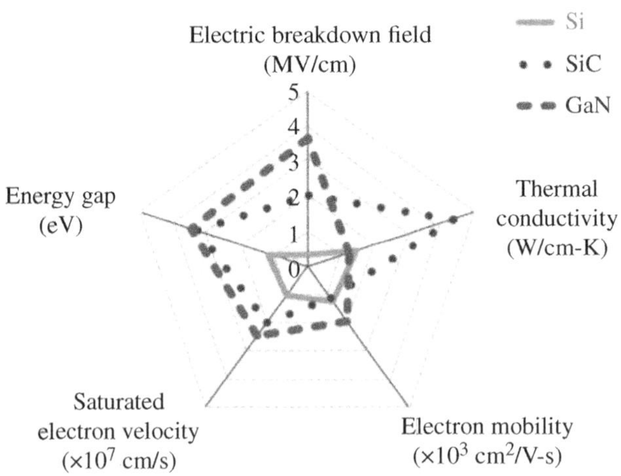

Figure 2.5 Performance comparison among silicon, SiC, and GaN switches [1]/with permission of IEEE.

2.1.5 GaN HEMTs

GaN HEMT is another type of emerging WBG semiconductor switches commercialized in recent years, which have been increasingly used in power conversion systems in industries. Similar to a silicon MOSFET, GaN HEMT is a three-terminal device, including a gate terminal, a drain terminal, and a source terminal. The majority of the present commercial GaN HEMTs are lateral devices, and the cross-sectional structure is shown in Figure 2.6. Vertical GaN transistors are still under development for medium-voltage applications, and the challenges of material quality and doping processing have not been solved [1]. GaN transistors can be classified into two categories, namely enhancement mode (E-mode) and depletion mode (D-mode) transistors. An E-mode transistor is a normally off device which can be turned on by providing a positive gate-to-source (V_{gs}) voltage. On the contrary, a D-mode transistor is normally on, and a negative gate-to-source (V_{gs}) voltage is required to turn it off. D-mode transistors are usually connected and packaged in cascode with low-voltage silicon MOSFETs to change them into normally off devices, and the circuit is shown in Figure 2.7. As a result, the D-mode GaN transistor is turned on and off by controlling the silicon MOSFET in the cascode package.

Compared to the conventional silicon MOSFETs, GaN HEMTs have much faster switching speed, higher blocking voltage capability, lower on-state conduction losses and switching losses, and higher temperature operation capability. Figure 2.5 shows the performance comparison between silicon, SiC, and GaN devices. Furthermore, unlike a MOSFET, there is no P–N junction within the lateral structure of a GaN HEMT. Thus, there is no body diode, eliminating the associated reverse recovering losses, which enables efficient high-frequency switching operation in hard-switched power converter topologies. Compared to SiC MOSFETs, the current technology status of GaN HEMTs is more suitable for high-frequency low-power operation, such as up to a few hundreds of kilowatts. The blocking voltage rating of the present GaN HEMTs in the market is up to 650 V, and the rating of continuous drain current is up to 100 A approximately.

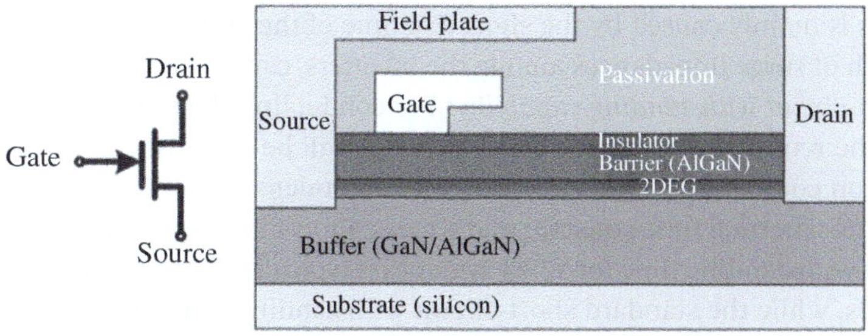

Figure 2.6 Symbol and basic lateral structure of a GaN HEMT.

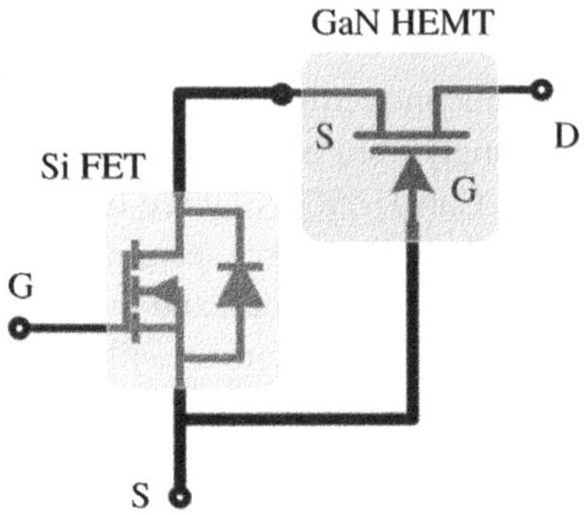

Figure 2.7 Cascode normally-off GaN transistor based on a D-mode GaN HEMT.

Nevertheless, as new types of semiconductor devices, the integration of WBG transistors into power converter systems also introduces challenges due to ultrafast switching of power semiconductors, which require special attention during the design process, which can be summarized as follows:

- *Increased electromagnetic interference (EMI)*: Specifically, the high common-mode dv/dt resulting from the fast switching coupled with the module-to-heatsink stray capacitance as well as the gate driver transformer primary–secondary capacitance ($C * dv/dt$) will generate significant common-mode current during switching, which may trip the system microcontrollers and saturate the common-mode filter inductors. Under such a scenario, common-mode EMI filters need to be properly designed to attenuate the high dv/dt.
- *Increased surge voltage stress on motor windings*: The high differential-mode dv/dt generated from the fast switching of WBG switches, if not mitigated properly, will cause significant surge voltage on the motor terminals due to surge impedance mismatch, especially if there are long cables interconnected between the WBG inverters and electric motors. Specifically, the first few winding turns closer to the inverter side might be subject to severe insulation stress, in which the high-frequency voltage spikes might double the DC-bus voltage. This is mainly caused by the short rise time of the WBG switches and the mismatch of surge impedances among the inverters, cables, and machines.
- *Weaker short-circuit withstanding capability*: For conducting the same amount of current, the required die size for WBG switches will be much smaller than that for silicon counterparts. Nevertheless, such high-density WBG chips also indicate shorter thermal time constant and weaker short-circuit capability. The short-circuit withstanding time for WBG switches is typically shorter than a few microseconds, while the standard short-circuit withstanding time for Si IGBTs is normally 10 μs. This may pose a challenge for the short-circuit protection due to the requirement of fast fault detection speed.

- *Higher turn-off voltage overshoot*: Due to the large di/dt of WBG switches during switching, the turn-off voltage overshoot could be considerable, especially in high-power applications where the parasitic inductance of the high-voltage SiC modules and the commutation loops in converters might be significant. For instance, with a typical *di*/*dt* of 20 kA/μs for SiC MOSFETs, the turn-off voltage overshoot can reach 1 kV superposed on the DC-bus voltage if the parasitic inductance of the commutation loops is 50 nH ($V = L * di/dt$). This requires a very dedicated design of the bus bars and circuit layout of the converters.

2.2 DC–DC Converters

2.2.1 Buck Converter

A buck converter is a DC–DC step-down converter which changes the input DC voltage to stabilized lower values of output DC voltage. A basic buck converter circuit generally consists of a semiconductor switch such as an IGBT or MOSFET, a power diode, and a passive energy storage component such as an inductor, as well as an output filter capacitor to smooth out the output voltage ripples, as shown in Figure 2.8. The operating principle of a buck converter can be summarized as follows:

(1) *Switch turned on*: When the switch *S* is turned on, the voltage across the inductor will be $V_L = V_{in} - V_{out}$, where V_{in} is the input voltage and V_{out} is the output voltage of the buck converter. In this scenario, the inductor will be charged and the current through the inductor will be increased linearly approximately. No current will flow through the diode since it is reverse biased, as shown in Figure 2.9a.

(2) *Switch turned off*: When the switch is turned off, the diode will be forward biased due to the induced voltage in the inductor. In this scenario, the voltage across the inductor is $V_L = -V_{out}$, if the voltage drop across the diode is neglected. Accordingly, the current flowing through the inductor will decrease, as shown in Figure 2.9b.

Depending on whether the inductor current drops to zero, the operation modes of the buck converter can be either in a continuous mode where the inductor

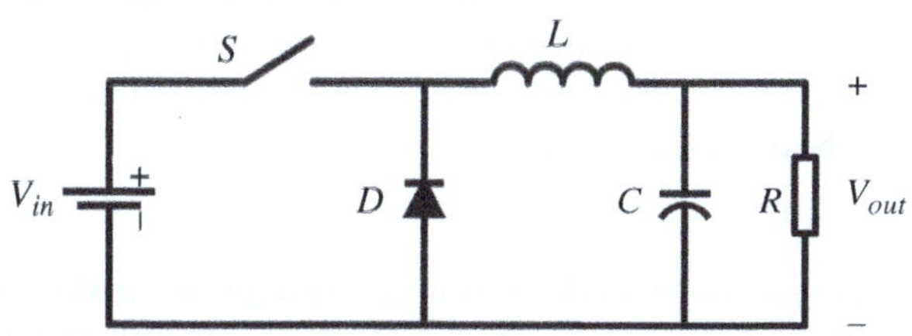

Figure 2.8 Basic circuit of a buck converter.

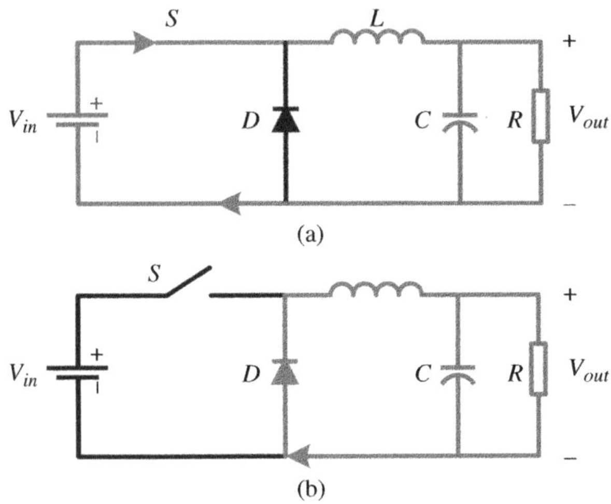

Figure 2.9 Current flow of the buck converter when (a) the switch is turned on and (b) the switch is turned off.

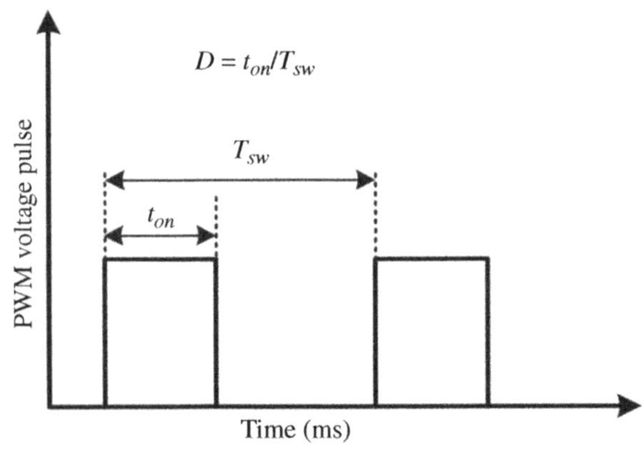

Figure 2.10 Illustration of turn-on time, a switching period, and duty cycle.

current always stays positive or in a discontinuous mode where the inductor current drops to zero for a short duration during a commutation cycle. In a continuous mode, the relationship between the output voltage and the input voltage can be expressed as follows:

$$V_{out} = \frac{t_{on}}{T_{sw}} V_{in} \tag{2.1}$$

where t_{on} is the turn-on duration of a semiconductor switch and T_{sw} is the switching period, the reciprocal of the switching frequency. The ratio of t_{on} and T_{sw} is defined as the duty cycle of a switch, as depicted in Figure 2.10. Therefore, Eq. (2.1) can be expressed as follows:

$$V_{out} = D * V_{in} \tag{2.2}$$

Considering that the duty cycle D is typically in the range of $0 < D < 1$, the output voltage of a buck converter will be lower than the input voltage.

Under some scenarios, if the required load power is too low, the inductor current falls to zero for a short period during the operation of the Buck converter, which is called a discontinuous mode. Such a discontinue mode differs from the continuous mode in that, the inductor is completely discharged in the end of a switching cycle. Following the inductor "volt-second balance" principle where the inductor positive voltage during the turn-on period equals the negative voltage during the turn-off period, one can derive that, the output voltage of a buck converter is a function consisting of not only the input voltage and duty cycle, but also the inductance, switching frequency, and the output load current. More detailed operating principle for buck converters operating in a discontinue mode is elaborated in many power electronics textbooks such as [2], and thus will not be repeated here.

In motor-drive systems, a buck converter is often cascaded with a current source inverter (CSI) to provide a lower DC input voltage for the CSI because the CSI converts the input DC voltage into a higher magnitude output AC voltage, which may sometimes pose challenges on the dielectric insulation of the stator windings of an electric motor if a buck converter is unavailable.

2.2.2 Boost Converter

A boost converter is a DC–DC step-up converter that is intended to boost (or increase) the input voltage of an unregulated DC supply to a stabilized higher output voltage. Similar to a buck converter, a basic boost converter circuit consists of an inductor, a power switch (e.g., IGBT or SiC MOSFET), a power diode, and a filtering capacitor, while the boost circuit topology is different from that for a buck converter, as shown in Figure 2.11.

When the switch is turned on, the current will flow through the inductor and the inductor will be charged, which stores some energy, and the polarity of the left side of the inductor is positive. In this scenario, the power diode is reverse biased, so there will be no current flowing through the load resistor, *R*, while charging the capacitor, *C*, as shown in Figure 2.12a.

When the switch is turned off, the current flowing in the circuit will be reduced due to the involvement of a higher impedance (i.e., the load resistance, *R*). The magnetic field created in the inductor during the previous turn-on mode will decrease to maintain the current flowing through the load (i.e., inductor discharging). Accordingly, the induced voltage across the inductor will be reversed,

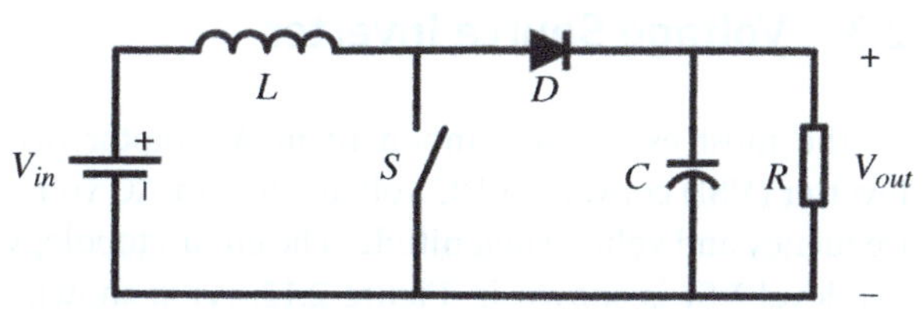

Figure 2.11 Basic circuit of a boost converter.

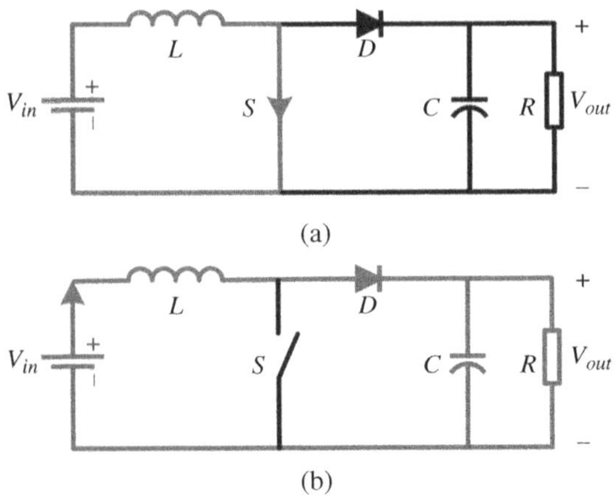

Figure 2.12 Current flow path of a boost converter under the condition (a) when the switch is turned on and (b) when the switch is turned off.

enabling a higher voltage to charge the capacitor through the power diode, D, as shown in Figure 2.12b.

If the switching period is short enough (i.e., high switching frequency), the inductor in the boost circuit will not be able to fully discharge, so the load will always see a voltage greater than that of the input DC source when the switch is turned off. The relationship between the output voltage and the input voltage of a boost converter is described as follows:

$$V_{out} = \frac{V_{in}}{1 - D} \tag{2.3}$$

where V_{out} and V_{in} refer to the output voltage and the input voltage of the boost converter, respectively, and D denotes the duty cycle of the semiconductor switch. Considering that the duty cycle is in the range of $0 < D < 1$, the output voltage will be always higher than the input voltage.

Boost converters are commonly cascaded with voltage source inverters (VSI) for electric motor-drive systems, since a VSI is a step-down of the input DC voltage into AC voltages with lower magnitude. For instance, in the traction drive system of an electric vehicle, the boost converter is used to step up the battery voltage to a higher DC voltage, so lower number of battery cells can be used to save cost and space. Afterwards, the boosted DC voltage is inverted to a desired AC voltage with a VSI to energize the stator windings of an electric motor.

2.3 Voltage Source Inverter

As the most essential component in AC motor drive systems, voltage source inverter (VSI) converts a DC voltage into an AC voltage with controllable output frequency and voltage magnitude. The circuit topology of a very basic three-phase two-level VSI is shown in Figure 2.13. As is shown, each phase leg of the VSI

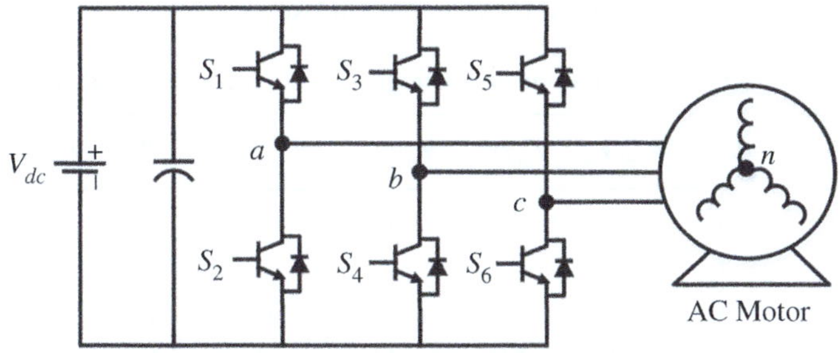

Figure 2.13 Three-phase PWM voltage source inverter.

consists of two complementary switches, i.e., one upper switch and one lower switch, which can be an IGBT, MOSFET, or other types of active switches. Each switch is configured with an anti-parallel diode for circulating reverse current when the complementary switch is turned off. The two switches can not be turned on at the same time, to avoid any DC-bus shoot-through fault. In practical implementation, a short duration called deadtime (e.g., one or two microseconds), is generally set up between the switching actions of the two complementary switches, to eliminate any DC-bus short-circuit fault. For the whole 3-phase VSI, there are six switches in total, S_1 to S_6, which are controlled by the associated pulse width modulation (PWM) strategy to synthesize 3-phase sinusoidal waveforms. Here, the switching function of the VSI is defined as S_k, ($k = a$, b, or c), given in Equation (2.4) below:

$$S_k = \begin{Bmatrix} 1 & \textit{The upper switch is on} \\ 0 & \textit{The upper switch is off} \end{Bmatrix} \tag{2.4}$$

Since there are two switching states for each phase, there are $2^3 = 8$ possible switching patterns (i.e., voltage vectors) for the 3-phase VSI. As shown in Table 2.1, the 8 voltage vectors, including two zero vectors and 6 active vectors. At zero vectors, the output voltage of the VSI is zero. At any of the six active vectors, the values of the line-to-neutral voltages (i.e., phase voltage) and line-to-line voltages, both normalized by the VSI DC-bus voltage, are given in Table 2.1.

2.4 Pulse Width Modulation (PWM)

There are many PWM methods proposed in the literature over the past decades. Here, in this section, only a basic and fundamental PWM method, i.e., sinusoidal PWM (SPWM) will be introduced due to its simplicity and wide utilization. SPWM has been intensively used in AC motor drives and general-purpose inverters in industries. In this modulation method, the three-phase sinusoidal reference modulating signals are compared with a common triangular carrier,

Table 2.1 Switching patterns and output voltages of the 3-phase VSI.

Voltage vectors	Switching pattern			Line-to-neutral voltages			Line-to-line voltages		
	A	B	C	V_{an}/V_{dc}	V_{bn}/V_{dc}	V_{cn}/V_{dc}	V_{ab}/V_{dc}	V_{bc}/V_{dc}	V_{ca}/V_{dc}
$\vec{V}_0$	0	0	0	0	0	0	0	0	0
$\vec{V}_1$	1	0	0	2/3	−1/3	−1/3	1	0	−1
$\vec{V}_2$	1	1	0	1/3	1/3	−2/3	0	1	−1
$\vec{V}_3$	0	1	0	−1/3	2/3	−1/3	−1	1	0
$\vec{V}_4$	0	1	1	−2/3	1/3	1/3	−1	0	1
$\vec{V}_5$	0	0	1	−1/3	−1/3	2/3	0	−1	1
$\vec{V}_6$	1	0	1	1/3	−2/3	1/3	1	−1	0
$\vec{V}_7$	1	1	1	0	0	0	0	0	0

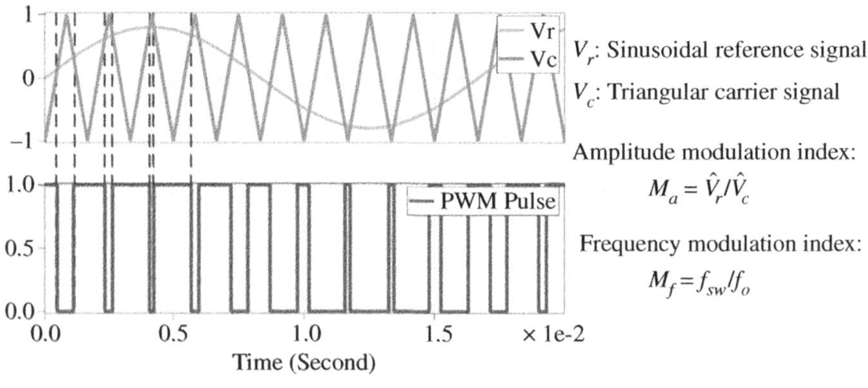

Figure 2.14 Illustration of the sinusoidal PWM (SPWM).

and their intersections determine the switching actions of the power devices in the inverter, as shown in Figure 2.14. Specifically, the sinusoidal reference signal with an amplitude of $\widehat{V_r}$ compares to the real-time value of the carrier signal V_c that has an amplitude of 1. At any time, if the sinusoidal reference signal, V_r, is larger than the carrier signal, V_c, the top switch of an inverter phase leg will be turned on, and otherwise the bottom switch will be turned on. It should be noted that the top switch and the bottom switch on the same phase leg of the inverter can never be turned on at the same time, to avoid a DC-bus short-circuit fault. The ratio between the amplitudes of the reference signal and the carrier signal is defined as the amplitude modulation index, M_a, expressed as

$$M_a = \widehat{V_r}/\widehat{V_c} = \widehat{V_r}/1 = \widehat{V_r} \tag{2.5}$$

where $\widehat{V_r}$ and $\widehat{V_c}$ refer to the amplitude (peak) of the sinusoidal reference signal and amplitude (peak) of the carrier signal, respectively, and $\widehat{V_c}$ is generally equal to 1. Thus, M_a is actually only determined by $\widehat{V_r}$. M_a is typically in the range of $0 \leq M_a \leq 1$. During the control of a VSI, M_a is used to regulate the output voltage amplitude of the inverter.

Furthermore, the frequency, f_o, of the sinusoidal reference signal, V_r, determines the fundamental output frequency of the inverter, and the frequency, f_{sw}, of the triangular carrier signal, V_c, determines the switching frequency of the power transistors. The higher the value of f_{sw}, the better the output waveform quality of the inverter, but the inverter will dissipate more switching losses and result in lower efficiency. The ratio of carrier frequency, f_{sw}, to the reference signal frequency, f_o, is defined as the frequency modulation index, M_f:

$$M_f = f_{sw}/f_o \tag{2.6}$$

In practice, the frequency modulation index, M_f, should be at least 10, to ensure sufficient sampling of the reference signal. Additionally, the value of M_f is recommended to be an odd integer multiple of three (triplen), to minimize the harmonics in the inverter output waveforms.

Based on the Fourier analysis, the peak value (i.e., amplitude) of the line-to-neutral voltage $V_{ln}(t)$ of the 3-phase SPWM inverter can be calculated as:

$$\widehat{V}_{ln} = \frac{V_{dc}}{2} m_a \tag{2.7}$$

Accordingly, the RMS value of the line-to-neutral voltage can be expressed as:

$$V_{ln(RMS)} = \frac{V_{dc}}{2\sqrt{2}} m_a \tag{2.8}$$

Thus, the RMS value and peak value of the line-to-line voltage V_{ll} (e.g., V_{ab}) can be derived as follows:

$$V_{ll(RMS)} = \frac{\sqrt{3}V_{dc}}{2\sqrt{2}} m_a \approx 0.612\, m_a V_{dc} \tag{2.9}$$

$$\widehat{V}_{ll} = \frac{\sqrt{3}V_{dc}}{2} m_a \approx 0.866\, m_a V_{dc} \tag{2.10}$$

With the SPWM method, if the amplitude modulation index, m_a, reaches the maximum value of 1, the peak value of the line-to-line voltage will be $0.866V_{dc}$.

To better utilize the DC-bus voltage, zero-sequence components (i.e., triplen harmonics) can be injected into the sinusoidal reference signals, which can increase the amplitude modulation index m_a up to $\frac{2}{\sqrt{3}}$ (i.e., increased by approximately 15%), which indicates that the output line-to-line voltage amplitude is almost equal to the DC-bus voltage. Such a harmonic injection approach in SPWM is equivalent to a well-known more advanced modulation strategy, namely, space vector PWM (SVPWM), as detailed in [3].

If the amplitude modulation index is further increased, the sinusoidal reference signals will not intersect the triangular carrier signal during the instants when

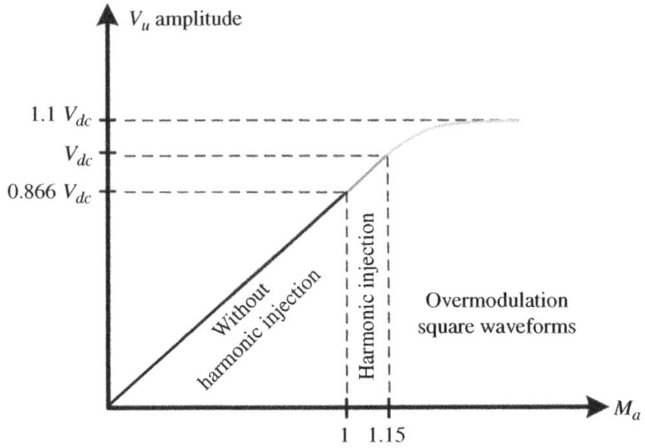

Figure 2.15 Relationship between the line-to-line voltage amplitude and the modulation index of the PWM inverter.

the amplitude of the reference signals is constantly larger than that of the carrier signals, resulting in square voltage output waveforms for a certain period and consequently low-order harmonics (e.g., 5th and 7th) in the inverter output voltages. Such a modulation phenomenon is called overmodulation, which is also known as six-step operation due to the presence of six voltage steps in its output voltage waveforms. The advantage with the overmodulation is mainly the increased inverter output voltage, with an output line-to-line voltage amplitude of 1.1 times of the DC-bus voltage (see Chapter 6.7 for details), which can be leveraged for motor start-up and high-speed operation modes, but at the cost of much increased low-order harmonic components. Figure 2.15 shows the variations of the magnitude of the line-to-line voltage with respect to the increase of the amplitude modulation index.

Homework Problems

2.1 For semiconductor devices, developed from the conventional thyristors, to modern IGBTs and wide bandgap switches, the switching speed of these devices becomes faster and faster (e.g., from millisecond scale to nanosecond scale). What are the advantages and disadvantages of utilizing such increasingly fast switching devices in power converters?

2.2 An IGBT typically requires a turn-on gate voltage of 15 V. If an IGBT switch is provided with a 10 V turn-on gate voltage, what would be the impact on the IGBT short-circuit current and inverter efficiency?
(Hint: please refer to the IGBT output characteristics)

2.3 In Figure P3.1, a half-bridge inverter circuit is connected to an inductive load. In the half-bridge circuit, why is an antiparallel diode used for each IGBT?

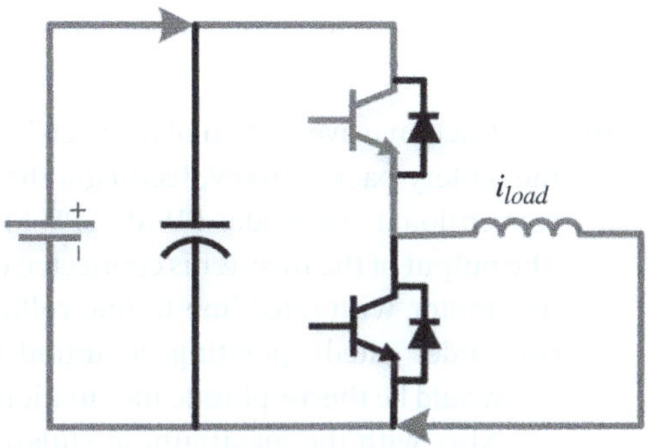

Figure P3.1 An IGBT half-bridge circuit connected to an inductive load.

2.4 In a DC–DC buck converter as shown in Figure P4.1, the input DC voltage $V_{\text{in}} = 200\,\text{V}$, the duty cycle of the switch S is 0.5, the inductor is 200 μH, the output capacitor is 200 μF, the load resistor is 10 Ω, and the switching frequency is 30 kHz. Assuming that the voltage drop across the power diode D is negligible, please calculate the following variables:

(a) The output voltage of the buck converter

(b) The output voltage ripple percentage

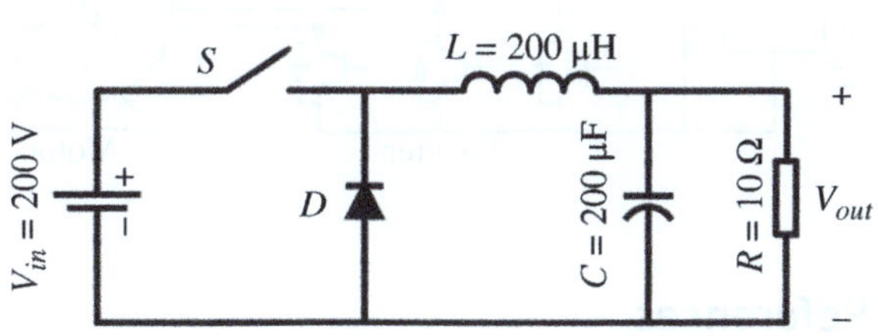

Figure P4.1 A DC–DC buck converter.

2.5 In a battery charging system of an electric vehicle (EV), an external 400 V DC supply needs to be stepped up through a boost converter to charge a 800 V battery pack in the vehicle. As shown in Figure P5.1, the inductor in the boost converter is 200 μH, the output capacitor is 200 μF, the load resistor is 10 Ω, and a 1200 V SiC MOSFET is used as the switch.

(a) To boost the output voltage to 800 V, what would be the duty cycle of the switch?

(b) To obtain an output voltage ripple of 2%, what is the switching frequency, f_{sw}?

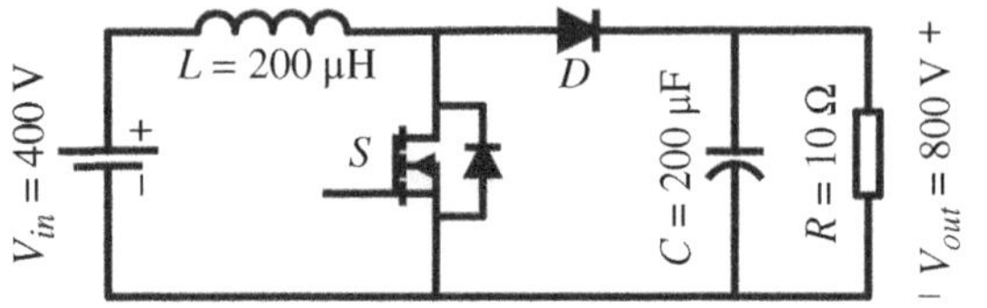

Figure P5.1 A DC–DC boost converter for an EV battery charging system.

2.6 In a traction drive system of an electric vehicle (EV), the rated DC voltage of the battery pack is 800 V, feeding a three-phase two-level inverter with the conventional sinusoidal PWM (SPWM) strategy, as shown in Figure P6.1. The output of the inverter is connected directly to a three-phase 100 kW traction motor, with rated line-to-line voltage of 500 V RMS.

(a) Under rated operating condition of the traction drive system, what would be the amplitude modulation index for the SPWM of the inverter? Also, with the maximum modulation index, what would be the output line voltage of the inverter?

(b) If the motor rated at 500 V RMS is replaced with a higher voltage motor rated at 800 V RMS of the line-to-line voltage, what would be the recommended change in the power converter circuit to match the increased motor terminal voltage rating? Draw the updated power converter circuit for the EV traction drive system.

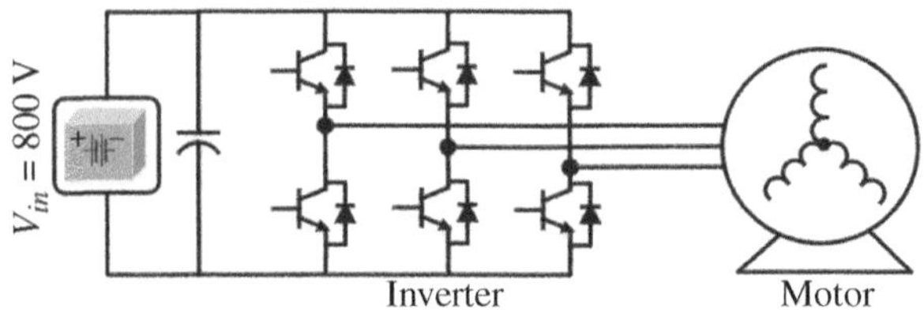

Figure P6.1 A battery-powered inverter-motor traction drive system for EVs.

References

1 Jones, E.A., Wang, F.F., and Costinett, D. (2016). Review of commercial GaN power devices and GaN-based converter design challenges. *IEEE Journal of Emerging and Selected Topics in Power Electronics* 4 (3): 707–719.

2 Erickson, R.W. and Maksimovic, D. (2020). Fundamentals of Power Electronics, Springer.

3 Holmes, D.G. and Lipo, T. A. (2003). Pulse Width Modulation for Power Converters: Principles and Practice, Wiley-IEEE Press.

3

Review of Magnetostatics, Magnetic Circuits, and Fundamentals of Electromechanical Energy Conversion

In this chapter, we review the basics of magnetostatic fields and associated notations and fundamentals followed throughout this book. Naturally, magnetostatics implies nonradiating electromagnetic fields. The well-known right-handed rule of relationship between electric current flow in conductors and the resulting magnetic field intensity vector is adopted throughout this book. The well-known Maxwell's equations are the governing formulations. In Section 3.1, we review Ampere's law, Faraday's law, the constitutive relations, and the adopted definition of the magnetic vector potential (MVP).

3.1 Magnetostatic Fields Notations and Ampere's Law

The first of Maxwell's equations is Ampere's law, which can be written in differential form as follows:

$$\nabla \times \overline{H} = Curl(\overline{H}) = \overline{J} \tag{3.1}$$

where $\overline{H}$ is the vector of magnetic field intensity and $\overline{J}$ is the vector of current density, and the second of Maxwell's equations is Faraday's law, which can be written in differential form as follows:

$$\nabla \times \overline{E} = Curl(\overline{E}) = -\partial\overline{B}/\partial t \tag{3.2}$$

where $\overline{E}$ is the vector of electric field intensity and $\overline{B}$ is the vector of magnetic flux density.

Here, the constitutive relations are as follows:

$$\overline{J} = \sigma\overline{E} \tag{3.3}$$

or

$$\overline{E} = \rho\overline{J} \tag{3.4}$$

Electric Machinery and Drives: An Electromagnetics Perspective, First Edition.
Nabeel A. O. Demerdash, JiangBiao He, and Hao Chen.

Companion website: www.wiley.com/go/demerdash/electricmachineryanddrives

where σ is the electrical conductivity and ρ is the electrical resistivity of the material and

$$\overline{B} = \mu \overline{H} \tag{3.5}$$

or

$$\overline{H} = \nu \overline{B} \tag{3.6}$$

where μ is the magnetic permeability and ν is the magnetic reluctivity of the material.

Throughout this book, the MVP, $\overline{A}$, is defined in terms of the magnetic flux density, $\overline{B}$, as follows:

$$\overline{B} = \nabla \times \overline{A} = Curl(\overline{A}) \tag{3.7}$$

Hence, one can write the following:

$$\overline{H} = \nu \overline{B} = \nu(\nabla \times \overline{A}) = \nu \, Curl(\overline{A}) \tag{3.8}$$

Accordingly, based on the formulation of Ampere's law in Eqs. (3.1) and (3.8), one can write the following important relationship, which will reappear in later chapters of this book:

$$\nabla \times \overline{H} = \nabla \times [\nu(\nabla \times \overline{A})] = \overline{J} \tag{3.9}$$

or

$$Curl(\overline{H}) = Curl[\nu \, Curl(\overline{A})] = \overline{J} \tag{3.10}$$

At this stage, it is useful to restate Ampere's and Faraday's laws in integral form because of the fact that this form is the most useful in applications to be introduced in the early part of this book. Accordingly, Ampere's law in integral form can be written as follows:

$$\oint_L \overline{H} \cdot d\overline{l} = \sum_{enclosed} I = \text{the sum of the currents enclosed within the closed path, } L \tag{3.11}$$

and Faraday's law in integral form can be written as follows:

$$V_L = \oint_L \overline{E} \cdot d\overline{l} = \frac{\partial}{\partial t}\left(\int_S \overline{B} \cdot d\overline{S}\right) = \frac{\partial}{\partial t}(\Phi) \tag{3.12}$$

where V_L is the induced voltage in the closed path, L, surrounding the open surface area, S, enclosed within this path, see Figures 3.1 and 3.2; $d\overline{l}$ is the differential vector length along the closed path, L; and $d\overline{S}$ is the differential vector surface area on the open surface area, S, enclosed within the closed path, L.

Figure 3.1 Ampere's law in integral form.

$$\oint_L \bar{H} \cdot d\bar{l} = \sum_{\text{Enclosed within integral path}} I$$

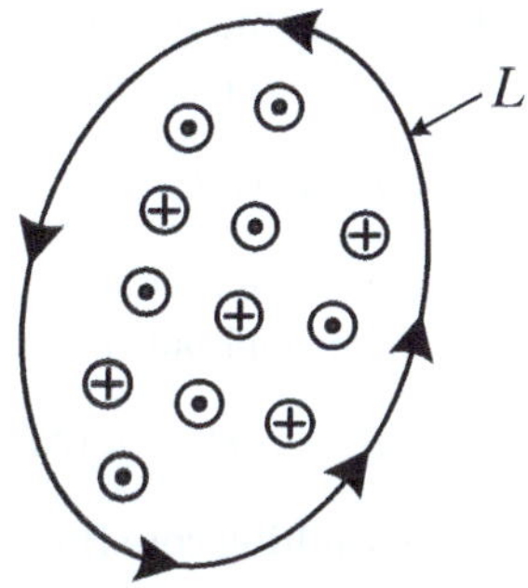

Figure 3.2 Faraday's law in integral form.

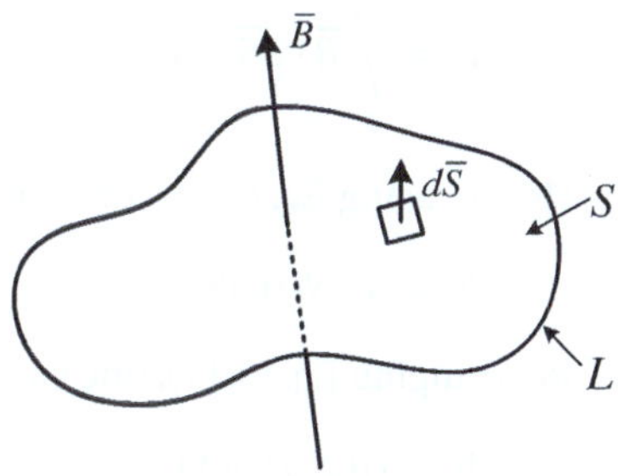

3.2 Magnetic Circuits with Ferrous Magnetic Cores

Consider the ferrous core shown in Figure 3.3 with a single excitation coil with a number of turns, N, wound around the core in the counterclockwise direction as shown in this figure in elevation and in "bird's eye-view" cross-section. With a current, i, passing through the coil, applying Ampere's law along the mean path in the core, l_c, with a cross-sectional area, S_c, one can approximate the line integral, $\oint \overline{H} \cdot d\bar{l}$, along this mean path as follows:

$$\oint \overline{H} \cdot d\bar{l} = H_c l_c = Ni \tag{3.13}$$

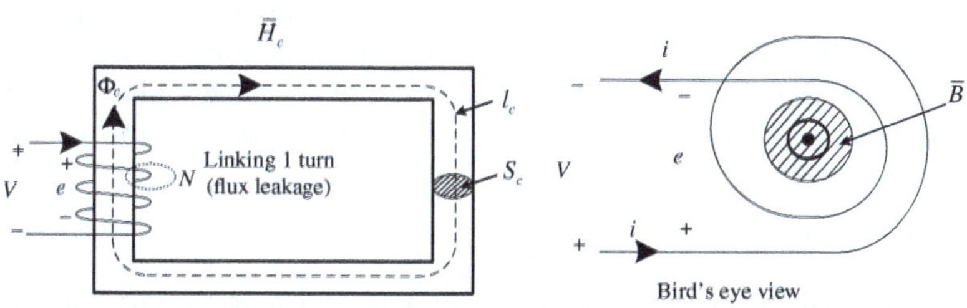

Figure 3.3 A magnetic circuit with ferrous core and one winding.

where H_c is the mean (average) value of the magnetic field intensity throughout the ferrous core; hence,

$$H_c = Ni/l_c \tag{3.14}$$

Therefore, for a core magnetic permeability, μ_c, one can express the mean flux density throughout this core, B_c, as follows:

$$B_c = \mu_c H_c = \mu_c(Ni/l_c) \tag{3.15}$$

The resulting core flux, Φ_c, from an integral ($\int_{S_c} \overline{B} \cdot d\overline{S}$) over the cross-section of the core, S_c, can be approximated as follows:

$$\Phi_c = \int_{S_c} \overline{B} \cdot d\overline{S} = B_c S_c \tag{3.16}$$

Hence, using Eq. (3.15), one can write the following:

$$\Phi_c = \mu_c(Ni/l_c)S_c \tag{3.17}$$

Rearranging Eq. (3.17), one can write the following:

$$\Phi_c = (\mu_c S_c/l_c)(Ni) \tag{3.18}$$

The term $(\mu_c S_c/l_c)$ is the magnetic permeance, of this single-core magnetic circuit, and the term (Ni) is the magnetomotive force (MMF) of the exciting coil, . Hence, one can express the core flux, Φ_c, as follows:

$$\Phi_c = \mathbf{P}_c \mathbf{F}_c \tag{3.19}$$

where

$$\mathbf{P}_c = \mu_c S_c/l_c \tag{3.20}$$

and

$$\mathbf{F}_c = Ni \tag{3.21}$$

The reader should observe the magnetic and electric circuit analogies depicted in Figure 3.4, in which the permeance, $\mathbf{P}_c$, is analogous to the conductance, G, in the electric circuit; the MMF, $\mathbf{F}_c = Ni$, is analogous to the voltage source, E; and the magnetic flux, Φ_c, is analogous to the electric current, I. Notice that the inverse of the permeance $(1/\mathbf{P}_c)$ is the magnetic reluctance, $\mathbf{R}_c$, where

$$\mathbf{R}_c = 1/\mathbf{P}_c = l_c/(\mu_c S_c) \tag{3.22}$$

The reluctance term, $\mathbf{R}_c$, is analogous to the resistance, R, in the electric circuit, which is the inverse of the conductance, G.

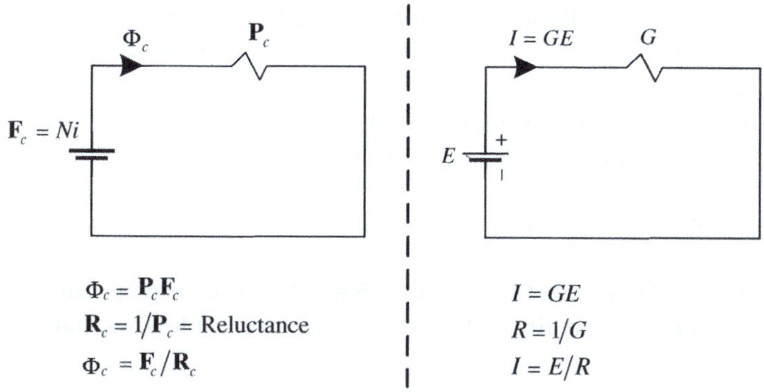

Figure 3.4 The magnetic and electric circuits analogy.

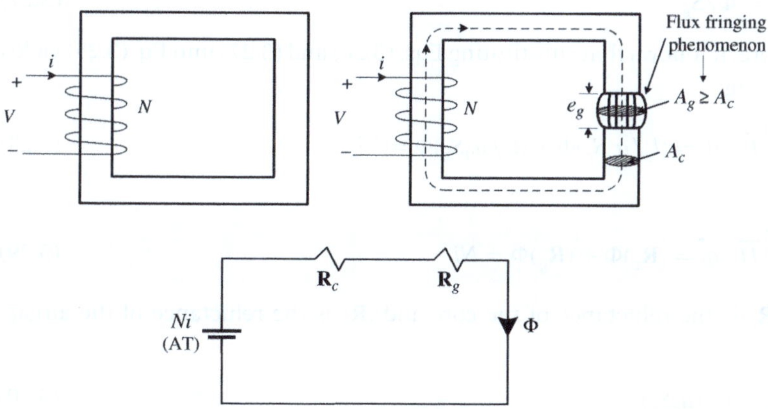

Figure 3.5 A ferrous core with an airgap and an *N*-turns excitation coil.

Now, consider the magnetic circuit with a ferrous core and an airgap of length, l_g, shown in Figure 3.5. Applying Ampere's law along the mean path throughout this magnetic circuit, one can write the following:

$$\oint \overline{H} \cdot d\overline{l} = \int_{core(l_c)} \overline{H} \cdot d\overline{l} + \int_{airgap(l_g)} \overline{H} \cdot d\overline{l} = Ni \tag{3.23}$$

where N is the number of turns in the exciting coil and i is the current passing through the coil, see Figure 3.5.

Now, an approximation of Eq. (3.23) can be expressed as follows:

$$\oint \overline{H} \cdot d\overline{l} = H_c l_c + H_g l_g = Ni \tag{3.24}$$

where H_c and H_g are the magnitudes of the average magnetic field intensity in the core and airgap, respectively, while l_c and l_g are the mean lengths in the core and airgap, respectively.

For a core permeability, μ_c, and airgap permeability, μ_0, Eq. (3.24) can be written in terms of the core and airgap flux densities, B_c and B_g, as follows:

$$\oint \overline{H} \cdot d\bar{l} = (B_c/\mu_c)l_c + (B_g/\mu_0)l_g = Ni \tag{3.25}$$

Now, for an amount of magnetic flux, Φ, and cross-sectional areas perpendicular to the flux flow for the core and airgap, S_c and S_g, respectively, it follows that

$$B_c = \Phi/S_c \tag{3.26}$$

and

$$B_g = \Phi/S_g \tag{3.27}$$

Therefore, it follows that substituting Eqs. (3.26) and (3.27) into Eq. (3.25) yields the following:

$$\oint \overline{H} \cdot d\bar{l} = (l_c/\mu_c S_c)\Phi + (l_g/\mu_0 S_g)\Phi = Ni \tag{3.28}$$

or

$$\oint \overline{H} \cdot d\bar{l} = (\mathbf{R}_c)\Phi + (\mathbf{R}_g)\Phi = Ni \tag{3.29}$$

Here, $\mathbf{R}_c$ is the reluctance of the core and .$\mathbf{R}_g$ is the reluctance of the airgap, where

$$\mathbf{R}_c = l_c/(\mu_c S_c) \tag{3.30}$$

and

$$\mathbf{R}_g = l_g/(\mu_0 S_g) \tag{3.31}$$

See Figure 3.5.

3.3 Magnetic Flux Linkages, Inductance, and Electromotive Forces

Examining the coil depicted in Figure 3.6 wrapped around a portion of a magnetic core, through which an amount of flux, Φ_c, threads through each turns of this N-turns coil, the flux linkage, λ_c, of this coil can be expressed as follows:

$$\lambda_c = N\Phi_c \tag{3.32}$$

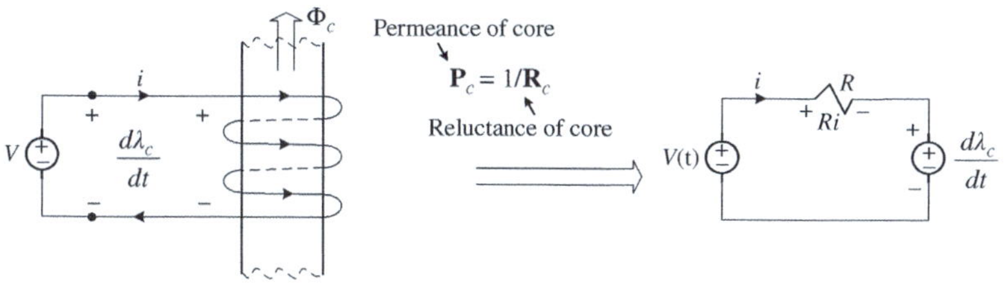

Figure 3.6 Flux linkage and electromotive force.

where

$$\Phi_c = \mathbf{P}_c(Ni) \tag{3.33}$$

in which $\mathbf{P}_c$ is the permeance of the core and i is the current exciting the N-turns coil.

Accordingly, substituting Eq. (3.33) into (3.32) yields the following:

$$\lambda_c = N^2\mathbf{P}_c i \tag{3.34}$$

Therefore, the inductance of this coil, L_c, which is defined as the flux linkage per ampere of the excitation current, i, can be expressed as follows:

$$L_c = \lambda_c/i = (N^2\mathbf{P}_c i)/i = N^2\mathbf{P}_c \tag{3.35}$$

or

$$L_c = N^2\mathbf{P}_c = N^2/\mathbf{R}_c \tag{3.36}$$

where the reluctance $\mathbf{R}_c = 1/\mathbf{P}_c$.

Now, considering the B–H nonlinear characteristic of ferrous cores, shown in Figure 3.7, the core permeability, μ_c, is as depicted in this figure, so is the inverse of permeability, the reluctivity, ν_c. Accordingly, the inductance, as defined by Eqs. (3.35) and (3.36), indicates that as magnetic cores saturate, inductances decrease.

If the flux linkage, λ_c, of a coil varies with time, according to Faraday's law, there will be an induced electromotive force (EMF), e_c, that can be expressed as follows:

$$e_c = d\lambda_c/dt \tag{3.37}$$

whose polarity is as depicted in Figure 3.6. In Figure 3.6, the voltage source, $V(t)$, supplying this N-turns coil with an ohmic resistance, R, is related to the induced EMF, e_c, as follows:

$$V(t) = Ri + e_c = Ri + (d\lambda_c/dt) \tag{3.38}$$

or

$$d\lambda_c/dt = V - Ri \tag{3.39}$$

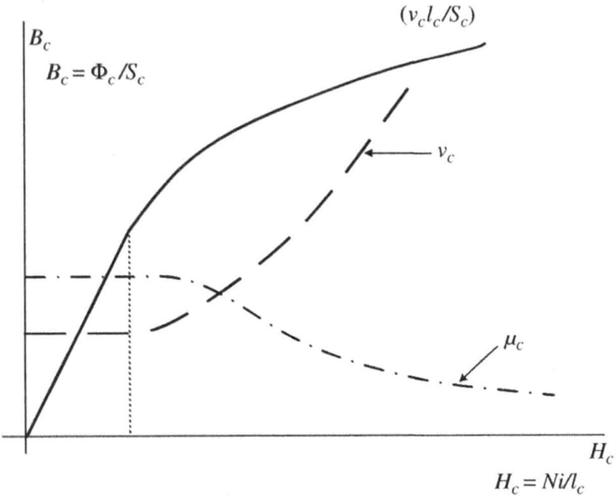

Figure 3.7 *B–H* characteristic nonlinearity.

However, from Eq. (3.35), the current, i, can be expressed as follows:

$$i = \lambda_c/L_c \tag{3.40}$$

Hence, Eq. (3.39) can be rewritten as follows:

$$d\lambda_c/dt = V - R(\lambda_c/L_c) \tag{3.41}$$

or

$$d\lambda_c/dt = V - (R/L_c)\lambda_c \tag{3.42}$$

that is,

$$\dot{\lambda}_c = -(R/L_c)\lambda_c + V \tag{3.43}$$

which is in the so-called state-space form $\dot{x} = ax + bu$, as in many control systems state-space formulations, which will appear in later chapters of this book. More specifically, Eq. (3.43) is the state-space formulation in the so-called flux linkage frame of reference throughout this book.

Now, consider the magnetic circuit that includes moving parts such as shown in Figure 3.8. This magnetic circuit consists of two reluctance paths: one path is entirely in a ferrous core of reluctance, $\mathbf{R}_c$, and a second path in series with the ferrous core which encompasses an airgap and a moving ferrous piece, the position of which is given by the position variable, x, whose reluctance, $\mathbf{R}_g(x)$, is a function of (x). Therefore, the total reluctance of the magnetic circuit is $(\mathbf{R}_c + \mathbf{R}_g(x))$, and here the N-turns coil inductance, $L_c(x)$, can be expressed as follows:

$$L_c(x) = N^2/(\mathbf{R}_c + \mathbf{R}_g(x)) \tag{3.44}$$

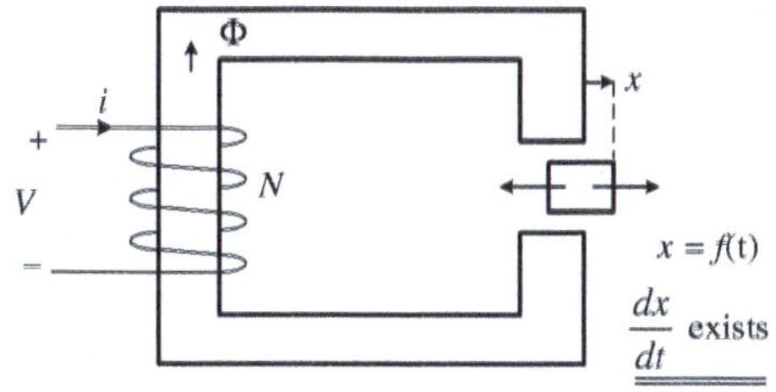

Figure 3.8 A magnetic circuit with an airgap and moving ferrous part.

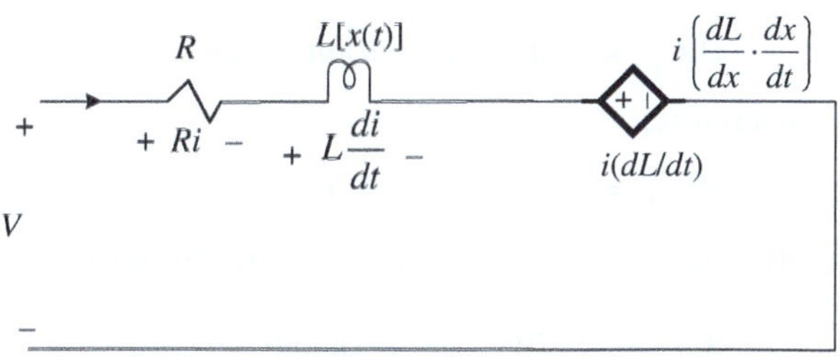

Figure 3.9 The electrical equivalent circuit with ohmic, transformer, and motional/translational voltage drops.

Hence, the flux linkage, λ_c, of this N-turns coil can be written as follows:

$$\lambda_c = L_c(x) \cdot i \tag{3.45}$$

and the voltage, V, at the terminals of the coil can be expressed as follows:

$$V = Ri + (d\lambda_c/dt) = Ri + d(L(x) \cdot i)/dt \tag{3.46}$$

which can be expanded as follows:

$$V = Ri + L(x) \cdot (di/dt) + i \cdot (dL(x)/dx) \cdot (dx/dt) \tag{3.47}$$

Equation (3.47) is depicted in the equivalent circuit shown in Figure 3.9, in which (Ri) is the "ohmic voltage drop" term, $[L(x) \cdot (di/dt)]$ is the "transformer voltage drop" term, and $[i \cdot (dL(x)/dx) \cdot (dx/dt)]$ is the "motional or translational voltage drop" term.

3.4 Energy Storage, Motion, and Forces in Magnetic Circuits

Referring to Figure 3.8, the terminal voltage, V, of the N-turns excitation coil can be expressed as stated earlier in Eq. (3.46) as follows:

$$V = Ri + e_c \tag{3.48}$$

where $e_c = d\lambda_c/dt$.

Therefore, the input power into the coil is

$$Vi = Ri^2 + (d\lambda_c/dt)i = Ri^2 + e_c i \tag{3.49}$$

The (Ri^2) term represents the ohmic resistive power dissipation, while the $[(d\lambda_c/dt)i] = e_c i$ term represents the power injected for energy storage into the magnetic field, or

$$e_c i = (d\lambda_c/dt)i \tag{3.50}$$

Hence, the differential input energy, $dW_{elec.}$, over a differential time period, dt, can be expressed as follows:

$$dW_{elec.} = e_c idt = (d\lambda_c/dt)idt = id\lambda_c \tag{3.51}$$

Therefore,

$$dW_{elec.} = id\lambda_c = id(L_c i) = (L_c i)di \tag{3.52}$$

or, the energy storage, $W_{elec.}$, can be expressed as follows:

$$W_{elec.} = \int dW_{elec.} = \int_0^i (L_c i)di = L_c i^2/2 \tag{3.53}$$

Meanwhile, for a mechanical force, F, exerted on the moving part of the magnetic circuit in Figure 3.8 and from basic principles of conservation of energy, one can express the mechanical work (energy) done by the force, F, exerted through a displacement, dx, as $dW_{mech.} = Fdx$ and state that

$$dW_{mech.} + dW_{elec.} = 0 = Fdx + d(L_c i^2/2) \tag{3.54}$$

or

$$F = -d(L_c i^2/2)/dx \tag{3.55}$$

That is, if the motion is rotational through an angular displacement, $d\theta$, the mechanical torque, T, can be expressed similarly as follows:

$$T = -d(L_c i^2/2)/d\theta \tag{3.56}$$

where the inductance term, L_c, and current, i, may be functions of the motion variable, x, or θ, and time, t.

Now, let us proceed to the development of a simplified method to express energy storage in a magnetic field of toroidal ferrous core excited by an N-turns idealized coil carrying an electric current, i. The toroidal ferrous core has a cross-sectional area, S, and a mean magnetic field path of length, l. Let the time-varying flux passing though this toroidal core be Φ. With a negligible resistance for the winding of this idealized coil, the induced voltage (EMF), e, at the terminals of this coil can be expressed based on Faraday's Law as follows:

$$e = N(d\Phi/dt) \tag{3.57}$$

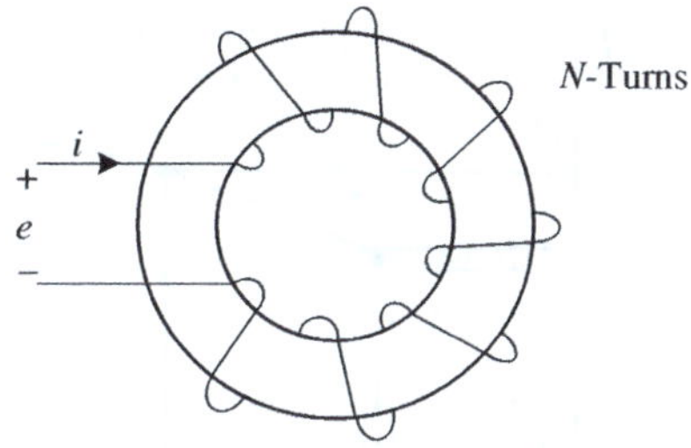

Figure 3.10 Toroidal N-turns-coil ferrous core device.

Hence, the instantaneous input power, P, can be written as follows:

$$P = ei = N(d\Phi/dt)i \tag{3.58}$$

and the input differential energy, $dW = Pdt$, becomes $dW = (ei)dt = N(d\Phi/dt)idt$, which yields the following:

$$dW = (Ni)d\Phi \tag{3.59}$$

However, for this simple toroidal geometry (Figure 3.10), and on the basis of Ampere's law, $\oint \overline{H} \cdot d\overline{l} = H_c l_c = Ni$. Furthermore, $d\Phi = d(SB) = S(dB)$. Accordingly, dW, in Eq. (3.12) can be rewritten as follows:

$$dW = (Hl)(SdB) = (Sl)(HdB) \tag{3.60}$$

Since (Sl) is the volume of the ferrous core, it follows that the term (HdB) is the stored energy density, dW, in joules per unit volume. Accordingly, in this case, the stored energy, W, in the magnetic field can be written as follows:

$$W = \text{Volume}\left(\int_0^{B_1} HdB\right) \tag{3.61}$$

which is depicted graphically in the nonlinear B–H magnetization characteristic of Figure 3.11, showing the concepts of energy per unit volume, $\left(\int_0^{B_1} HdB\right)$, and co-energy $\left(\int_0^{H_1} BdH\right)$ per unit volume.

3.4.1 Ferromagnetic Material *B*–*H* Characteristics

Consider the toroidal magnetic material core with a magnetization coil of N-turns, shown in Figure 3.10, with a mean path of length, l. Let the excitation current, $i(t)$, be an AC current of sinusoidally time-varying nature as shown in Figure 3.12. Consider the toroidal magnetic circuit of Figure 3.10, from its raw unmagnetized status at point "0" in the time domain instantaneous sinusoidally time-varying current $i(t)$ waveform of Figure 3.12 and in the B–H plane of Figure 3.13. The instantaneous value of the excitation current $i(t)$ varies throughout the AC cycles from point "0" in Figure 3.12 to points "1," "3," "4," "5," "7," "8," and point "1"

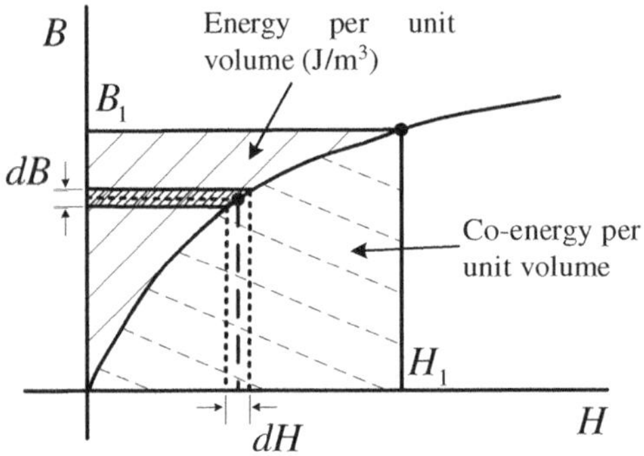

Figure 3.11 The concept of energy and co-energy density per unit volume.

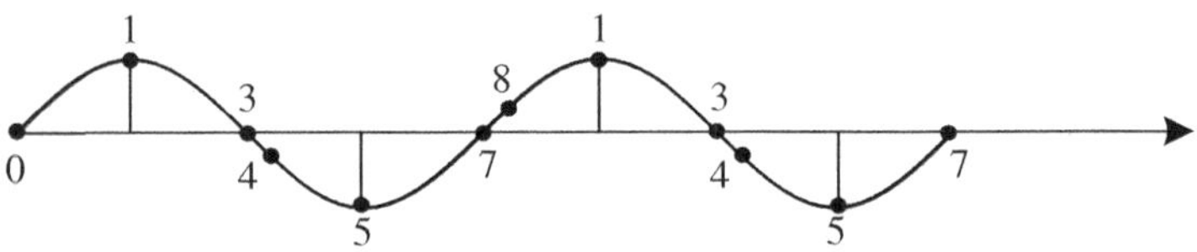

Figure 3.12 A sinusoidally time-varying excitation current, $i(t)$.

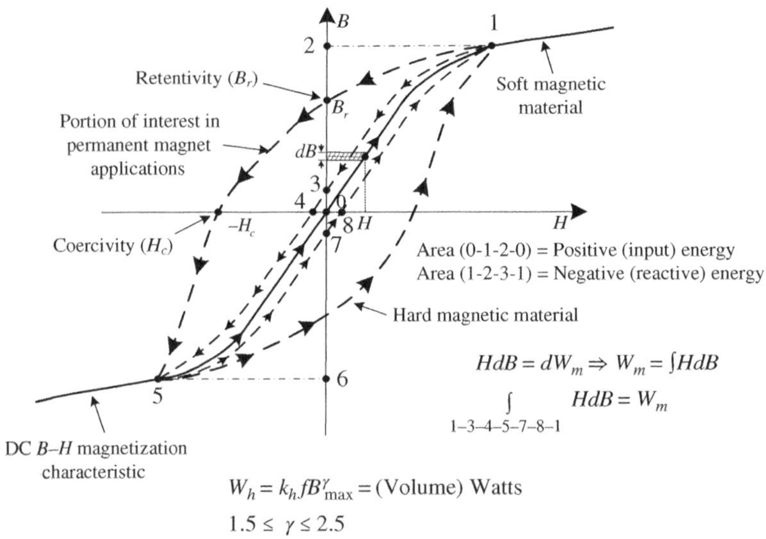

Figure 3.13 The B–H hysteresis phenomena for soft and hard magnetic materials and associated energy losses.

again in this figure. Meanwhile, in the *B–H* plane of Figure 3.13, the corresponding (*B*, *H*) coordinate points start at point "0" in this figure, followed by the sequence of points "1," "3," "4," "5," "7," "8," and back to point "1," to form the "narrow" hysteresis loop for soft magnetic materials, as depicted in Figure 3.13. The area comprising this hysteresis loop represents an energy loss, and hence a power loss in such material, the so-called hysteresis iron or ferrous losses, which can be expressed based on empirical practice by, W_h, the hysteresis power loss as

$$W_h = k_h f (B_{\max})^{\gamma} \tag{3.62}$$

where W_h is the loss in watts per unit of core volume, k_h is a hysteresis core loss factor function of the material, f, of the AC excitation cycle frequency in Hertz, $B_{\max}$ is the maximum flux density of the AC excitation in Tesla, while γ is a dimensionless exponent function of the shape of the hysteresis loop of the material, and in general, $1.5 \leq \gamma \leq 2.5$.

The typical shape of a wide hysteresis loop associated with hard magnetic materials is also shown in Figure 3.13, where the portion of such a loop occurring in the second quadrant of the *B–H* plane represents the portion of the magnetic *B–H* characteristic, which is most important in characterizing permanent magnet (PM) materials, finding exceeding importance in the design and performance of PM machines. See Figure 3.14 for the characteristics of commonly used PM materials.

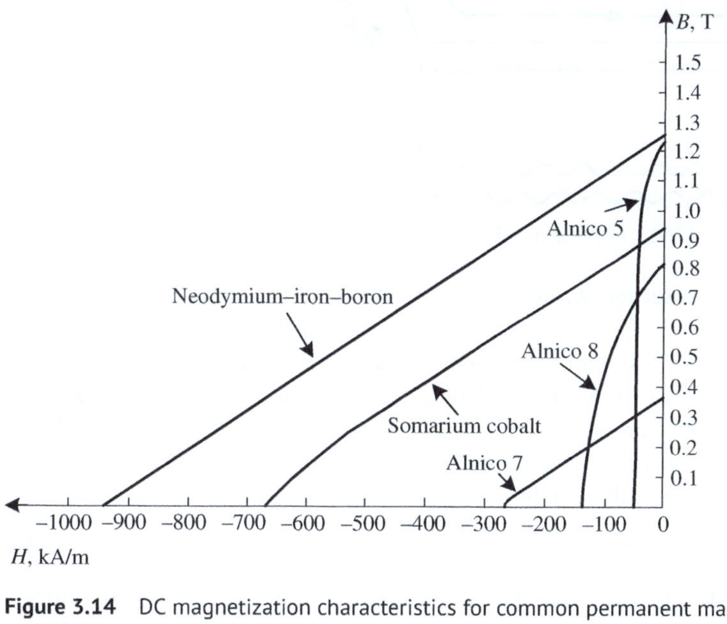

Figure 3.14 DC magnetization characteristics for common permanent magnet (PM) materials.

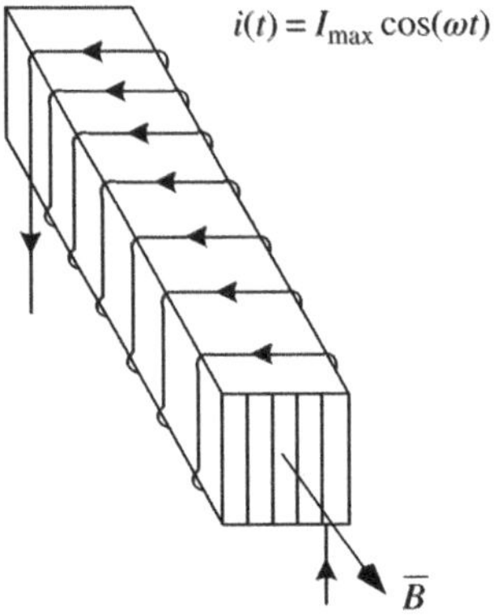

Figure 3.15 Laminated ferrous core excited by an AC current-carrying coil.

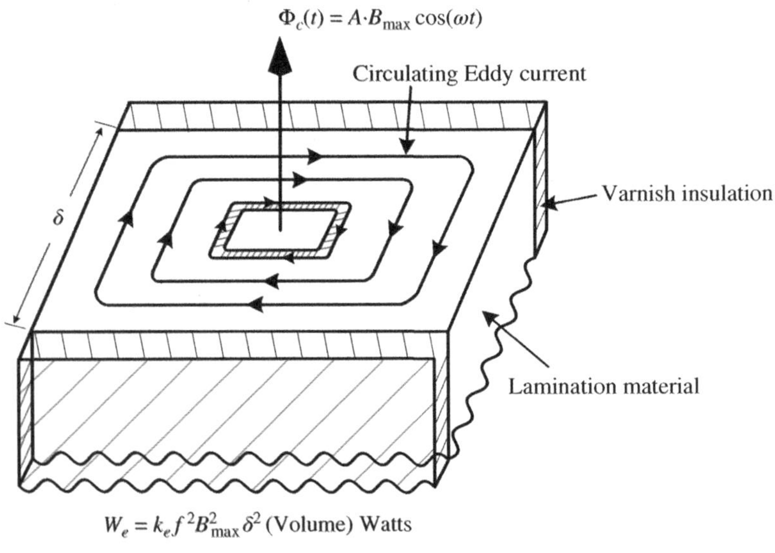

Figure 3.16 Cross-section of a lamination with varnish insulation.

Another type of losses in laminated ferrous magnetic cores, see Figures 3.15 and 3.16, is due to induced circulating eddy currents in the laminations, as depicted in a typical cross-section of a single lamination in Figure 3.16. The losses due to these circulating induced eddy currents per unit volume of laminated ferrous core material, W_e, can be expressed as follows:

$$W_e = k_e f^2 B_{max}^2 \delta^2 \tag{3.63}$$

where W_e is the loss in watts per unit volume of core, k_e is an eddy current loss factor function of the material conductivity of the lamination, f is the AC excitation cycle frequency in Hertz, B_{max} is the maximum flux density of the AC excitation in Tesla, and δ is the thickness of the lamination depicted clearly in Figure 3.16.

3.5 Lorentz's Law-Induced Voltage and Forces in Magnetic Fields

According to Lorentz's law, a force, $\overline{F}$, exerted on a moving charge, q, with a velocity vector, $\overline{v}$, in the presence of a magnetic field flux density, $\overline{B}$, and an electric field intensity, $\overline{E}$, can be expressed as follows:

$$\overline{F} = q(\overline{E} + \overline{v} \times \overline{B}) \tag{3.64}$$

from which it follows that if a conductor of length, $\overline{l}$, see Figure 3.17, is moving with a velocity vector, $\overline{v}$, in the presence of a magnetic flux density, $\overline{B}$, the induced EMF, e, in this conductor can be expressed as follows:

$$e = \overline{l} \cdot (\overline{v} \times \overline{B}) \tag{3.65}$$

However, for the most pertinent example shown in Figure 3.17, one can write the following:

$$\overline{l} = (l)\hat{z}, \quad \overline{v} = (v)\hat{x}, \quad \overline{B} = (B)\hat{y} \tag{3.66}$$

Here,

$$\overline{v} \times \overline{B} = (vB)(\hat{x} \times \hat{y}) = (vB)\hat{z} \tag{3.67}$$

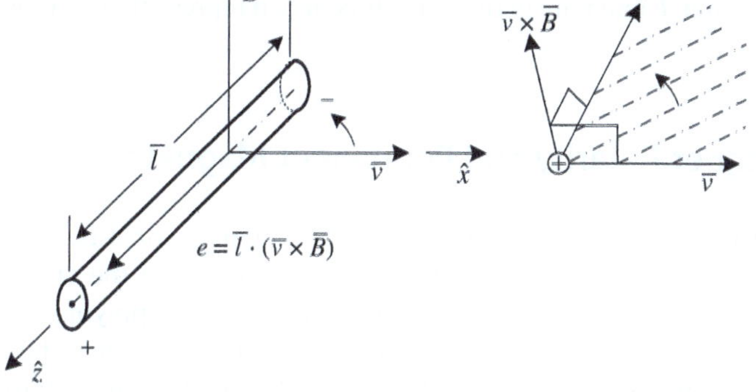

Figure 3.17 Induced EMF in a conductor moving with a velocity, $\overline{v}$, in the presence of a magnetic flux density, $\overline{B}$.

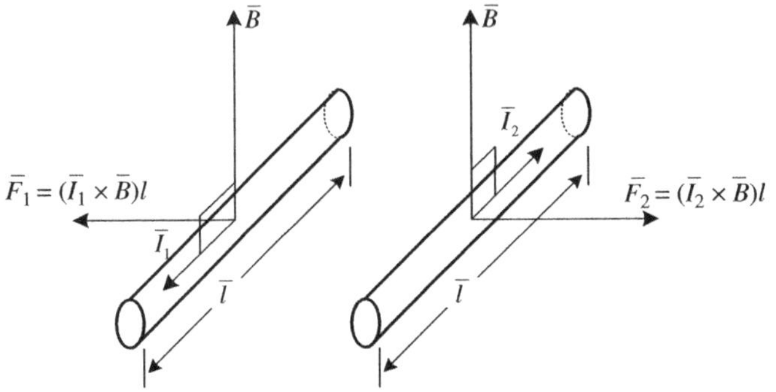

Figure 3.18 The orientation of forces, flux densities, and current flow in straight conductors.

Therefore, it follows that the EMF, e, can be expressed as follows:

$$e = (vB)\hat{z} \cdot (l)\hat{z} = Blv \tag{3.68}$$

The case depicted in Figure 3.17 is most commonly encountered for EMFs induced in electrical conductors and windings in electric machinery.

Based on Lorentz's law, we now formulate expressions of forces experienced by current-carrying electric conductors in the presence of a magnetic field flux density, $\overline{B}$, see Figure 3.18. If we represent the directional current flow vector, $\overline{I}$, the force, $\overline{F}$, exerted on a conductor of length, l, one can write the following:

$$\overline{F} = (\overline{I} \times \overline{B})l \tag{3.69}$$

The reader should note the orientations of the current vectors, $\overline{I}_1$ and $\overline{I}_2$, and the corresponding exerted forces, $\overline{F}_1$ and $\overline{F}_2$, in Figure 3.18, respectively. Meanwhile, based on the relative direction of current flow, $\overline{I}$, mechanical force, $\overline{F}$, and velocity, $\overline{v}$, orientation, the case of $\overline{F}$ and $\overline{v}$ in opposite direction represents a "generator case" and, vice versa, when $\overline{F}$ and $\overline{v}$ are in the same direction, it represents a "motor case," see Figure 3.19.

3.6 Electromagnet Polarity and Permanent Magnets

Consider the electromagnet illustrated in Figure 3.20, which consists of a current-carrying coil wrapped around a magnetic material core with the orientations of the $\overline{B}$ field surrounding the core as shown in this figure. Applying the right-handed rule principles, the $\overline{B}$ field exits outward of the top surface, thus creating the appearance of a situation of a north pole of a PM, while it enters into the bottom surface, thus creating the appearance of a situation of a south pole of a PM. Meanwhile, in plane (bird's eye-view), the $\overline{B}$ field is oriented outward of

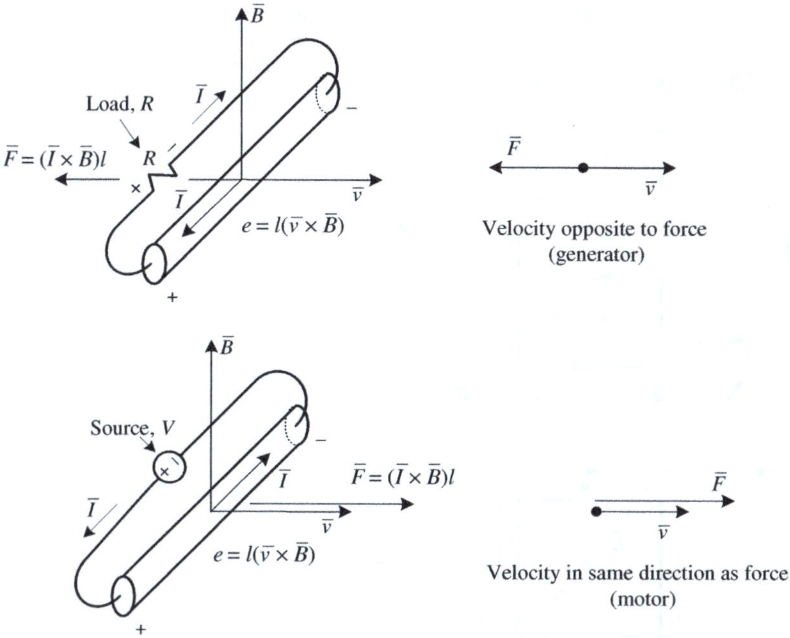

Figure 3.19 The cases of a "generator" and a "motor" for a straight current-carrying conductor in the presence of a magnetic field, $\bar{B}$.

the page within the core and surrounding coil and inward into the page outside the core and surrounding coil. Therefore, one can imagine replacing the core and coil, which is this electromagnet, with a PM of the same shape, while orienting the north pole of the PM upwards along the *y*-axis orientation in Figure 3.20, creating the same $\bar{B}$-field pattern as this electromagnet. We will use this analogy between electromagnets and PMs to develop a basic and simple understanding of torque production in electric machinery with stator and rotor cores of cylindrical geometry, referring to such polarities as virtual "North" and "South" PM poles.

3.7 Elementary Electromechanical Torque Production and Magnetomotive Force (MMF) Distribution Concepts

Consider the cylindrical device shown in Figure 3.21a. The outer stationary cylinder (the stator) is lined with an inner layer of four cylindrical segments of PM material, whose inner surfaces facing a cylindrical airgap shell alternately have North, South, North, South poles, of the four respective PMs, as depicted in Figure 3.21a. Meanwhile, an inner cylindrical shaft is covered with four

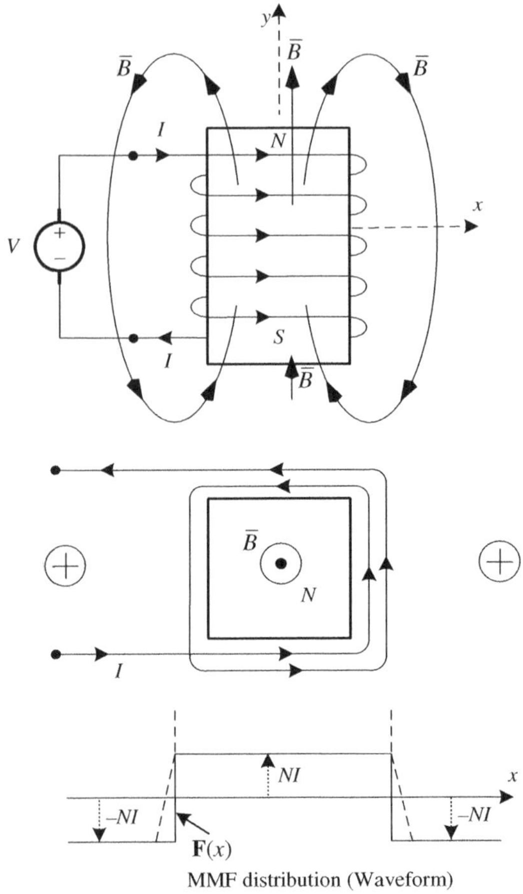

Figure 3.20 An electromagnet shown in elevation and plane including MMF spatial distribution (waveform).

quadrants of PM material cylindrical shells facing the airgap alternately having sequential PM polarities of North, South, North, South poles, shown in this same figure. The corresponding flattened diagram of this device is shown in part (b) of this figure, Figure 3.21b. Also, shown in Figure 3.21b are the corresponding MMF waveforms of the stationary and rotating portions of this device, $\vec{\mathbf{F}}_{stator}$ and $\vec{\mathbf{F}}_{rotor}$, respectively, which are shifted in space from each other by an angular shift, δ, the so-called torque angle.

Carefully examining the relative positions of the PM "North" and "South" polarities facing the airgap in Figure 3.21a,b, one should realize that there will be mechanical forces of attraction between the "North" and "South" polarities on both the stationary and rotating portions of this device attempting to line up each "North" with "South" poles in a respective manner in this device.

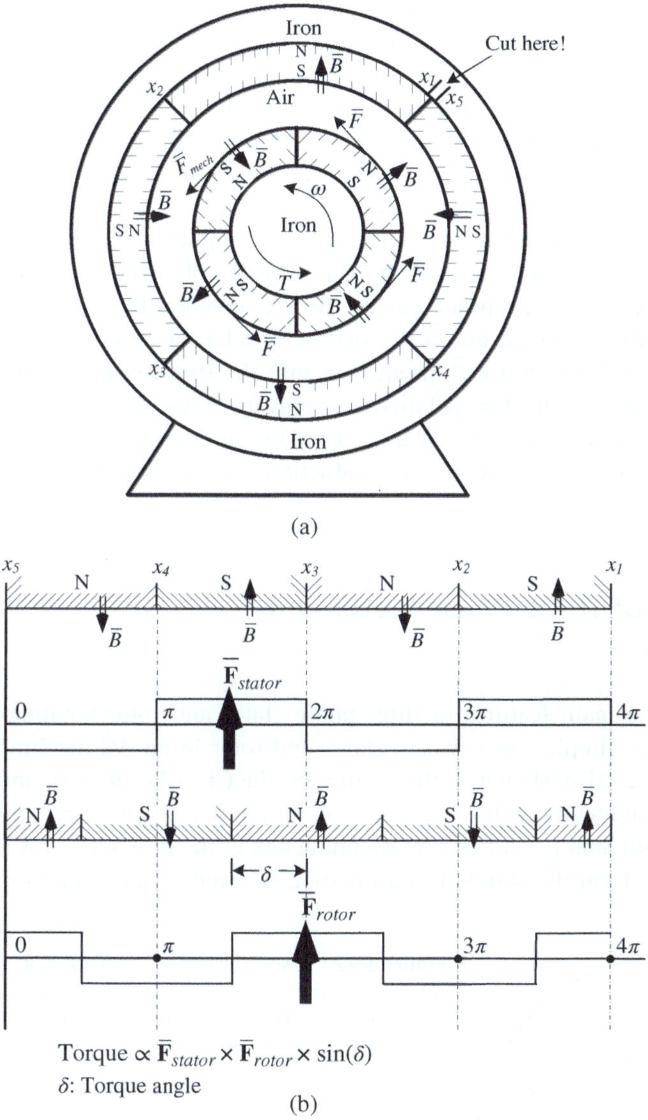

Figure 3.21 Part (a) A cylindrical device with ferrous cores and permanent magnets. Part (b) A flattened redraw of the device and its corresponding MMF waveforms.

This constitutes a mechanical force, and hence torque, on the rotor of this device, in the counterclockwise direction as depicted in part (a) of Figure 3.21, and a corresponding mechanical force to the left on the rotating member of this device in part (b) to line up each "North" with each "South" on the stationary portion of this device. These forces of attraction of each "North" of a PM (or virtual PM) with each "South" of a PM (or virtual PM) on the rotating and stationary portions of such devices cause the production of a mechanical torque in virtually most rotating machines. Here, we mean by a "virtual PM" being of a "North" or a "South" PM polarity that may be created by the presence of current-carrying coils or a winding with an appropriate current flow orientation that creates a flux density, $\overline{B}$, which is oriented out of the surface of a ferrous material for a virtual "North" PM polarity, and a $\overline{B}$ orientation into the surface of a ferrous material for a virtual "South" PM polarity, respectively. These concepts and principles will be applied in Sections 3.9 and 3.10 to explain mechanical torque production in three-phase AC synchronous, induction, and brush-commutator DC machines.

3.8 Elements of Three-Phase Stator and Rotor Windings

Here, we discuss the main features of three-phase stator and rotor windings in AC machines. The simplest is a six-slot stator and rotor in an AC machine shown in Figure 3.22. Also, shown in this figure are the (A, $-A$), (B, $-B$), and (C, $-C$) coil-sides/conductors of the A, B, and C stator phase coils/turns, as well as the (a, $-a$), (b, $-b$), and (c, $-c$) coil-sides/conductors of the a, b, and c rotor phase coils/turns, to form the simplest of a three-phase winding configuration.

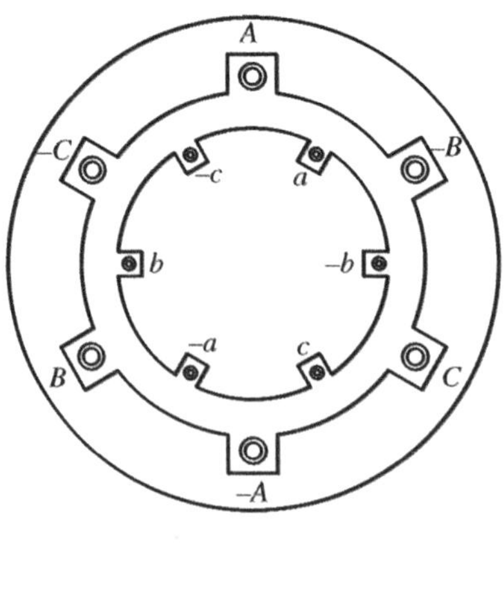

Figure 3.22 A six-slot stator and a six-slot rotor including the three-phase A, B, and C stator phase coils/turns and three-phase a, b, and c rotor phase coils/turns.

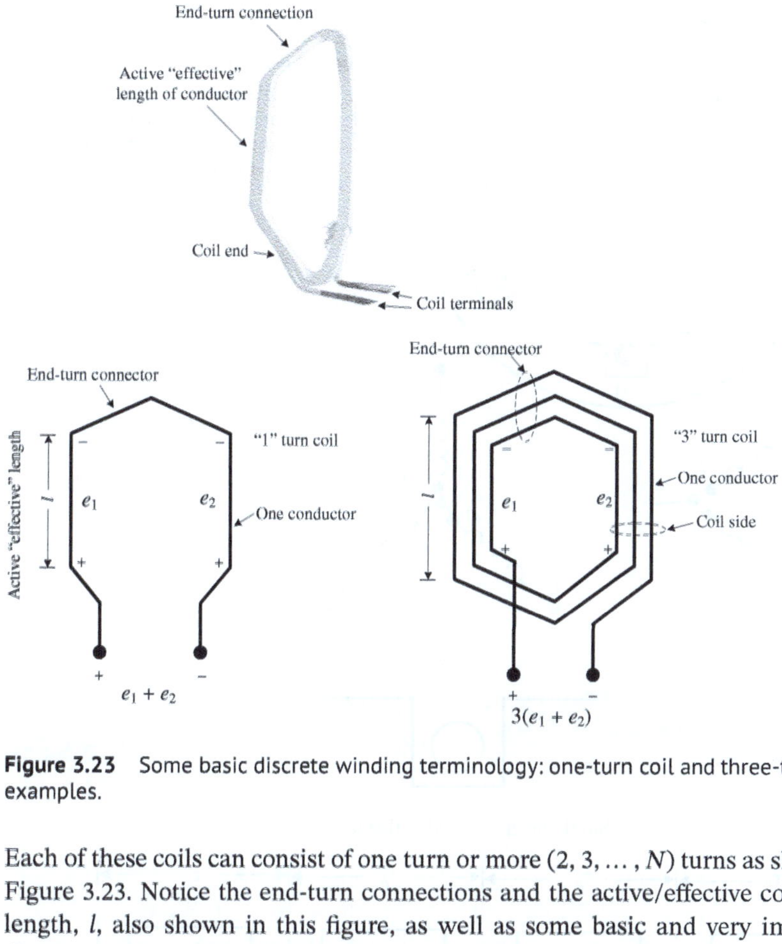

Figure 3.23 Some basic discrete winding terminology: one-turn coil and three-turns coil examples.

Each of these coils can consist of one turn or more (2, 3, … , N) turns as shown in Figure 3.23. Notice the end-turn connections and the active/effective conductor length, l, also shown in this figure, as well as some basic and very important discrete winding/coil terminology used throughout this book.

In Figure 3.24, we depict a stator with a three-phase set of coils (A, $-A$), (B, $-B$), and (C, $-C$), which can be connected in either a Delta or Wye configuration as shown in this figure. Once more, these coils can be single-turn (one-turn) or multiturns coils, see Figure 3.23.

We now present AC winding layout types and coil types in Figures 3.25–3.28, with accompanying symbolisms and terminologies used throughout this book. Here, P is the number of poles, while N_s is the number of stator slots housing the winding and m is the number of phases, hence, $m = 3$, for three-phase windings. Hence, the number of slots per pole per phase, q, can be expressed as follows:

$$q = N_s/(m \times P) \tag{3.70}$$

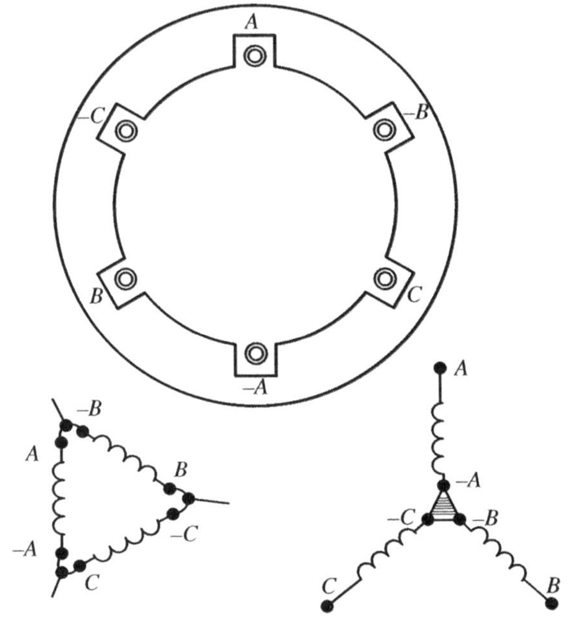

Figure 3.24 A six-slot stator including the three-phase *A*, *B*, and *C* phase coils/turns and six coil-sides/conductors connected in either Delta or Wye fashion.

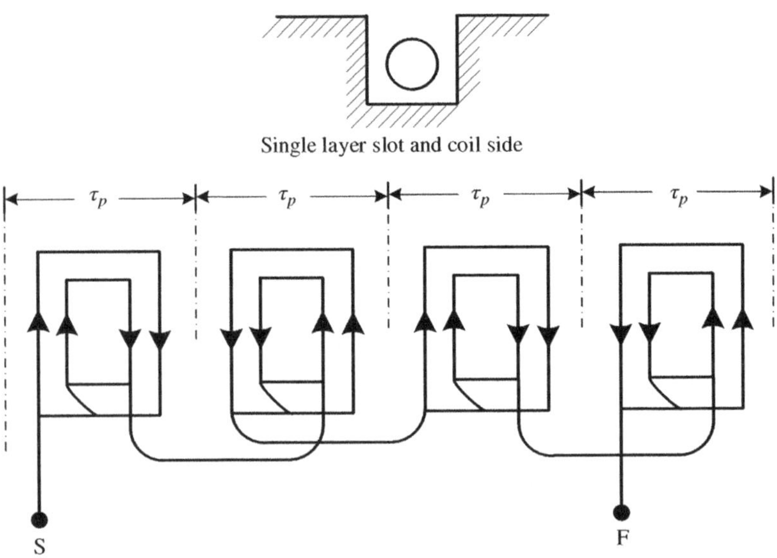

Figure 3.25 Single-layer winding with dissimilar coils and ends (concentric-coils) $P = 4$ and $q = 4$.

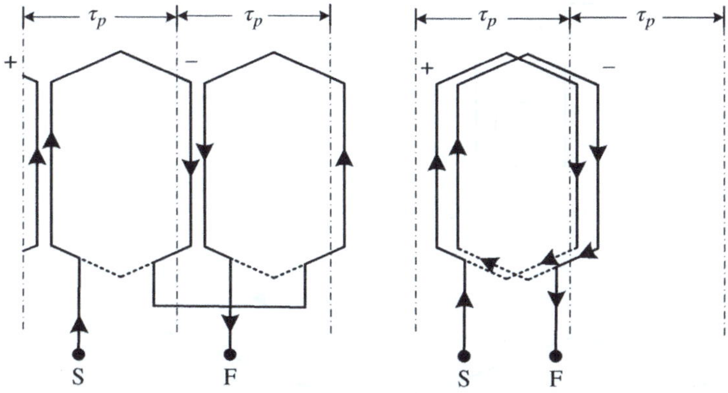

Figure 3.26 Single-layer winding with similar coils and ends (lattice-coils or grid-coils) $P = 2$ and $q = 2$, q = slots per pole per phase (nonfractional), $q = N_s/(m \times P)$.

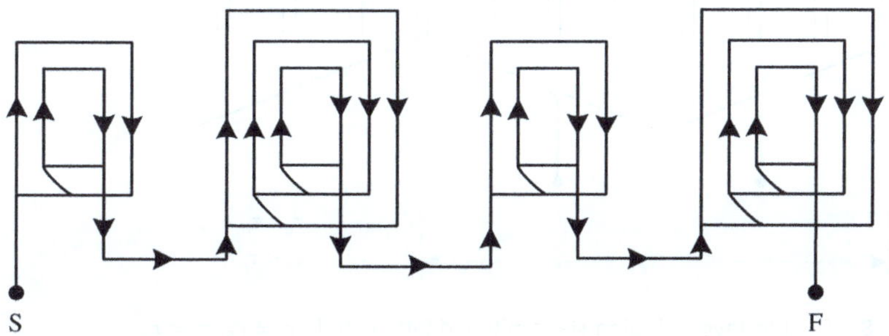

Figure 3.27 Single fractional slot winding with dissimilar coils and coil ends (concentric-coils) $P = 8$ and $q = 2\frac{1}{2}$.

Here, in Figures 3.25–3.28, τ_P is the pole-pitch or pole-span, expressed in number of slots per pole, or in meters at mid-airgap diameter. In electrical radians, $\tau_P = \pi$, or in electrical degrees, $\tau_P = 180°e$, or in mechanical radians, $\tau_P = 2\pi/P$, or in mechanical degrees, $\tau_P = 360°/P$, mech. deg.

In case the slots per-pole per-phase, Eq. (3.70). is a fractional number, the winding is referred to as a tractional-slot winding, see Figure 3.27 as an example. In the concentric, dissimilar coils with single-layer windings, an example with $P = 4$ and $q = 4$ is shown in Figure 3.25. Meanwhile, a single-layer winding with similar coils of the lattice-coils or grid-coils is shown in Figure 3.26, for an example with $P = 2$ and $q = 2$. Recall that the number of slots per pole per phase is given by Eq. (3.70).

A case of double-layer, short-pitched winding, where the coil-pitch or span, $\tau_c < \tau_P$, of lattice-coils-type, is shown in Figure 3.28. Whenever $\tau_c < \tau_P$, we call

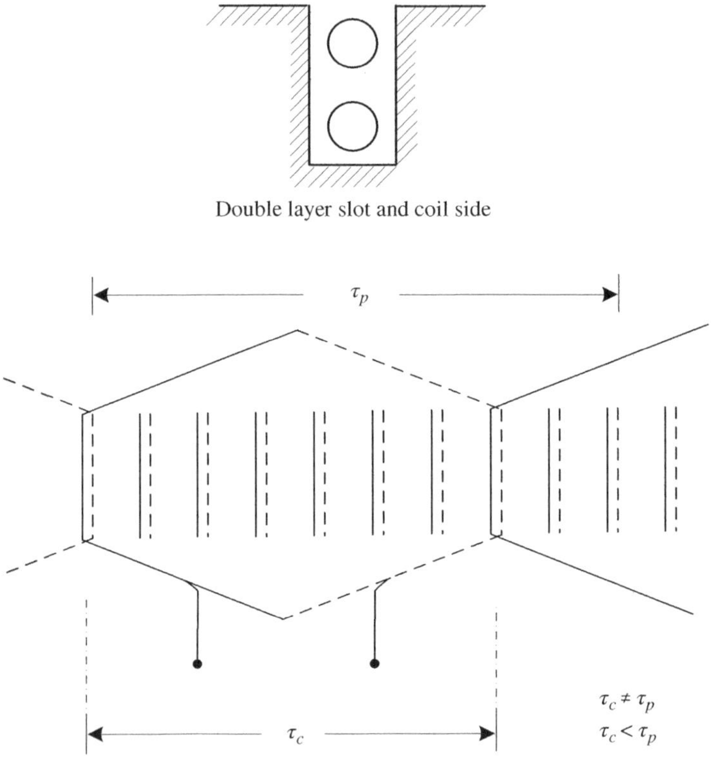

Figure 3.28 Double-layer winding short-pitched lattice-coils $\tau_c \neq \tau_p$, $\tau_c < \tau_p$.

such a winding "short-pitched." Meanwhile, whenever $\tau_c > \tau_p$, such a winding is referred to as "long-pitched" and whenever $\tau_c = \tau_p$, such a winding is referred to as a "full-pitched" winding.

3.9 Elementary Three-Phase Synchronous and Induction Machines

In this section, we present the magnetic circuit and winding spatial layout schematics of synchronous and induction machines in Figures 3.29–3.33, while Figure 3.34 shows the photograph of the actual coil embedding of coils/coil-sides in the stator of an AC machine such as in the synchronous and induction types.

Here, a simple six-slot, two-pole, salient-pole, wound-field synchronous machine schematic is shown in Figure 3.29, with the circuit schematic of

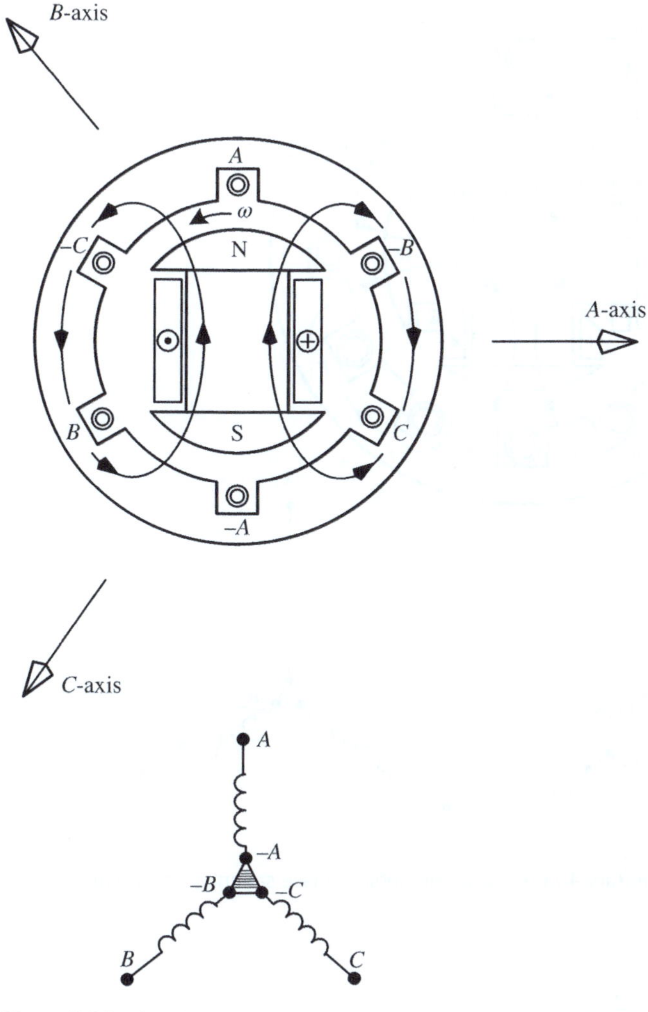

Figure 3.29 An elementary two-pole, three-phase salient-pole synchronous machine.

a Wye-connected stator winding. Meanwhile, a 12-slot, 4-pole, salient-pole, wound-field synchronous machine is depicted in Figure 3.30 with its possible circuit schematic of two Wye-connection stator winding arrangements. Meanwhile, Figures 3.31 and 3.32 depict in isometric three-dimensional configuration a 24-slot, 4-pole, salient-pole, wound-field synchronous machine, including the details of the *A*, *B*, and *C* stator slot labeling. Meanwhile, a simple 6-slots stator and a squirrel-cage rotor are depicted in Figure 3.33, including an isometric configuration of a 12-bars squirrel-cage.

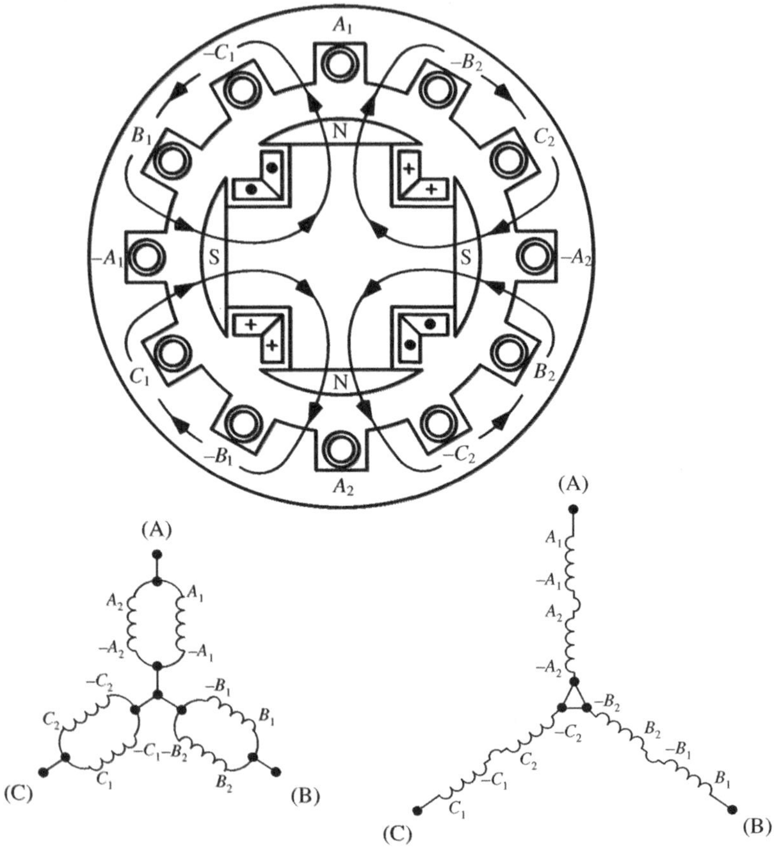

Figure 3.30 An elementary 4-pole, 12 stator slots, 3-phase salient-pole synchronous machine.

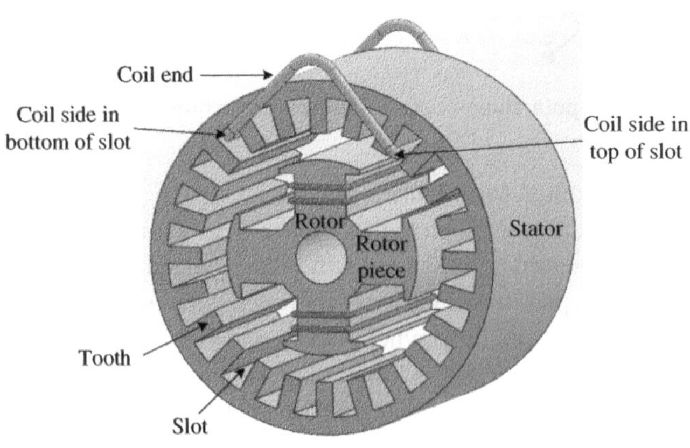

Figure 3.31 The magnetic circuit of a 4-pole, salient-pole, 24 stator slots synchronous machine.

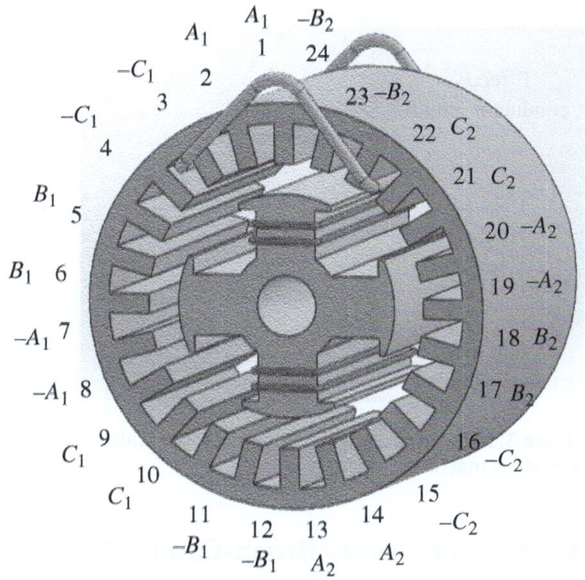

Figure 3.32 The magnetic circuit of a 4-pole, salient-pole, 24 stator slots synchronous machine with phases *A*, *B*, and *C* slot labelings.

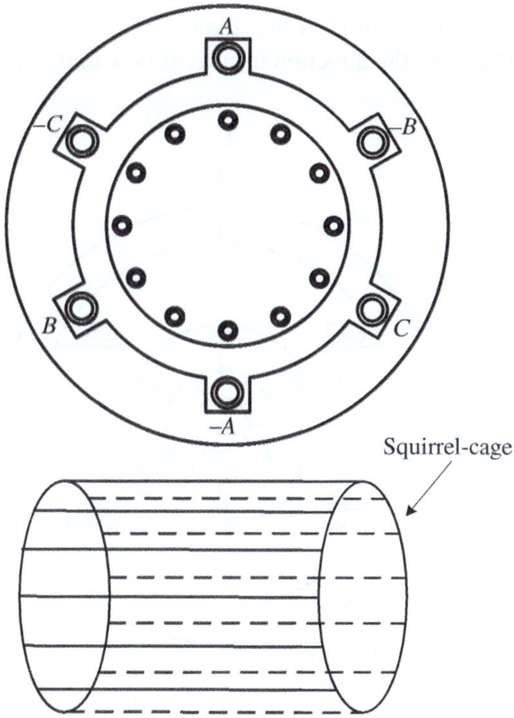

Figure 3.33 An elementary two-pole, three-phase squirrel-cage induction machine.

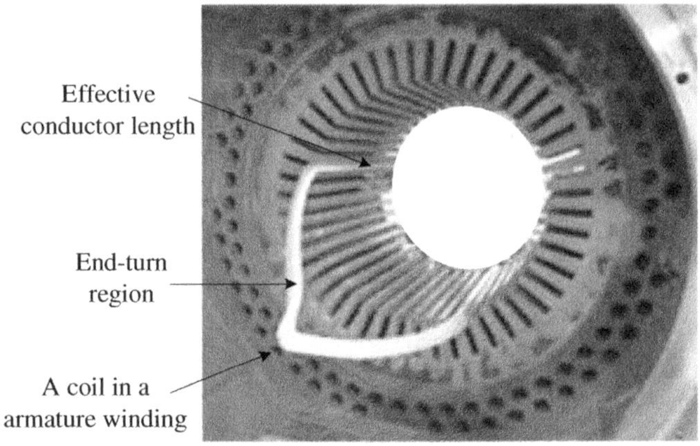

Figure 3.34 Armature coils embedded in a stator-slotted core of an AC synchronous or induction machine.

3.10 Elementary Brush-Commutator DC Machines

In Figure 3.35, we show a simple two-pole, six-slot armature (rotor), cross-section of a DC machine, while indicating in this figure the orientation, in-and-out of the page, the direction of current flow in the field winding surrounding the poles

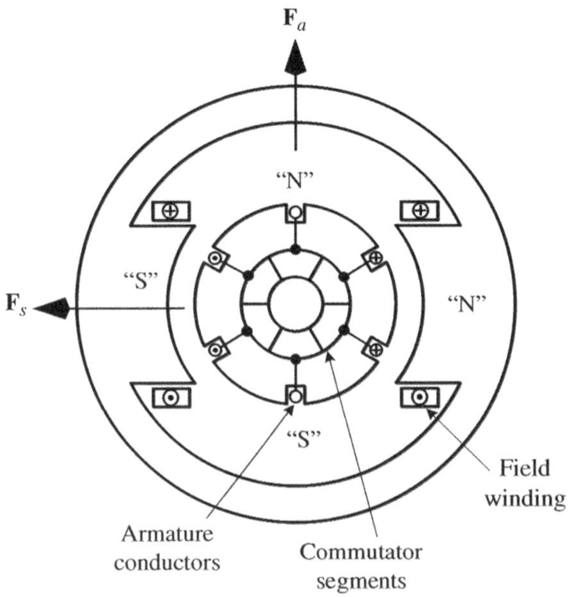

Figure 3.35 An elementary two-pole, six-armature slots DC machine and commutator segments as well as field coils structure.

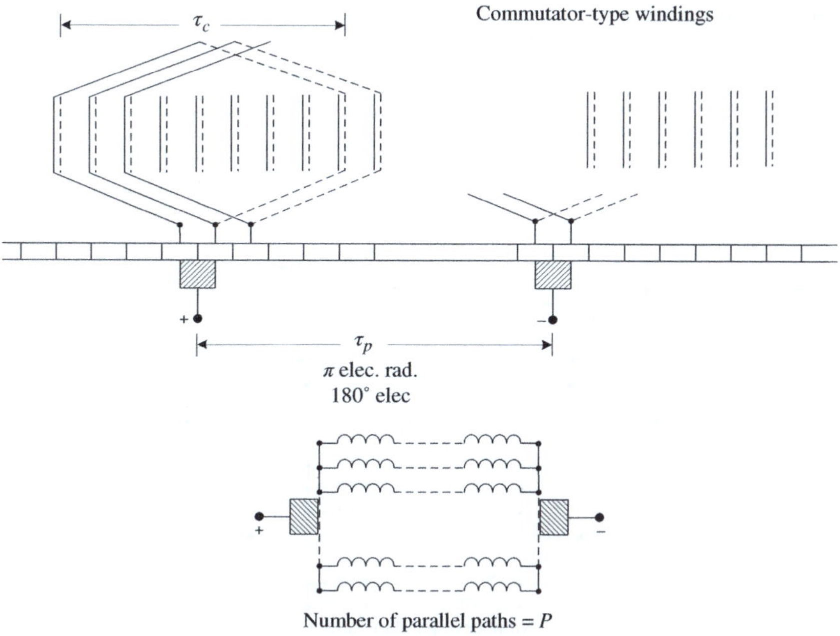

Figure 3.36 Lap-type lattice DC armature coils layout.

(on the stator) and current flow, in-and-out of the page, in the armature winding (on the rotor). Shown also in this figure are the commutator segments to which the armature coils (turns) are connected. Not shown here are the brushes which rest on the commutator segments' outer cylindrical surface.

Meanwhile, a "lap-type" lattice DC armature coils (turns) layout is shown in Figure 3.36 and a "wave-type" lattice DC armature coils (turns) layout is shown in Figure 3.37. Details on these windings will be discussed in a later chapter of this book on DC machines.

3.11 Mechanical Torque Production in AC and DC Machines

In this section, we will discuss some elementary aspects of electromechanical torque production, based on the simple principle that north "N" and south "S" poles of PMs produce a force of attraction and like poles "N" and "N" or "S" and "S" poles produce a force of repulsion. These "N" and "S" poles are due to PM materials or virtual ones resulting from electromagnets produced by the flow

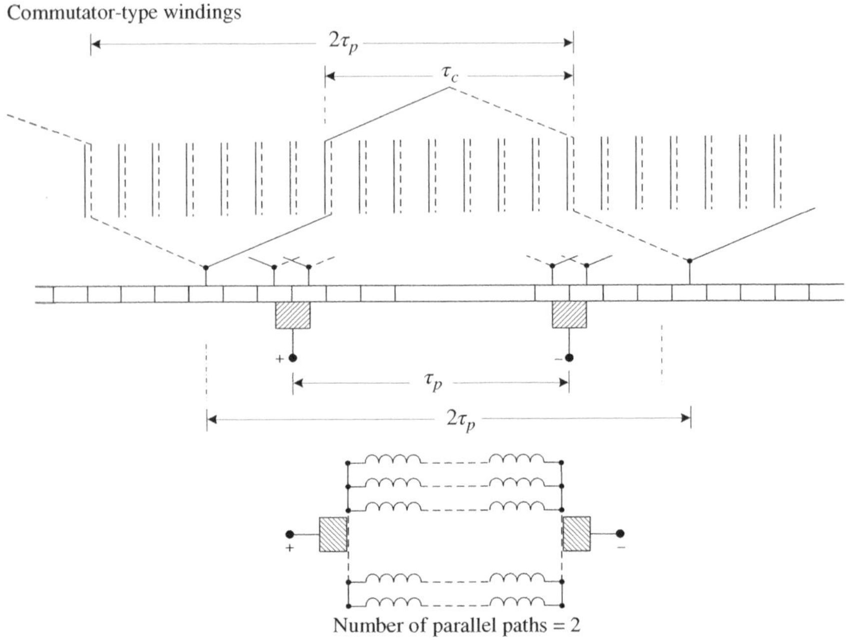

Figure 3.37 Wave-type lattice DC armature coils layout.

of currents in windings, and the resulting establishment of MMFs. In this vein, we invoke a discussion we presented earlier in Section 3.7, and specifically with reference to Figure 3.21, with its two parts, part (a) and part (b).

Accordingly, regarding electromechanical torque production in DC machines, we refer to the cross-section of an elementary two-pole, six armature slots DC machine shown in Figure 3.35. The orientation of the MMF of the field winding, $\mathbf{F}_f$, and orientation of the MMF of the armature (rotor) winding, $\mathbf{F}_a$, produced two virtual north "N" poles on the stator and rotor, respectively, and produced two virtual south poles on the stator and rotor, respectively. Examining Figure 3.35 closely, one can argue that the two "N" poles on the stator and rotor repel, and the two "S" poles on the stator and rotor also repel, thus creating/establishing a torque in the counterclockwise direction. This machine will be said to operate in a "motoring" mode, that is, when the torque and rotation are in the same direction. Meanwhile, if the machine is forced by an external prime mover to rotate in opposition to the orientation/direction of the torque, the machine will be said to operate in a "generating" mode. It should be stated that a "motoring" mode means that electrical energy is being converted to mechanical, and vice versa for a "generating" mode when the torque and rotation are in opposite orientation/direction.

Furthermore, with regard to electromechanical torque production in AC machines, that is, synchronous and induction machines, we refer to Figure 3.38a–d for the synchronous machine and Figures 3.39 and 3.40 for the wound-rotor and squirrel-cage induction machines, respectively. We start our discussion with the case of the synchronous machine. In this case, the A, B, and C, three-phase stator winding positive sequence phase currents can be expressed in the time-domain as follows:

$$i_A(t) = I_{\max} \cos(\omega t) \tag{3.71}$$

$$i_B(t) = I_{\max} \cos(\omega t - 2\pi/3) \tag{3.72}$$

$$i_C(t) = I_{\max} \cos(\omega t - 4\pi/3) \tag{3.73}$$

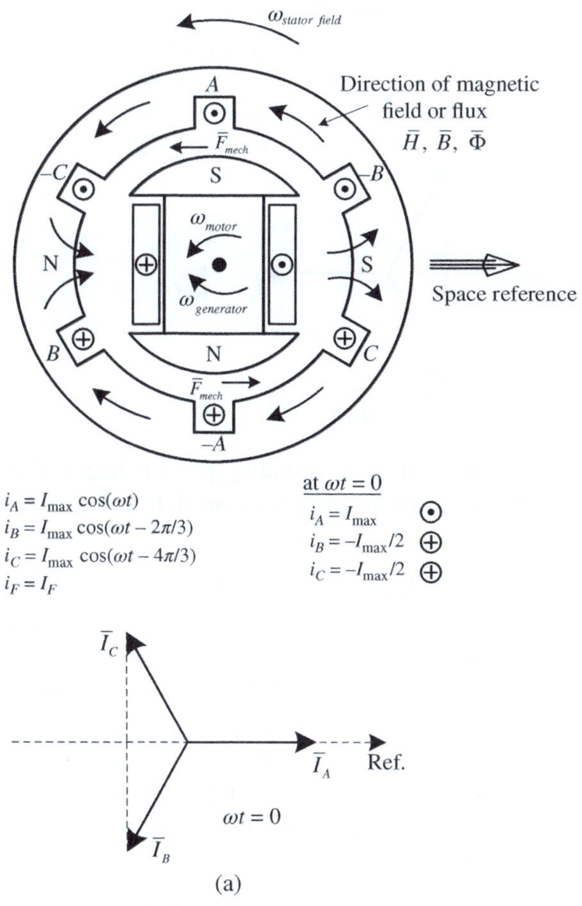

Figure 3.38 (a) Rotating field in the stator of a two-pole, salient-pole synchronous machine at $\omega t = 0$. (b) Rotating field in the stator of a two-pole, salient-pole synchronous machine at $\omega t = 2\pi/3$. (c) Rotating field in the stator of a two-pole, salient-pole synchronous machine at $\omega t = 4\pi/3$. (d) Rotating field in the stator of a two-pole, salient-pole synchronous machine at $\omega t = 2\pi$.

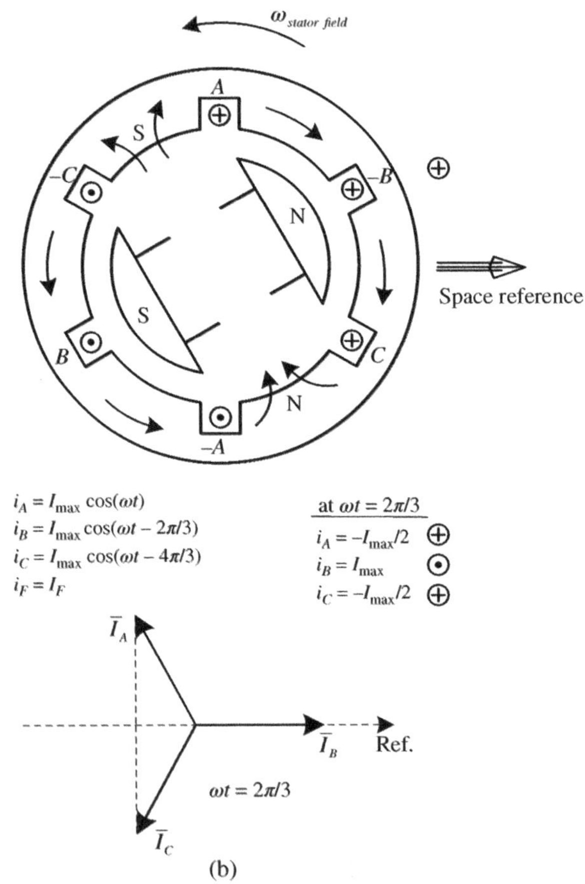

Figure 3.38 (*Continued*)

At $\omega t = 0$, the corresponding phasor diagram representation is as shown in Figure 3.38a, and the instantaneous A, B, and C current values are correspondingly as follows:

$$i_A(0) = I_{\max}, \quad i_B(0) = -I_{\max}/2, \quad i_C(0) = -I_{\max}/2 \tag{3.74}$$

Shown in the cross-section of the machine in Figure 3.38a by "thick" dots and crosses for ($I_{\max}$) and ($-I_{\max}$), respectively, and "thin" dots and crosses for ($I_{\max}/2$) and ($-I_{\max}/2$), respectively. This results in the magnetic field orientation/pattern shown in the synchronous machine cross-section of Figure 3.38a, which results in the virtual "N" and "S" poles shown in this figure.

At $\omega t = 2\pi/3$, the corresponding phasor diagram representation is as shown in Figure 3.38b, and the instantaneous A, B, and C current values are correspondingly as follows:

$$i_A(2\pi/3) = -I_{\max}/2, \quad i_B(2\pi/3) = I_{\max}, \quad i_C(2\pi/3) = -I_{\max}/2 \tag{3.75}$$

Figure 3.38 (*Continued*)

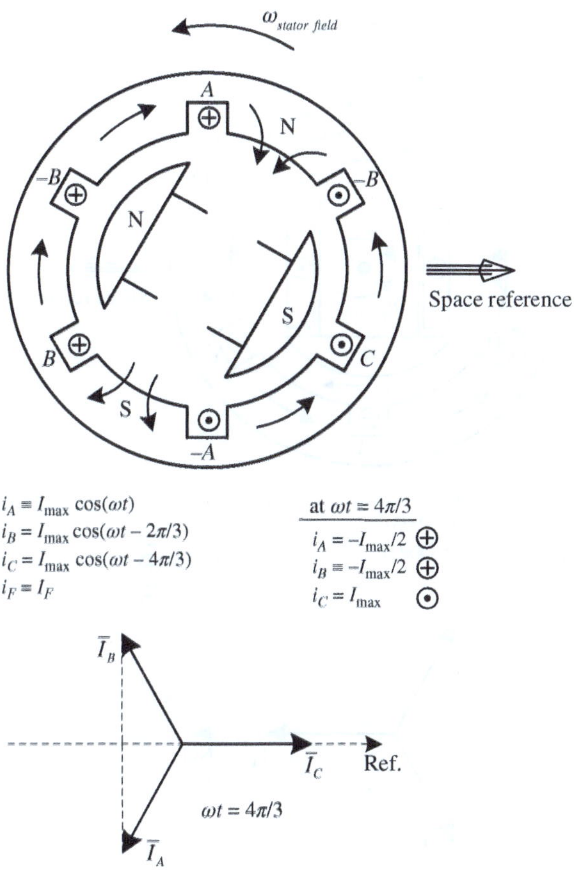

(c)

Figure 3.38b shows the cross-section of the machine by "thick" and "thin" dots and crosses, as explained earlier in Figure 3.38a. This results in the virtual "N" and "S" poles shown in Figure 3.38b.

Similarly, at $\omega t = 4\pi/3$, we have the resulting phasor diagram of Figure 3.38c, and correspondingly,

$$i_A(4\pi/3) = -I_{max}/2, \qquad i_B(4\pi/3) = -I_{max}/2, \qquad i_C(4\pi/3) = I_{max} \qquad (3.76)$$

Consequently, this results in the virtual "N" and "S" poles shown in Figure 3.38c.

Finally, at $\omega t = 2\pi$, we have an identical current values and field pattern to those at $\omega t = 0$, resulting in the "N" and "S" poles shown in Figure 3.38d.

Examining Figure 3.38a–d, one can see clearly that the field pattern rotates in a similar manner as the rotation of the phasor diagram of the A, B, and C phase currents. Hence, if the rotor is to rotate at the same speed of rotation as the field pattern, later referred to as the synchronous speed, the virtual "N" and "S" poles

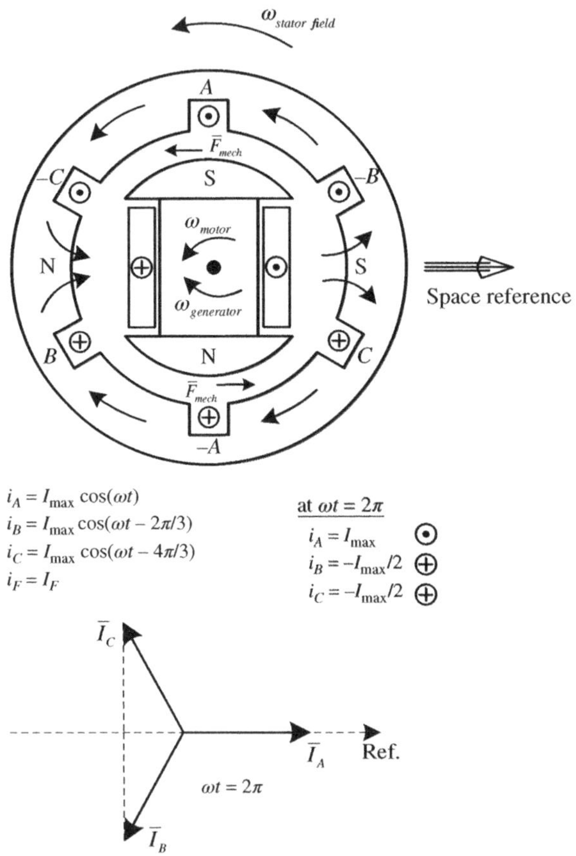

(d)

Figure 3.38 *(Continued)*

on the stator will always be pulling on the opposite "S" and "N" poles on the rotor, respectively, thus producing a counterclockwise torque. Hence, it should be pointed out that the angular speed of rotation of this rotating field is the synchronous speed, n_{syn}, given in rev/min by the following expression:

$$n_{syn} = 120f/P \tag{3.77}$$

where f is the frequency of the AC current in Hertz and P is the number of poles in an AC machine. The proof of the expression of Eq. (3.77) will be revisited in later chapters on synchronous and induction machines. It should also be pointed out that

$$\omega = 2\pi f \tag{3.78}$$

where ω is the angular frequency in electrical radians per second.

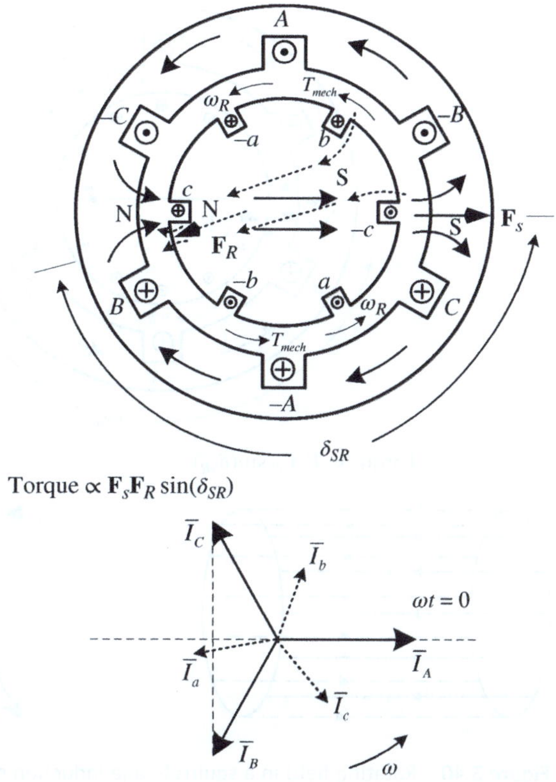

Figure 3.39 Rotating field in a wound-rotor induction machine at $\omega t = 0$.

Now, we discuss the case of wound-rotor and squirrel-cage induction machines, illustrated in Figures 3.39 and 3.40, respectively. In such induction machines, the current flowing in the rotor conductors is induced in such a manner as to establishing a demagnetizing MMF opposing the magnetizing MMF established by the current flowing in the *A*, *B*, and *C* stator phase windings. Hence, notice the obtuse angle, δ_{SR}, between the stator MMF space vector, $\overline{\mathbf{F}}_s$, and the rotor MMF space vector, $\overline{\mathbf{F}}_R$, in both the wound-rotor case, Figure 3.39, and the squirrel-cage case, Figure 3.40, induction machine. Also, notice the near πelectrical radians out-of-phase nature of the angle between the currents, $\overline{I}_A$ and $\overline{I}_a$, $\overline{I}_B$ and $\overline{I}_b$, $\overline{I}_C$ and $\overline{I}_c$, respectively, as depicted in the wound-rotor induction machine in case of Figure 3.39.

These MMF stator and rotor space vectors shown in Figures 3.39 and 3.40 lead to the induced virtual "N" and "S" poles established on the inner stator rim and outer rotor rim, shown in these two figures. Examining the positions of each respective pair of near-proximity north "N" and near-proximity south "S" virtual poles on the stator and rotor, shown in Figures 3.39 and 3.40, the repulsion

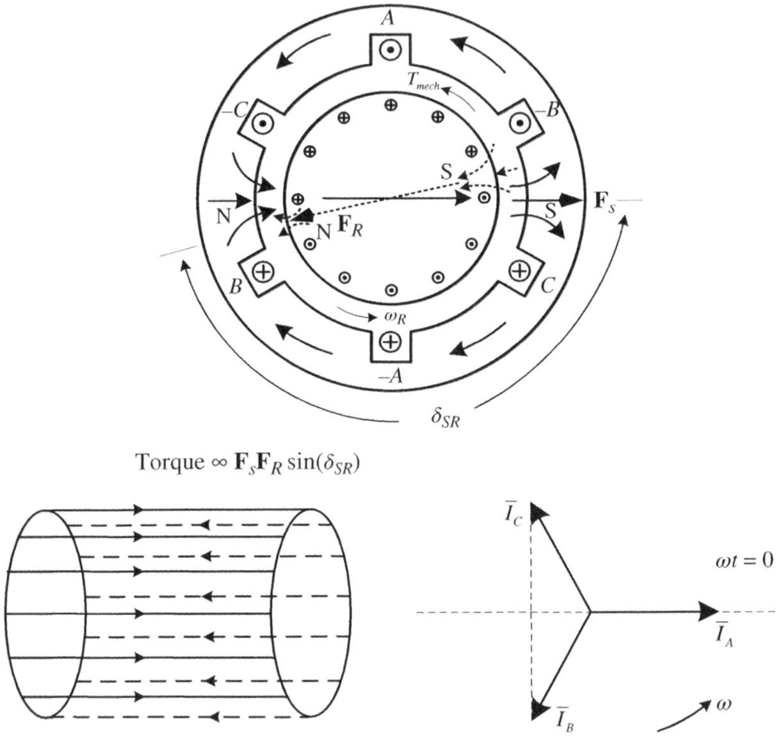

Figure 3.40 Rotating field in a squirrel-cage induction machine at $\omega t = 0$.

forces between each pole of like poles exert a counterclockwise mechanical torque pulling these two rotors in the counterclockwise direction. These concepts discussed earlier explain the process of mechanical torque production in both the wound-rotor and squirrel-cage induction machines operating in a motoring mode with rotor rotation and mechanical torque orientation in the same direction, here in this case in the counterclockwise direction. Again, the reader should closely examine Figures 3.39 and 3.40 for a fuller grasp of these processes of producing/developing torque.

In all the cases discussed in this section, the developed mechanical torque, T_{dev}, is proportional to the product of the amplitude of the stator MMF, $\overline{\mathbf{F}}_s$, the rotor MMF, $\overline{\mathbf{F}}_R$, and $\sin(\delta_{SR})$, δ_{SR} being the torque angle subtended between $\overline{\mathbf{F}}_s$ and $\overline{\mathbf{F}}_R$. That is,

$$T_{dev} \propto \mathbf{F}_s \cdot \mathbf{F}_R \cdot \sin(\delta_{SR}) \tag{3.79}$$

Explanations on this and other key matters on MMF and EMF are the subject of Chapter 5 in this book.

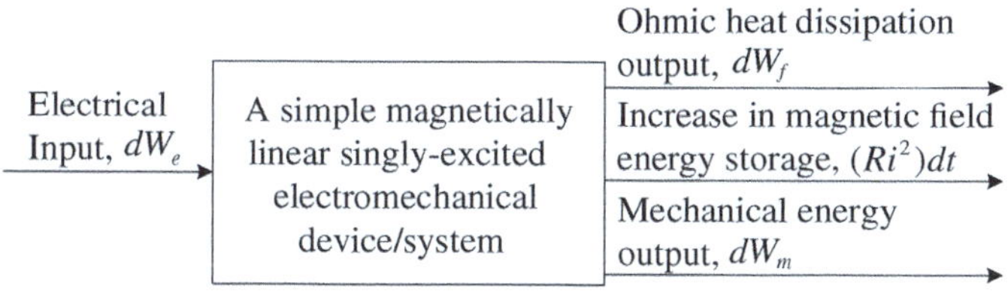

Figure 3.41 A single-excited electromechanical system.

Meanwhile, in Chapter 4, we discuss some basic electromechanical concepts in electric machines and drives, including mechanical load dynamics. This will include Newtonian rotational load/prime mover and motor/generator dynamics.

3.12 Forces and Torques in Magnetically Linear Singly Excited Electromechanical Devices and Systems

From basic principles of conservation of energy, one can write the following:

$$\sum \text{Input energy} = \sum \text{Output energy} \tag{3.80}$$

Hence,

$$dW_e = (Ri^2)dt + dW_f + dW_m \tag{3.81}$$

In a singly excited system, the electrical input energy, dW_e, can be expressed using Faraday's law as follows:

$$dW_e = vidt = (Ri + d\lambda/dt)idt \tag{3.82}$$

$$dW_e = (Ri^2)dt + id\lambda \tag{3.83}$$

Substituting Eq. (3.83) into (3.81) yields the following:

$$(Ri^2)dt + id\lambda = (Ri^2)dt + dW_f + dW_m \tag{3.84}$$

Hence, it follows that

$$dW_f + dW_m = id\lambda \tag{3.85}$$

However, for rapidly moving electromechanical systems, a differential displacement (mechanical motion), *dx*, happens so fast in an infinitesimally small differential time that the flux linkage of the system, λ, can be considered to have remained almost constant throughout that displacement. That is, one can write the following:

$$d\lambda = 0 \tag{3.86}$$

Hence, from Eqs. (3.85) and (3.86), one can write the following:

$$dW_f + dW_m = 0 \tag{3.87}$$

or

$$dW_f = -dW_m \tag{3.88}$$

However, for a system force, F_x, moving through a displacement, dx, one can write the following:

$$dW_m = (F_x)dx \tag{3.89}$$

Hence, from Eqs. (3.88) and (3.89), we have

$$dW_m = (F_x)dx = -dW_f \tag{3.90}$$

Accordingly, in general, Eq. (3.90) gives F_x, as follows:

$$F_x = -\partial W_f / \partial x \tag{3.91}$$

The partial derivative notation is necessary because the magnetically stored energy, W_f, is a function of several variables (here i and x).

For the case of rotational motion, one deals with torque, T_θ, and angular displacement, $d\theta$, and Eq. (3.91) becomes as follows:

$$T_\theta = -\partial W_f / \partial \theta \tag{3.92}$$

Meanwhile, the magnetically stored energy, W_f, in singly excited systems can be expressed as follows:

$$W_f = \frac{1}{2} L i^2 = \frac{1}{2} \frac{(Li)^2}{L} = \frac{1}{2} \frac{\lambda^2}{L} \tag{3.93}$$

where λ = constant and L is the inductance.

Hence,

$$\begin{aligned} \frac{\partial W_f}{\partial x} &= \frac{\partial}{\partial x}\left(\frac{\lambda^2}{2} \cdot L^{-1}\right) = \frac{\lambda^2}{2} \frac{\partial}{\partial x}(L^{-1}) \\ &= \frac{\lambda^2}{2} \frac{d}{dx}(L^{-1}) = \frac{\lambda^2}{2} \frac{d}{dL}(L^{-1}) \cdot \frac{dL}{dx} \end{aligned} \tag{3.94}$$

That is,

$$\frac{\partial W_f}{\partial x} = \frac{\lambda^2}{2}(-L^{-2}) \cdot \frac{dL}{dx} \tag{3.95}$$

or

$$\frac{\partial W_f}{\partial x} = -\frac{\lambda^2}{2L^2}\left(\frac{dL}{dx}\right) \tag{3.96}$$

Substituting Eq. (3.95) into (3.91), one obtains force, F_x, as follows:

$$F_x = -\left[-\frac{\lambda^2}{2L^2}\left(\frac{dL}{dx}\right)\right] \tag{3.97}$$

or

$$F_x = \frac{\lambda^2}{2L^2}\left(\frac{dL}{dx}\right) = \frac{i^2}{2}\left(\frac{dL}{dx}\right) \tag{3.98}$$

Similarly, for torque, T_θ, we have the following:

$$T_\theta = \frac{\lambda^2}{2L^2}\left(\frac{dL}{d\theta}\right) = \frac{i^2}{2}\left(\frac{dL}{d\theta}\right) \tag{3.99}$$

If the stored energy in the magnetic field, W_f, is to be preferably expressed in terms of total system flux, Φ, and system magnetic reluctance, $\mathbf{R}$, one can write the following:

$$\left.\begin{aligned} \lambda &= N\Phi \\ L &= N^2/\mathbf{R} \\ \Phi &= \mathbf{F}/\mathbf{R} \end{aligned}\right\} \tag{3.100}$$

where $\mathbf{F}$ is the MMF of the singly excited system.

If $d\lambda = 0$, Eq. (3.86), it follows that

$$d\lambda = d(N\Phi) = N(d\Phi) = 0 \tag{3.101}$$

or

$$d\Phi = 0 \tag{3.102}$$

and

$$W_f = \frac{1}{2}Li^2 = \frac{1}{2}\frac{\lambda^2}{L} = \frac{1}{2}\frac{(N\Phi)^2}{N^2/\mathbf{R}} = \frac{1}{2}\mathbf{R}\Phi^2 \tag{3.103}$$

Hence,

$$W_f = \frac{1}{2}\mathbf{R}\Phi^2 \tag{3.104}$$

Hence, with $d\Phi = 0$ and $\Phi =$ constant, it follows that

$$\frac{\partial W_f}{\partial x} = \frac{1}{2}\Phi^2\left(\frac{\partial \mathbf{R}}{\partial x}\right) \tag{3.105}$$

Therefore,

$$F_x = -\frac{\partial W_f}{\partial x} = -\frac{1}{2}\Phi^2\left(\frac{\partial \mathbf{R}}{\partial x}\right) \tag{3.106}$$

Similarly, for torque, T_θ, we have the following:

$$T_\theta = -\frac{\partial W_f}{\partial \theta} = -\frac{1}{2}\Phi^2\left(\frac{\partial \mathbf{R}}{\partial \theta}\right) \tag{3.107}$$

This is the so-called reluctance torque, and will exist if and only if $(\partial \mathbf{R}/\partial\theta)$ is not equal to zero.

Homework Problems

3.1 An inductor is to be designed using a magnetic core such as shown in Figure P1.1. The core is of a uniform cross-sectional area, $A_c = 5\times10^{-4}\ \text{m}^2$, and is of mean length, $l_c = 0.20\ \text{m}$. It has an airgap of length, g, and a winding of N turns. Both g and N have to be obtained in the design process. Neglect fringing and leakage effects.

(a) For the direction of current shown in the winding, indicate on the magnetic circuit's diagram the direction of flow of flux, Φ, and the direction of the flux density, B, in the core and airgap.

(b) The relative permeability of the core $\mu_r = 5000$, and the core material barely begins to saturate if the flux density begins to exceed 1.2 T. Find the values of N and g if the inductor core reaches a flux density of 1.2 T when the current in the coil is 5 A. (Hint: $\lambda = \text{Li} = N\Phi$, $L = 15\ \text{mH}$.)

(c) State the orientation of the force acting on the two surfaces of the iron core facing the airgap and whether it tends to close or open such an airgap. Also, compute the magnitude of this force in Newton.

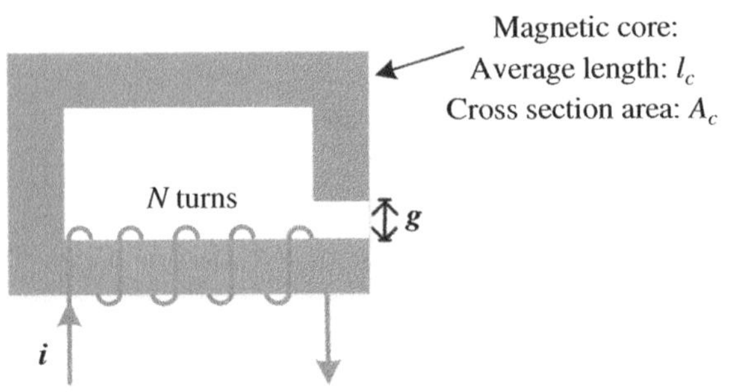

Figure P1.1 Geometry of an inductor.

3.2 In the electric machine shown in Figure P2.1, please perform the following tasks:

(a) Show the direction of magnetic flux and induced apparent "north" and "south" magnetic poles on the outer surfaces of the rotors and inner surfaces of the stators of each.

(b) State the type of machine, whether it is DC brush-commutator type, PM brushless DC type, permanent magnet AC synchronous type, field excitation AC synchronous type, or AC induction machine.

(c) Mark the appropriate boxes (use "Y" for yes and "N" for no) applicable to the status of torque, motor rotation, generator rotation, and presence of armature winding on the stator or rotor or both, for the machine shown in Figure P2.1.

The torque is: Clockwise_____ Counterclockwise_____
Motor rotation is: Clockwise_____ Counterclockwise_____
Generator rotation is: Clockwise_____ Counterclockwise_____
Armature winding is on: Stator_____ Rotor_____ Both stator and rotor_____

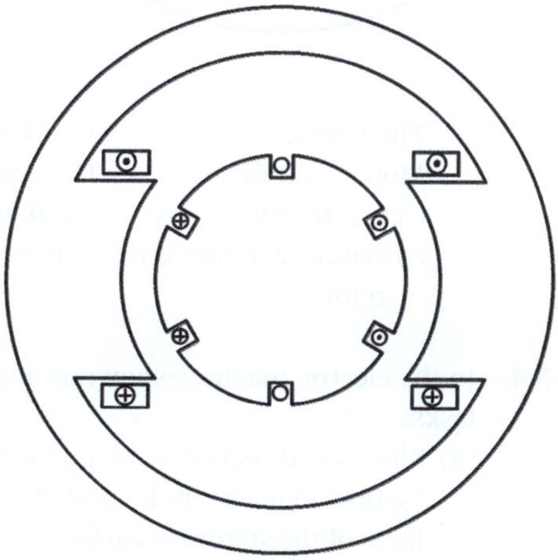

Figure P2.1 Cross-section of an elementary 2-pole DC machine.

3.3 In the electric machine shown in Figure P3.1, please perform the following tasks:

(a) Show the direction of the magnetic flux and induced apparent "north" and "south" magnetic poles on the outer surfaces of the rotors and inner surfaces of the stators of each.

(b) State the type of machine, whether it is DC brush-commutator type, PM brushless DC type, PM AC synchronous type, field excitation AC synchronous type, or AC induction machine.

(c) Mark the appropriate boxes (use "Y" for yes and "N" for no) applicable to the status of torque, motor rotation, generator rotation, and presence of armature winding on the stator or rotor or both, for the machine shown in Figure P3.1.

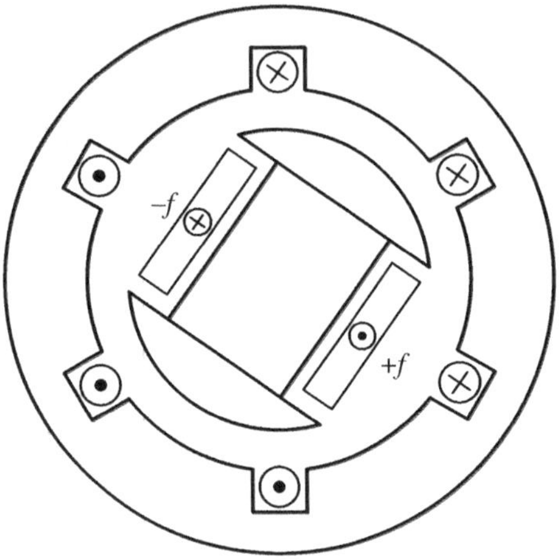

Figure P3.1 Cross-section of an elementary 2-pole 3-phase synchronous machine.

The torque is: Clockwise_____ Counterclockwise_____
Motor rotation is: Clockwise_____ Counterclockwise_____
Generator rotation is: Clockwise_____ Counterclockwise_____
Armature winding is on: Stator_____ Rotor_____ Both stator and rotor_____

3.4 In the electric machine shown in Figure P4.1, please perform the following tasks:

(a) Show the direction of magnetic flux and induced apparent "north" and "south" magnetic poles on the outer surfaces of the rotors and inner surfaces of the stators of each.

(b) State the type of machine, whether it is DC brush-commutator type, PM brushless DC type, PM AC synchronous type, field excitation AC synchronous type, or AC induction machine.

(c) Mark the appropriate boxes (use "Y" for yes and "N" for no) applicable to the status of torque, motor rotation, generator rotation, and presence of armature winding on the stator or rotor or both, for the machine shown in Figure P4.1.

The torque is: Clockwise_____ Counterclockwise_____
Motor rotation is: Clockwise_____ Counterclockwise_____
Generator rotation is: Clockwise_____ Counterclockwise_____
Armature winding is on: Stator_____ Rotor_____ Both stator and rotor_____

Figure P4.1 Cross-section of an elementary 2-pole 3-phase wound rotor induction machine.

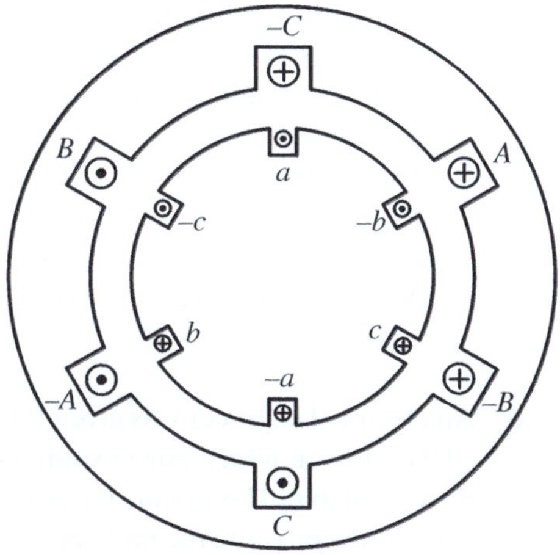

3.5 In the electric machine shown in Figure P5.1, please perform the following tasks:

(a) Show the direction of magnetic flux and induced apparent "north" and "south" magnetic poles on the outer surfaces of the rotors and inner surfaces of the stators of each.

(b) State the type of machine, whether it is DC brush-commutator type, PM brushless DC type, PM AC synchronous type, field excitation AC synchronous type, or AC induction machine.

Figure P5.1 Cross-section of an elementary 2-pole 3-phase PM machine.

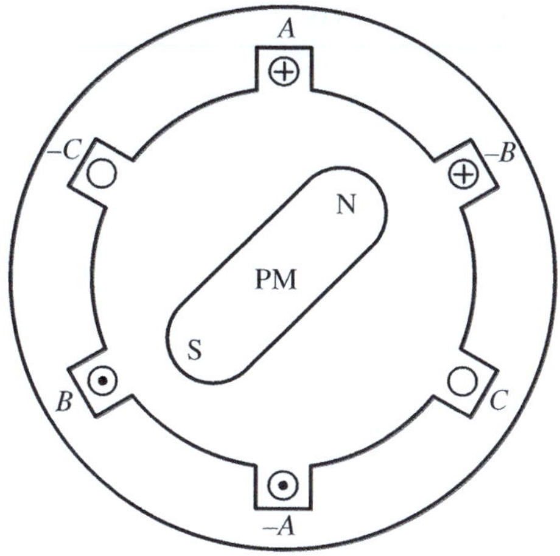

(c) Mark the appropriate boxes (use "Y" for yes and "N" for no) applicable to the status of torque, motor rotation, generator rotation, and presence of armature winding on the stator or rotor or both, for the machine shown in Figure P5.1.
The torque is: Clockwise_____ Counterclockwise_____
Motor rotation is: Clockwise_____ Counterclockwise_____
Generator rotation is: Clockwise_____ Counterclockwise_____
Armature winding is on: Stator_____ Rotor_____ Both stator and rotor_____

3.6 An electrical engineer was given three AC motors (*A*), (*B*), and (*C*) to test in a laboratory at which power supplies with frequencies of 50, 60, and 400 Hz were available. The motors (*A*), (*B*), and (*C*) were tested under no-load and load conditions, and the no-load and load speeds in rev/min (revolutions per minute) were recorded as given in Table P6.1:
Please find the test frequency in Hertz, the number of poles, and the type of motor. Find whether synchronous or induction motor for all three (*A*), (*B*), and (*C*) motors and why.

Table P6.1 Test results for motors (*A*), (*B*), and (*C*).

AC motor	No-load speed (rev/min)	Load speed (rev/min)
(*A*)	749	734
(*B*)	8,000	8,000
(*C*)	719	710

4

Electromechanical Concepts in Electric Machines

4.1 Introductory Discussion

Naturally, the main purpose of electric machines is the conversion of electric energy into mechanical energy (motors) and vice versa mechanical energy into electric energy (generators). The dynamics of the electrical system constituting the machine and all its electrical attachments are the subject of all the other chapters of this textbook. However, in this chapter, we concentrate on the dynamics of the mechanical system: machine-mechanical load for motors or machine-mechanical prime mover for generators. The machine (motor)-mechanical load as a "mechanical system" is the main thrust of the derivations and developments which are to follow in this chapter, because motors-drives-mechanical loads are much more ubiquitous in our engineering work than prime movers-generators-electrical loads. Motor electric power and mechanical load (consumer) system notations are adopted here and throughout this textbook. Reference [1] is an important source for the materials which is to follow in this chapter.

4.2 Motor-Mechanical Load Dynamics

Figure 4.1a depicts an electrical source (supply)-motor-mechanical load system, and Figure 4.1b depicts a mechanical prime mover-generator-electrical load (sink) system [1]. These figures give an idea about the various torques and rotational speed orientations. Again, the motor (consumer) notation is adopted throughout, with counterclockwise rotation looking into the cross-section of a motor from the mechanical load side as the positive orientation of rotation.

For a motor, the equation of motion for the rotating masses (from Newton's law) can be written as follows:

$$T_{dev} - T_{RL} - T_m = J(d\omega_{rm}/dt) \tag{4.1}$$

Electric Machinery and Drives: An Electromagnetics Perspective, First Edition.
Nabeel A. O. Demerdash, JiangBiao He, and Hao Chen.

Companion website: www.wiley.com/go/demerdash/electricmachineryanddrives

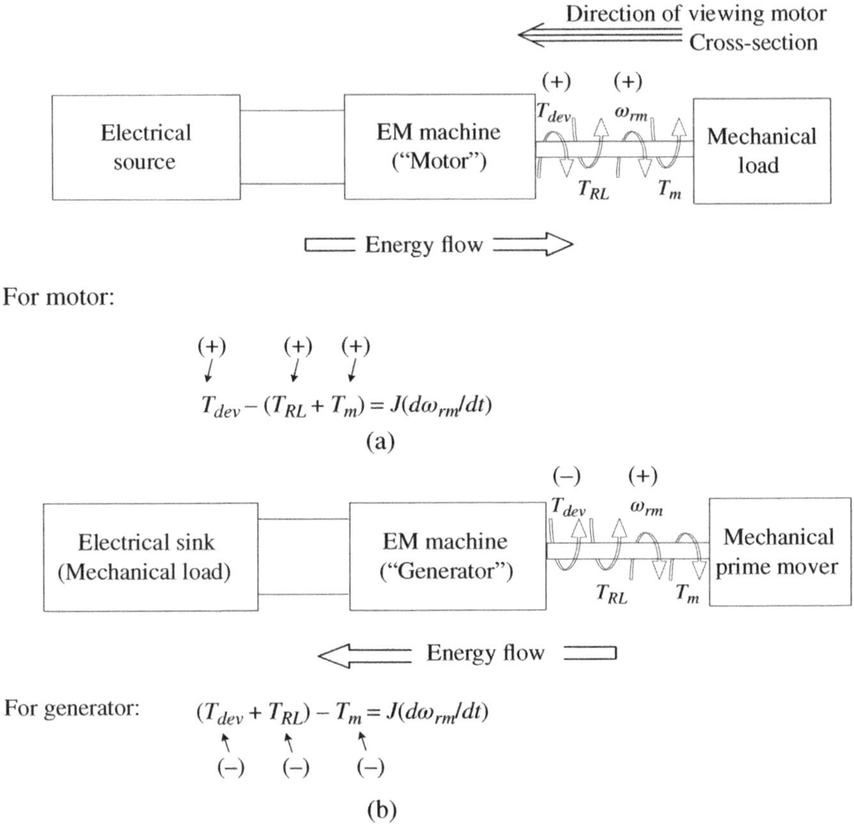

Figure 4.1 Electromechanical system. (a) The EM rotational machine: motor convention. (b) The EM rotational machine: generator convention.

where T_{dev} is the electromechanical torque developed by the motor, that is, the developed motor torque, in Nm. T_{RL} is the rotation loss torque (friction, windage, and magnetic core losses) internal to the motor, where the output motor torque $T_{out} = (T_{dev} - T_{RL})$, Nm. T_m is the torque produced by the mechanical load in opposition to the developed motor torque. J is the polar mass moment of inertia of all the rotating masses (parts) including the motor's rotor, shaft, coupling or gears, and load, in kg m^2. ω_{rm} is the angular velocity of the rotating parts in mechanical radians per second, mech. rad/s.

It is found to be more convenient to lump the motor's rotational loss torque, T_{RL}, with the mechanical load torque, T_m, to form the equivalent (gross mechanical) load torque $T_{gm} = (T_m + T_{RL}) = T'_m$.

Hence, for motors:

$$T_{gm} = T'_m = (T_m + T_{RL}) \tag{4.2}$$

and Eq. (4.1) can be rewritten as follows:

$$T_{dev} - T_{gm} = T_{dev} - T'_m = J(d\omega_{rm}/dt) \tag{4.3}$$

again, J is the sum of the moments of inertia of all the rotating masses of the motor, coupling, and mechanical load, in kg m^2.

Now, consider the associated system power components in the motor-mechanical load system.

The electric power converted to mechanical form in the motor is the developed power, P_{dev}, in Watts (W):

$$P_{dev} = T_{dev}\omega_{rm} \tag{4.4}$$

The rotational power losses internal to the motor, which consist of friction, windage, core, and stray losses, P_{RL} in W, are:

$$P_{RL} = T_{RL}\omega_{rm} \tag{4.5}$$

The power absorbed (consumed) by the mechanical load which consists of the mechanical load losses and the useful mechanical load power, P_m, in W, is:

$$P_m = T_m\omega_{rm} \tag{4.6}$$

Here, it should be pointed out that:

$$P_{dev} = P_{RL} + P_m \tag{4.7}$$

or

$$P_{dev} - P_{RL} = P_m \tag{4.8}$$

Consider an example to illustrate the above-mentioned concepts:

Example 4.1

Consider a motor-mechanical load system, whose (P_{dev} – speed) and ($T_{gm} = T'_m$ – speed) characteristics are shown in Figure 4.2a and whose total moment of inertia, $J = 2$ kg m^2.

Find:

(a) the system speed transient starting from zero rotor speed, $\omega_{rm}(t = 0) = 0$.
(b) the system speed transient starting from 300 rad/s rotor speed.
(c) the steady-state speed in rad/s and rev/min.
(d) the mechanical load power and torque in kW and hp, and Nm and ft. lbs, respectively.

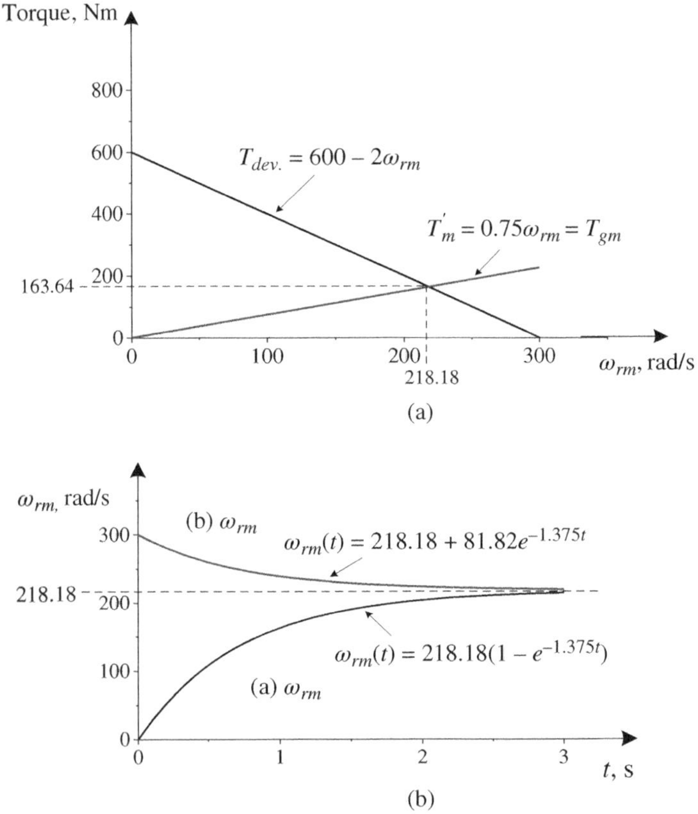

Figure 4.2 Motor and mechanical load torque–speed characteristics and speed transient. (a) Motor and load torque–speed characteristics. (b) Speed transients.

Solution

From Figure 4.2a, T_{dev} can be written as follows:

$$T_{dev} = (600 - 2\omega_{rm}) \quad \text{Nm} \tag{4.9}$$

Hence, for $\omega_{rm} = 0$ rad/s $\Rightarrow T_{dev} = 600$ Nm

for $\omega_{rm} = 300$ rad/s $\Rightarrow T_{dev} = 0$ Nm

and

$$T_{gm} = T'_m = 0.75\omega_{rm} \text{ Nm} \tag{4.10}$$

Hence, from Eq. (4.3), one can write,

$$J(d\omega_{rm}/dt) = T_{dev} - T'_m = 600 - 2\omega_{rm} - 0.75\omega_{rm}$$

Hence,

$$2(d\omega_{rm}/dt) = 600 - 2.75\omega_{rm}$$

Hence,

$$(d\omega_{rm}/dt) + 1.375\omega_{rm} = 300$$

or

$$\frac{8}{11}(d\omega_{rm}/dt) + \omega_{rm} = 218.18 \tag{4.11}$$

Laplace transforming Eq. (4.11) gives:

$$\frac{8}{11}S\Omega_{rm} - \frac{8}{11}\Omega_{rm}(0) + \Omega_{rm} = \frac{218.18}{S} \tag{4.12}$$

(a) For $\omega_{rm}(0) = 0 \Rightarrow \Omega_{rm}(0) = 0$
Hence,

$$\frac{8}{11}S\Omega_{rm} + \Omega_{rm} = \frac{218.18}{S}$$

Hence,

$$\left(\frac{8}{11}S + 1\right)\Omega_{rm} = \frac{218.18}{S}$$

Hence,

$$\Omega_{rm} = \frac{218.18}{S\left(\frac{8}{11}S + 1\right)} = \frac{300}{S(S + 1.375)} \tag{4.13}$$

Hence,

$$\Omega_{rm} = \frac{k_1}{S} + \frac{k_2}{S + 1.375} = \frac{300}{S(S + 1.375)}$$

Using the well-known algebra of partial fractions, one has the following:

$$k_1 = \left.\frac{300}{S + 1.375}\right|_{S=0} = \frac{300}{1.375} = 218.18$$

and

$$k_2 = \left.\frac{300}{S}\right|_{S=-1.375} = \frac{300}{-1.375} = -218.18$$

Hence,

$$\Omega_{rm} = \frac{218.18}{S} - \frac{218.18}{S + 1.375} \tag{4.14}$$

Inverse Laplace transforming Eq. (4.14) gives

$$\omega_{rm}(t) = (218.18 - 218.18e^{-1.375t}) = 218.18(1 - e^{-1.375t})\ \text{rad/s} \tag{4.15}$$

See the plot in Figure 4.2b.

(b) For $\omega_{rm}(0) = 300$ rad/s, from Eq. (4.12), it follows that

$$\frac{8}{11}S\Omega_{rm} - \frac{8}{11}(300) + \Omega_{rm} = \frac{218.18}{S}$$

Hence,

$$\left(\frac{8}{11}S + 1\right)\Omega_{rm} = \frac{218.18}{S} + 218.18$$

Therefore,

$$\Omega_{rm} = \frac{218.18}{S\left(\frac{8}{11}S + 1\right)} + \frac{218.18}{\left(\frac{8}{11}S + 1\right)}$$

Hence,

$$\Omega_{rm} = \frac{218.18}{S} - \frac{218.18}{S + 1.375} + \frac{300}{S + 1.375}$$

That is,

$$\Omega_{rm} = \frac{218.18}{S} + \frac{81.82}{S + 1.375} \tag{4.16}$$

Inverse Laplace transforming Eq. (4.16) gives

$$\omega_{rm}(t) = (218.18 + 81.82e^{-1.375t})\text{rad/s} \tag{4.17}$$

See the plot in Figure 4.2b.

(c) ω_{rm} (steady-state, as $t \to \infty$) = 218.18 rad/s for both (a) and (b) of Figure 4.2 at which condition:

$$\text{as } t \to \infty, \text{ we must have } T_{dev} = T'_m = T_{gm}$$

which is "the steady-state equilibrium point" for the motor-mechanical load system.

$$\omega_{rm}(t \to \infty) = 218.18 \times (60/2\pi) = 2083 \text{ rev/min}$$

At $\omega_{rm} = 218.18$ rad/s = 2083 rev/min

Equations (4.9) and (4.10) give

$$T_{dev} = 600 - 2 \times 218.18 = T'_m = 0.75 \times 218.18 = 163.64 \text{ Nm}$$

That is,

$$T_{dev} = T'_m = T_{gm} = 163.64 \text{ Nm}$$

or

$$T_{dev} = T'_m = 163.64/(1.356 \text{ Nm/ft.lb}) = 120.68 \text{ ft.lb}$$

Hence,

$$P_{dev} = T_{dev}\omega_{rm} = P'_m = T'_m\omega_{rm} = 163.64 \times 218.18 = 35703 \text{ W} \approx 35.70 \text{ kW}$$

That is,

$$P_{dev} = P'_m = (35703/746) = 47.86 \text{ hp}$$

Examining the results of this example, see Figure 4.2a,b, and Eqs. (4.15) and (4.17), one can state that whenever $T_{dev} > T'_{m,}$ the system accelerates, and whenever $T_{dev} < T'_{m,}$ the system decelerates.

In both conditions, the motor-mechanical load system sought out the equilibrium condition of

$$T_{dev} = T'_m \tag{4.18}$$

where the system torques are in balance.

Hence, the steady-state speed for any motor-mechanical load system occurs at the intersection of the motor and mechanical load torque–speed characteristics, $T_{dev} = T'_m$.

Also, whenever $T_{dev} > T'_m \Rightarrow$ the system is accelerating, and whenever $T_{dev} < T'_m \Rightarrow$ the system is decelerating.

4.3 Mechanical Load Torque–Speed Characteristics

The dynamics of the mechanical behavior of a motor-mechanical load system depend heavily on the nature of the mechanical load's torque–speed characteristic. That is, the relationship between the load torque, T_L, and the load speed, ω_L. Usually, T_L is in Nm, and ω_L is in mech. rad/s. A useful general form of this $T_L - \omega_L$ relationship is represented imperially by the following expression, which is a polynomial in ω_L at order n, where n can be equal to "0, 1, 2, 3, 4, … , etc.":

$$T_L = A_0 + A_1\omega_L + A_2\omega_L^2 + A_3\omega_L^3 + \cdots + A_n\omega_L^n \tag{4.19}$$

where again:

$$n = 0, 1, 2, 3, \cdots = \text{order of load torque relation.}$$

For applications in which the motor shaft is directly connected to the load shaft "direct drive," one writes

$$T_m = T_L \text{ and } \omega_{rm} = \omega_L \tag{4.20}$$

Examples of mechanical loads are

(1) Constant torque loads: ($n = 0$, $T_L = A_0$)
These are loads such as conveyer belts, cranes (hoists), elevators, looms, rolling mills, and saws.

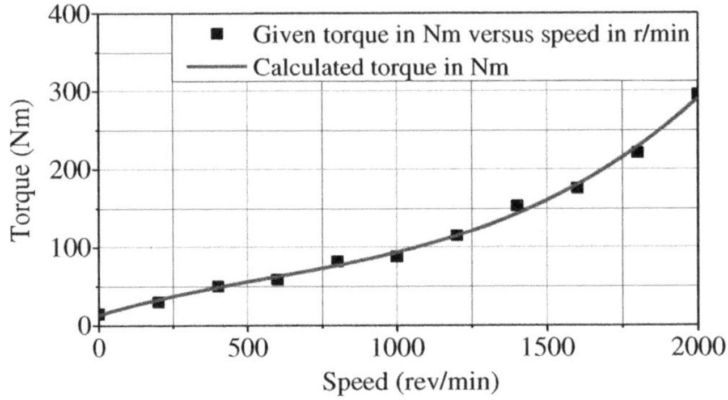

Figure 4.3 Third-order torque load example with least mean-squared curve fitting polynomial.

(2) Linear (first-order) torque loads: ($n = 1$, $T_L = A_1\omega_L$, or $T_L = A_0 + A_1\omega_L$)
These are viscous friction and friction-type loads generally.
(3) Parabolic (second-order) torque loads: ($n = 2$, $T_L = A_2\omega_L^2$)
These are loads such as centrifugal pumps, fans, blowers, and rotary (not reciprocating) compressors.
(4) General-type torque loads (given in tabulated torque–speed data):

$$T_L = A_0 + A_1\omega_L + A_2\omega_L^2 + \cdots + A_n\omega_L^n$$

See Figure 4.3 for an example of the third-order nature, $n = 3$, for the torque–speed tabulation and a least mean-squared third-order curve-fit polynomial for the data.

Example 4.2

Speed (rev/min)	Torque (Nm)
0	15
200	30
400	50
600	59
800	82
1000	88
1200	115
1400	153
1600	176
1800	221
2000	295

$$T_L = A_0 + A_1\omega_L + A_2\omega_L^2 + A_3\omega_L^3$$

A_0	$1.32E{+}01$
A_1	$1.12E{-}01$
A_2	$-7.692E{-}05$
A_3	$4.511E{-}08$

$$T_L = 13.2 + 0.112\omega_L - 0.00007692\omega_L^2 + 0.00000004511\omega_L^3$$

Constant Power Loads: (P_L = constant = $T_L\omega_L$)

These are mechanical loads such as take-up or rewind reels, or what one would encounter in paper, textile, fabric, and tape products industries, see the reel schematic of Figure 4.4. Here, one can write:

$$T_L = \left(\frac{\text{constant}}{\omega_L}\right) \tag{4.21}$$

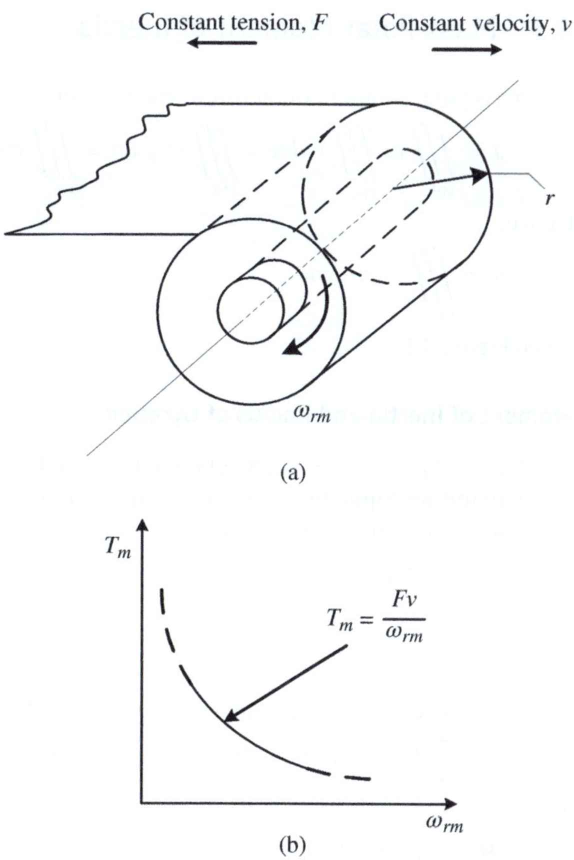

Figure 4.4 Constant power load. (a) Take-up reel. (b) Torque–speed characteristic.

To keep the process of rewinding or take-up moving properly, it is highly desirable, if not often absolutely required, to keep the tangential force, F, in Figure 4.4 and velocity, v, both constant and independent of the roll radius, r.

Hence,

$$T_L = F \times r \tag{4.22}$$

and

$$\omega_L = \frac{v}{r} \tag{4.23}$$

Accordingly,

$$P_L = T_L \omega_L = F \times r \times \frac{v}{r} = Fv$$

$$P_L = T_L \omega_L = Fv = \text{constant} \tag{4.24}$$

Hence,
$$T_L = \left(\frac{Fv}{\omega_L} \right) \tag{4.25}$$

4.4 Mass Polar Moment of Inertia

The mass polar moment of inertia, J, can be expressed as follows:

$$J = \iiint_{Vol.} dJ \iiint_{Vol.} r^2 dm = \iiint_{Vol.} r^2 (\rho dv) = \iiint_{Vol.} r^2 (\rho r d\theta dz dr)$$

Hence,

$$J = \iiint_{Vol.} \rho r^3 dr dz d\theta \tag{4.26}$$

See Figure 4.5.

Moment of Inertia and Radius of Gyration

The radius of gyration, K, in meters is defined as the distance such that when it is squared and multiplied by the mass of the rotating body, M, in kilogram, results in the mass polar moment of inertia, J, in kg m^2. Hence, one can write

$$J = M(K^2) \tag{4.27}$$

or

$$K = \sqrt{J/M} \tag{4.28}$$

Hence, for the hollow cylinder (rotor) of Figure 4.5

$$J = \frac{\pi}{2} \rho l \left(r_2^4 - r_1^4 \right) = \frac{\pi}{2} \rho l \left(r_2^2 - r_1^2 \right) \left(r_2^2 + r_1^2 \right)$$

and,

$$M = \frac{\pi}{2} \rho l \left(r_2^2 - r_1^2 \right)$$

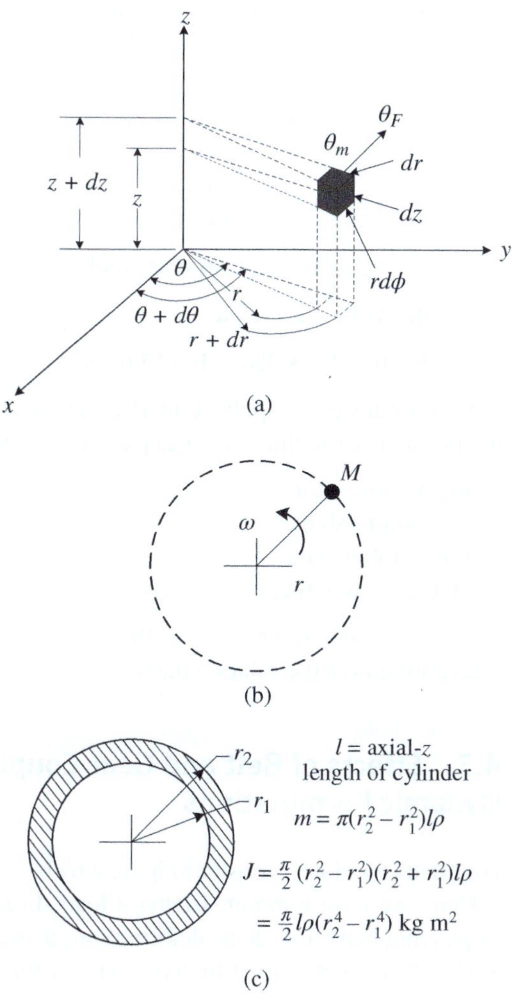

Figure 4.5 Moment of inertia. (a) *dm* in a cylindrical coordinate system. (b) $J = mr^2$ kg m² for point mass *M*. (c) $J = \frac{1}{2}m\left(r_2^2 + r_1^2\right)$ for annular cylinder.

Hence,

$$K = \sqrt{\frac{\pi}{2}\rho l\left(r_2^2 - r_1^2\right)\left(r_2^2 + r_1^2\right) \bigg/ \frac{\pi}{2}\rho l\left(r_2^2 - r_1^2\right)}$$

Therefore,

$$K = \sqrt{\left(r_2^2 + r_1^2\right)}$$

and

$$J = K^2 M = \frac{\pi}{2}\rho l\left(r_2^4 - r_1^4\right) \quad \text{kg m}^2$$

Sometimes in the United States, "WK^2" is used to mean "MK^2."

For example, if "WK^2" = 250 Lb ft^2, then it follows that one can reason the following relationship:

$$1 \text{ kg} = 2.188 \text{ Lb}_{\text{mass}} \text{ and } \ 1 \text{ m} = 3.281 \text{ ft}$$

$$J(\text{in kg m}^2) = \frac{WK^2(\text{in Lb}_{\text{mass}} \text{ ft}^2)}{2.188 \times (3.281)^2}$$

$$J(\text{in kg m}^2) = 0.04246\, WK^2(\text{in Lb}_{\text{mass}} \text{ ft}^2) \tag{4.29}$$

Hence, for $WK^2 = 250$ Lb ft^2

$$J = 0.04246 \times 250 = 10.615 \text{ kg m}^2$$

Also, for any given application, the total moment of inertia, J, in Eq. (4.26) must include all the rotating (moving) parts which would include the following:

1. the motor's rotor,
2. the motor's shaft,
3. the coupling, and
4. all the parts of the mechanical load.

For rigid bodies (all rotating at the same angular mechanical speed), one simply adds all the Js of the rotating parts.

4.5 Effects of Belt and Gear Couplings on Mechanical Dynamic Formulations

For "Belt Coupling," consider Figure 4.6a,b.

Here, we have a motor driving a load through a belt coupling arrangement. Neglecting belt slippage or deformation, if the motor "pulley #1" turns through angle, θ_1, a length equal to "$r_1\theta_1$" enters the pulley's groove. The same length "$r_1\theta_1$" must leave the load's "pulley #2" grove, hence,

$$r_1\theta_1 = r_2\theta_2 \tag{4.30}$$

or,

$$\theta_2 = (r_1/r_2)\theta_1 \tag{4.31}$$

Differentiating (4.31) with respect to time, we obtain:

$$(d\theta_2/dt) = (r_1/r_2)(d\theta_1/dt)$$

That is,

$$\omega_2 = (r_1/r_2)\omega_1 \tag{4.32}$$

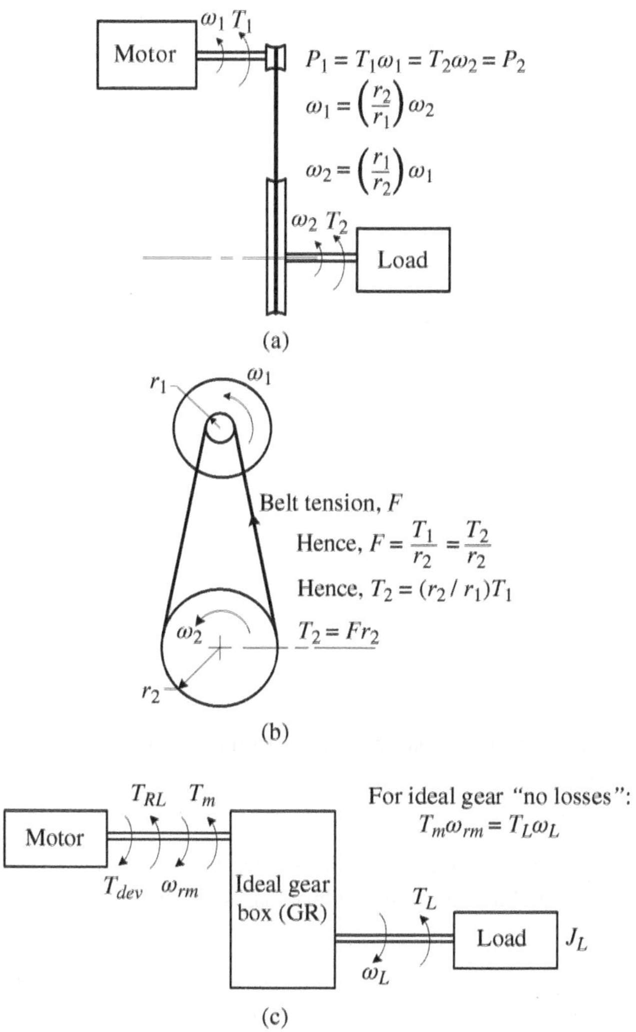

Figure 4.6 Belt coupling (a and b) and gear coupling (c). (a) End view. (b) Side view. (c) Ideal gear box.

The motor exerts a torque, T_1, on pulley #1, and the belts tension force is the same throughout the belt.

Hence,

$$T_1 = Fr_1 \quad \text{and} \quad T_2 = Fr_2$$

Hence,

$$F = \frac{T_1}{r_1} = \frac{T_2}{r_2}$$

Therefore,

$$T_2 = (r_2/r_1)T_1 \tag{4.33}$$

and,

$$P_1 = \omega_1 T_1 = \left(\frac{r_2}{r_1}\right)\omega_2 \times \left(\frac{r_1}{r_2}\right) T_2 = \omega_2 T_2 = P_2$$

or,

$$P_1 = \omega_1 T_1 = \omega_2 T_2 = P_2 \tag{4.34}$$

Now, if we consider a pure inertia load,

$$T_2 = J_2(d\omega_2/dt)$$

Hence, from Eq. (4.33), it follows that

$$(r_2/r_1)T_1 = J_2(d[(r_1/r_2)\omega_1]/dt)$$

Hence,

$$(r_2/r_1)T_1 = J_2(r_1/r_2)(d\omega_1/dt)$$

or,

$$T_1 = \left\{(r_1/r_2)^2 J_2\right\}(d\omega_1/dt) = J_1(d\omega_1/dt) \tag{4.35}$$

Hence,

$$J_1 = (r_1/r_2)^2 J_2$$

Therefore, the effective inertia of the load presented to the motor through the pulley system is as follows:

$$J_{load\ at\ motor} = (r_1/r_2)^2 J_{load} \tag{4.36}$$

In a more general ideal "Gear Coupling," consider Figure 4.6c.

Let GR be the primary motor shaft speed/secondary load shaft speed.

Hence,

$$GR = \omega_{rm}/\omega_L \tag{4.37}$$

Because the gear box is ideal with no losses, it follows that

$$P_m = T_m\omega_{rm} = P_L = T_L\omega_L \tag{4.38}$$

Hence,

$$T_m = (\omega_L/\omega_{rm})T_L \tag{4.39}$$

However,

$$\omega_L/\omega_{rm} = 1/GR \tag{4.40}$$

Hence,

$$T_m = (1/GR)T_L \tag{4.41}$$

However,

$$T_L = J_L(d\omega_L/dt) \tag{4.42}$$

Hence,

$$T_L = J_L(d(\omega_{rm}/GR)/dt)$$

That is,

$$T_L = (J_L/GR)(d\omega_{rm}/dt) \tag{4.43}$$

Hence,

$$T_m = (1/GR)(J_L/GR)(d\omega_{rm}/dt)$$

That is,

$$T_m = (J_L/(GR)^2)(d\omega_{rm}/dt) \tag{4.44}$$

Hence,

$$J_m = J_L/(GR)^2 \tag{4.45}$$

J_m is the load inertia viewed from the motor's side.

4.6 Operating Modes of Electric Machines

Figure 4.7 is a good summary of the four basic operating modes of an electric machine. Notice that

$$P_{dev} = T_{dev}\omega_{rm} \tag{4.46}$$

Here, for motor modes, $P_{dev} > 0$ and for generator modes, $P_{dev} < 0$.

Let us apply this concept to the induction machine's well-known typical torque–speed (slip) characteristic, see Figures 4.8 and 4.9 for details.

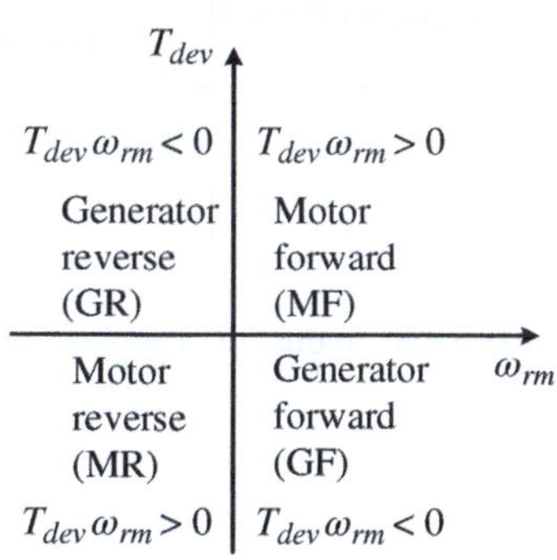

Figure 4.7 The four basic operating modes of an electric machine.

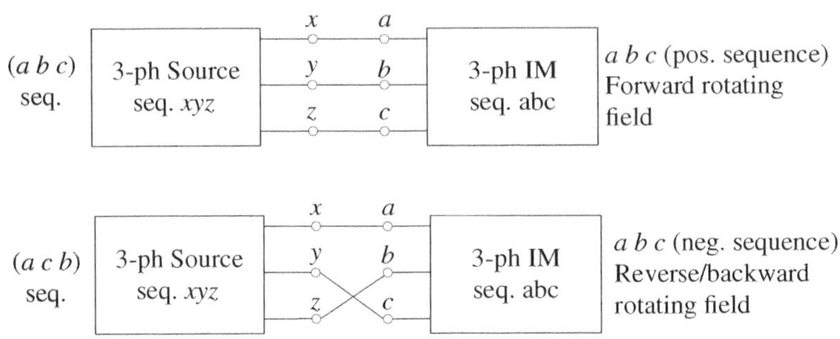

Figure 4.8 Induction machine phase sequencing.

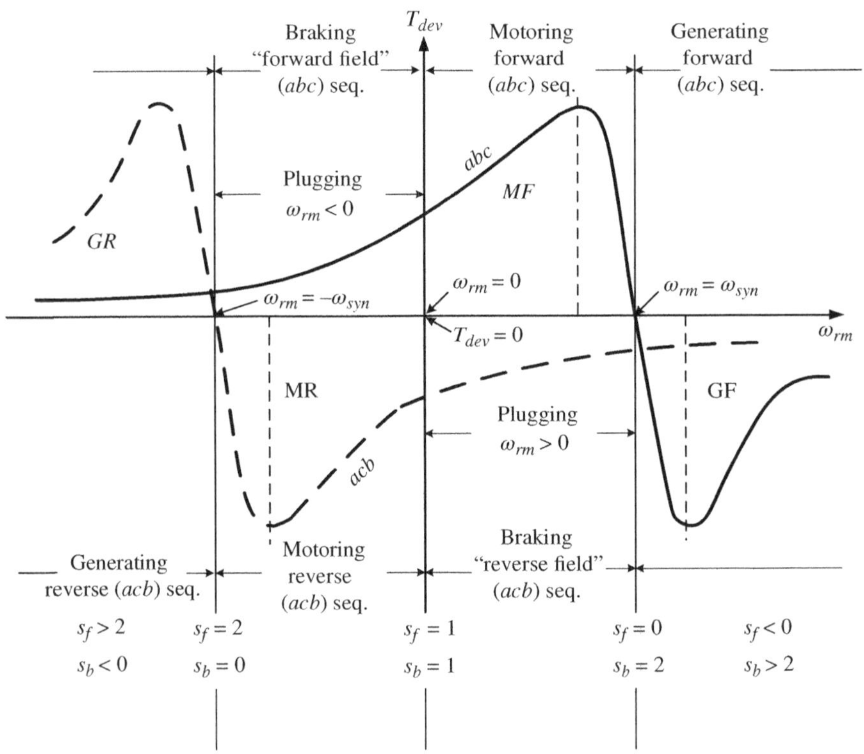

Figure 4.9 Induction machine operating modes.

4.7 On Time Constants of the Mechanical Dynamics of Motor-Mechanical Load Systems

Here, we have (Figure 4.10):

$$T_m = k_m - k_m\left(\frac{\omega_{rm}}{\omega_0}\right) = k_m - \left(\frac{k_m}{\omega_0}\right)\omega_{rm} \tag{4.47}$$

$$T_L = \left(\frac{k_L}{\omega_0}\right)\omega_{rm} \tag{4.48}$$

The equation of motion of the rotating masses is

$$J\frac{d\omega_{rm}}{dt} = T_m - T_L = k_m - \left(\frac{k_m}{\omega_0}\right)\omega_{rm} - \left(\frac{k_L}{\omega_0}\right)\omega_{rm}$$

Hence,

$$J\frac{d\omega_{rm}}{dt} = k_m - \left(\frac{k_m + k_L}{\omega_0}\right)\omega_{rm}$$

That is,

$$J\frac{d\omega_{rm}}{dt} + \left(\frac{k_m + k_L}{\omega_0}\right)\omega_{rm} = k_m \tag{4.49}$$

Laplace transforming Eq. (4.49), assuming $\omega_{rm}(t = 0) = 0$, one writes:

$$JS\Omega_{rm} + \left(\frac{k_m + k_L}{\omega_0}\right)\Omega_{rm} = \frac{k_m}{S}$$

Hence,

$$[JS + (k_m + k_L)/\omega_0]\Omega_{rm} = \frac{k_m}{S}$$

Therefore,

$$\Omega_{rm} = \frac{k_m}{S[JS + (k_m + k_L)/\omega_0]}$$

Figure 4.10 Motor and load torque–speed characteristics.

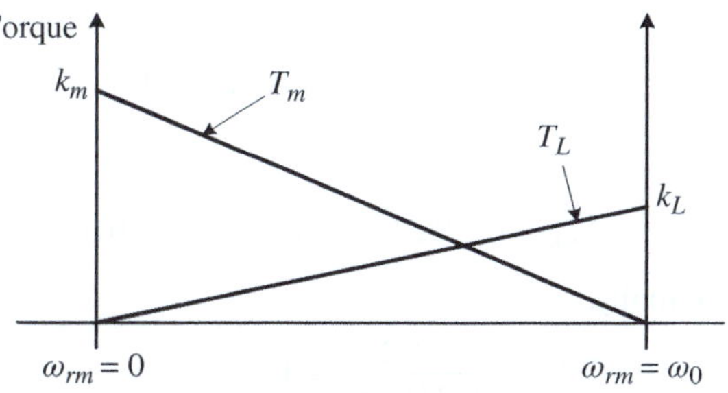

or,

$$\Omega_{rm} = \frac{k_m/J}{S\left[S + \left(\frac{k_m+k_L}{\omega_0 J}\right)\right]} \tag{4.50}$$

Hence,

$$\Omega_{rm} = \frac{k_1}{S} + \frac{k_2}{S + \left(\frac{k_m+k_L}{\omega_0 J}\right)} \tag{4.51}$$

Accordingly, the transient involves a time term, which can be expressed as follows:

$$e^{-\left(\frac{k_m+k_L}{\omega_0 J}\right)t} = e^{-t/\tau_{\text{mech}}} \tag{4.52}$$

Hence, the mechanical time constant, τ_{mech}, can be estimated as follows:

$$\tau_{mech} = \left(\frac{\omega_0 J}{k_m + k_L}\right) \tag{4.53}$$

τ_{mech} is in seconds.

Accordingly, the larger the inertia J and speed ω_0, the longer the time constant. Also, the larger the k_m and k_L, the shorter the time constant.

Example 4.3

Consider the motor-mechanical load system in Example 4.1. In this system, $k_m = 600$ Nm, $k_L = 0.75 \times 300 = 225$ Nm, and $\omega_0 = 300$ rad/s. Also, in this system, $J = 2$ kg m^2. See Figure 4.2.

Hence,

$$\tau_{mech} = \left(\frac{\omega_0 J}{k_m + k_L}\right) = \frac{300 \times 2}{600 + 225}$$

That is,

$$\tau_{mech} = \frac{600}{825} = 0.72727 \text{ seconds}$$

This should be in comparison with an AC cycle's time, T, in any a 60-Hz AC system:

$$T = \frac{1}{f} = \frac{1}{60} = 0.0166667 \text{ seconds}$$

Here,

$$\tau_{mech} = 0.72727 \text{ seconds} \gg T = 0.0166667 \text{ seconds}$$

Actually,

$$\frac{\tau_{mech}}{T} = \frac{0.72727}{0.0166667} \approx 44$$

This justifies many more integration time samplings of the electrical system dynamics for every one single integration sampling of the mechanical system dynamics, as will be demonstrated in later chapters of this textbook.

4.8 Modeling and Simulation of Motor Starting Transients

We will treat induction motor starting transients, through the use of equivalent circuit steady state AC circuit models and torque computation, using what was discussed in Sections 4.2–4.7 and Chapter 6. We will also revisit this starting transient, using time-domain state-space modeling of induction motors in Chapter 10.

Meanwhile, in the next chapter, Chapter 5, we will discuss electric machinery windings and their associated electromotive forces (EMFs) and magnetomotive forces (MMFs), including the "all-important" rotating MMFs produced by AC currents flowing in poly-phase windings of AC machines. We also derive torque and volt–ampere ratings/power ratings in relation to machine volumes based on mid-airgap diameters and effective stator/rotor conductors axial lengths.

Homework Problems

4.1 A flywheel with a moment of inertia of 4 kg m^2 is initially at rest. If a torque of 6 Nm (counterclockwise) is suddenly applied to the flywheel, what will be the speed of the flywheel after five seconds? Express that speed in both radians per second and revolutions per minute.

4.2 A motor is supplying 50 Nm of torque to its load. If the motor's shaft is turning at 1500 rev/min, what is the mechanical power supplied to the load in watts? In horsepower?

Reference

1 Gross, C.A. (2006). *Electric Machines*, Edition 1. CRC Press.

5

Electric Machinery Windings and Associated Electromotive and Magnetomotive Forces

In this chapter, we present the analytical side of what was qualitatively presented and discussed in Chapter 3, with regard to AC machines, Section 3.8, and DC commutator-brush-type machines, Section 3.10. This will necessarily involve deriving analytical expressions for induced electromotive forces (EMFs) and magneto motive forces (MMFs) established by the flow of currents in these machine windings and associated mechanical torques developed through the interaction of the winding currents and the magnetic fields established in these devices.

5.1 AC Winding Layouts

The nature of various types of coils/turns used in the construction of single-layer and double-layer windings in AC machines is discussed in Chapter 3, Section 3.8. Here, we elaborate specifically on the layout of the more complex cases of double-layer windings through a practical example of a 4-pole, 48-stator slots, and 3-phase stator (Figure 5.1), which could belong to any 4-pole, synchronous (wound-field or permanent-magnet) machine, or induction (wound-rotor or squirrel-cage) machine. We use short-pitched coils, in this specific case of Figures 5.2–5.6, where the number of stator slots per pole-pitch is $\tau_p = 12$ slots, and the number of stator slots per coil-pitch is $\tau_c = 10$ slots. The number of stator slots is $N_s = 48$ slots, while in electrical measure, $\tau_p = \pi$ electrical radians = 180° electrical, and, in the coil span, $\beta = (\tau_c/\tau_p)\pi = (5/6)\pi$ electrical radians = 150° electrical. Meanwhile, the angular pitch between two adjacent slots can be calculated as: $\alpha = p(\pi)/N_s = (4\pi/48) = (\pi/12)$ electrical radians = $(4\times180°/48) = 15°$ electrical. Here, in this case, the number of slots per pole per phase is $q = N_s/(3\times p) = (48/3\times4) = 4$. With all the aforementioned particulars, the starting point/slots, for phase (A), S_{a1} and S_{a2}, for phase (B), S_{b1}

Electric Machinery and Drives: An Electromagnetics Perspective, First Edition.
Nabeel A. O. Demerdash, JiangBiao He, and Hao Chen.

Companion website: www.wiley.com/go/demerdash/electricmachineryanddrives

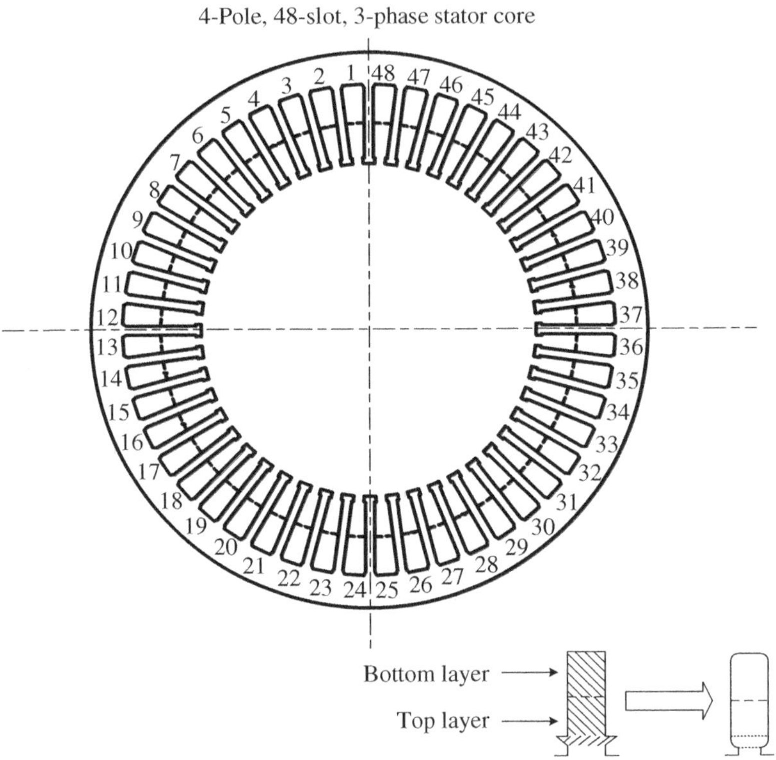

Figure 5.1 A double-layered 48 slots stator.

and S_{b2}, as well as for phase (C), S_{c1} and S_{c2}, are clearly depicted in slots, #1 and #25, for phase (A), slots, #9 and #33, for phase (B), and slots, #17 and #41, for phase (C), see Figure 5.2. Meanwhile, the slots that belong to phase (A) coils/turns are clearly depicted in Figure 5.3, in addition to the four separate sets of coils that belong to phase (A), as well as two possible ways to connect these four "Separate Sets" of coils, with the option of the number of parallel paths, $m = 1$, $m = 2$, respectively. Furthermore, the slots that belong to phase (B) and phase (C) are depicted in Figures 5.4 and 5.5, respectively. Meanwhile, the entire phase (A), phase (B), and phase (C) slots are shown in Figure 5.6, respectively.

In addition, the connection diagram, including the slot numbers that belong to the top (solid line) and the bottom (dotted line) layers of this set of coils belonging to phase (A), is shown in Figure 5.7. In Figure 5.7, all coils belonging to this phase are connected in series, in one path, $m = 1$, with the start and finish of each set of coils, $a_1(+)$ and $a_1(-)$, $a_2(+)$ and $a_2(-)$, $a_3(+)$ and $a_3(-)$, as well as $a_4(+)$ and $a_4(-)$, respectively.

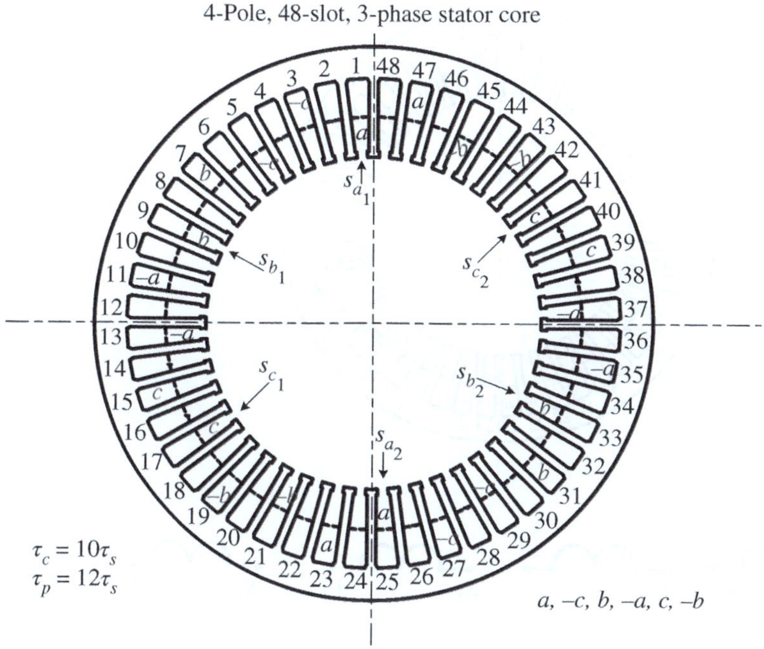

$$\alpha = \frac{p \times 180^\circ}{N_s} = \frac{4 \times 180^\circ}{48} = 15^\circ e$$

$$\beta = \frac{\tau_c}{\tau_p} \times 180^\circ = \frac{10\tau_s}{12\tau_s} \times 180^\circ = 150^\circ e$$

Hence, $120^\circ e = 8 \times 15^\circ e = 8\alpha$.

Figure 5.2 The start of coils of phases (*A*), (*B*), and (*C*).

5.2 DC Winding Layouts

The nature of the always double-layer winding layouts of conventional commutator-brush-type DC machines is presented in Chapter 3, Section 3.10. Specifically, the "Lap-Type Lattice" DC winding layout is shown in Figure 3.36, where the number of parallel paths, m, is equal to the number of poles, p, in simple-lap types shown in this figure and where the separation between a positive (+) brush and a next negative (−) brush is a pole-pitch, τ_p, in the number of commutator segments in-between the (+) and the (−) brushes.

Also, a "Wave-Type Lattice" DC winding layout of the simple wave type is shown in Section 3.10 (Figure 3.37), where the separation between the (+) brush and the next (−) brush is again a pole-pitch, τ_p, in the number of commutator segments

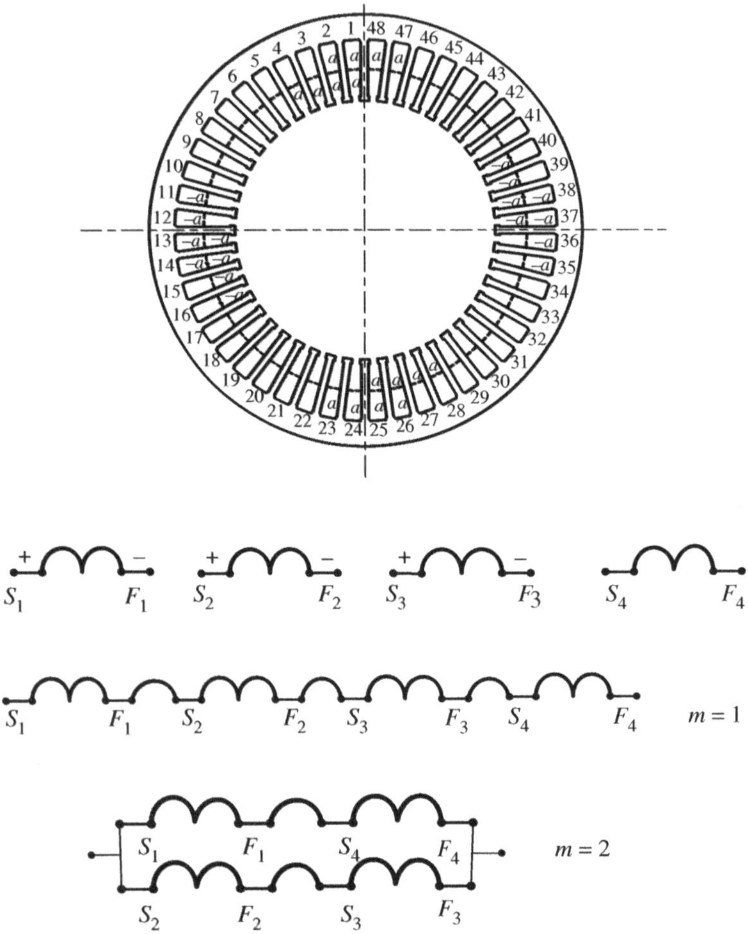

Figure 5.3 The phase (*A*) slots and possible connections.

in-between those two (+) and (−) brushes. In this case, the number of parallel paths, *m*, is always equal to 2, that is, $m = 2$. More details are given in Chapter 8 on conventional brush-commutator-type DC machines.

5.3 Induced Electromotive Forces in Single-Phase and Poly-Phase AC Windings

Consider the field excitation coil (f, $-f$) mounted on a cylindrical rotor as shown in Figure 5.8 and a stator-mounted turn/coil (a, $-a$) mounted on the inner bore

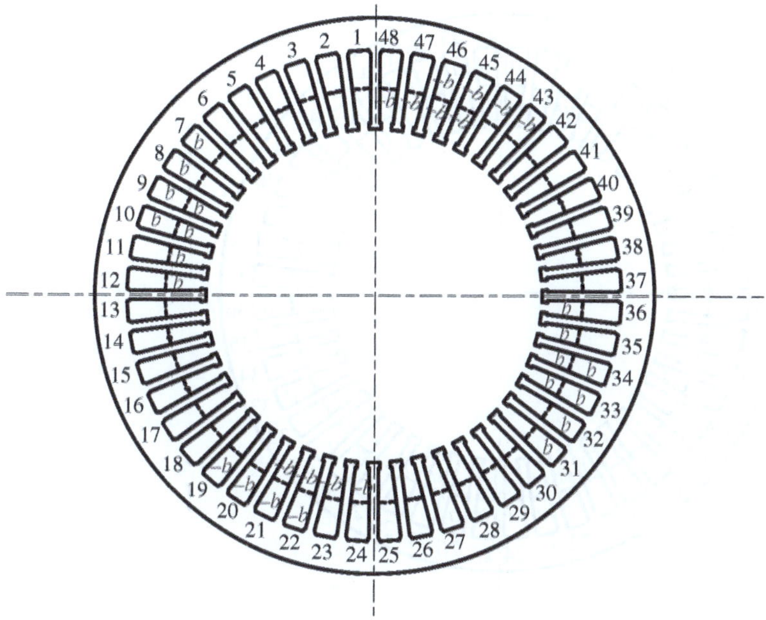

Figure 5.4 The phase (*B*) slots.

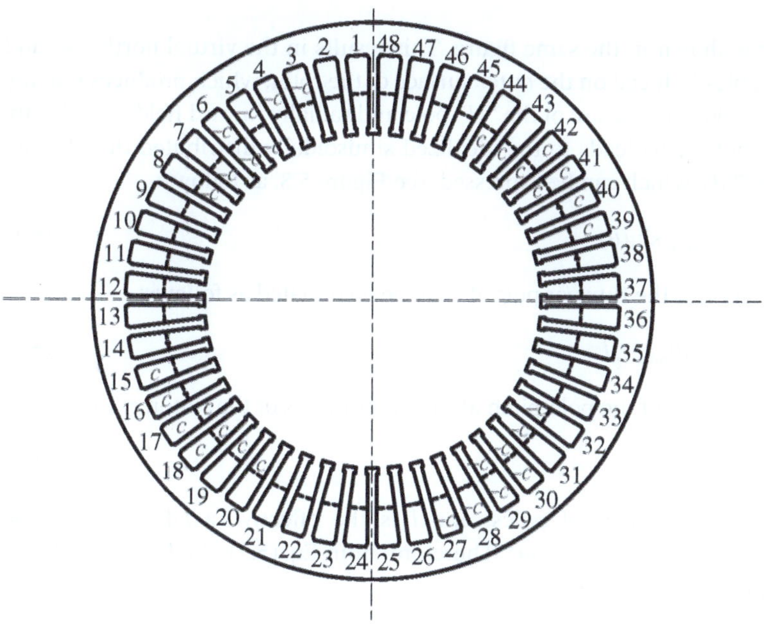

Figure 5.5 The phase (*C*) slots.

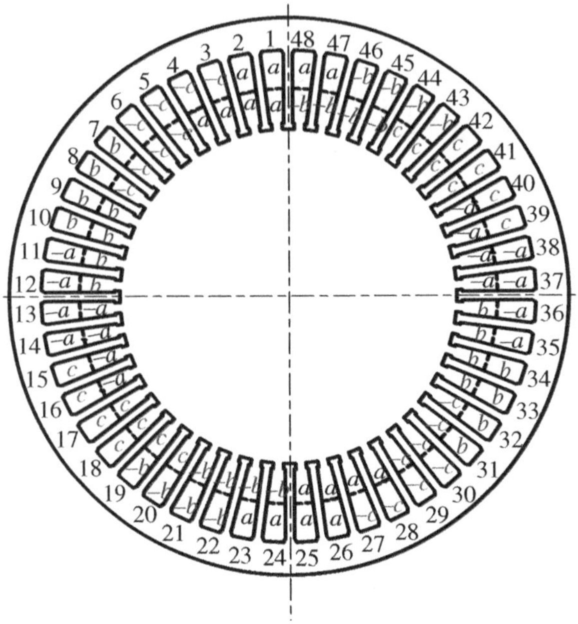

Figure 5.6 The entire phase (*A*), (*B*), and (*C*) coils in this four Pole 48 slots stator.

of a stator is shown in the same figure. This results in the virtual north "N" and south "S" poles induced on the outer surface of the rotor, which produces radially outward $\overline{B}$ field over the arc of the "N" pole and radially inward field over the arc of the "S" pole. This leads to the idealized sinusoidally distributed flux density waveform, $B(\theta)$, which can be expressed, see Figure 5.8, as follows:

$$B(\theta) = B_{\max} \sin(\theta) \tag{5.1}$$

which means that the flux per pole, Φ, can be formulated as follows:

$$\Phi = (2/\pi)B_{\max}\,\tau_p l \tag{5.2}$$

where τ_p is the arc of the pole-pitch at the inner radius of the stator, r. Hence,

$$\tau_p = 2\pi r/p \tag{5.3}$$

where p is the number of poles, and l is the effective axial length of the conductor/coil-sides (a) and $(-a)$. Therefore, from Eq. (5.2), the flux per pole, Φ, can be expressed as follows:

$$\Phi = (2/\pi)B_{\max}(2\pi r/p)l = 4B_{\max}(rl/p) \tag{5.4}$$

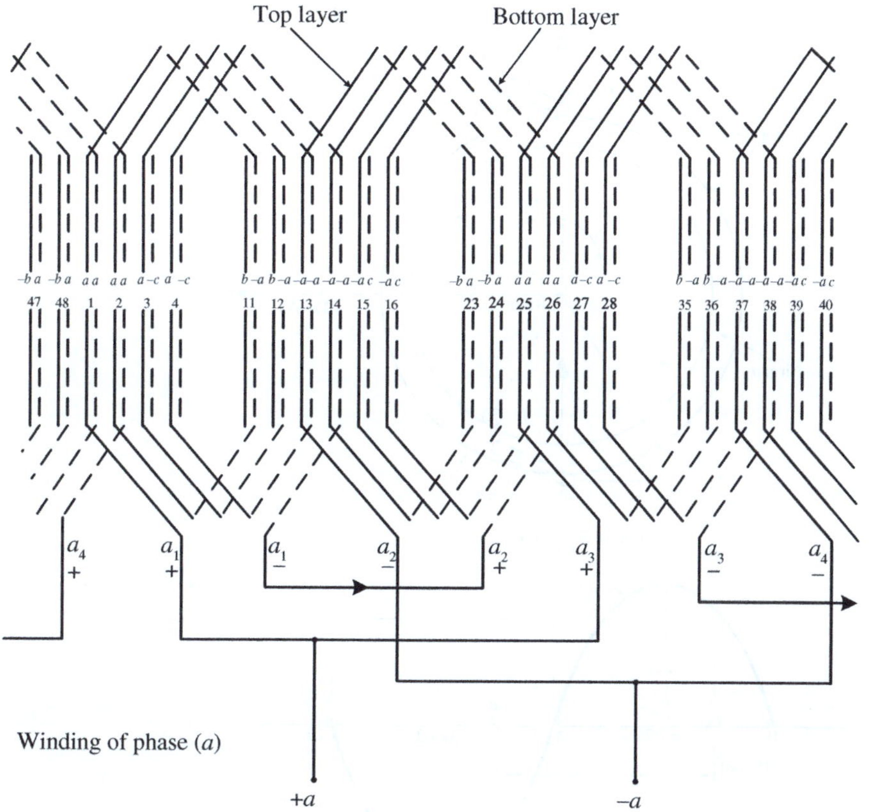

Figure 5.7 The top and bottom coil sides of phase (*A*) – one parallel path.

Once the rotor in Figure 5.8 starts to rotate in the counterclockwise direction, it means that there exists a velocity vector, $\overline{v}$, of relative motion between the conductors (a) and ($-a$) as shown in the figure. This relative motion takes place in the presence of a radially oriented $\overline{B}$ field, which is also shown in this figure. This results in induced voltages due to the well-known ($\overline{v} \times \overline{B}$) volts/unit length, which can be expressed as follows:

$$e_{cond} = B_{conductor} \times v \times l \tag{5.5}$$

where

$$v = \omega_{mech}(r) \text{ and, } B_{conductor} = B_{\max} \sin\theta \tag{5.6}$$

Here, $\omega_{mech} = (2/p)\omega$, where $\omega = 2\pi f$. Accordingly, from Eqs. (5.1) and (5.6), one can express (e_{cond}) as follows:

$$e_{cond} = (2/p)\omega r l B_{\max} \sin\theta \tag{5.7}$$

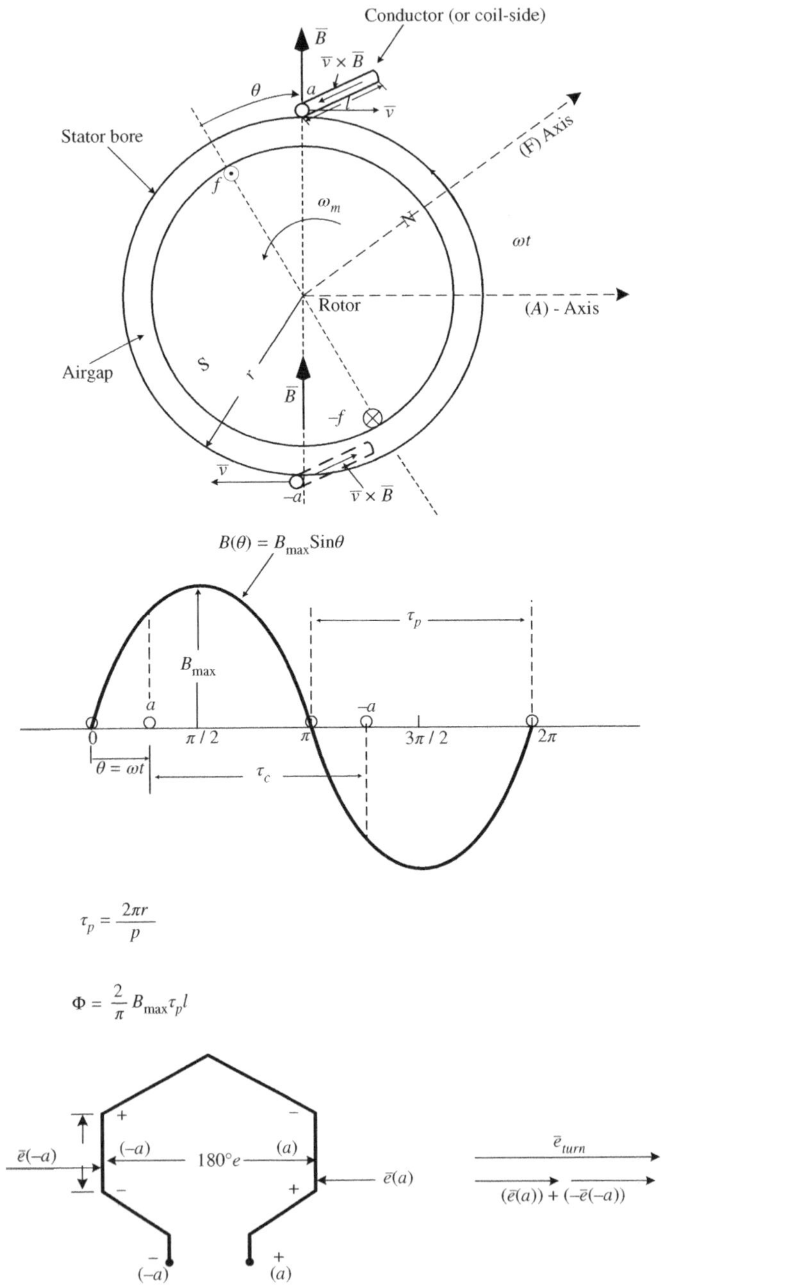

Figure 5.8 Rotor with field excitation (f) and ($-f$) and a one-turn coil (a) and ($-a$). A quasi-isometric view with a sinusoidal flux density waveform, $B(\theta)$.

and an EMF per turn becomes

$$e_{turn} = 2e_{cond} = (4/p)\omega rlB_{\max} \sin\theta \tag{5.8}$$

Equation (5.8) can be rewritten as follows:

$$e_{turn} = (4B_{\max} rl/p)\omega \sin\theta \tag{5.9}$$

However, recalling from Eq. (5.4) that the flux per pole, $\Phi = (4B_{\max}rl/p)$, it follows that

$$e_{turn} = \omega\Phi \sin\theta \tag{5.10}$$

However, $\theta = \omega t$ and $\omega = 2\pi f$, where ω is the electrical angular frequency in electrical radians per second and f is the frequency in Hertz. Hence,

$$e_{turn} = \omega\Phi \sin(\omega t) = 2\pi f\Phi \sin(2\pi ft) \tag{5.11}$$

or

$$e_{turn} = E_{\max} \sin(\omega t) \tag{5.12}$$

where

$$E_{\max} = 2\pi f\Phi \tag{5.13}$$

and the RMS value of the maximum EMF in Eq. (5.13), $E_{rms} = E_{\max}/\sqrt{2}$, hence,

$$E_{rms} = (2\pi/\sqrt{2})f\Phi = 4.44f\Phi \tag{5.14}$$

For a coil with N-turns, the RMS value of its EMF

$$E_{rms}(N - turns) = 4.44f\Phi N \tag{5.15}$$

or

$$E_{rms}(N - turns) = (2\pi/\sqrt{2})f\Phi N \tag{5.16}$$

Hence, for a phase winding with N_{ph}-turns per phase distributed in slotted cores with a number of slots per pole per phase, q, and coils with coil-pitch, $\tau_c \neq \tau_p$, see Figure 5.9; electrical angles between two adjacent slots, α, see Figure 5.10; and coil skew angle, ρ, see Figure 5.11, the EMF per phase, E_{ph}, is given by

$$E_{ph} = (2\pi/\sqrt{2})f\Phi N_{ph}k_w \tag{5.17}$$

where the winding factor, k_w, is given by

$$k_w = k_p k_d k_s \tag{5.18}$$

Here, the pitch factor, k_p, see Figure 5.9, is given by

$$k_p = \sin(\beta/2) = \sin[(\tau_c/\tau_p)\pi/2] \tag{5.19}$$

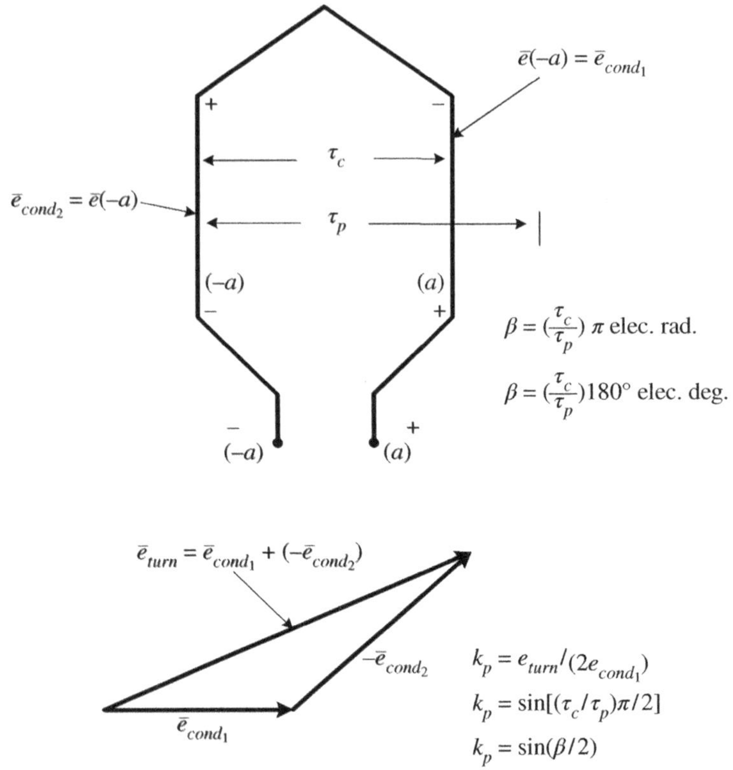

Figure 5.9 Effects of short-pitched turns/coils.

and the distribution factor, k_d, see Figure 5.10, is given by

$$k_d = \sin(q\alpha/2)/[q\sin(\alpha/2)] \tag{5.20}$$

and the skew factor, k_s, see Figure 5.11, is given by

$$k_s = \sin(\rho/2)/(\rho/2) \tag{5.21}$$

where ρ is the skew angle in electric radians.

For a slot pitch angle, α ° elec., and the number of slots per pole per phase, q, we can have the following:

For a flux density distribution with harmonic components higher than the fundamental component, the Fourier Series form, $B(\theta)$ can be expressed as follows (see Figure 5.12):

$$B(\theta) = \sum_{h=1,3,5\cdots}^{\infty} B_{m_h} \sin(h\theta + \psi_h) \tag{5.22}$$

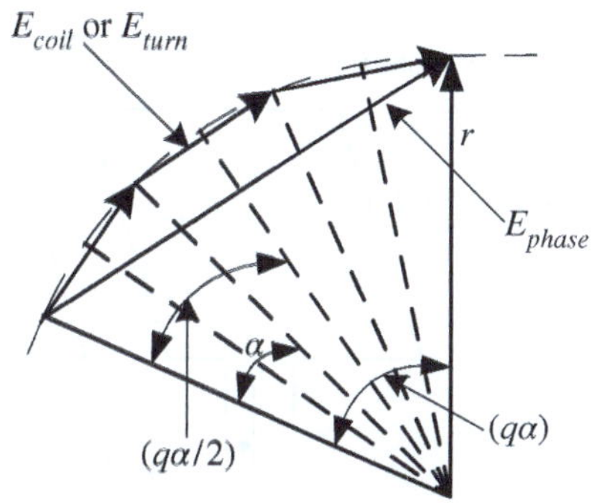

Figure 5.10 Distribution factor – effects of distributed windings.

$$\frac{E_{phase}}{qE_{coil}} = k_d = \text{Distribution factor}$$

$$k_d = \frac{2r\sin(q\alpha/2)}{q(2r\sin(\alpha/2))} = \frac{\sin(q\alpha/2)}{q\sin(\alpha/2)}$$

Here, $(q\alpha)$ = Phase belt angle

where the harmonic flux per pole for the "h" harmonic is

$$\Phi_h = \frac{2}{\pi} \cdot B_{m_h} \cdot \left(\frac{\tau_p}{h}\right) \cdot l \tag{5.23}$$

and the harmonic winding factor, k_{w_h}, is

$$K_{w_h} = \sin(h\beta/2) \cdot \frac{\sin(qh\alpha/2)}{q\sin(h\rho/2)} \cdot \frac{\sin(h\rho/2)}{(h\rho/2)} \tag{5.24}$$

and the harmonic EMF per phase, E_{ph_h}, is as follows:

$$E_{ph_h} = 4.44(hf)\Phi_h \cdot (N_{ph}/m) \cdot k_{w_h} \tag{5.25}$$

and the total phase RMS, EMF, E_{ph}, is

$$E_{ph}(rms) = \sqrt{E_{ph_1}^2 + E_{ph_3}^2 + \cdots + E_{ph_k}^2 + \cdots} \tag{5.26}$$

where N_{ph} is the number of turns per phase, m is the number of parallel paths per phase, and h is the harmonic component order. Therefore, for the fundamental component, $h = 1$,

$$E_{ph_1} = 4.44 f\Phi_1(N_{ph}/m)k_{w_1} \tag{5.27}$$

An important observation regarding the relationship between the fundamental flux per phase, Φ_1, AC frequency, f, and fundamental phase EMF, E_{ph_1}, which is approximately close to the phase voltage, V_{ph_1}, is that the following can be written as

$$V_{ph_1} \simeq E_{ph_1} = 4.44 f\Phi_1(N_{ph}/m)k_{w_1} \tag{5.28}$$

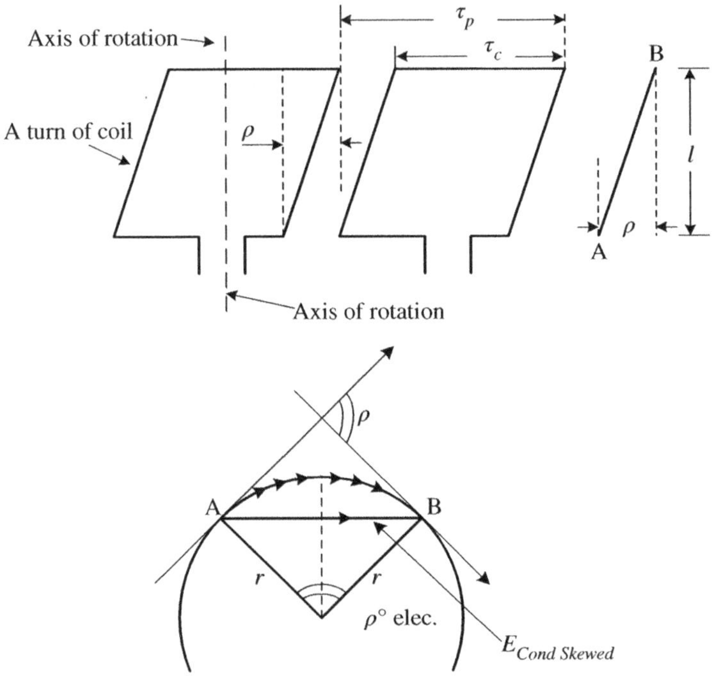

$$\frac{E_{Cond\ Skewed}}{E_{Cond\ Unskewed}} = \frac{E_{Turn\ Skewed}}{E_{Turn\ Unskewed}} = \frac{2r\sin(\rho/2)}{\rho r} = k_s$$

$$\text{Skew factor} = k_s = \frac{\sin(\rho/2)}{(\rho/2)}$$

Figure 5.11 Effects of skewing winding turns/coils.

Therefore, an AC machine is connected to a controller/drive that preserves the ratio (V_{ph_1}/f) to a constant value, that is,

$$(V_{ph_1}/f) = \text{constant} \tag{5.29}$$

It therefore follows from Eq. (5.28) that

$$(V_{ph_1}/f) = 4.44\Phi_1(N_{ph} \cdot k_{w_1}/m) = \text{constant} \tag{5.30}$$

which means that the fundamental flux per pole, Φ_1, will remain constant as long as the ratio (V_{ph_1}/f) remains constant. This concept is known in adjustable-speed drives (ASDs) as the concept of constant volts per Hertz, a key principle in AC machine ASD systems and technologies. This concept of constant volts per Hertz will be discussed in more detail in later chapters of this book.

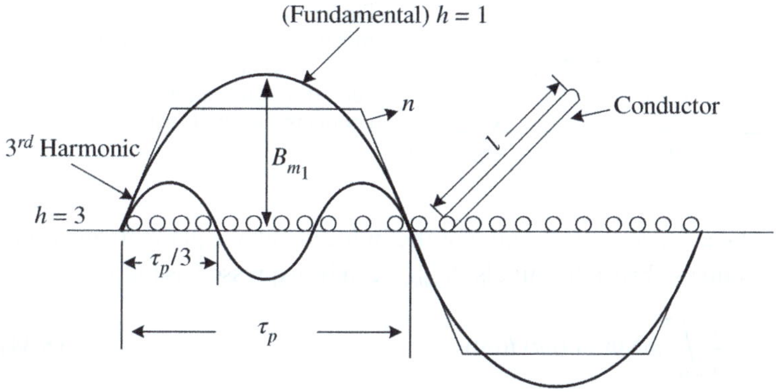

Figure 5.12 Flux density waveform with harmonics.

5.4 Induced Electromotive Forces in Brush-Commutator-Type DC Windings

In any turn in the armature winding (rotor winding) of a DC machine, see Figure 3.35, depicting a two-pole machine, the generated/induced voltage is an AC-type, whose fundamental component per turn, e_{turn}, can be expressed, as derived in Eq. (5.11), as follows:

$$e_{turn} = \omega\Phi \sin(\omega t) \tag{5.31}$$

Here, ω is the angular speed of the rotor (armature winding) in electrical radians per second, that is,

$$\omega = (p/2)\omega_m \tag{5.32}$$

where ω_m is the angular speed in mechanical radians per second and p is the number of field winding poles on the stator. Therefore, for an N-turns coil, the induced voltage in the coil, e_{coil}, can be expressed as follows:

$$e_{coil} = Ne_{turn} = \omega N\Phi \sin(\omega t) \tag{5.33}$$

However, through the construct of the armature coils connections to the commutator segments, and the fixed position of the brushes pressing against the cylindrical surface of the commutator segments, see the simplified example two-pole DC machine of Figure 3.35, a mechanical rectification (commutation) process takes place that renders the induced EMF waveform, seen through the terminals connected to the brushes as shown in Figure 5.13.

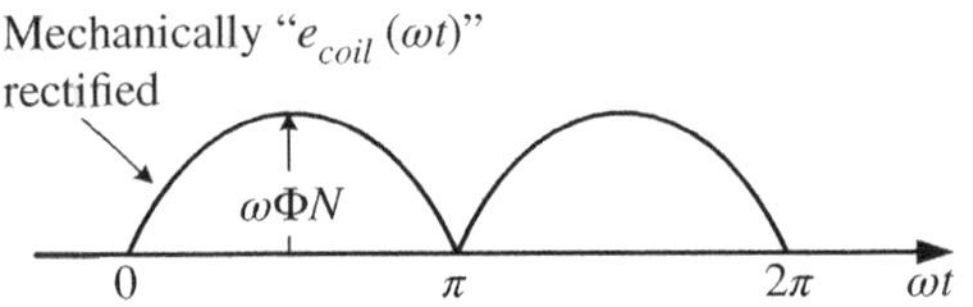

Figure 5.13 Waveform of mechanical rectified induced voltage in an armature coil in a DC machine seen through the terminals connected to the brushes.

Therefore, the average, or DC, value of the induced voltage per coil appearing across the (+) and (−) brush terminals, $E_{coilav.}$, can be expressed as follows:

$$E_{coilav.} = \frac{1}{\pi}\int_0^{\pi} \omega N\Phi \sin(\omega t)d(\omega t) \tag{5.34}$$

which yields

$$E_{coilav.} = (2/\pi)\omega N\Phi \tag{5.35}$$

However, substituting for ω from Eq. (5.32) into Eq. (5.35) yields

$$E_{coilav.} = (\omega_m pN\Phi/\pi) \tag{5.36}$$

For a DC machine armature with armature turns, N_a, and a number of parallel paths, m, see Figures 3.36 and 3.37 for lap and wave-type windings, respectively; the number of armature turns connected in series per parallel path becomes (N_a/m); the induced armature voltage appearing between the (+) and (−) brushes of the DC machine, E_a, can therefore be expressed as follows:

$$E_a = [\omega_m p(N_a/m)\Phi]/\pi \tag{5.37}$$

However, for a rotational speed, n, in rev/min, the mechanical speed, ω_m, can be written as follows in mechanical radians per second:

$$\omega_m = (n/60)(2\pi) = \pi n/30 \tag{5.38}$$

Substituting from Eq. (5.38) into Eq. (5.37) yields the following expression for the induced voltage (EMF) in a DC armature

$$E_a = np(N_a/m)\Phi/30 \tag{5.39}$$

For a number of armature conductors, C_a, the number of armature turns, $N_a = C_a/2$, therefore, E_a, can be rewritten as follows:

$$E_a = np(C_a/m)\Phi/60 \tag{5.40}$$

Next, we introduce MMFs in AC and DC machines.

5.5 MMF in Distributed and Concentrated AC and DC Machine Windings and Associated Flux Density Distributions/Waveforms in Airgaps

First, consider the single-coil, single-phase, cross-sectional model of a laminated stator steel core, with two slots (a) and ($-a$), Figure 5.14, housing a coil with two coil sides in slots (a) and ($-a$). The coil is made of N-turns. The stator core surrounds an unslotted laminated-steel cylinder, separated from the rotor core by an airgap of radial height, h_g (or g), that is, h_g and g may be used interchangeably throughout this book. There is a depiction of the expected shape of flux density, $\overline{B}$, lines, to be established by a current, i, with an orientation designated by the dot and cross-notations in slots (a) and ($-a$), respectively, as shown in Figure 5.14. A corresponding developed (flattened) depiction of the airgap, the inner slotted surface of the stator, and the unslotted surface of the rotor, with an in-between airgap of height h_g (or g), are also shown at the bottom of Figure 5.14.

Also shown in Figure 5.14 is the current distribution over this airgap circumference, with (Ni) ampere-conductors in (a) slot and ($-Ni$) ampere-conductors in slot ($-a$). Applying Ampere's law of $\oint \overline{H} \cdot d\overline{l} = \sum$ enclosed currents to the current distribution shown in this figure, see the closed path integral for ($\oint \overline{H} \cdot d\overline{l}$) shown in the figure, we obtain the trapezoidal-profiled MMF waveform, $\mathbf{F}(\theta)$, also shown in the figure. This is in addition to the rectangular idealization of this trapezoidally profiled waveform, also shown in the figure. Examining Figure 5.14 closely, one notices that the "$\overline{B}$" or "$\overline{H}$" lines cross the airgap twice to form the closed loop in $\oint \overline{H} \cdot d\overline{l} = \sum$ enclosed currents. Hence, the flat part of the trapezoidal waveform is ($Ni/2$) and so is the height of the rectangular waveform idealization in this figure. Considering the idealized rectangular waveform, according to Fourier series analysis, the fundamental component, $\mathbf{F}_1(\theta)$, of this waveform can be analytically written as follows:

$$\mathbf{F}_1(\theta) = (4/\pi)(Ni/2)\cos(\theta) \tag{5.41}$$

where $\mathbf{F}_1(\theta)$ is the ampere-turns (AT) per airgap, that is, AT/pole.

If we consider the laminated ferrous core to be idealized with very large (or nearly infinite) permeabilities, one can express the fundament flux density, $B_1(\theta)$, waveform, assuming $\mathbf{F}_1(\theta)$ to be spent to push through the "B" field only across the airgap, as follows:

$$B_1(\theta) = \mu_0[\mathbf{F}_1(\theta)/h_g] \tag{5.42}$$

That is,

$$B_1(\theta) = (4/\pi)\mu_0(Ni/2h_g)\cos(\theta) \tag{5.43}$$

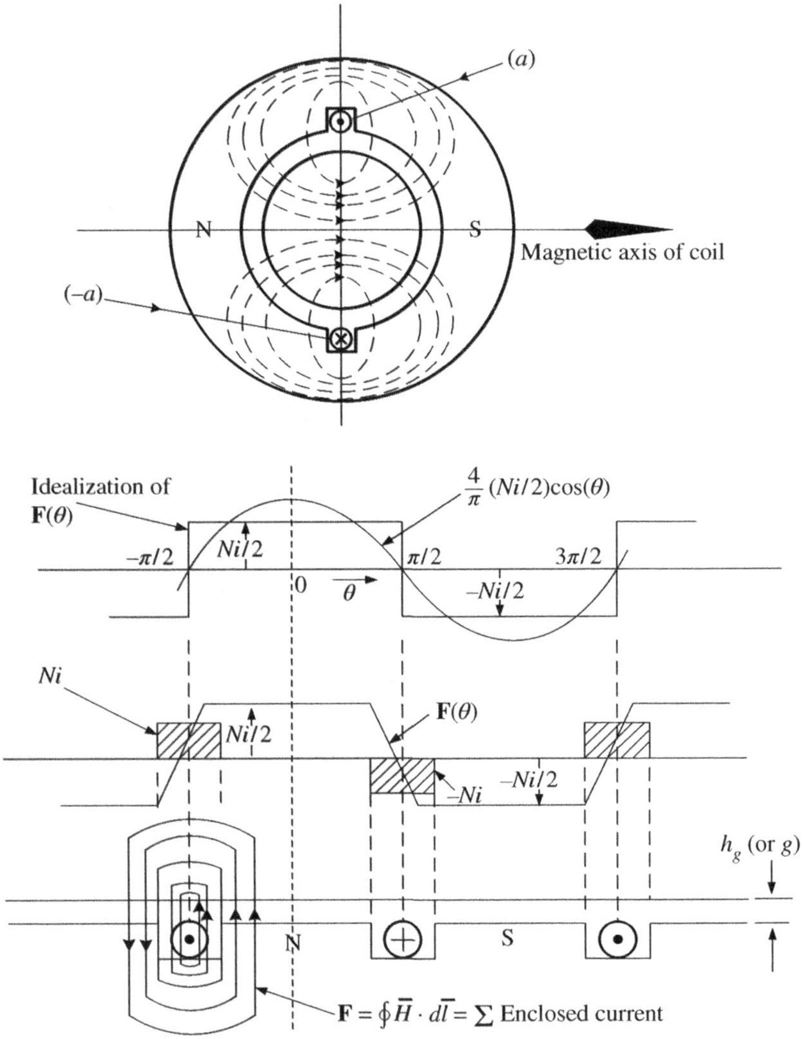

Figure 5.14 A single-coil model of a single-phase AC armature surrounding a cylindrical unslotted rotor with an airgap of radial height, h_g (or g).

where the peak of the flux density waveform, $B_{1\max}$, becomes

$$B_{1\max} = (4/\pi)\mu_0(Ni/2h_g) \tag{5.44}$$

where $B_{1\max}$ is in Tesla, and the corresponding peak of the MMF, $\mathbf{F}_1(\theta)$, $F_{1\max}$, in AT per pole or AT per airgap is as follows:

$$F_{1\max} = (4/\pi)(Ni/2) \tag{5.45}$$

Furthermore, a close examination of Figure 5.14 reveals that the $\overline{B}$ flux lines have an axis of symmetry, which is the magnetic axis of the coil [(a) & $(-a)$]. Along this magnetic axis, the $\overline{B}$ flux lines radially exit the inner boundaries of the stator core into the airgap, thus creating a "virtual" north "N" pole facing the airgap on the left side of this cross-sectional figure. Meanwhile, the opposite phenomenon takes place on the right side of this figure, thus creating a "virtual" south "S" pole facing the airgap, where the $\overline{B}$ flux lines radially enter the stator core along the magnetic axis of the excitation coil. See the "N" pole and "S" pole designations in the developed diagram at the bottom of Figure 5.14.

Second, we will analytically develop the MMF waveform of only one phase in a three-phase, a, b, and c, stator winding in an AC three-phase machine, see Figure 5.15. This is a 24-stator slots, 2-pole, and 3-phase machine. That is, the number of stator slots per pole per phase, q, is equal to four slots/pole/phase. The a, $-a$, b, $-b$, c, and $-c$ slots are clearly designated in Figure 5.15, including the phase axes, Axis (a), Axis (b), and Axis (c). Notice that, in a counterclockwise rotational direction, the (a), (b), and (c) phase sequence is preserved. Also, shown in Figure 5.15 is the a, b, and c phase coil connection schematic in a series "one-path" connection per phase, $m = 1$, and the schematic of phase (a) connection for a "two parallel-path" connection, that is, $m = 2$, is also shown in the figure.

Meanwhile, Figure 5.16 depicts the developed diagram of the slots of phase (a) of the machine previously shown in Figure 5.15, all-be-it with a double-layer winding housed in the (a) and $(-a)$ slots, four (a) and four $(-a)$ slots, that is, $q = 4$ slots per pole per phase.

For a number of conductors, n_c, per coil side, and parallel paths per phase, m, the ampere-conductors per slot are $2n_c(i_a/m)$, for a phase current, i_a. This results in the current ampere-conductor distribution, as shown in Figure 5.16, and the resulting MMF waveform is the two-trapezoidal profiles shown in the figure. Therefore, this results in an MMF with a fundamental component, $\mathbf{F}_{a1}(\theta)$, that can be expressed as follows:

$$\mathbf{F}_{a1}(\theta) = [(4/\pi)k_w(N_{ph}/p)(i_a/m)]\cos\theta \tag{5.46}$$

where k_w is the winding factor, including the pitch factor, k_p, the distribution factor, k_d, and the skew factor, k_s, Eqs. (5.19), (5.20), and (5.21). Equation (5.46) can be rewritten as follows:

$$\mathbf{F}_{a1}(\theta) = \mathbf{F}_{a1\max}\cos(\theta) \tag{5.47}$$

where

$$\mathbf{F}_{a1\max} = (4/\pi)k_w(N_{ph}/p)(i_a/m) \tag{5.48}$$

where N_{ph} is the number of turns per phase, p is the number of poles, i_a is the current in phase (a), and m is the number of parallel paths.

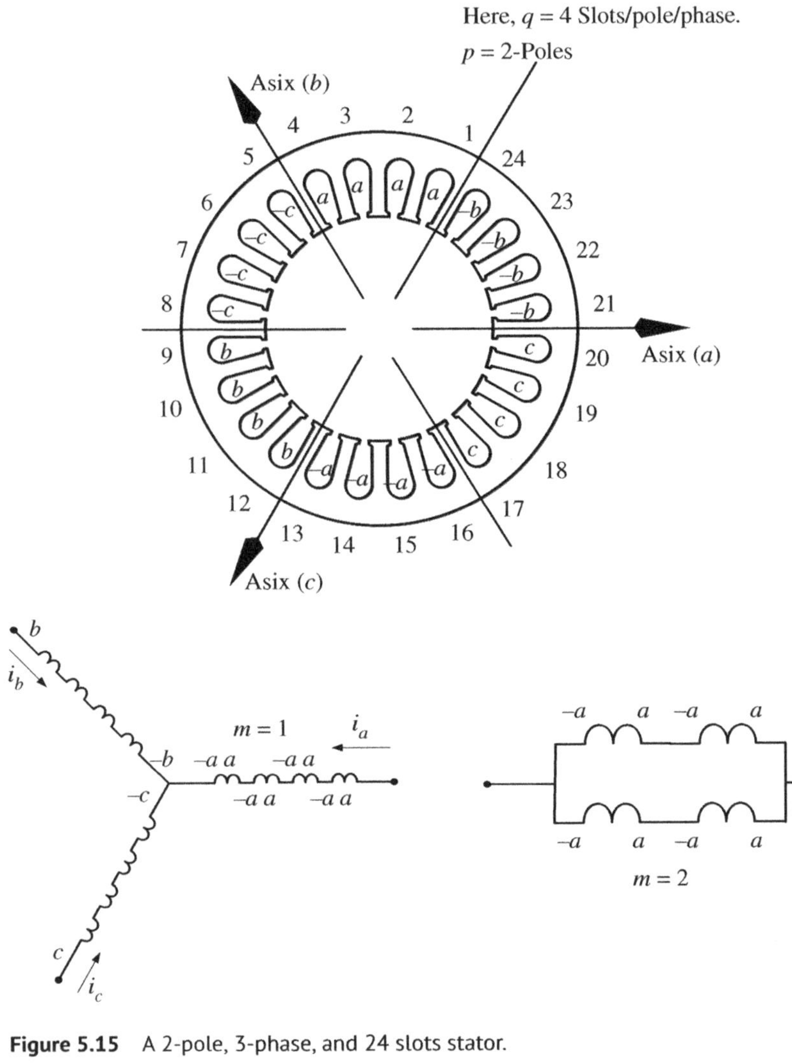

Figure 5.15 A 2-pole, 3-phase, and 24 slots stator.

Accordingly, the corresponding flux density waveform, $B_{a1}(\theta)$, can be expressed as follows:

$$B_{a1}(\theta) = \mu_0[\mathbf{F}_{a1}(\theta)/h_g] \tag{5.49}$$

assuming stator and rotor ferrous lamination with very large permeabilities. Hence, one can write the following:

$$B_{a1}(\theta) = \mu_0(\mathbf{F}_{a1\,\max}(\theta)/h_g)\cos(\theta) \tag{5.50}$$

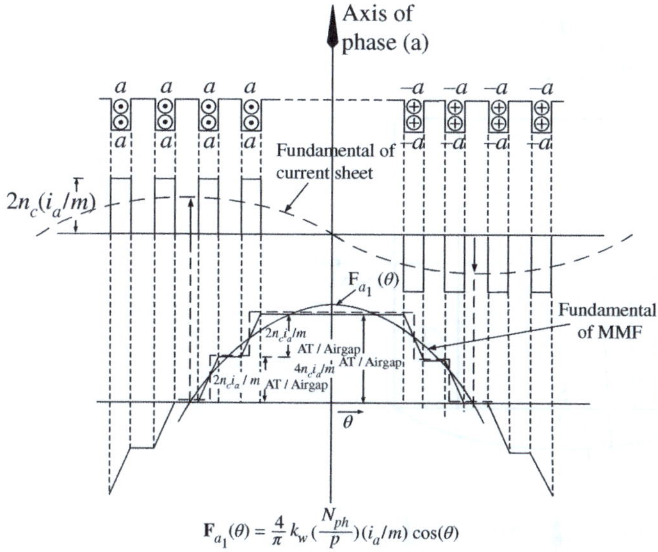

Figure 5.16 The MMF waveform of phase (a) for the 24 slots, 2-pole, and three-phase winding of Figure 5.15.

That is,

$$B_{a1}(\theta) = [(\mu_0/h_g)(4/\pi)k_w(N_{ph}/p)(i_a/m)]\cos(\theta) \quad (5.51)$$

or

$$B_{a1}(\theta) = B_{a1\,\max}\cos(\theta) \quad (5.52)$$

where

$$B_{a1\,\max}(\theta) = [(\mu_0/h_g)(4/\pi)k_w(N_{ph}/p)(i_a/m)] \quad (5.53)$$

Third, we will consider the MMF waveform and fundamental harmonic flux associated with salient-pole rotors in AC machines. Consider the two-pole, salient-pole rotors shown in Figure 5.17 and associated flattened/developed diagram is shown in this figure. Applying Ampere's law of $\oint \overline{H}\cdot d\bar{l} = \sum$ enclosed currents, starting from the center of the positive current distribution along the closed paths shown in the figure, results in the trapezoidal MMF waveform shown in the figure. The fundamental component of this MMF waveform, $\mathbf{F}_{r1}(\theta)$, can be expressed as follows:

$$\mathbf{F}_{r1}(\theta) = (4/\pi)k_{f1}(N_f/p)(i_f)\cos(\theta) \quad (5.54)$$

where N_f is the total number of field winding/rotor field winding turns, p is the number of salient poles, i_f is the field winding current passing through the turns

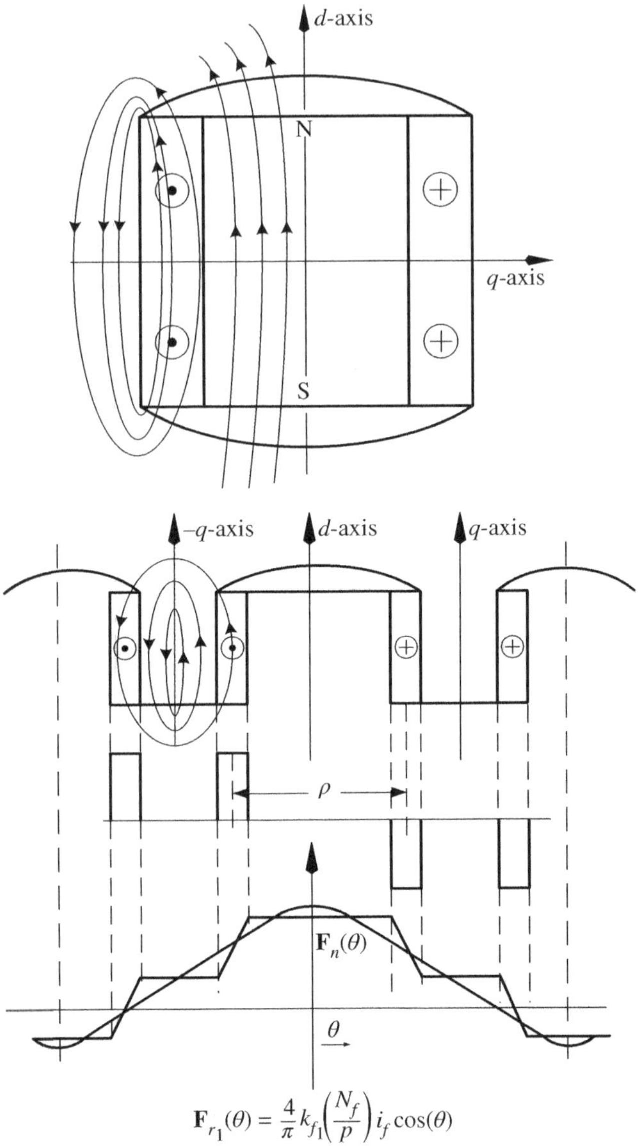

Figure 5.17 A two-pole salient field winding and resulting MMF waveform.

of the field winding, and k_{f1} is a fundamental space harmonic factor given by the following equation:

$$k_{f1} = \sin(\rho/2) \tag{5.55}$$

where ρ is the angular spread in electrical measure as defined by the angular spread between the two sides of the field excitation winding surrounding the poles.

Equation (5.54) can be rewritten as follows:

$$\mathbf{F}_{r1}(\theta) = \mathbf{F}_{r1\,\max}\cos(\theta) \tag{5.56}$$

where

$$\mathbf{F}_{r1\,\max} = (4/\pi)k_{f1}(N_f/p)(i_f) \tag{5.57}$$

and the corresponding maximum flux density, $B_{r1\max}$, in an airgap of height, h_g, with very large (or nearly infinite) ferrous lamination permeabilities becomes

$$B_{r1\,\max} = \mu_0\mathbf{F}_{r1\,\max}/h_g \tag{5.58}$$

That is,

$$B_{r1\,\max} = (\mu_0/h_g)(4/\pi)k_{f1}(N_f/p)i_f \tag{5.59}$$

and $B_{r1}(\theta_e) = B_{r1\max}\sin(\theta_e)$

Fouth, in order to compute the flux per pole, Φ_{pole}, resulting from a field winding surrounding a salient-pole, the reader should refer to Figure 5.18 of a sample four-pole rotor. The diameter at mid-airgap is D, and, therefore, the pole arc, τ_p, can be expressed as follows:

$$\tau_p = (\pi D/p) \tag{5.60}$$

where p is the number of poles.

Examining Figure 5.18, one can write

$$\Phi_{pole}\int_{S_{pole}}\overline{B}\cdot d\overline{S} = \int_{S_{pole}} B_{r_{1\max}}\sin(\theta_e)dS \tag{5.61}$$

where, in this case, dS can be expressed, see Figure 5.18, as follows:

$$dS = l(D/2)d\theta_m = l(D/2)(2/p)d\theta_e = (Dl/p)d\theta_e \tag{5.62}$$

Substituting for (dS) from Eq. (5.62) into Eq. (5.61), the flux per pole, Φ_{pole}, can be expressed as follows, where l is the rotor axial effective length:

$$\Phi_{pole} = \int_0^{\pi}(B_{r_{1\max}}\sin\theta_e)(Dl/p)d\theta_e \tag{5.63}$$

Therefore,

$$\Phi_{pole} = B_{r_{1\max}}(Dl/p)\int_0^{\pi}(\sin\theta_e)d\theta_e \tag{5.64}$$

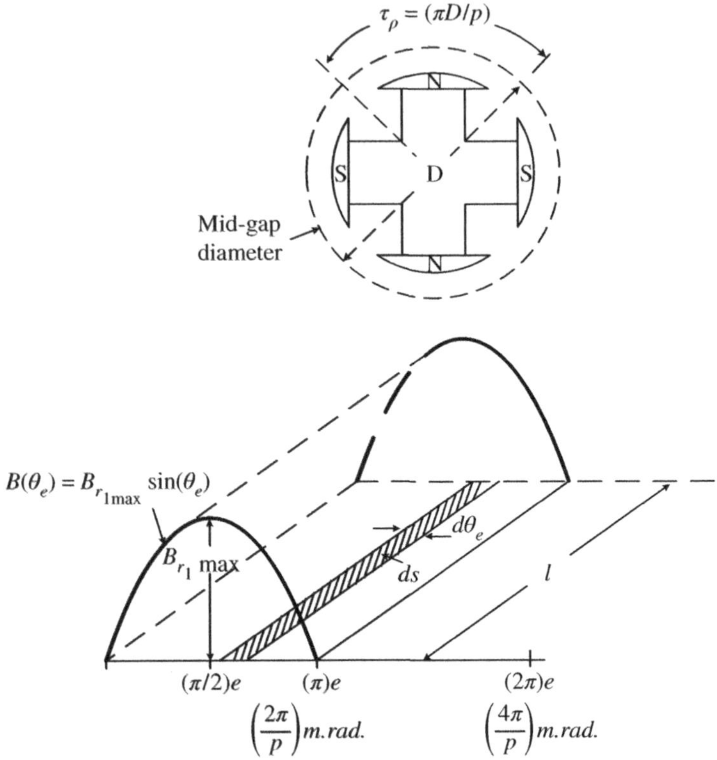

$$\Phi_{pole} = \int_{S_{pole}} \bar{B} \cdot d\bar{S} = \int_{S_{pole}} B(\theta_e) \cdot dS$$

$$dS = l\left(\frac{D}{2}\right)d\theta_m = l\left(\frac{D}{2}\right)\cdot\left(\frac{2}{p}\right)d\theta_e = l\left(\frac{D}{p}\right)d\theta_e$$

$$\theta_e = \left(\frac{p}{2}\right)\theta_m$$

$$d\theta_e = \left(\frac{p}{2}\right)d\theta_m$$

$$d\theta_m = \left(\frac{2}{p}\right)d\theta_e$$

Figure 5.18 Computation of sinusoidally distributed flux.

which yields

$$\Phi_{pole} = (2Dl/p)B_{r_{1\,max}} \tag{5.65}$$

However, from Eq. (5.60), it follows that $(D/p) = (\tau_p/\pi)$, therefore,

$$\Phi_{pole} = (2/\pi)\tau_p l B_{r_{1\,max}} \tag{5.66}$$

Next, we give a numerical/practical example:

Example 5.1 ***A Salient Pole Synchronous Machine Has the Following Design Features***

$$p = 6$$

$$N_f = 114$$

$$h_g = 0.032\ \text{m}$$

$$k_r = 0.92$$

$$B_{r_{1\max}} = 1.3\ \text{T}$$

$$D_r = 1.1\ \text{m}$$

$$D = D_{midgap} = 1.132\ \text{m}$$

$$l = 1.82\ \text{m}$$

$$F_{r_1}(\theta) = F_{r_{1\max}} \cos(\theta)$$

$$\begin{aligned} F_{r1\max} &= \frac{4}{\pi} k_r \left(\frac{N_f}{p}\right) i_f \\ &= \frac{4}{\pi} \times 0.92 \times \left(\frac{114}{6}\right) i_f\ \text{AT/p} \end{aligned}$$

$$\begin{aligned} B_{r_{\max 1}} &= \mu_0 \left(\frac{F_{r_{1\max}}}{h_g}\right) \\ &= \frac{4\pi \times 10^{-7}}{0.032} \left[\frac{4}{\pi} \times 0.92 \times \left(\frac{114}{6}\right) i_f\right] \\ &= 1.3\ \ \text{Tesla} \end{aligned}$$

$$i_f = \frac{1.3 \times 6 \times 0.032 \times 10^7}{16 \times 0.92 \times 114} = 1487.4\ \text{A}$$

$$\Phi_{pole} = \frac{2}{\pi} B_{r_{1\max}} \cdot \tau_p \cdot l$$

$$\tau_p = \frac{\pi D}{p} = \frac{\pi(1.1 + 0.032)}{6} = 0.5927\ \text{m}$$

$$\Phi_{pole} = \frac{2}{\pi} \times 1.3 \times 0.5927 \times 1.82 = 0.8928\ \text{Wb}$$

Fifth, consider the two-pole, eight slots armature DC commutator-brush-type machine shown in the cross-section of Figure 5.19. For a field-winding with N_f turns and a field current, i_f, there are (N_f/p) turns per pole and $(N_f/p)i_f$

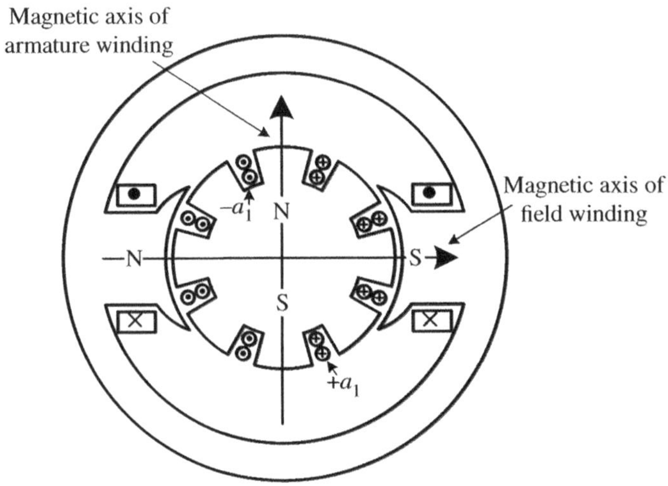

Figure 5.19 A two-pole 12 slots armature DC commutator-brush-type machine.

ampere-turns per pole distributed over the width of the field coils to render the trapezoidally shaped field winding MMF, $\mathbf{F}_f(\theta)$, as shown in Figure 5.20, whose fundamental harmonic component, $\mathbf{F}_{f_1}(\theta)$, can be expressed as follows:

$$\mathbf{F}_{f_1}(\theta) = (4/\pi)k_{f_1}(N_f/p)i_f\cos(\theta) \tag{5.67}$$

where k_{f_1} is a fundamental harmonic space factor derived from Fourier analysis and is given by

$$k_{f_1} = \sin(\rho/2) \tag{5.68}$$

where the space angle, ρ, in electrical measure is defined in Figure 5.20 as the spread from center-to-center of the field coils surrounding each pole.

Meanwhile, the ampere-conductor distribution in the armature, for a number turns-per coil/conductors-per coil-side, n_c, and current per conductor, i_a, are also shown in Figure 5.20. Furthermore, this armature MMF, $\mathbf{F}_a(\theta)$, is idealized by a triangular waveform in Figure 5.21, with a peak, F_a, which can be expressed, based on the ampere-conductor distribution shown in Figure 5.21, as follows:

$$F_a(\theta) = (1/2)(2N_a/p)(i_a/m) = (N_a i_a/mp) \tag{5.69}$$

where N_a is the number of armature turns, i_a is the armature current, m is the number of parallel paths in the armature, and p is the number of poles. The fundamental component, $\mathbf{F}_{a_1}(\theta)$, of the armature MMF, $\mathbf{F}_a(\theta)$, based on Fourier analysis of this triangular waveform, can be expressed as follows:

$$\mathbf{F}_{a_1}(\theta) = (8/\pi^2)F_a\cos(\theta) \tag{5.70}$$

Figure 5.20 Profiles of field and armature winding waveforms in DC and AC commutator-brush machines.

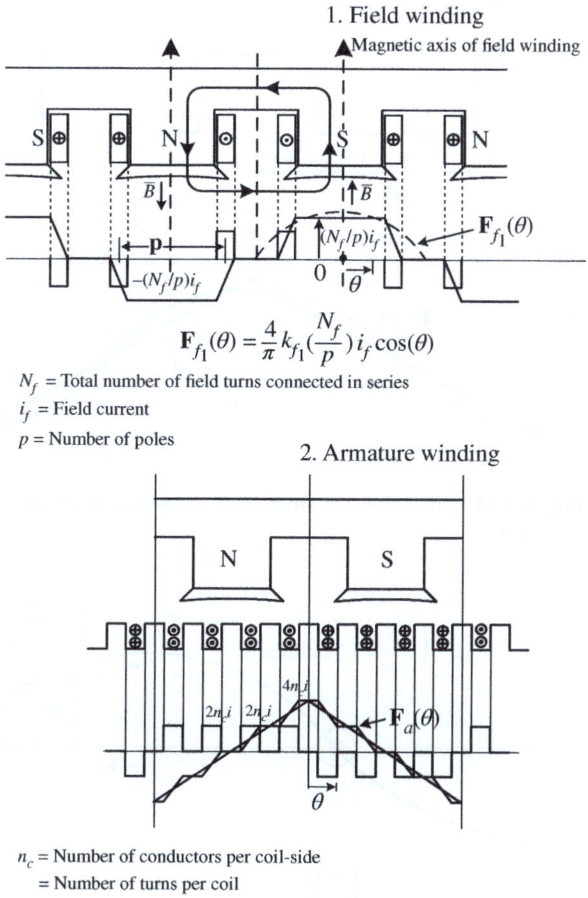

Thus, substituting from Eq. (5.69) into Eq. (5.70) yields the following for $\mathbf{F}_{a_1}(\theta)$:

$$\mathbf{F}_{a_1}(\theta) = (8/\pi^2)(N_a i_a/mp)\cos(\theta) \tag{5.71}$$

Here, the number of parallel paths, m, is equal to the number of poles for simple-lap windings and is equal to 2 for simple-wave windings.

Sixth, we now consider the MMFs resulting from single-phase, two-phase, balanced, and unbalanced winding, as well as balanced three-phase windings carrying AC currents.

(a) Now, we start with the single-phase case. The fundamental MMF component of a single-phase distributed winding, with q slots per pole per phase, and a winding factor, k_w, with a number of turns per phase, N_{ph}, and a number of poles, p, see Figure 5.22, can be written on the basis of Eq. (5.41) with a current,

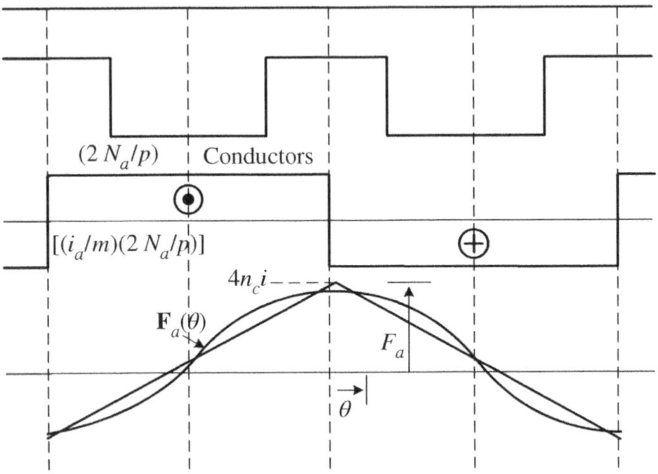

Figure 5.21 Idealized armature MMF waveform in DC and AC commutator-brush machines.

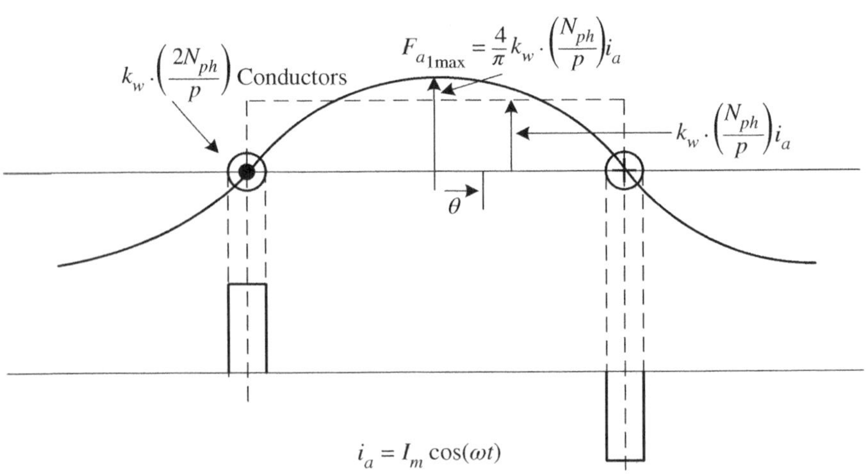

Figure 5.22 The fundamental harmonic component of the MMF, $\mathbf{F}_{a_1}(\theta)$, of a single-phase winding.

i_a, flowing through the phase winding, as follows:

$$\mathbf{F}_{a_1}(\theta) = (4/\pi)k_w(N_{ph}/p)i_a(t)\cos(\theta) \tag{5.72}$$

If i_a is an AC current that can be expressed as

$$i_a(t) = \sqrt{2}I\cos(\omega t) \tag{5.73}$$

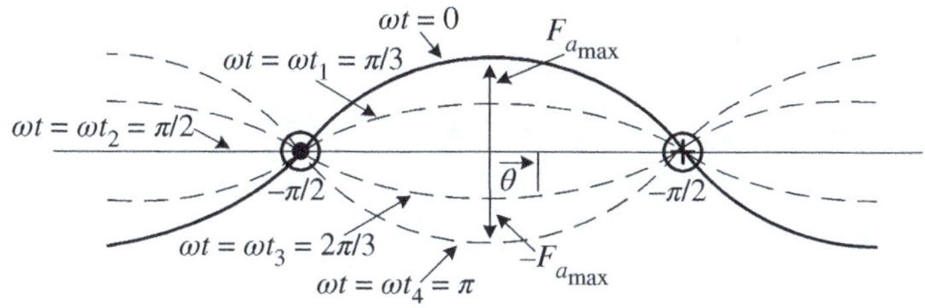

Figure 5.23 Standing pulsating MMF waveform of a single-phase winding at $\omega t = 0$, $\pi/3$, $\pi/2$, $2\pi/3$, and $\omega t = \pi$.

where I is the RMS magnitude of this AC current, $i_a(t)$, and, hence, $\mathbf{F}_{a_1}(\theta)$ can be rewritten as

$$\mathbf{F}_{a_1}(\theta) = \left[(4/\pi)(k_w N_{ph}/p)(\sqrt{2}I)\right]\cos(\theta)\cos(\omega t) \tag{5.74}$$

The MMF, $\mathbf{F}_{a_1}(\theta)$, in Eq. (5.74), is depicted in Figure 5.23 at $\omega t = 0$, $\pi/3$, $\pi/2$, $2\pi/3$, and $\omega t = \pi$.

This is the so-called standing wave. Defining $F_{\max}$ as follows:

$$F_{\max} = \left[(4/\pi)(k_w N_{ph}/p)(\sqrt{2}I)\right] \tag{5.75}$$

The MMF, $\mathbf{F}_{a_1}(\theta)$, in Eq. (5.74) can be rewritten as follows:

$$\mathbf{F}_{a_1}(\theta) = F_{\max}\cos(\theta)\cos(\omega t) \tag{5.76}$$

Utilizing the trigonometric identity, $\cos\alpha\cos\beta = [\cos(\alpha-\beta)+\cos(\alpha+\beta)]/2$, Eq. (5.76) can be rewritten as follows:

$$\mathbf{F}_{a_1}(\theta) = (F_{\max}/2)\cos(\theta - \omega t) + (F_{\max}/2)\cos(\theta + \omega t) \tag{5.77}$$

The first term in Eq. (5.77), $\mathbf{F}^{(+)}_{a_1}(\theta, \omega t)$, is

$$\mathbf{F}^{(+)}_{a_1}(\theta, \omega t) = (F_{\max}/2)\cos(\theta - \omega t) \tag{5.78}$$

and the second term in Eq. (5.77), $\mathbf{F}^{(-)}_{a_1}(\theta, \omega t)$, is

$$\mathbf{F}^{(-)}_{a_1}(\theta, \omega t) = (F_{\max}/2)\cos(\theta + \omega t) \tag{5.79}$$

Here, $\mathbf{F}^{(+)}_{a_1}(\theta, \omega t)$ constitutes a forward rotating (traveling) wave, while $\mathbf{F}^{(-)}_{a_1}(\theta, \omega t)$ constitutes a backward (reverse) rotating (traveling) wave, which are depicted graphically in Figure 5.24, in space waveform profiles and forward as well as backward (reverse) space vectors. Here, forward means space-vector rotation in the counterclockwise direction, and backward (reverse) means rotation in the clockwise direction, see the space-vector graphical representation in Figure 5.24.

(b) We now discuss the case of a balanced two-phase machine, see the schematic of Figure 5.25, and an actual two-phase, phase (a) and phase (b), embedded in a 24-stator slot lamination of a two-pole machine given in Figure 5.26.

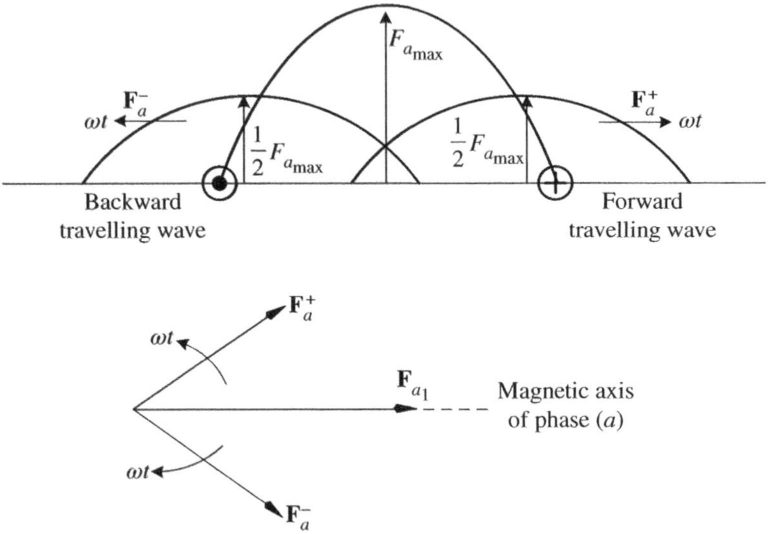

Figure 5.24 Graphical depiction of the forward and backward (reverse) traveling waveforms and rotating space vectors produced by the passage of AC current in a single-phase winding.

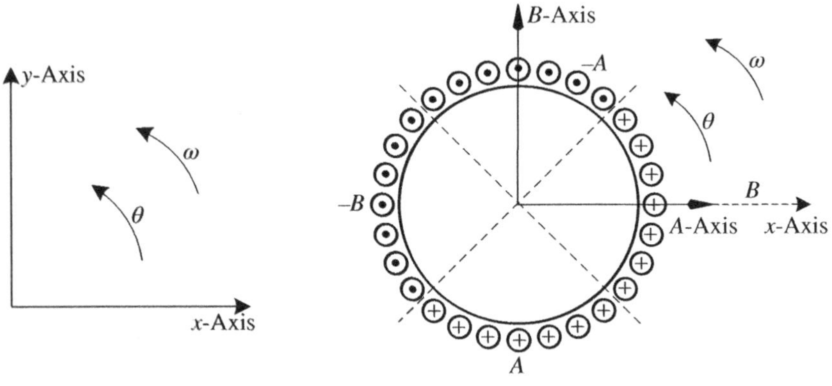

Figure 5.25 The schematic layout of a two-phase two-pole stator for phases (*A*) and (*B*).

If the two phases (*A*) and (*B*) in Figure 5.25 are carrying balanced currents, it follows that

$$i_a(t) = \sqrt{2}I\cos(\omega t) \tag{5.80}$$

and

$$i_b(t) = \sqrt{2}I\cos(\omega t - \pi/2) \tag{5.81}$$

Figure 5.26 Phases (*a*) and (*b*) in a 24-slots lamination of a 2-pole balanced 2-phase stator armature.

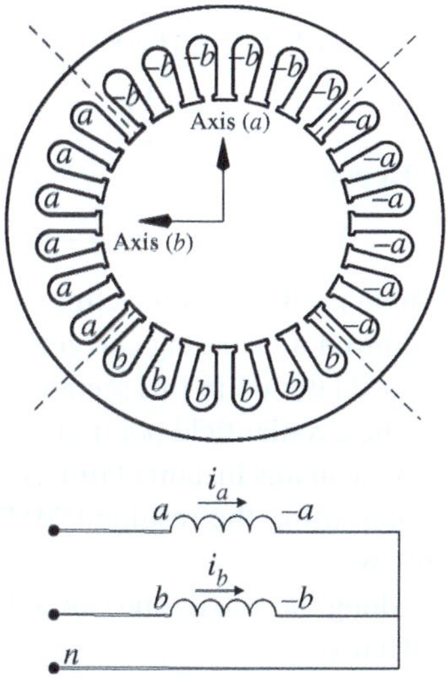

Hence, the MMFs produced by phases (*A*) and (*B*) can be expressed as follows for phases (*A*) and (*B*):

Phase (*A*):

$$\mathbf{F}_{a_1}(\theta, t) = \left[\frac{4}{\pi}k_{w_1}\left(\frac{N_{ph}}{p}\right)\sqrt{2}I\right]\cos\theta\cos(\omega t) \tag{5.82}$$

Hence,

$$\mathbf{F}_{a_1}(\theta, t) = F_m\cos(\theta)\cos(\omega t) \tag{5.83}$$

That is,

$$\mathbf{F}_{a_1}(\theta, t) = \frac{1}{2}F_m\cos(\theta - \omega t) + \frac{1}{2}F_m\cos(\theta + \omega t) \tag{5.84}$$

Phase (*B*):

$$\mathbf{F}_{b_1}(\theta, t) = \left[\frac{4}{\pi}k_{w_1}\left(\frac{N_{ph}}{p}\right)\sqrt{2}I\right]\cos\left(\theta - \frac{\pi}{2}\right)\cos\left(\omega t - \frac{\pi}{2}\right) \tag{5.85}$$

Hence,

$$\mathbf{F}_{b_1}(\theta, t) = F_m\cos\left(\theta - \frac{\pi}{2}\right)\cos\left(\omega t - \frac{\pi}{2}\right)$$

Therefore,

$$\mathbf{F}_{b_1}(\theta, t) = \frac{1}{2}F_m\cos(\theta - \omega t) + \frac{1}{2}F_m\cos(\theta + \omega t - \pi) \tag{5.86}$$

The resultant (total) MMF effect of the two phases (*A*) and (*B*), $\mathbf{F}_1(\theta, t)$, can therefore be expressed as follows:

$$\mathbf{F}_1(\theta, t) = \mathbf{F}_{a_1}(\theta, t) + \mathbf{F}_{b_1}(\theta, t) \tag{5.87}$$

$$\mathbf{F}_1(\theta,t) = \left[\frac{1}{2}F_m\cos(\theta-\omega t) + \frac{1}{2}F_m\cos(\theta+\omega t)\right] + \left[\frac{1}{2}F_m\cos(\theta-\omega t) + \frac{1}{2}F_m\cos(\theta+\omega t-\pi)\right]$$

Hence,

$$\mathbf{F}_1(\theta,t) = F_m\cos(\theta-\omega t) \tag{5.88}$$

which is a forward traveling (rotating) MMF wave as in the case of a three-phase balanced ac winding carrying three-phase-balanced currents, which will be discussed later on in this section.

The circular field point of view:

Now, at any instant of time, t, and angle, θ, the locus of the tip of the space vector representing the resultant MMF, $\mathbf{F}_1(\theta,\, t) = F_m\cos(\theta - \omega t)$, can be developed as follows:

Along the x-axis, one can write $\theta = 0$.

Hence,

$$x = \mathbf{F}_1(0,t) = F_m\cos(0-\omega t) = F_m\cos(\omega t)$$

That is,

$$x = F_m\cos(\omega t)$$

Hence,

$$\frac{x}{F_m} = \cos\omega t \tag{5.89}$$

Along the y-axis, one can write $\theta = \frac{\pi}{2}$.

Hence,

$$y = \mathbf{F}_1\left(\frac{\pi}{2},t\right) = F_m\cos\left(\frac{\pi}{2}-\omega t\right) = F_m\sin(\omega t)$$

That is,

$$y = F_m\sin(\omega t)$$

Hence,

$$\frac{y}{F_m} = \sin(\omega t) \tag{5.90}$$

Therefore, one can write the following:

$$\left(\frac{x}{F_m}\right)^2 + \left(\frac{y}{F_m}\right)^2 = \cos^2(\omega t) + \sin^2(\omega t) = 1 \tag{5.91}$$

which is the locus of the tip of the MMF space vector of magnitude, F_m, yielding the following locus:

$$x^2 + y^2 = F_m^2 \tag{5.92}$$

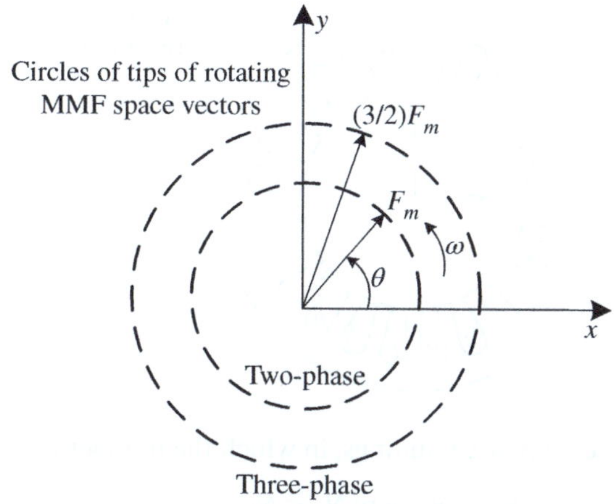

Figure 5.27 Locus of tip of space vectors representing the resultant MMFs of two-phase and three-phase cases.

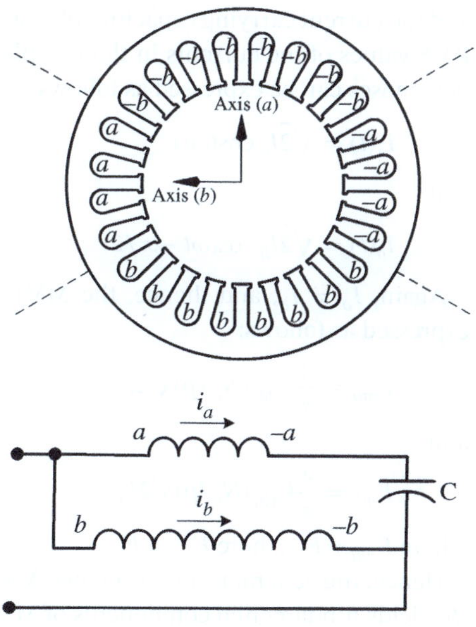

Figure 5.28 The case of an unbalanced two-phase winding, phases (*a*) and (*b*), where $N_a < N_b$, with a possible split-phase capacitor connection that is permanent.

See the circular locus of the MMF space vector in Figure 5.27. We will next discuss the unbalanced two-phase case, which leads to an MMF space vector whose tip will trace an ellipse, the so-called elliptic MMF space vector.

(c) Here, we analyze the unbalanced two-phase case, in which the number of turns for phases (*A*) and (*B*), N_a and N_b, is not equal. That is, $N_a \neq N_b$, such as shown in (*a*) and (*b*) winding layouts in Figures 5.28 and 5.29.

This is a case which is very similar to the balanced two-phase case of Section 6.2 except that the windings (*A*) and (*B*) normally represent a main and an auxiliary

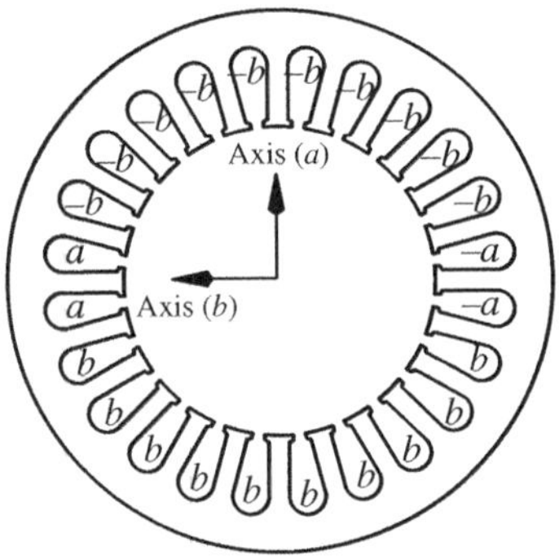

Figure 5.29 Another case of an unbalanced two-phase winding, where $N_a \ll N_b$.

set of phase windings, in which the number of turns per phase

$$N_a \neq N_b, \text{here } N_a < N_b$$

and the current carrying capacities of phases (A) and (B) are not equal. That is, the RMS values of the currents in the two phases, $I_a \neq I_b$. Hence, the currents in the two phases can be expressed as follows:

$$i_a(t) = \sqrt{2}I_a \cos(\omega t) \tag{5.93}$$

and

$$i_b(t) = \sqrt{2}I_b \cos(\omega t - \pi/2) \tag{5.94}$$

Again, $I_a \neq I_b$; and, hence, the MMF amplitudes of the two phases can be expressed as follows:

$$F_{ma} = \frac{4}{\pi} k_{wa_1} (N_a/p)\sqrt{2}I_a \tag{5.95}$$

and

$$F_{mb} = \frac{4}{\pi} k_{wb_1} (N_b/p)\sqrt{2}I_b \tag{5.96}$$

where $F_{ma} \neq F_{mb}$, here $F_{ma} > F_{mb}$.

Hence, the general form for the two MMFs produced by the two phases (a) and (b) yields fundamental components of MMFs, $\mathbf{F}_{a_1}(\theta, t)$ and $\mathbf{F}_{b_1}(\theta, t)$, which can be expressed as follows:

$$\mathbf{F}_{a_1}(\theta, t) = F_{ma} \cos(\theta) \cos(\omega t) \tag{5.97}$$

and

$$\mathbf{F}_{b_1}(\theta, t) = F_{mb} \cos\left(\theta - \frac{\pi}{2}\right) \cos\left(\omega t - \frac{\pi}{2}\right) \tag{5.98}$$

Therefore,

$$\mathbf{F}_{a_1}(\theta, t) = \frac{1}{2}F_{ma} \cos(\theta - \omega t) + \frac{1}{2}F_{ma} \cos(\theta + \omega t) \tag{5.99}$$

and

$$\begin{aligned}\mathbf{F}_{b_1}(\theta,t) &= \frac{1}{2}F_{mb}\cos(\theta-\omega t)+\frac{1}{2}F_{mb}\cos(\theta+\omega t-\pi)\\ &= \frac{1}{2}F_{mb}\cos(\theta-\omega t)-\frac{1}{2}F_{mb}\cos(\theta+\omega t)\end{aligned} \tag{5.100}$$

Hence, the total (resultant) MMF has a fundamental component, $\mathbf{F}_1(\theta, t)$, as follows:

$$\mathbf{F}_1(\theta,t) = \mathbf{F}_{a_1}(\theta,t)+\mathbf{F}_{b_1}(\theta,t) \tag{5.101}$$

$$\mathbf{F}_1(\theta,t) = \underbrace{\frac{1}{2}(F_{ma}+F_{mb})\cos(\theta-\omega t)}_{\text{Forward travelling wave}}+\underbrace{\frac{1}{2}(F_{ma}-F_{mb})\cos(\theta+\omega t)}_{\text{Reverse (backward) travelling wave}} \tag{5.102}$$

Equation (5.102) constitutes a forward-traveling MMF wave of amplitude $\frac{1}{2}(F_{ma}+F_{mb})$ and a reverse (backward)-traveling MMF wave of amplitude $\frac{1}{2}(F_{ma}-F_{mb})$. Thus, there could be forward- and reverse-oriented torque components produced. Depending on the degree of unbalance between phases (A) and (B), the forward torque and MMF component should be stronger than the reverse component. If phase (B) is disconnected, the aforementioned equation reduces to that of a single-phase case presented in Section 6.1.

The elliptic field point of view:

Now, at any instant of time, t, and angle, (θ), the locus of the tip of the space vector representing the resultant MMF, $\mathbf{F}_1(\theta, t)$, where

$$\mathbf{F}_1(\theta,t) = \frac{1}{2}(F_{ma}+F_{mb})\cos(\theta-\omega t)+\frac{1}{2}(F_{ma}-F_{mb})\cos(\theta+\omega t) \tag{5.103}$$

can be developed in the x–y plane as follows:

Along the x-axis, one can write $\theta = 0$, hence,

$$x = \mathbf{F}_1(0,t) = \frac{1}{2}(F_{ma}+F_{mb})\cos(\omega t)+\frac{1}{2}(F_{ma}-F_{mb})\cos(\omega t) \tag{5.104}$$

Hence,

$$x = F_{ma}\cos(\omega t)$$

That is,

$$\frac{x}{F_{ma}} = \cos(\omega t) \tag{5.105}$$

Along the y-axis, one can write $\theta = \pi/2$, hence,

$$y = \mathbf{F}_1\left(\frac{\pi}{2},t\right) = \frac{1}{2}(F_{ma}+F_{mb})\cos\left(\frac{\pi}{2}-\omega t\right)+\frac{1}{2}(F_{ma}-F_{mb})\cos\left(\frac{\pi}{2}+\omega t\right) \tag{5.106}$$

$$y = \frac{1}{2}(F_{ma} + F_{mb})\sin(\omega t) - \frac{1}{2}(F_{ma} - F_{mb})\sin(\omega t) \tag{5.107}$$

Hence,

$$y = F_{mb}\sin(\omega t)$$

or,

$$\frac{y}{F_{mb}} = \sin(\omega t) \tag{5.108}$$

That is,

$$\left(\frac{x}{F_{ma}}\right)^2 + \left(\frac{y}{F_{mb}}\right)^2 = \cos^2(\omega t) + \sin^2(\omega t) = 1 \tag{5.109}$$

or,

$$\left(\frac{x}{F_{ma}}\right)^2 + \left(\frac{y}{F_{mb}}\right)^2 = 1 \tag{5.110}$$

which is the locus of the tip of the MMF space vector, that is, an ellipse, as shown in Figure 5.30.

The elliptic field represents the most general case of magnetic field (MMF waveform), which can exist in an electric machine. The resultant field (MMF) space vector changes both its magnitude and spatial position (phase) with time in a continuous fashion such that its endpoint (tip locus) describes an ellipse as shown in Figure 5.30.

(c) We now analyze the MMF resulting from a distributed three-phase winding carrying three-phase-balanced AC currents, such as the 24-slots stator shown in Figure 5.31. This stator is wound for a two-pole, three-phase, and single-layer winding, that is a winding configuration with, $q = 4$ slots per pole per phase, and magnetic phase axes, Axis (a), Axis (b), and Axis (c) arranged for an a, b,

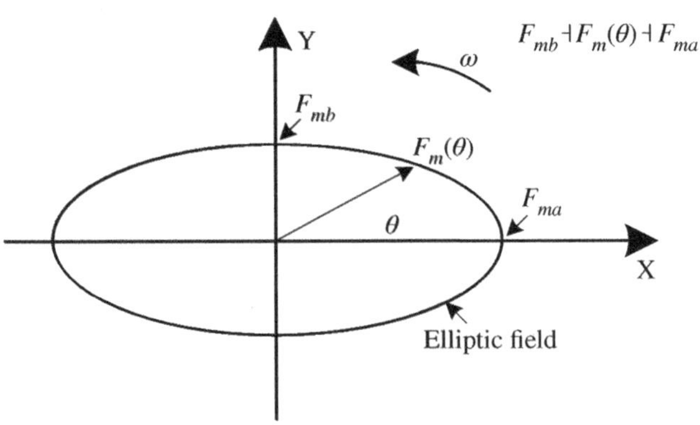

Figure 5.30 The elliptic field case.

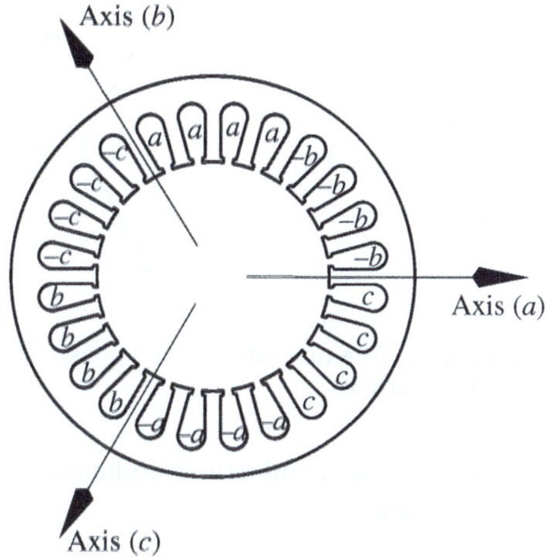

Figure 5.31 A 24-slot stator wound for a three-phase and two-pole armature of an AC machine.

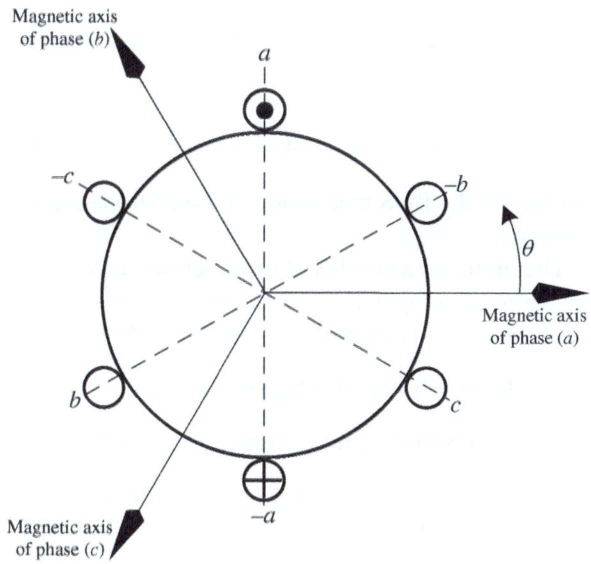

Figure 5.32 A schematic simplified representation of a three-phase and two-pole AC winding.

and *c* sequence for a counterclockwise rotating field. A simplified schematic representation of this winding by three equivalent $(a, -a)$, $(b, -b)$, and $(c, -c)$ coils is shown in Figure 5.32, with (a), (b), and (c) being the magnetic axes sequence in the counterclockwise rotational orientation. The time-domain and phasor representations of the balanced three-phase excitation AC currents are depicted in Figure 5.33. Notice the (a), (b), and (c) sequences for a counterclockwise rotation of the phasor diagram of the currents shown in Figure 5.33.

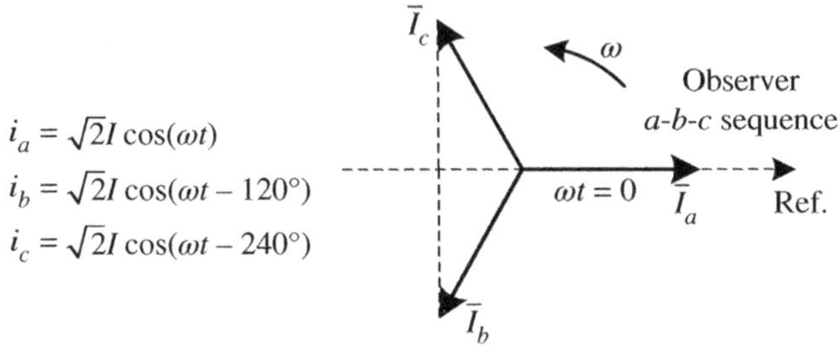

Figure 5.33 The time-domain and phasor representation of the three-phase AC excitation currents.

Hence, for a, b, and c time-domain currents, $i_a(t)$, $i_b(t)$, and $i_c(t)$, one can write the following:

$$i_a(t) = \sqrt{2}I\cos(\omega t) \tag{5.111}$$

$$i_b(t) = \sqrt{2}I\cos(\omega t - 2\pi/3) = \sqrt{2}I\cos(\omega t - 120°) \tag{5.112}$$

and,

$$i_c(t) = \sqrt{2}I\cos(\omega t - 4\pi/3) = \sqrt{2}I\cos(\omega t - 240°) \tag{5.113}$$

where I is the RMS magnitude of the phase currents, and $\omega = 2\pi f$ are electrical radians.

Therefore, for a number of turns per phase, N_{ph}, and a winding factor, k_w, based on earlier developments in Eqs. (5.47) and (5.48), one can write the following for the fundamental component of the MMF, $\mathbf{F}_{a_1}(\theta, t)$, of phase ($a$):

$$\mathbf{F}_{a_1}(\theta, t) = (4/\pi)k_w(N_{ph}/p)i_a(t)\cos\theta \tag{5.114}$$

Hence, substituting for $i_a(t)$ from Eq. (5.111) yields the following:

$$\mathbf{F}_{a_1}(\theta, t) = (4/\pi)k_w(N_{ph}/p)\sqrt{2}I\cos\theta\cos(\omega t) \tag{5.115}$$

or

$$\mathbf{F}_{a_1}(\theta, t) = F_{\max}\cos\theta\cos(\omega t) \tag{5.116}$$

where

$$F_{\max} = (4/\pi)\sqrt{2}k_w(N_{ph}/p)I \tag{5.117}$$

Using trigonometric identifies, $\mathbf{F}_{a_1}(\theta, t)$ in Eq. (5.116) can be rewritten as follows:

$$\mathbf{F}_{a_1}(\theta, t) = (F_{\max}/2)\cos(\theta - \omega t) + (F_{\max}/2)\cos(\theta + \omega t) \tag{5.118}$$

Similarly, the fundamental component of the MMF, $\mathbf{F}_{b_1}(\theta, t)$, of phase (b), can be written as follows:

$$\mathbf{F}_{b_1}(\theta, t) = F_{\max} \cos(\theta - 2\pi/3)\cos(\omega t - 2\pi/3) \tag{5.119}$$

That is,

$$\mathbf{F}_{b_1}(\theta, t) = (F_{\max}/2)\cos(\theta - \omega t) + (F_{\max}/2)\cos(\theta + \omega t - 4\pi/3) \tag{5.120}$$

Also, similarly, for the fundamental component of the MMF, $\mathbf{F}_{c_1}(\theta, t)$, of phase (c), can be written as follows:

$$\mathbf{F}_{c_1}(\theta, t) = F_{\max} \cos(\theta - 4\pi/3)\cos(\omega t - 4\pi/3) \tag{5.121}$$

That is,

$$\mathbf{F}_{c_1}(\theta, t) = (F_{\max}/2)\cos(\theta - \omega t) + (F_{\max}/2)\cos(\theta + \omega t - 2\pi/3) \tag{5.122}$$

Accordingly, the resultant fundamental component of the MMF, $\mathbf{F}_1(\theta, t)$, of the entire (a), (b), and (c) phases of the windings in Figures 5.31 and 5.32, can be expressed as follows:

$$\mathbf{F}_1(\theta, t) = \mathbf{F}_{a_1}(\theta, t) + \mathbf{F}_{b_1}(\theta, t) + \mathbf{F}_{c_1}(\theta, t) \tag{5.123}$$

Substituting for $\mathbf{F}_{a_1}(\theta, t)$, $\mathbf{F}_{b_1}(\theta, t)$, and $\mathbf{F}_{c_1}(\theta, t)$ from Eqs. (5.118), (5.120), and (5.122) into Eq. (5.123) yields the following:

$$\begin{aligned}\mathbf{F}_1(\theta, t) &= (3/2)F_{\max} \cos(\theta - \omega t) \\ &\quad + (1/2)F_{\max}[\cos(\theta + \omega t) + \cos(\theta + \omega t - 4\pi/3) + \cos(\theta + \omega t - 2\pi/3)]\end{aligned} \tag{5.124}$$

However, the term $[\cos(\theta + \omega t) + \cos(\theta + \omega t - 4\pi/3) + \cos(\theta + \omega t - 2\pi/3)]$ is equal to zero, based on well-known trigonometric identities. Hence, the resultant MMF, $\mathbf{F}_1(\theta, t)$, in Eq. (5.124), can be rewritten as follows:

$$\mathbf{F}_1(\theta, t) = (3/2)F_{\max} \cos(\theta - \omega t) \tag{5.125}$$

Accordingly, for an observer starting from an initial position, θ_0, and moving in the counterclockwise direction at a rate of ω electrical radians per second, his/her position, $\theta(t)$ at any time, t, can be written as follows:

$$\theta(t) = \theta_0 + \omega t \tag{5.126}$$

Such an observer will see a value of the resultant MMF, $\mathbf{F}_1(\theta, t)$, that can be expressed, based on Eq. (5.125), as follows:

$$\mathbf{F}_1(\text{observer}) = (3/2)F_{\max} \cos(\theta_0 + \omega t - \omega t) = (3/2)F_{\max} \cos(\theta_0) \tag{5.127}$$

which is a fixed value. This implies that the MMF as a rotating (traveling) MMF waveform is rotating (traveling) with the observer. Thus, the MMF, $\mathbf{F}_1(\theta, t)$, is

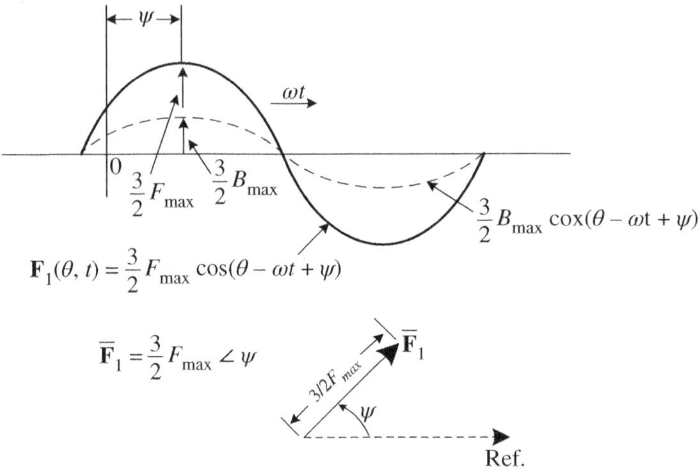

Figure 5.34 The resultant fundamental MMF, $F_1(\theta, t)$, as a rotating/traveling waveform and the corresponding space-vector representation, $\overline{\mathbf{F}}_1$.

a rotating waveform, rotating at a speed, ω, electrical radians per second, see Figure 5.34 for a generalized graphical representation of this rotating MMF waveform, and its corresponding "Space-Vector" representation. Also, the reader should examine Figure 5.27 for the locus of the tip of the space vector, $(3/2)\mathbf{F}_m$, as depicted in Figure 5.27.

A further graphical depiction of this rotating MMF and its magnetic flux lines orientation is given in Figure 5.35 at times, $\omega t = 0$, $\omega t = 2\pi/3$, $\omega t = 4\pi/3$, and $\omega t = 2\pi$.

The corresponding rotating flux density waveform, $B_1(\theta, t)$, in a machine with a uniform airgap of radial height, h_g, can therefore be written as follows for core materials with very large magnetic permeabilities:

$$B_1(\theta, t) = (3/2)(F_{max}/h_g)\mu_0 \cos(\theta - \omega t) \tag{5.128}$$

or,

$$B_1(\theta, t) = (3/2)B_{max} \cos(\theta - \omega t) \tag{5.129}$$

where

$$B_{max} = \mu_0(F_{max}/h_g) \tag{5.130}$$

In Section 5.6, we will analyze the interaction between the stator and rotor MMFs to produce a torque in electric machines and the conditions for such torque production.

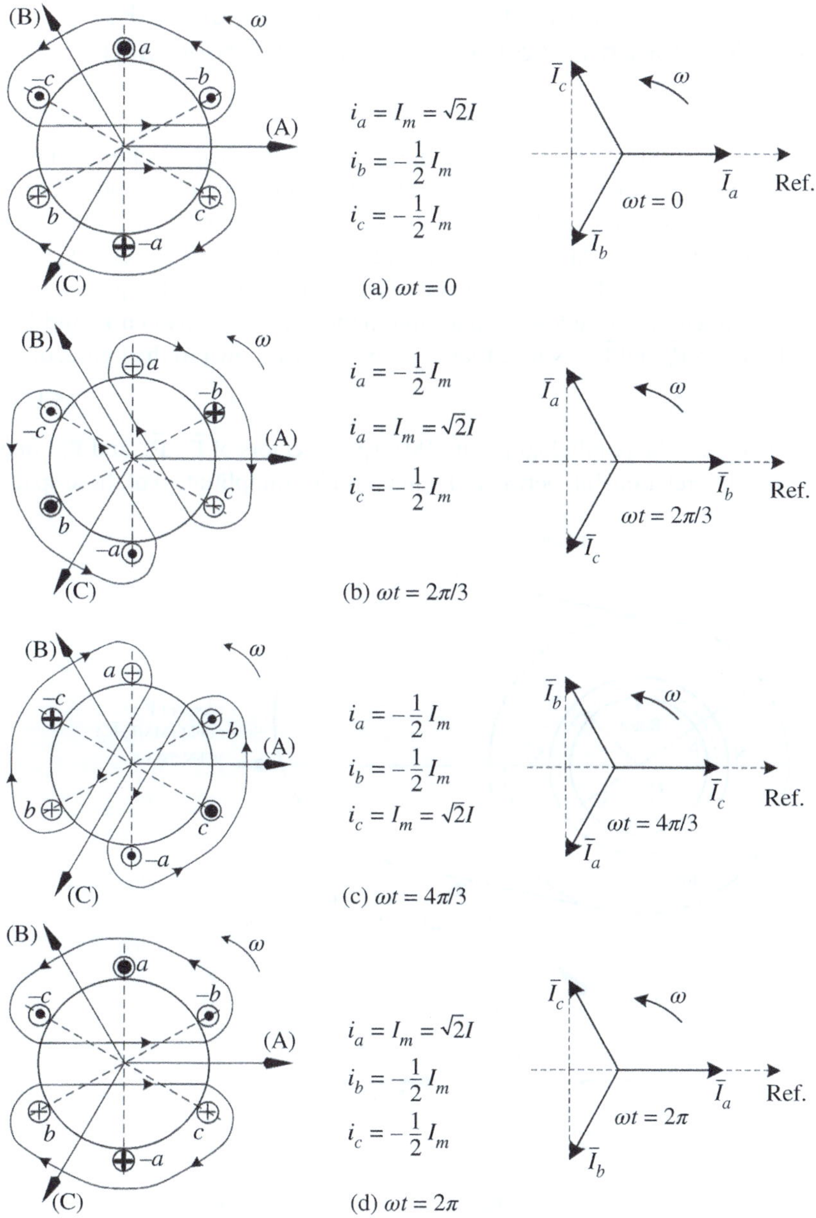

Figure 5.35 A qualitative graphic view of the rotating MMF/field produced by balanced three-phase winding.

5.6 Relationships Among Magnetomotive Forces, Flux Density Distributions, Volume, Developed Torque, and Power

Consider the simplified cylindrical geometry of an electrical machine's stator, rotor, and airgap of radial height, $h_g = g$, and mid-airgap diameter, D, and effective axial core length (stator stack length), l, as shown in Figure 5.36. Also, shown in Figure 5.36 are the stator MMF space vector, $\overline{\mathbf{F}}_s$, rotor MMF space vector, $\overline{\mathbf{F}}_r$, and the angles, δ_{sr} between $\overline{\mathbf{F}}_s$ and $\overline{\mathbf{F}}_r$. Meanwhile, the resultant MMF space vector, $\overline{\mathbf{F}}_{sr}$, is also shown in Figure 5.36, in addition to the angles δ_s, between $\overline{\mathbf{F}}_s$ and $\overline{\mathbf{F}}_{sr}$, and δ_r, between $\overline{\mathbf{F}}_r$ and $\overline{\mathbf{F}}_{sr}$. Notice that, $\delta_{sr} = \delta_s + \delta_r$, as shown in the figure, or

$$\delta_{sr} = \delta_s + \delta_r \tag{5.131}$$

Here, based on the parallelogram of MMF space vectors of $\overline{\mathbf{F}}_s$, $\overline{\mathbf{F}}_r$, and $\overline{\mathbf{F}}_{sr}$, one can express the relationship between the magnitudes (amplitudes) of these space

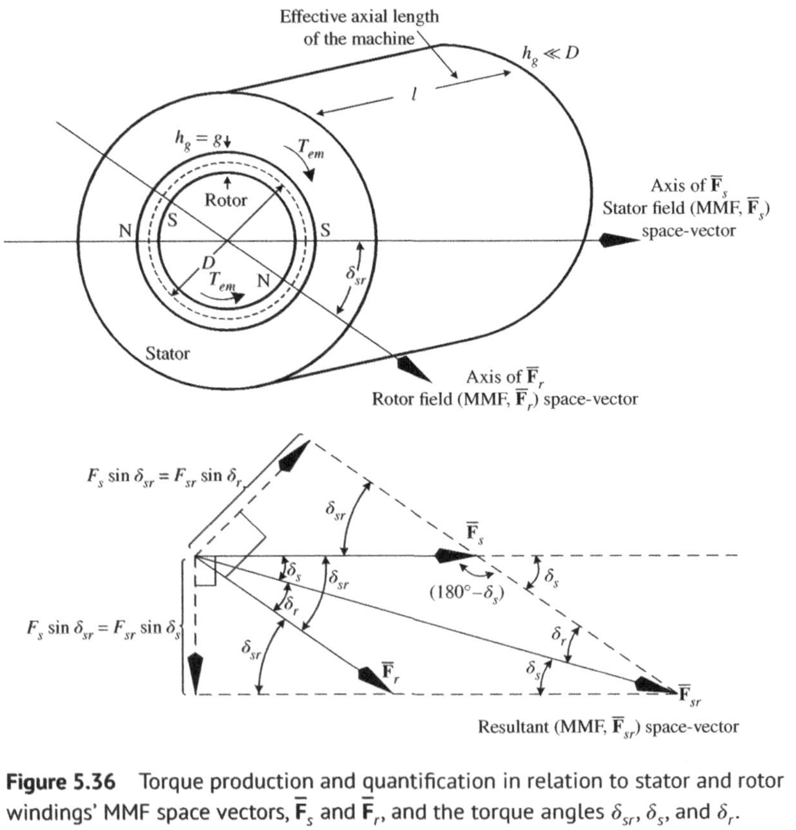

Figure 5.36 Torque production and quantification in relation to stator and rotor windings' MMF space vectors, $\overline{\mathbf{F}}_s$ and $\overline{\mathbf{F}}_r$, and the torque angles δ_{sr}, δ_s, and δ_r.

vectors as follows:

$$(F_{sr})^2 = (F_s)^2 + (F_r)^2 + 2F_sF_r\cos\delta_{sr} \tag{5.132}$$

That is,

$$F_{sr} = \sqrt{(F_s)^2 + (F_r)^2 + 2F_sF_r\cos\delta_{sr}} \tag{5.133}$$

The reader should be forewarned that the angle, δ_{sr}, shown as an acute angle in Figure 5.36 will turn-out to be an obtuse angle, as we detail the coverage of AC induction, synchronous, and brushless DC machines in later Chapters 6–10.

Also, in this case, the maximum flux density, B_{sr}, produced by the MMF space-vector amplitude, F_{sr}, can be expressed as follows:

$$B_{sr} = \mu_0(F_{sr}/h_g) \tag{5.134}$$

where the flux density waveform, see Figure 5.37, $B_g(\theta)$, can be expressed as follows:

$$B_g(\theta) = B_{sr}\cos\theta = \mu_0(F_{sr}/h_g)\cos\theta \tag{5.135}$$

Hence, the average $B_g^2(\theta)$, $B_{g_{av}}^2(\theta)$, over one pole-pitch, τ_p, can be expressed as follows:

$$B_{g_{av}}^2(\theta) = \frac{1}{\pi/2}\int_0^{\pi/2} B_{sr}^2(\cos\theta)^2 d\theta = B_{sr}^2/2 \tag{5.136}$$

$$B_{g_{av}}^2(\theta) = (\mu_0F_{sr}/h_g)^2/2 \tag{5.137}$$

Hence, the average co-energy in the airgap, w_{coen}, can be expressed as follows:

$$w_{coen} = B_{g_{av}}^2/(2\mu_0) = \mu_0(F_{sr}/h_g)^2/4 \tag{5.138}$$

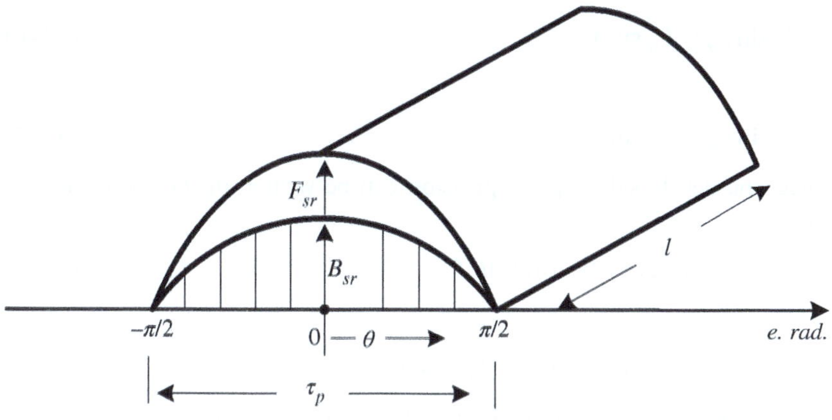

Figure 5.37 Relationship among resultant MMF space vector, $\bar{\mathbf{F}}_{sr}$, peak resultant flux density, B_{sr}, and machine dimensions at mid-airgap diameter, D.

Substituting for F_{sr} from Eq. (5.133) into Eq. (5.138) yields

$$w_{coen} = \mu_0/\left(4h_g^2\right)\left(F_s^{\,2} + F_r^{\,2} + 2F_sF_r\cos\delta_{sr}\right) \tag{5.139}$$

Hence, the total coenergy, W_{coen}, stored in the airgap volume of a cylindrical shell, (πDlh_g), can be written as follows:

$$W_{coen} = (\pi Dlh_g)w_{coen} = \pi\mu_0 Dl\left(F_s^{\,2} + F_r^{\,2} + 2F_sF_r\cos\delta_{sr}\right)/4h_g \tag{5.140}$$

Now, based on earlier developments in Section 3.4, the developed electromechanical torque, T_{dev}, can be expressed as follows, using Eq. (5.140):

$$T_{dev} = \frac{\partial W_{coen}}{\partial\theta} = \frac{\partial W_{coen}}{\partial\delta_{sr}} \tag{5.141}$$

$$T_{dev} = \frac{\partial\left[(\pi\mu_0 Dl/4h_g)\left(F_s^{\,2} + F_r^{\,2} + 2F_sF_r\cos\delta_{sr}\right)\right]}{\partial\delta_{sr}} \tag{5.142}$$

Therefore, T_{dev} is given by the following per pair of poles:

$$T_{dev} = -(\mu_0\pi Dl/2h_g)F_sF_r\sin\delta_{sr} \tag{5.143}$$

That is, for a p-pole machine, the developed torque, T_{dev}, can be expressed as follows:

$$T_{dev} = -(p/2)(\mu_0\pi Dl/2h_g)F_sF_r\sin\delta_{sr} \tag{5.144}$$

Hence, the developed torque, T_{dev}, can be rewritten as follows:

$$T_{dev} = -(\mu_0/2)(p/2)(\pi Dl/h_g)F_sF_r\sin\delta_{sr} \tag{5.145}$$

Meanwhile, upon a careful examination of the parallelogram in Figure 5.36, one can write the following:

$$F_r\sin\delta_{sr} = F_{sr}\sin\delta_s \tag{5.146}$$

and

$$F_s\sin\delta_{sr} = F_{sr}\sin\delta_r \tag{5.147}$$

Hence, the developed torque expression can be written in the following two forms:

$$T_{dev} = -(\mu_0/2)(p/2)(\pi Dl/h_g)F_sF_{sr}\sin\delta_s \tag{5.148}$$

and

$$T_{dev} = -(\mu_0/2)(p/2)(\pi Dl/h_g)F_rF_{sr}\sin\delta_r \tag{5.149}$$

It should be pointed out that the negative sign in all the developed torque Eqs. (5.143)–(5.149) indicates that the orientation of the torque is in a direction that tends to shrink all the torque angles δ_{sr}, δ_s, and δ_r. Hence, there are two equal and opposing torques acting on the rotor and stator of any electrical rotating machine.

Now, we will derive the developed torque expression in terms of the resultant flux per pole, Φ_{sr}, in an electric machine. Referring to Figure 5.37, this resultant flux per pole can be expressed as follows:

$$\Phi_{sr} = (2/\pi)B_{sr}\tau_p l \tag{5.150}$$

where τ_p is the arc of the pole-pitch, that is,

$$\tau_p = \pi D/p \tag{5.151}$$

$$B_{sr} = \mu_0(F_{sr}/h_g) \tag{5.152}$$

Accordingly,

$$\Phi_{sr} = (2/\pi)(\mu_0 F_{sr}/h_g)(\pi D/p)l \tag{5.153}$$

That is,

$$\Phi_{sr} = [2\mu_0 Dl/(h_g p)]F_{sr} \tag{5.154}$$

Hence,

$$F_{sr} = [h_g p/(2\mu_0 Dl)]\Phi_{sr} \tag{5.155}$$

Substituting for F_{sr} from Eq. (5.155) into the two developed torque expressions in Eqs. (5.148) and (5.149) results in the following:

$$T_{dev} = -(\pi/2)(p/2)^2\Phi_{sr}F_s \sin\delta_s \tag{5.156}$$

$$T_{dev} = -(\pi/2)(p/2)^2\Phi_{sr}F_r \sin\delta_r \tag{5.157}$$

Now, for the developed mechanical power, P_{dev}, we first will treat two cases of AC machines, namely, the synchronous and induction machines.

For AC poly-phase synchronous machines, the mechanical rotational speed is the synchronous speed given in Eq. (3.74) in rev/min as, $n_{syn} = 120f/p$. Hence, the mechanical speed, ω_m, in mech. rad/s, can be expressed as follows:

$$\omega_m = (120f/p)(2\pi/60) = 4\pi f/p \tag{5.158}$$

Hence, P_{dev} can be expressed based on T_{dev} in Eqs. (5.156) and (5.157) as follows:

$$P_{dev} = \|\omega_m T_{dev}\| = (p/2)(\pi^2 f)\Phi_{sr}F_s \sin\delta_s \tag{5.159}$$

or

$$P_{dev} = (p/2)(\pi^2 f)\Phi_{sr}F_r \sin\delta_r \tag{5.160}$$

For AC poly-phase induction machines briefly discussed in Section 3.11, we introduce the slip, s, defined in per unit of the synchronous speed in rev/min and the actual rotor mechanical speed, n, also in rev/min as follows:

$$s = (n_{syn} - n)/n_{syn} \tag{5.161}$$

Therefore, the rotor speed, n, can be expressed in terms of the slip, s, and the synchronous speed, n_{syn}, in rev/min, as follows:

$$n = (1 - s)n_{syn} \tag{5.162}$$

or

$$n = \frac{120f(1 - s)}{p} \tag{5.163}$$

Hence, the mechanical angular speed, ω_m, in mech. rad/s, can be written as follows:

$$\omega_m = [120f(1 - s)/p](2\pi/60)$$

or

$$\omega_m = 4\pi f(1 - s)/p \tag{5.164}$$

Accordingly, the developed mechanical power in such an induction machine can be expressed as follows:

$$P_{dev} = \|\omega_m T_{dev}\| = (p/2)(\pi^2 f)(1 - s)\Phi_{sr} F_s \sin \delta_s \tag{5.165}$$

Finally, in this chapter, we will relate the volume of an AC or a DC machine based on the cylindrical volume of a mid-airgap diameter, D, and a stack or effective/active conductor axial length, l, that is $(\pi D^2 l/4)$ to both the developed torque, T_{dev}, and the developed power, P_{dev}.

We first define the magnetic loading as the maximum value of the resultant airgap flux density waveform, B_{sr}, in Tesla, and, here, the resultant flux per pole, Φ_{sr}, is related to the magnetic loading through Eq. (5.150). Therefore, one can write the following:

$$\Phi_{sr} = (2/\pi)B_{sr}(\pi D/p)l = 2B_{sr}Dl/p \tag{5.166}$$

We also define the ampere loading, A, as the ampere-conductors per unit length of the circumference at mid-airgap diameter, D. Therefore, in a three-phase winding with a number of turns per phase, N_{ph}, and a number of parallel-paths per phase, m, and an RMS current per phase, I_{ph}, the total number of ampere-conductors in such a machine is given by $(3 \times N_{ph} \times 2 \times I_{ph}/m)$, that is, $(6N_{ph}I_{ph}/m)$. Hence, the ampere-loading, A, can be expressed as follows:

$$A = (6N_{ph}I_{ph}/m)/(\pi D) \tag{5.167}$$

Therefore,

$$I_{ph}/m = \pi DA/(6N_{ph}) \tag{5.168}$$

However, the stator MMF, $\mathbf{F}_s$, is $(3/2)F_{max}$, see Eq. (5.125), where F_{max} is given by Eq. (5.117), where $I = I_{ph}/m$. Hence,

$$\mathbf{F}_s = (3/2)[(4/\pi)\sqrt{2}k_w(N_{ph}/p)(I_{ph}/m)] \tag{5.169}$$

Substituting for (I_{ph}/m) from Eq. (5.167) into Eq. (5.169) yields the following:

$$\mathbf{F}_s = (6\sqrt{2}/\pi)k_w(N_{ph}/p)[\pi DA/(6N_{ph})] \tag{5.170}$$

Hence,

$$\mathbf{F}_s = \sqrt{2}k_w DA/p \tag{5.171}$$

Therefore, substituting from Eqs. (5.166) and (5.171) for the resultant flux per pole, Φ_{sr}, and the stator MMF, F_s, respectively, into the torque expression, Eq. (5.156), yields the following for AC synchronous and induction machines:

$$T_{dev} = -(\pi/2)(p/2)^2(2B_{sr}Dl/p)(\sqrt{2}k_w DA/p)\sin\delta_s \tag{5.172}$$

which reduces to the following form:

$$T_{dev} = -(\pi\sqrt{2}/4)(B_{sr}AD^2l)\sin\delta_s \tag{5.173}$$

But the volume of the mid-airgap cylinder of axial length, l, *Vol.*, is

$$Vol. = \pi D^2 l/4 \tag{5.174}$$

which yields the following:

$$D^2 l = (4/\pi)Vol. \tag{5.175}$$

Hence,

$$T_{dev} = \sqrt{2}AB_{sr}(Vol.)\sin\delta_s \tag{5.176}$$

Consequently, the developed power in an AC synchronous machine,

$$P_{dev} = T_{dev}(4\pi f/p) = \sqrt{2}AB_{sr}(Vol.)(4\pi f/p)\sin\delta_s \tag{5.177}$$

That is,

$$P_{dev} = [4\pi\sqrt{2}AB_{sr}f(Vol.)/p]\sin\delta_s \tag{5.178}$$

Meanwhile, in AC three-phase induction machines, the developed power is

$$P_{dev} = [4\pi\sqrt{2}AB_{sr}f(1-s)(Vol.)/p]\sin\delta_s \tag{5.179}$$

Next, in Section 5.7, we develop the relationship between magnetic loading, B_{sr}, ampere loading, A, volume, $Vol. = \pi D^2 l/4$, and volt-ampere rating, S, of AC machines.

5.7 Relationship Among Electric Machinery Volume and Developed Torque and/or Volt–Ampere Capabilities

As presented in Section 5.6, the relationship between developed power, P_{dev}, and volume in a three-phase synchronous machine is given by Eq. (5.178), restated here for convenience,

$$P_{dev} = [4\pi\sqrt{2}AB_{sr}f(Vol.)/p]\sin\delta_s \tag{5.180}$$

where A is the ampere-loading defined in Eq. (5.167), B_{sr} is the magnetic loading, which is the maximum value of the resultant flux density waveform in the airgap, $(Vol.)$ is the mid-airgap cylindrical volume, that is, $(\pi D^2 l/4)$, see Figure 5.36, p is the number of poles in the machine, and δ_s is a torque angle and also defined in Figure 5.36.

Meanwhile, in a three-phase induction machine, the relationship between P_{dev} and volume, Eq. (5.179), is restated here also for convenience,

$$P_{dev} = [4\pi\sqrt{2}AB_{sr}f(1-s)(Vol.)/p] \tag{5.181}$$

where s is the slip defined in Eq. (5.161).

The volt–ampere capability, S, of both three-phase synchronous and induction machines is defined as follows:

$$S = 3I_{ph}E_{ph} \tag{5.182}$$

where I_{ph} is the RMS current per phase and E_{ph} is the inducted EMF per phase due to the resultant airgap flux per pole, Φ_{sr}, Eq. (5.166).

Here, one can write the following for E_{ph}:

$$E_{ph} = (2\pi/\sqrt{2})f\Phi_{sr}(N_{ph}/m)k_{w_1} \tag{5.183}$$

where N_{ph} is the number of turns per phase, m is the number of parallel paths per phase, and k_{w_1} is the winding factor for the fundamental component of airgap flux density waveform.

Hence, one can write the following for S:

$$S = 3I_{ph}E_{ph} = 3I_{ph}[(2\pi/\sqrt{2})f\Phi_{sr}(N_{ph}/m)k_{w_1}] \tag{5.184}$$

However, from earlier developments in Section 5.6, Eqs. (5.167) for (I_{ph}/m), and (5.150) for, Φ_{sr}, we have

$$(I_{ph}/m) = \pi DA/(6N_{ph}) \tag{5.185}$$

and

$$\Phi_{sr} = (2/\pi)B_{sr}(\pi D/p)l = (2DlB_{sr}/p) \tag{5.186}$$

Substituting from Eqs. (5.185) for (I_{ph}/m) and (5.186) for Φ_{sr} into Eq. (5.184) yields the following expression for the volt–ampere capability:

$$S = (6\sqrt{2}/\pi)f(2DlB_{sr}/p)(I_{ph}/m)k_{w_1}[\pi mDA/(6N_{ph})] \tag{5.187}$$

which is algebraically reduced to the following:

$$S = (2\pi^2/\sqrt{2})(f/p)B_{sr}Ak_{w_1}(D^2l) \tag{5.188}$$

However, since the volume $(Vol.)$ is

$$(Vol.) = \pi D^2 l/4 \tag{5.189}$$

It follows that

$$D^2 l = 4(Vol.)/\pi \tag{5.190}$$

Substituting for (D^2l) from Eq. (5.190) into Eq. (5.188) yields the following for S:

$$S = (8\pi/\sqrt{2})(f/p)B_{sr}Ak_{w_1}(Vol.) \tag{5.191}$$

The developed power, P_{dev}, Eqs. (5.180) and (5.181), and the volt–ampere capability, S, expressed in Eq. (5.191), all point to the fact that the higher the magnetic-loading, B_{sr}, the higher the ampere-loading, A, the lower the required volume of the machine for the same P_{dev} rating and S capability. Also, Eqs. (5.180), (5.181), and (5.191) point to the fact that the higher the rated frequency, f, in Hertz, and synchronous speed, $n_{syn} = (120f/p)$ rev/min, the lower the volume, $(Vol.)$, for the same developed power, P_{dev}, and volt–ampere capability, S.

Furthermore, based on Eqs. (5.180), (5.181), and (5.191), the power factors, PF, for the synchronous and induction machines can be approximated, or estimated, by the following expressions for the synchronous and induction machines in Eqs. (5.192) and (5.193), respectively:

(1) For the synchronous machine, $PF_{syn.}$ is given as follows:

$$PF_{syn.} \simeq \frac{[4\pi\sqrt{2}AB_{sr}(f/p)(Vol.)\sin\delta_s]}{[(8\pi/\sqrt{2})AB_{sr}(f/p)(Vol.)k_{w_1}]} = \frac{\sin\delta_s}{k_{w_1}} \tag{5.192}$$

Hence,

$$PF_{syn.} \simeq \sin\delta_s/k_{w_1} \tag{5.193}$$

(2) For the induction machine, $PF_{ind.}$ is as follows:

$$PF_{ind.} \simeq \frac{[4\pi\sqrt{2}AB_{sr}(f/p)(1-s)(Vol.)\sin\delta_s]}{[(8\pi/\sqrt{2})AB_{sr}(f/p)(Vol.)k_{w_1}]} = \frac{\sin\delta_s}{k_{w_1}} \tag{5.194}$$

Hence,

$$PF_{ind.} \simeq (1-s)\sin\delta_s/k_{w_1} \tag{5.195}$$

In other words, for synchronous machines, one can state, based on Eq. (5.193), that

$$\delta_s = \sin^{-1}(k_{w_1} \cdot PF_{syn.}) \tag{5.196}$$

and for induction machines, one can state, based on Eq. (5.195), that

$$\delta_s = \sin^{-1}[k_{w_1} \cdot PF_{ind.}/(1 - s)] \tag{5.197}$$

Next, in Section 5.8, we present some numerical application examples to the material presented in Section 5.7.

5.8 Worked Examples

We present in this section four examples: Example 5.1 on synchronous machines, specifically, a 4-pole turbogenerator of a volt–ampere rating, $S = 659$ MVA. Example 5.2 on a 5-Hp, six-pole induction motor, Example 5.3 on the derivation of the relationship between volt–ampere/power and volume in a DC commutator-brush type of machine, and finally Example 5.4 on EMF computation in a synchronous machine.

Example 5.1 *Turbogenerator/Wound Rotor Synchronous Machine*

$S = 659$ MVA, 60 Hz, 0.9 PF, 20 kV, 3-phase, 4-pole, 48-stator slots, Dr = 60 in., $g = 2.5$ in., peak airgap flux density = 70,000 lines/in.^{2}, coil span is 10 slots, $\tau_c = 10$, and double-Y-connected turbogenerator with double-layer lap stator winding.

$$E_{ph} = \frac{20,000}{\sqrt{3}} = 11,547 \text{ V}$$

$$N_{ph} = \left(\frac{48 \times 2}{2}\right) \Big/ 3 = 16 \text{ turns/path/phase}$$

$$m(\text{number of parallel paths}) = 2$$

$$N_{ph}/m = 8 \text{ turns/path/phase}$$

$$\alpha = \frac{4 \times 180^\circ}{48} = \frac{180^\circ}{12} = 180^\circ e$$

$$\beta = \frac{\tau_c}{\tau_p}(180^\circ) = \frac{10}{12} \times 180^\circ = 150^\circ \text{e}$$

$$q = \frac{48}{3 \times 4} = 4 \text{ slots/pole/phase}$$

$$k_{d_1} = \frac{\sin(q\alpha/2)}{q\sin(\alpha/2)} = \frac{\sin(4\times 15^\circ/2)}{4\sin(15^\circ/2)} = \frac{0.5}{0.5221} = 0.9577$$

$$k_{p_1} = \sin(\beta/2) = \sin(150^\circ/2) = 0.9659$$

$$k_{w_1} = k_{d_1}\cdot k_{p_1} = 0.9251$$

$$E_{ph} = \frac{2\pi}{\sqrt{2}} f\Phi_1 k_{w_1}(N_{ph}/m) = 4.44\times f\times \Phi_1(k_{w_1}N_{ph}/m)$$

$$\Phi_1 = \frac{11,547}{4.44\times 60\times 0.9251\times 8} = 5.85674\ \text{Wb}$$

$$D = 60 + 0.25 = 62.5\ \text{in} = \frac{62.5\times 2.54}{100} = 1.5875\ \text{m}$$

$$\Phi_1 = \frac{2DlB_{m_1}}{p} = \frac{2\times(60+2.5)\times l\times 70,000}{4} = 5.85674\times 10\ \text{lines}$$

$$l = \frac{5.85674\times 10^8\times 4}{2\times 62.5\times 7\times 10^4} = 267.74\ \text{in.} = 6.80\ \text{m}$$

$$1\ \text{Tesla} = 64,500\ \text{lines/in}^2$$

$$B_{m_1} = 70,000/64,500 = 1.0853\ \text{Tesla}$$

$$\left\{\begin{array}{l}\text{if } B_{m_1} = 75,000\ \text{lines/in.}^2 = 1.1628\ \text{Tesla}\\ \qquad l\ \text{would be} = 249.9\ \text{in.}\end{array}\right.$$

$$S = \left(\frac{2\pi^2}{\sqrt{2}}\right)\left(\frac{f}{p}\right) B_{m_1} A k_{w_1} D^2 l\ \text{VA}$$

Hence,

$$\frac{2\pi^2}{\sqrt{2}}\left(\frac{60}{4}\right)\times 1.0853\times A\times 0.9251\times(1.5875)^2\times 6.8 = 659\times 10^6$$

$$A\times 13.5977\times 15\times 1.0854\times 0.9251\times 2.52016\times 6.8 = 659\times 10^6$$

$$A = \frac{659\times 10^6}{3509.713} = 187,765\ \text{A/m}$$

$$A = 4769.2\ \text{A/in.}$$

However, based on the "nameplate design rating" of this machine, we have

$$I_{ph} = 659\times 10^6/(\sqrt{3}\times 20,000)$$

$$I_{ph} = 19,023.7\ \text{A}$$

$$I_{ph}/m = 9511.85\ \text{A}$$

$$A = \frac{6N_{ph}(I_{ph}/m)}{\pi D} = \frac{6 \times 16 \times 9511.85}{\pi \times 62.5} = 4650.6 \quad \text{A/in.} = 183,094.5 \quad \text{A/m}$$

Therefore, it follows that

$$B_{m_1} \text{(actual design peak flux density)} = \frac{4769.2}{4650.6} \times 70,000$$
$$= 71,785 \text{ lines/in.} = 1.113 \text{ Tesla}$$

All numbers mentioned earlier are within reasonable engineering approximations.

Example 5.2 *Induction Motor*

A 5-Hp, 60 Hz, 0.777 lagging PF, 460 V, 3-phase, 6-pole, 36-stator slots, D_r = 5.72 in., g = 0.015 in., coil span is five slots (τ_c = 5 slots and τ_p = 6 slots), Y-connected induction motor with a double-layer lap stator winding, and has 222 turns per phase all connected in series.

Design analysis computations are as follows:

(1) Estimate design analysis, we have

$$\tau_c = 5 \text{ slots and } \tau_p = 6 \text{ slots}$$

$$\beta = \frac{\tau_c}{\tau_p} \times 180°e = \frac{5}{6_p} \times 180°e = 150°e$$

$$k_p = \sin(\beta/2) = \sin(150°/2) = 0.966$$

$$\alpha = \text{slot pitch angle} = \frac{p \times 180°e}{36} = \frac{6 \times 180°e}{36} = 30°e$$

$$q(\text{slot/pole/phase}) = \frac{36}{6 \times 3} = 2 \text{ slots/pole/phase}$$

$$k_d = \frac{\sin(q\alpha/2)}{q\sin(\alpha/2)} = \frac{\sin(2 \times 30°/2)}{2\sin(30°/2)} = 0.966$$

$$k_w = k_d \times k_p = 0.966 \times 0.966 = 0.933$$

$$E_{ph} = 460/\sqrt{3} = 265 \text{ V}$$

$$E_{ph} = 4.44 f \Phi N_{ph} k_w$$

$$\Phi = E_{ph}/(4.44 f N_{ph} k_w)$$

$$\Phi = 265/(4.44 \times 60 \times 222 \times 0.933) = 0.0048 \text{ Wb} = 0.0048 \times 10^8 \text{ lines}$$

But,

$$\Phi = \frac{2}{\pi} B_m \tau_p l = \frac{2}{\pi} B_m \left(\frac{\pi D}{p} \right) l$$

Hence,

$$B_m = p\Phi/(2Dl)$$

$$D = D_r + g = 5.72 + 0.015 = 5.735 \text{ in.}$$

$$l = 5 \text{ in.}$$

Hence,

$$D = 5.735 \times 0.0254\, m = 0.14567 \text{ m}$$

and

$$l = 5 \times 0.0254\, m = 0.127 \text{ m}$$

Hence,

$$B_m = 6 \times 0.0048/(2 \times 0.14567 \times 0.127) = 0.7784 \text{ Wb/m}^2 = 0.7784 \text{ Tesla}$$

$$Vol. = \frac{\pi D^2 l}{4} = \frac{\pi (0.14567)^2 \times 0.127}{4} = 0.002116575 \text{ m}^3 = 2.1166 \times 10^{-3} \text{ m}^3$$

(2) Based on "nameplate ratings" of the machine, we have

$$P(\text{Power}) = 5 \times 746 = 3730 \text{ W}$$

$$S = \frac{P(\text{Power})}{PF} = \frac{3730}{0.777} = 4800.51 \text{ VA}$$

$$I_{ph} = S/(3E_{ph}) = 4800.51/(3 \times 265) = 6.0384 \text{ A}$$

$$\begin{aligned} A &= \frac{3(2 \times N_{ph}) \times I_{ph}}{\pi D} = \frac{6N_{ph} I_{ph}}{\pi D} \\ &= \frac{6 \times 222 \times 6.0384}{\pi \times 0.14567} = 17575.44 \text{ A/m} \end{aligned}$$

Hence,

$$A = 17575.44 \times 0.0254 = 446.42 \text{ A/in.}$$

But,

$$\begin{aligned} S &= (8\pi/\sqrt{2})(f/p) B_{sr} A k_{w_1} (Vol.) \\ &= (8\pi/\sqrt{2})(60/6) \times 0.7784 \times 17575.44 \times 0.933 \times 2.1166 \times 10^{-3} \\ &= 4801.25 \text{ VA} \end{aligned}$$

Checks with $S = 4800.51$ VA.

All numbers mentioned earlier are within reasonable engineering approximations.

Example 5.3 ***Volume to Power (Volt-Ampere) Relationship in DC Machines***

$$E = pn\Phi C_a/(60m)$$

$$\Phi = 2DlB_{m_1}/p$$

$$E = pn\left(\frac{2DlB_{m_1}}{p}\right)\frac{C_a}{60m} = n\left(\frac{2B_{m_1}C_a}{60m}\right)Dl$$

$$EI = P(\text{Power}) = n\left(\frac{2B_{m_1}C_a}{60m}\right)DlI$$

But, $C_a\left(\frac{I}{m}\right)/(\pi D) = A$ (ampere loading)
Hence,

$$I = \frac{\pi mDA}{C_a}$$

Hence,

$$EI = n\left(\frac{2B_{m_1}C_a}{60m}\right)Dl \times \frac{\pi mDA}{C_a}$$

$$EI = \frac{\pi}{30}B_{m_1}AnD^2l$$

But,

$$Vol. = \frac{\pi D^2 l}{4}$$

Hence,

$$D^2 l = \frac{4}{\pi}Vol.$$

Hence,

$$EI = \left(\frac{\pi}{30}B_{m_1}An\right)\left(\frac{4}{\pi}Vol.\right)$$

$$EI = \left(\frac{2}{15}B_{m_1}An\right)Vol.$$

$$EI = (2/15)B_{m_1}An(Vol.)$$

Example 5.4 ***On EMF Computation in a Synchronous Machine***

A 2-pole synchronous machine was designed so that its stator contained 24 slots, each slot containing two coil sides (each coil side consists of one conductor, i.e., there is one turn per coil). The three-phase winding of this machine was Y-connected, with one path per phase. The coils were designed to span 10 slot pitches.

(a) Calculate the pitch and distribution factors for the fundamental and fifth harmonics.
(b) Calculate the number of turns in series per phase.
(c) If the airgap flux density can be expressed as follows:

$$B(\theta) = 78,000\cos(\theta) + 16,000\cos(3\theta) + 14,000\cos(5\theta)$$

Lines/square inch and the machine diameter at mid-gap are 41 in. with an axial conductor length of 205 in. and calculate the fundamental and fifth harmonic flux per pole in Webers.
(d) For a fundamental frequency of 60 Hz, calculate the fundamental and fifth harmonic-induced EMFs per phase.
(e) Calculate the fundamental, third, and fifth harmonic-induced line-to-line EMFs.
(f) If the stator winding (armature) was connected in delta, what would the third harmonic-induced EMF cause? and why?

(a)

$$\alpha = \frac{2 \times 180^\circ e}{24} = 15^\circ e$$

$$\beta = \frac{\tau_c}{\tau_p}(180^\circ) = \frac{10}{12} \times 180^\circ = 150^\circ e$$

But,

$$k_{d_h} = \frac{\sin(qh\alpha/2)}{q\sin(h\alpha/2)} \text{ and } k_{p_h} = \sin(h\beta/2)$$

$$q(\text{slot/pole/phase}) = \frac{24}{2 \times 3} = 4\,\text{slots/pole/phase}$$

$$k_{d_1} = \frac{\sin(4 \times 15^\circ/2)}{4\sin(15^\circ/2)} = 0.9577$$

$$k_{p_1} = \cos(30^\circ/2) = 0.966$$

$$k_{d_3} = \frac{\sin(4 \times 3 \times 15^\circ/2)}{4\sin(3 \times 15^\circ/2)} = \frac{\sin(90^\circ)}{4\sin(22.5^\circ)} = 0.653$$

$$k_{p_3} = \cos(3 \times 30^\circ/2) = 0.7071$$

$$k_{d_5} = \frac{\sin(4 \times 5 \times 15^\circ/2)}{4\sin(5 \times 15^\circ/2)} = \frac{\sin(150^\circ)}{4\sin(37.5^\circ)} = 0.2053$$

$$k_{p_5} = \cos(5 \times 30^\circ/2) = 0.2588$$

(b) $\text{Total number of turns} = 24 \times 2/2\,\text{Turns}$

$\text{Turns/phase} = 24/3\ \text{Turns/phase}$

$N_{ph}/m = 8/1 = 8\,\text{Turns/path/phase}$

(c) $$\Phi_{pole_h} = \frac{2}{\pi} B_{\max_h} \left(\frac{\pi D}{hp}\right) \cdot l$$

$$\Phi_{pole_1} = \frac{2}{\pi} \times 78{,}000 \times \frac{\pi \times 41}{1 \times 2} \times 205 \times 10^{-8} = 6.556\ \text{Wb}$$

$$\Phi_{pole_3} = \frac{2}{\pi} \times 16{,}000 \times \frac{\pi \times 41}{3 \times 2} \times 205 \times 10^{-8} = 0.4483\ \text{Wb}$$

$$\Phi_{pole_5} = \frac{2}{\pi} \times 14{,}000 \times \frac{\pi \times 41}{5 \times 2} \times 205 \times 10^{-8} = 0.2353\ \text{Wb}$$

(d) $$E_{phase_h} = 4.44(fh) \cdot (N_{ph}/m) \cdot \Phi_{pole_h} \cdot (k_{d_h} k_{p_h})\,\text{V}$$

$$E_{ph_1} = 4.44 \times 60 \times 8 \times 6.556 \times 0.9577 \times 0.966 = 12{,}926.2\ \text{V}$$

$$E_{ph_3} = 4.44 \times 3 \times 60 \times 8 \times 0.4483 \times 0.653 \times 0.7071 = 1323.5\ \text{V}$$

$$E_{ph_5} = 4.44 \times 5 \times 60 \times 8 \times 0.2353 \times 0.2053 \times 0.2588 = 133.2\ \text{V}$$

(e) $$E_{L-L_1} = \sqrt{3} \times 12{,}926.2 = 22{,}389\ \text{V}$$

$$E_{L-L_3} = 0.000\ \text{V}$$

$$E_{L-L_5} = \sqrt{3} \times 133.2 = 230.7\ \text{V}$$

(f) E_{ph_3} would cause circulating current because E_{ph_3} for phases a, b, and c are in phase with one another. Hence, they add around the loop of a delta.

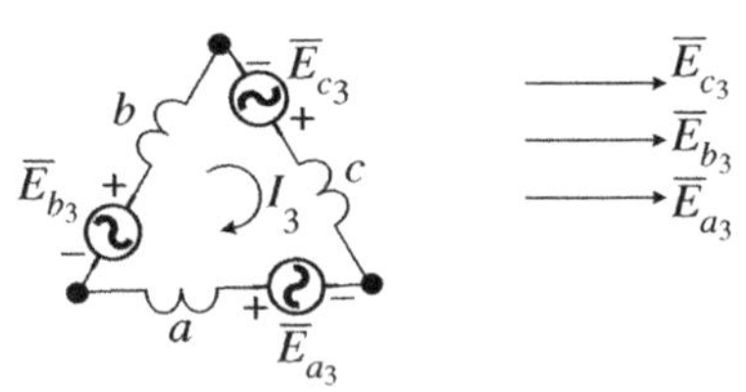

Homework Problems

5.1 In a balanced two-phase ac induction machine with a squirrel-cage rotor, the stator winding consists of two coils, the magnetic axes of which are displaced by 90° electrical ($\pi/2$ electrical radians) in space. For a number of series turns per phase, N_{ph}, a winding factor, k_w, and P-poles in such a machine

[Hint: $\cos\alpha \cos\beta = (1/2)\cos(\alpha+\beta) + (1/2)\cos(\alpha-\beta)$]:

(a) Derive an expression for the total (resultant) MMF produced by the two-phase stator winding, when the two-phase stator currents in amperes are

$$i_a = \sqrt{2}\,I\,\cos(\omega t)A \text{ and } i_b = \sqrt{2}\,I\,\cos(\omega t - \pi/2)A$$

State the nature of the resultant MMF produced by the two-phase stator winding.

(b) Can this machine function as a two-phase induction motor? Would such a motor have a nonzero starting torque? And why?

(c) If for some reason phase (b) was open circuited, that is, $i_b = 0$, write an expression for the resultant MMF produced by the stator winding. What is the nature of this resultant MMF? Would such a machine have a nonzero starting torque? And why?

(d) What is the effect on the direction of rotation of the motor if the connections of the (a) and (b) phases are interchanged? And why? Show your explanation both analytically and qualitatively.

(e) Based on the developments in (a) through (c), explain how does a single-phase squirrel-cage induction motor in a household appliance produce a nonzero starting torque and indicate in your answer what will have to constitute its stator windings? If such a machine loses its capability to start and operate normally, what is the most likely component to have failed in its stator winding circuits?

5.2 The design of a 4-pole, 3-phase, 460-V (line-to-line), Y-connected, 60-Hz induction motor is to be based on a stator inner diameter of 9.20 in., for an airgap height of 0.012 in., and a stator core length of 10.0 in. (1.0 in. = 0.0254 m). The stator winding has been chosen with a distribution factor, $k_w = 0.956$. The designer must pick the number of stator armature turns per phase so that the peak fundamental flux density at mid-airgap is 1.0T.

(a) Calculate the diameter at mid-airgap in meters, the resultant fundamental flux per pole, and the number of stator turns per phase. Also, calculate the peak value (amplitude) of the resultant MMF, F_{sr}, in AT/pole.

(b) If under the conditions in (a) the torque angles, $\delta_s = 70°e$, $\delta_{sr} = 170°e$, find the MMFs, F_s and F_r in AT/pole. Also, compute the stator phase current, I_s (or I_1), amperes, as well as the magnetizing current, $I_\emptyset$(or I_m), in amperes, and the magnetizing reactance, $x_\emptyset$, in Ohms.

(c) Calculate the developed torque in Nm and calculate the approximate value of developed power in kW and hp for the conditions in (a) and (b).

(d) Draw a neat MMF space-vector diagram showing F_s, F_r, F_{sr}, δ_s, δ_r, and δ_{sr} for a counterclockwise motor rotation. Indicate the direction of the motor's developed torque on this space-vector diagram.

(e) What is the relative angular speed in mechanical and electrical radians per second between the MMF, F_s, and the MMF, F_r, and what is the number of poles for each of the two MMFs in order to produce a nonzero developed average motor torque? And why?

5.3 In a balanced three-phase, P-pole, ac induction machine with a squirrel-cage rotor, the stator winding consists of the phases a, b, and c, housed in the normal counterclockwise sequence, $a, -c, b, -a, c, -b$, around the circumference of the airgap, with an angle $(2\pi/3)$ electrical radians between the a-axis and the b-axis, and an angle of $(4\pi/3)$ electrical radians between the a-axis and the c-axis, that is, in the normal a, b, and c, counterclockwise sequence manner. The number of series turns per phase is N_{ph}, with a winding factor, k_w. For the fundamental components of MMF:

(a) Derive an expression for the total (resultant) MMF produced by the three-phase stator winding, where the three-phase stator currents in amperes are

$$i_a = \sqrt{2}\ I\ \cos(\omega t)A,\ i_b = \sqrt{2}\ I\ \cos(\omega t - 2\pi/3)A, \text{ and}$$

$$i_c = \sqrt{2}\ I\ \cos(\omega t - 4\pi/3)A.$$

State the nature of the resultant MMF and its direction of rotation.

(b) Drive an expression for the total (resultant) MMF produced by the same three-phase stator winding, if the three-phase stator currents in amperes are

$$i_a = \sqrt{2}\ I\ \cos(\omega t)A,\ i_b = \sqrt{2}\ I\ \cos(\omega t - 4\pi/3)A, \text{ and}$$

$$i_c = \sqrt{2}\ I\ \cos(\omega t - 2\pi/3)A.$$

(c) What practical conclusion can you obtain from your derivations in (a) and (b) with regard to the operation of an actual three-phase squirrel-cage induction motor in a laboratory setting, if you have full access to the machine terminal wiring and the three-phase power supply?

6

The Poly-Phase Induction Machine

The poly-phase induction machine, and specifically three-phase induction motors and associated drives, constitutes what would be considered the main "work-horse" in industrial adjustable speed-drive (ASD) applications of various types. This includes industrial, transportation, heating and air-conditioning, public utility, manufacturing and automation applications, to name a few. These motors are rugged, robust, and very cost-effective when used in these various applications and are very versatile when used as prime movers in conjunction with ASDs in motor-drive systems. We will next describe their common construction features. These motors range from ratings as low as one or two hp up to tens of thousands of horsepower.

6.1 Main Constructional Features

Poly-phase induction machines consist of laminated and slotted stator and rotor cores, as shown in Figures 6.1–6.4. Example stators of such three-phase machines are shown in Figure 6.1 for a single-layer, 24-slots, 2-pole machine, including the A, B, and C labeling of the slots, and associated magnetic axes of the three phases. Such a stator winding is usually used for low horsepower (hp), up to 10–20 hp. Rotors of such machines can be of the wound-rotor variety with winding layout pattern, such as the stator of Figure 6.1, or the squirrel-cage variety, such as shown in Figure 6.2. Stators of larger horsepower variety of 30–40 hp and greater are usually wound with a double-layer pattern, such as shown in Figures 6.3 and 6.4, in which Part (I) and Part (II) of this double-layer winding are shown, respectively, for this 48-slot, 4-pole, and 3-phase stator winding. Furthermore, Figures 6.5–6.7 show physical photographs of stators, wound rotors with slip rings, and squirrel-cage rotors of three-phase induction machines, respectively.

Electric Machinery and Drives: An Electromagnetics Perspective, First Edition.
Nabeel A. O. Demerdash, JiangBiao He, and Hao Chen.

Companion website: www.wiley.com/go/demerdash/electricmachineryanddrives

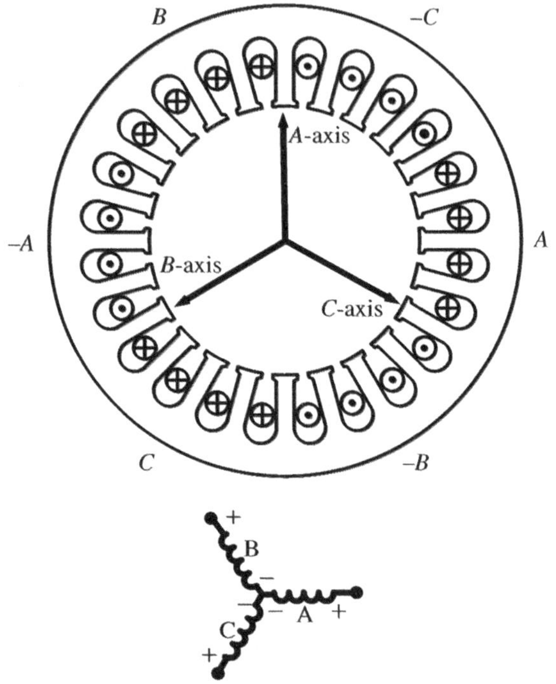

Figure 6.1 A stator lamination in a 24-slot, 3-phase, 2-pole single-layer winding layout with a schematic of a Wye-connection and designation of the magnetic axes of phases *A*, *B*, and *C*.

6.2 The *T*-Equivalent Circuit Model and Its Fundamental Formulation

The schematic diagram of Figure 6.8 is a simplified representation of the stators *A*, *B*, and *C*, phase windings, and the wound-rotors *a*, *b*, and *c*, phase windings (or equivalent representations of a squirrel-cage), in an induction machine. Here, the crescent-shaped distribution, attached to the rotor in this figure, represents half a radially outward-oriented airgap resultant flux density waveform, with a peak flux density, $B_{sr_{max}}$, leading to a resultant airgap flux per pole, Φ_{sr}, rotating (travelling) at the speed of rotation of the stator winding's MMF. Here, naturally, we have the rotating MMF established by the balanced three-phase stator currents traveling in a positive counterclockwise direction as indicated in Figure 6.8, at the synchronous speed, n_{syn}, in r/min, where

$$n_{syn} = 120f/p \tag{6.1}$$

Thus, the speed of rotation of Φ_{sr}, ω_{syn}, in electrical radians per second becomes

$$\omega_{syn} = (120f/p)(2\pi/60)(p/2) = 2\pi f \tag{6.2}$$

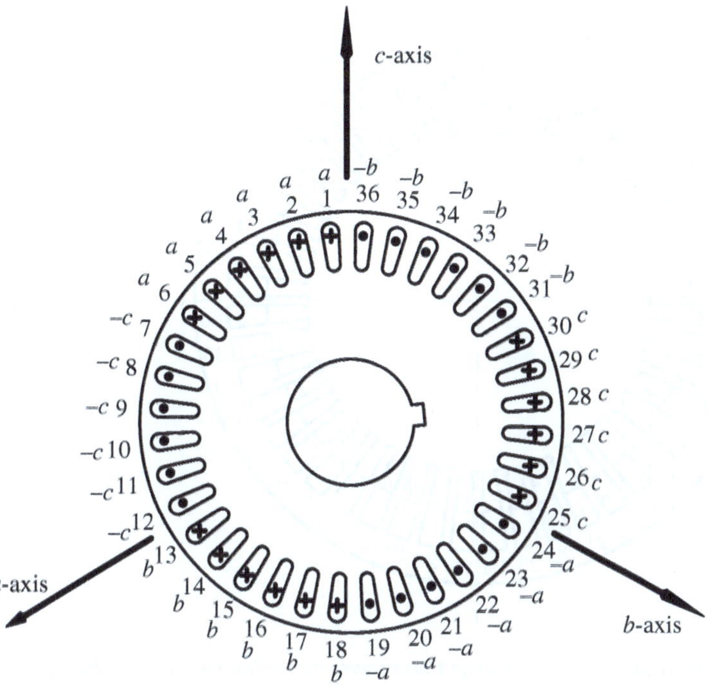

Figure 6.2 A squirrel-cage rotor lamination of an induction motor with 36-rotor slots for a 36-bar-cage.

We now define a quantity, called the slip, s, in per unit, as follows:

$$s = \left(\omega_{syn} - \omega_r\right) / \omega_{syn} = \left(n_{syn} - n_r\right) / n_{syn} \tag{6.3}$$

where n_r is the speed of rotation of the rotor in r/min; and ω_r is the speed of rotation of the rotor in electrical radians per second and given by

$$\omega_r = (1 - s)\,\omega_{syn} \tag{6.4}$$

Hence, n_r in r/min is given by

$$n_r = (1 - s)\,n_{syn} \tag{6.5}$$

and the rotor speed in mechanical radians per second, $\omega_{r\,mech.}$, is

$$\omega_{r\,mech.} = n_r\,(2\pi/60) = \pi n_r/30 \tag{6.6}$$

Therefore, the resultant flux is rotating past the rotor winding (or squirrel-cage bars) at a relative angular speed $(\omega_{syn} - \omega_r)$ in electrical radians per second, the

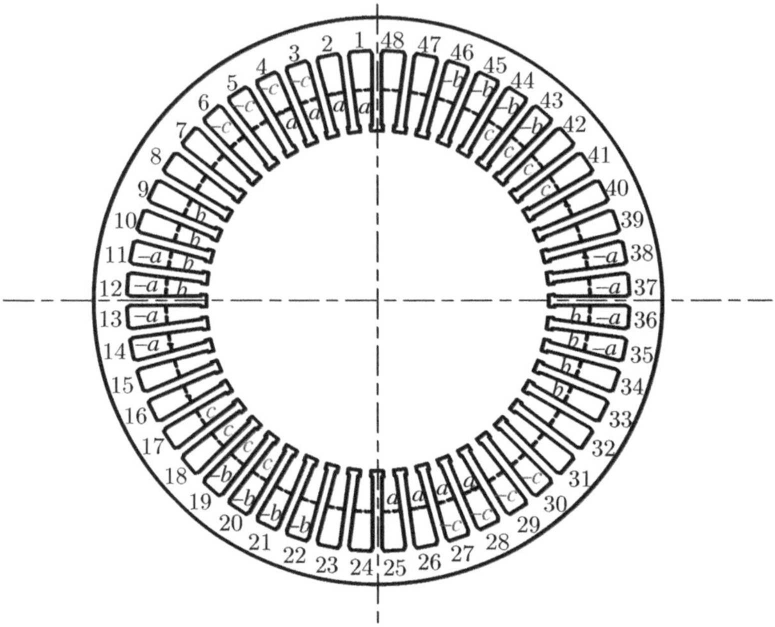

Figure 6.3 A stator lamination in a large horsepower motor for a 48-slot, 3-phase, 4-pole double-layer winding layout, with Part (I) of the double-layer winding with short-pitched coil span of 10-slots.

slip speed, leading to induced AC voltages and rotor currents at a frequency, f_r, given in Hertz by

$$f_r = \left(\left(\omega_{syn} - \omega_r\right) / \omega_{syn}\right) f = sf \tag{6.7}$$

which leads to the induced EMF amplitudes in phases (A) and (a) in the stator and rotor windings, $\left\|\overline{E}_A\right\|$ and $\left\|\overline{E}_a\right\|$, respectively, given as shown in Figure 6.9 by

$$\left.\begin{aligned} \left\|\overline{E}_A\right\| &= 4.44\,(f)\,\Phi_{sr} N_1 k_{w_1} \\ \left\|\overline{E}_a\right\| &= 4.44\,(sf)\,\Phi_{sr} N_2 k_{w_2} \end{aligned}\right\} \tag{6.8}$$

where N_1 and N_2 are the number of turns per phase, in the stator and rotor windings, respectively, k_{w_1} and k_{w_2} are the winding factors for the stator and rotor windings, respectively. It should be pointed out that the circuit schematic in Figure 6.9 presents the primary and secondary in a transformer, albeit with a variable turns ratio, $(a/s) : 1$, where,

$$((a/s) : 1) = \left(\left(N_1 k_{w_1} / N_2 k_{w_2}\right) / s \; : 1\right) \tag{6.9}$$

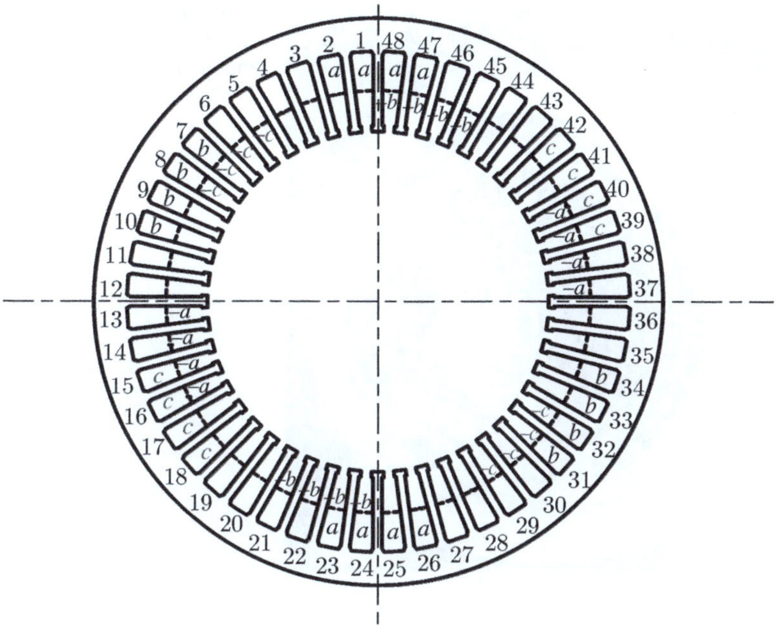

Figure 6.4 A stator lamination in a large horsepower motor for a 48-slot, 3-phase, 4-pole double-layer winding layout, with Part (II) of the double-layer winding with short-pitched coil span of 10-slots.

in which

$$a = \left(N_1 k_{w_1} / N_2 k_{w_2}\right) \tag{6.10}$$

This results in the variable turns-ratio equivalent circuit of Figure 6.10, where the frequency of the primary (stator) side is f, Hertz, and the frequency of the secondary (rotor) side is (sf). Dividing all the circuit elements on the rotor side by the slip, s, results in the equivalent circuit of Figure 6.11.

Here, in Figure 6.9, we have

$$E_2/E_1 = s\left(N_2 k_{w_2} / N_1 k_{w_1}\right) \tag{6.11}$$

That is,

$$E_2/E_1 = s/a \tag{6.12}$$

or

$$E_2 = sE_1/a \tag{6.13}$$

Also, the rotor currents, $\overline{I}_2$, can be expressed as follows:

$$\overline{I}_2 = \overline{E}_2/\left(R_2 + j2\pi\,(sf)\,L_2\right) = \overline{E}_2/\left(R_2 + jsX_2\right) \tag{6.14}$$

Figure 6.5 A stator of a three-phase induction motor. Source: Duke Electric.

Figure 6.6 A skewed wound-rotor of an induction motor, with slip rings and brushes.

where

$$X_2 = 2\pi f L_2 = \omega L_2 \tag{6.15}$$

Furthermore, in Figure 6.11, we have

$$\bar{I}_2 = \left(\bar{E}_2/s\right) / \left((R_2/s) + jX_2\right) = \left(\bar{E}_2/a\right) / \left((R_2/s) + jX_2\right) \tag{6.16}$$

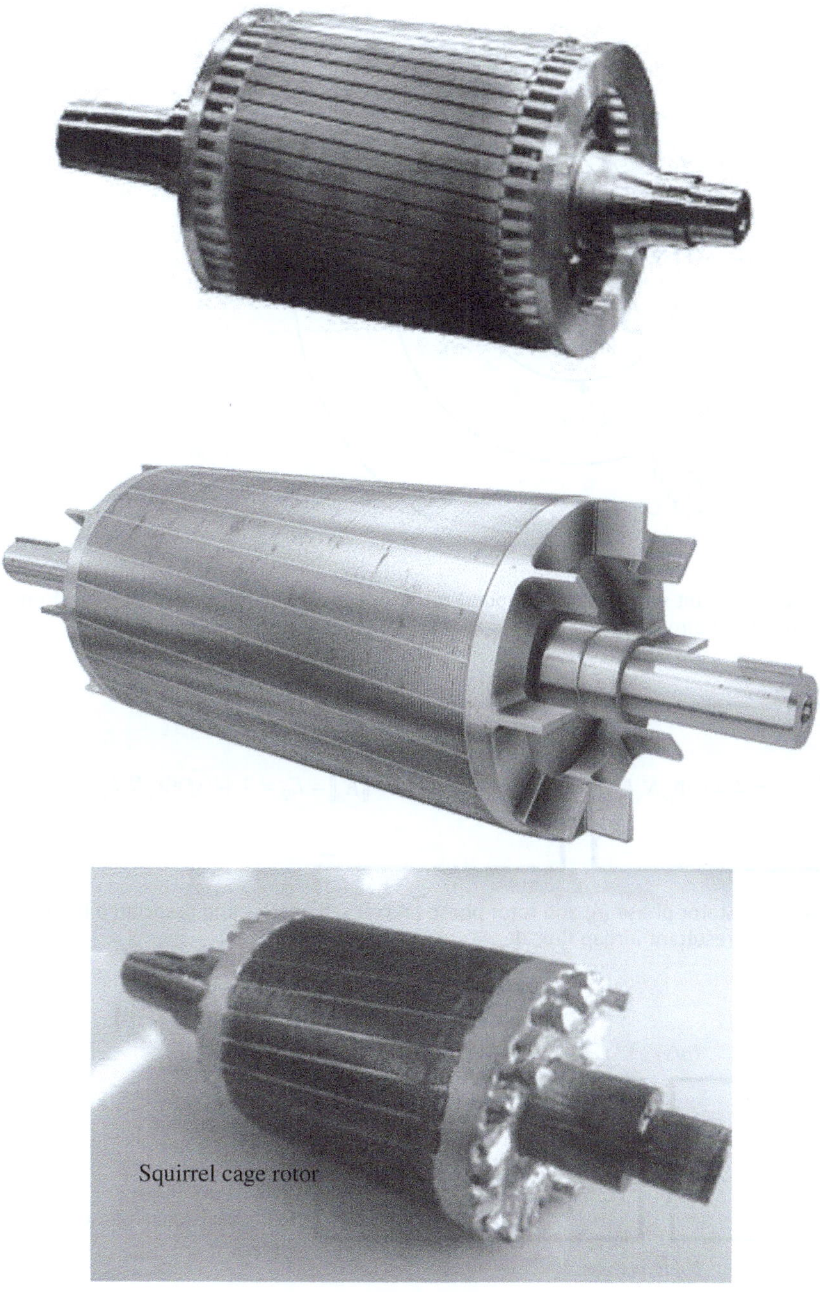

Figure 6.7 Three examples of skewed squirrel-cage rotors of induction motors.

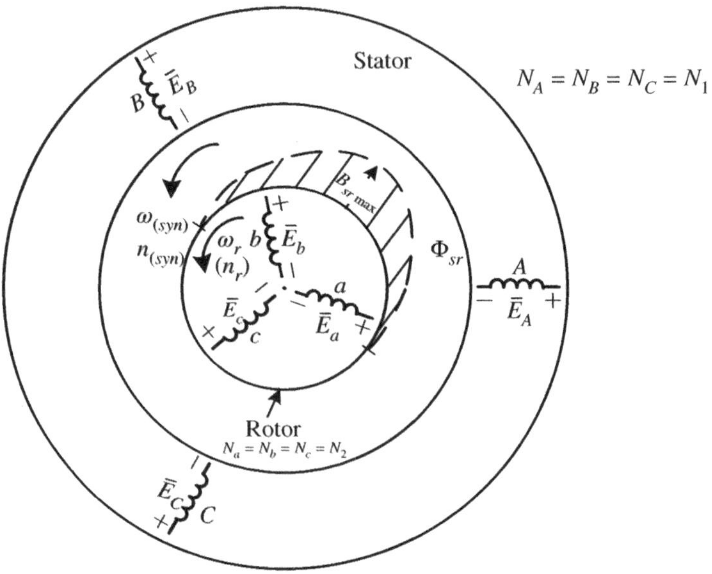

Figure 6.8 Circuit schematic of a wound-rotor induction machine and a half waveform of the resultant airgap flux.

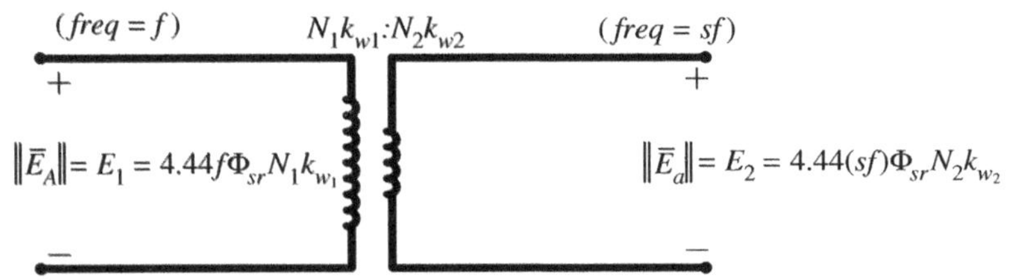

Figure 6.9 Stator phase (*A*) and rotor phase (*a*) coil schematics and associated induced EMFs due to resultant airgap flux, Φ_{sr}.

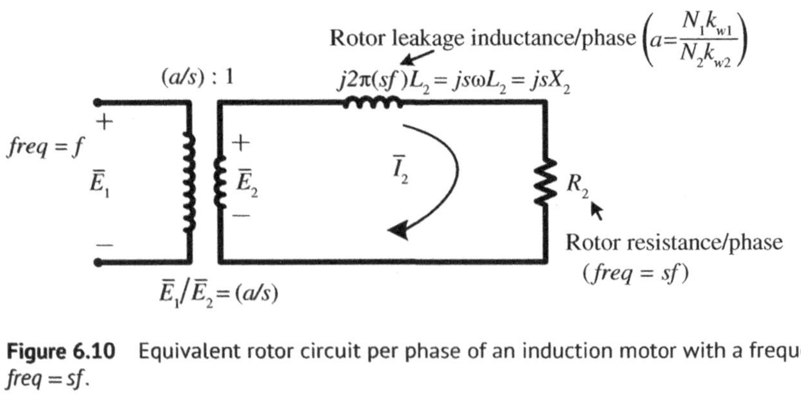

Figure 6.10 Equivalent rotor circuit per phase of an induction motor with a frequency, *freq* = *sf*.

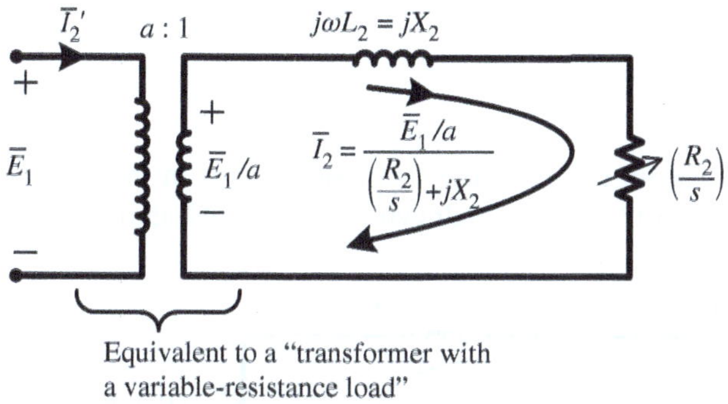

Figure 6.11 Equivalent rotor circuit per phase of an induction motor with a frequency, *freq* = *f*.

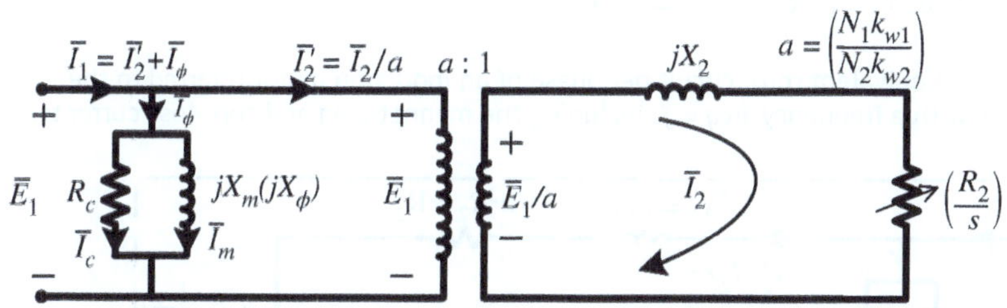

Figure 6.12 Equivalent rotor circuit per phase of an induction motor with a frequency, *freq* = *f*, including the magnetizing and core-loss components of currents per phase.

In Eq. (6.16), the frequency is f; hence, the frequencies of the primary and secondary sides in the equivalent circuit model of this figure are the same. Therefore, this circuit is identical to that of a regular transformer with a variable resistance load (R_2/s), function of the slip, s. Now, adding the effect of the core losses and magnetization MMF results in the equivalent circuit in Figure 6.12. In this figure, we have

$$\bar{I}_\phi = \bar{I}_c + \bar{I}_m \tag{6.17}$$

where

$$\bar{I}_c = \left(\bar{E}_1/R_c\right) \tag{6.18}$$

and

$$\bar{I}_m = \left(\bar{E}_1/jX_m\right) = \left(\bar{E}_1/jX_\phi\right) \tag{6.19}$$

Hence,

$$\bar{I}_\phi = \left(\bar{E}_1/R_c\right) + \left(\bar{E}_1/jX_m\right) = \left(\bar{E}_1/R_c\right) + \left(\bar{E}_1/jX_\phi\right) \tag{6.20}$$

Also, here the total core losses, $W_{core\ loss}$, are

$$W_{core\ loss} = 3\left(\frac{E_1^2}{R_c}\right) \tag{6.21}$$

Referring the rotor equivalent circuit to the stator side results in the equivalent circuit shown in Figure 6.13, which is redrawn in part (*A*) of Figure 6.14. In

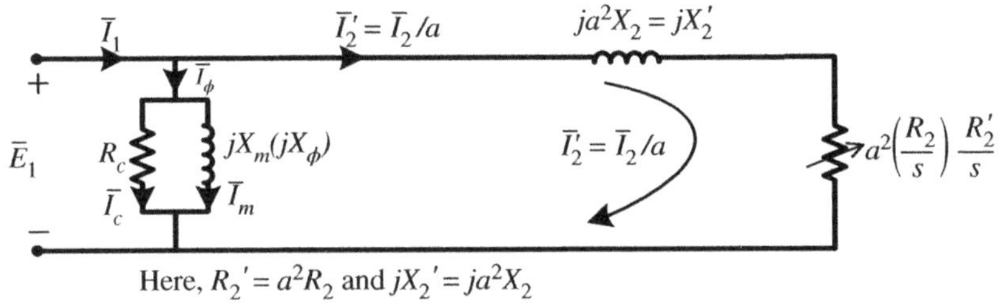

Figure 6.13 Equivalent rotor circuit per phase of an induction motor referred to the stator-side, with a frequency, *freq* = *f*, including the magnetizing and core-loss currents.

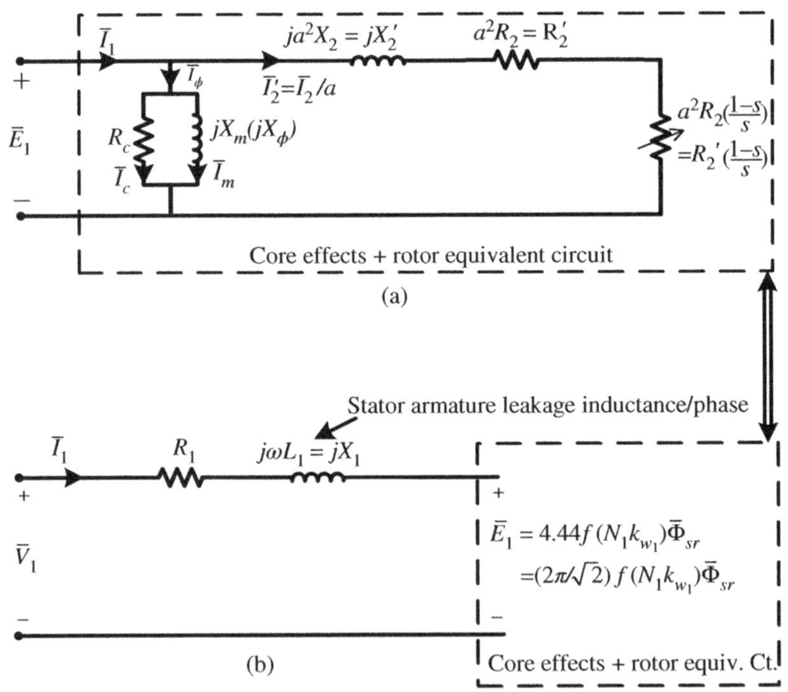

Figure 6.14 Equivalent rotor circuit per phase of an induction motor referred to the stator-side including the magnetizing branch, part (a), and the effect of stator winding leakage reactance and resistance, part (b).

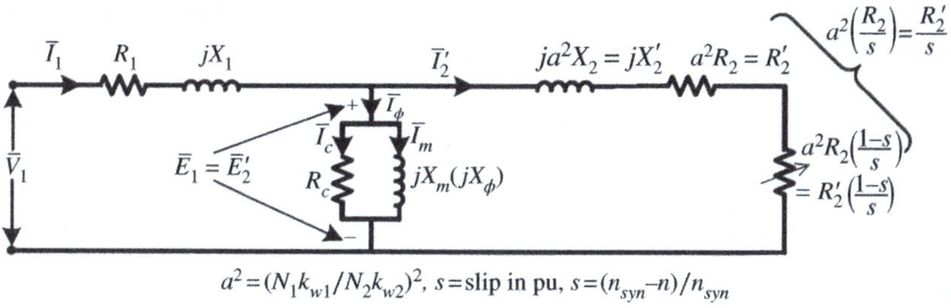

Figure 6.15 The *T*-equivalent circuit of a three-phase induction motor per phase referred to the stator side.

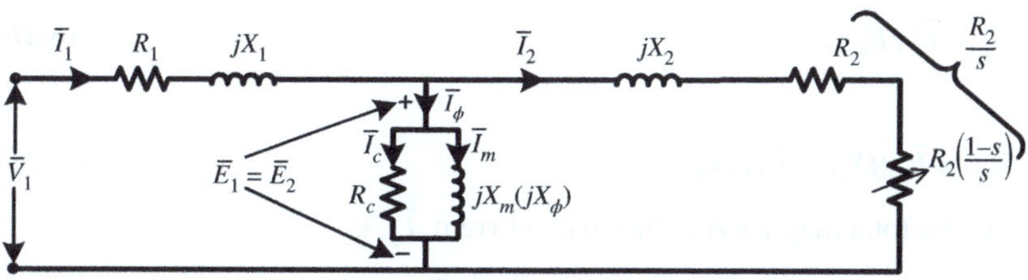

Figure 6.16 The *T*-equivalent circuit-all quantities referred to the stator side-redesignated.

this same figure, we merge the rotor equivalent circuit, referred to the stator-side, with the stator winding's resistance per phase, R_1, and stator leakage reactance, $j\omega L_1 = jX_1$, to yield part (*B*) of this figure. Hence, this leads to the two equivalent circuits of Figures 6.15 and 6.16, with redesignation of the referring process of all the parameters, eliminating the priming notation.

6.3 Performance Calculations Using the *T*-Equivalent Circuit Model

Examining the *T*-equivalent circuit per phase of a three-phase induction motor, Figure 6.16, the input power into the stator (armature #1) winding, P_{in}, can be expressed in phasor notation as follows:

$$P_{in} = \mathrm{Re}\left(3\overline{I}_1^* \cdot \overline{V}_1\right) \tag{6.22}$$

Meanwhile, ohmic losses in the stator winding can be expressed as follows:

$$P_{stator\_ohmic} = 3I_1^2R_1 \tag{6.23}$$

Furthermore, the power transferred across the airgap's magnetic field into the rotor circuit, P_{airgap}, can be expressed as follows:

$$P_{airgap} = \text{Re}\left(3\overline{I}_1^* \cdot \overline{V}_1\right) - 3I_1^2R_1 \tag{6.24}$$

or

$$P_{airgap} = \text{Re}\left(3\overline{I}_1^* \cdot \overline{E}_1\right) \tag{6.25}$$

where

$$\overline{E}_1 = \overline{V}_1 - (R_1 + jX_1)\overline{I} \tag{6.26}$$

Now, the currents, $\overline{I}_c$ and $\overline{I}_m$, see Figure 6.16, can be calculated as follows:

$$\overline{I}_c = \overline{E}_1/R_c \tag{6.27}$$

and

$$\overline{I}_m = \overline{E}_1/jX_m = \overline{E}_1/jX_\phi \tag{6.28}$$

Hence, the total magnetizing (no-load) current, $\overline{I}_\phi$, is

$$\overline{I}_\phi = \overline{I}_c + \overline{I}_m = \left(\overline{E}_1/R_c\right) + \left(\overline{E}_1/jX_m\right) = \left(\overline{E}_1/R_c\right) + \left(\overline{E}_1/jX_\phi\right) \tag{6.29}$$

Accordingly, the core-losses, eddy-current, and hysteresis losses, $P_{core\ loss}$, can be expressed as follows:

$$P_{core\ loss} = 3I_c^2R_c \tag{6.30}$$

or

$$P_{core\ loss} = 3E_1^2/R_c \tag{6.31}$$

Again, $P_{electromagnetic}$ (P_{em}), transferred for torque production into the rotor, can be expressed as follows:

$$P_{em} = \text{Re}\left(3\overline{I}_1^* \cdot \overline{E}_1\right) - 3E_1^2/R_c \tag{6.32}$$

or

$$P_{em} = 3I_2^2\left(R_2/s\right) \tag{6.33}$$

Meanwhile, the ohmic losses in the rotor (armature #2) winding, $P_{rotor\ ohmic}$, are

$$P_{rotor\ ohmic} = 3I_2^2R_2 \tag{6.34}$$

Accordingly, the developed power, $P_{dev.}$, which is also the so-called gross mechanical power converted from electrical energy to mechanical energy, $P_{gross\ mech.}$, can be written as follows:

$$P_{dev.} = P_{gross\ mech.} = 3I_2^2\left(R_2/s\right) - 3I_2^2R_2 \tag{6.35}$$

or

$$P_{dev.} = P_{gross\ mech.} = 3I_2^2R_2\left(\left(1-s\right)/s\right) \tag{6.36}$$

and the output mechanical power, P_{out}, is

$$P_{out} = P_{dev.} - P_{f\&W} = P_{gross\ mech.} - P_{f\&W} \tag{6.37}$$

where $P_{f\&W}$ is the mechanical friction and windage (rotational) losses.

Hence, the overall motor efficiency, η_{motor}, in *pu* becomes

$$\eta_{motor} = P_{out}/P_{in} \tag{6.38}$$

or in percent

$$\eta_{motor} = \left(P_{out}/P_{in}\right) \times 100 \tag{6.39}$$

Now, from the electromagnetic power, P_{em}, one can compute the electromagnetic torque, T_{em}, by dividing P_{em} by the synchronous speed, ω_{syn}, in mechanical radians per second. Here, ω_{syn}, in mech. rad/sec, is

$$\omega_{syn} = (120f/p)(2\pi/60) = 4\pi f/p \tag{6.40}$$

Hence, T_{em}, which is the developed mechanical torque, $T_{dev.}$, can be written as follows:

$$T_{em} = T_{dev.} = P_{em}/\omega_{syn} = \left(3I_2^2\left(R_2/s\right)\right)/\omega_{syn} \tag{6.41}$$

or

$$T_{em} = T_{dev.} = (p/4\pi f)\left(3I_2^2\left(R_2/s\right)\right) \tag{6.42}$$

Also, because of the rotor angular speed, $\omega_r = (1-s)\omega_{syn}$, one can rewrite the torque expression in Eq. (6.40) as follows:

$$T_{em} = T_{dev.} = T_{gross\ mech.} = \left(3I_2^2R_2\left(\left(1-s\right)/s\right)\right)/\omega_r \tag{6.43}$$

where again ω_r is the rotor speed in mech. rad/sec, that is,

$$\omega_r = (1-s)\,\omega_{syn} = (1-s)\,(4\pi f/p) \tag{6.44}$$

Accordingly, the output torque, T_{out}, which is the net mechanical torque, $T_{net\ mech.}$, can be expressed as follows:

$$T_{out} = T_{net\ mech.} = P_{out}/\omega_r = \left[3I_2^2R_2\left(\left(1-s\right)/s\right) - P_{f\&W}\right]/\omega_r \tag{6.45}$$

6.4 The Simplified T-Equivalent Circuit Model and Its Use in Performance Calculations and Associated Torque–Slip (Speed) Characteristics

Starting from the full T-equivalent circuit of Figure 6.16, where all rotor circuit parameters are referred to the stator-side, if we were to lump the core losses, $P_{core\ loss}$, Eqs. (6.30) and (6.31), with the friction and windage losses, $P_{f\,\&\,W}$, in Eqs. (6.36) and (6.45), the resulting losses will be referred to, henceforth, as the rotational losses, $P_{rotational}$. That is, one can write the following:

$$P_{rotational} = P_{core\ loss} + P_{f\&W} \tag{6.46}$$

The lumping of the above two loss components, $P_{core\ loss}$ and $P_{f\,\&\,W}$, into one loss, $P_{rotational}$, allows one to redraw the T-equivalent circuit referred to the stator side per phase as shown in Figure 6.17.

Now, the Thevenin equivalent to the portion of the circuit to the left of the line with arrows in this circuit is shown in Figure 6.18, in which it can be shown applying Thevenin's theorem analysis that in this figure, one can write the following for the voltage source, $\overline{V}_{1a}$, the resistance, R_{e1}, and reactance, jX_{e1}, respectively:

$$\overline{V}_{1a} = \left[\frac{jX_{\phi}}{R_1 + j(X_1 + X_{\phi})}\right]\overline{V}_1 \tag{6.47}$$

$$R_{e1} = \frac{R_1 X_{\phi}^2}{R_1^2 + (X_1 + X_{\phi})^2} \tag{6.48}$$

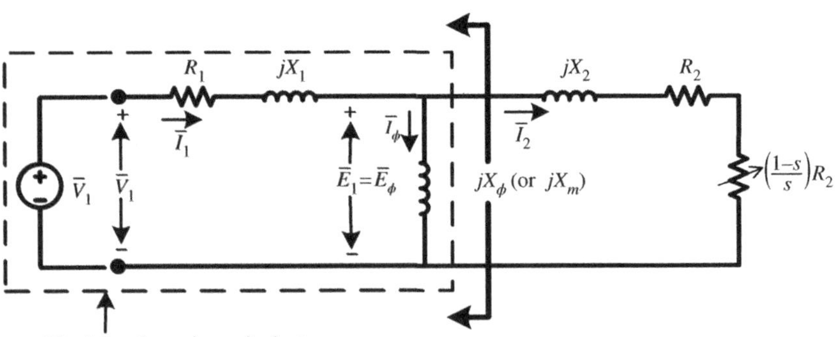

Figure 6.17 The simplified T-equivalent circuit per phase referred to the stator side (excluding the core losses) for an induction motor-here, $jX_{\phi} = jX_m$, is the magnetizing reactance.

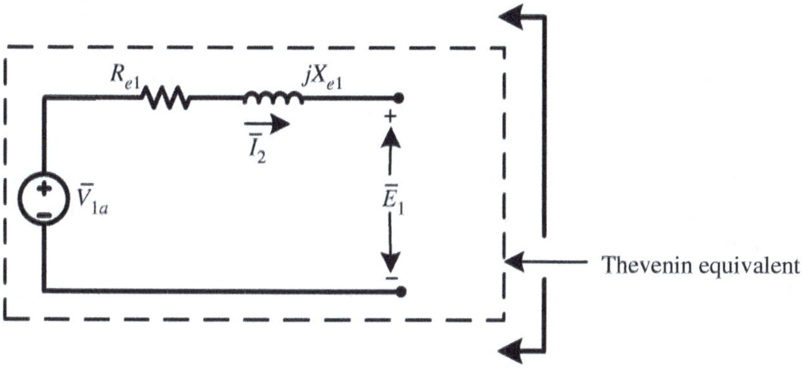

Figure 6.18 The Thevenin equivalent to the source, $\overline{V}_1$, and the stator-side up to the magnetizing reactance, $jX_\phi = jX_m$.

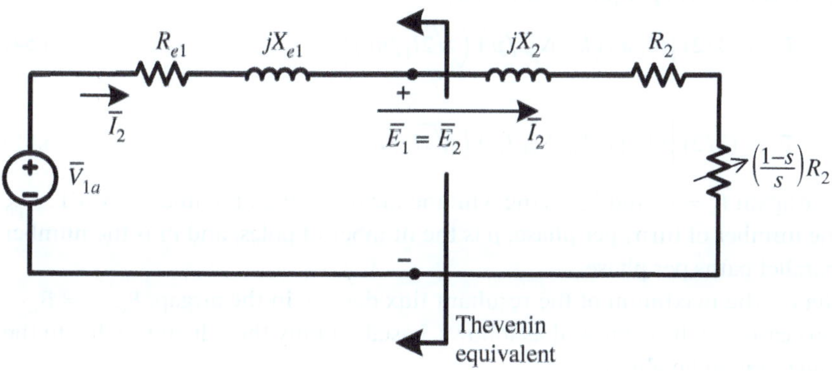

Figure 6.19 Thevenin equivalent of the T-equivalent circuit per phase referred to stator side (excluding core losses) of an induction motor.

and

$$jX_{e1} = j\left[\frac{X_\phi\left[R_1^2 + X_1\left(X_1 + X_\phi\right)\right]}{R_1^2 + \left(X_1 + X_\phi\right)^2}\right] \tag{6.49}$$

This results in the Thevenin equivalent circuit per phase depicted in Figure 6.19, in which the loop current, $\overline{I}_2$, flowing into the rotor side of the circuit can be expressed in RMS phasor form as follows:

$$\overline{I}_2 = \overline{V}_{1a}/\left[\left(R_{e1} + \left(R_2/s\right)\right) + j\left(X_{e1} + X_2\right)\right] \tag{6.50}$$

Hence, the amplitude of $\overline{I}_2$ can be expressed by

$$I_2 = \left\|\overline{I}_2\right\| = V_{1a}/\sqrt{\left[\left(R_{e1} + \left(R_2/s\right)\right)^2 + \left(X_{e1} + X_2\right)^2\right]} \tag{6.51}$$

Here, the Thevenin equivalent of the simplified T-equivalent circuit per phase, excluding core losses, is rendered as shown in Figure 6.19.

Referring back to Figure 6.17, if the induction machine is driven at the synchronous speed, while energized by the voltage, $\overline{V}_1$, per phase, the slip, s, will be zero. Hence, the rotor portion of the circuit in this figure will be open-circuited and, hence, passing no current. This means that under such a condition, the stator current, $\overline{I}_1$, will equal the current in the magnetizing branch. That is, one can write the following:

$$\overline{I}_\phi = \overline{I}_1 \tag{6.52}$$

and

$$\overline{I}_2 = 0 \tag{6.53}$$

Hence, one can express the amplitude of the resultant stator MMF, $\mathbf{F}_s$, in ampere-turns (AT) per pole as follows:

$$\mathbf{F}_s = (3/2)\left[(4/\pi)\left(k_{w_1}N_{ph}/p\right)\left(\sqrt{2}I_1/m\right)\right] \tag{6.54}$$

or

$$\mathbf{F}_s = (3/2)\left[(4/\pi)\left(k_{w_1}N_{ph}/p\right)\left(\sqrt{2}I_\phi/m\right)\right] \tag{6.55}$$

where again $\overline{I}_1 = \overline{I}_\phi$ and k_{w_1} is the winding factor for the fundamental MMF, N_{ph} is the number of turns per phase, p is the number of poles, and m is the number of parallel paths per phase.

Hence, the maximum of the resultant flux density in the airgap, $B_{sr_{max}} = B_{s_{max}}$, in this case, can be expressed as follows, based on only the reluctance due to the airgap of radial height, h_g:

$$B_{sr_{max}} = B_{s_{max}} = \mu_0\left(\mathbf{F}_s/h_g\right) \tag{6.56}$$

Meanwhile, the resultant flux per pole, Φ_{sr}, becomes

$$\Phi_{sr} = (2/\pi)\,\tau_p l B_{sr_{max}} = (2/\pi)(\pi D/p)\, l B_{sr_{max}} \tag{6.57}$$

Hence,

$$\Phi_{sr} = (2Dl/p)\,\mu_0\left(\mathbf{F}_s/h_g\right) = \left(2Dl\mu_0/h_g p\right)\mathbf{F}_s \tag{6.58}$$

while the EMF across the magnetizing reactance, $E_\phi = E_1$, see the electric circuit in Figure 6.17, can be expressed as follows in terms of Φ_{sr}:

$$E_\phi = \left(2\pi/\sqrt{2}\right) f\left(N_{ph}k_{w_1}/m\right)\Phi_{sr} \tag{6.59}$$

Substituting for Φ_{sr} from Eq. (6.58) into Eq. (6.59) yields the following:

$$E_\phi = \left(2\pi/\sqrt{2}\right) f\left(N_{ph}k_{w_1}/m\right)\left(2Dl\mu_0/h_g p\right)\mathbf{F}_s \tag{6.60}$$

And substituting for $\mathbf{F}_s$ from Eq. (6.55) into Eq. (6.60) gives the following for E_ϕ:

$$E_\phi = \left(2\pi/\sqrt{2}\right) f \left(N_{ph}k_{w_1}/m\right) \left(2Dl\mu_0/h_g p\right) \left[(6/\pi)\left(k_{w_1}N_{ph}/p\right)\left(\sqrt{2}I_\phi/m\right)\right] \tag{6.61}$$

The induced EMF, E_ϕ in Eq. (6.61), algebraically reduces to the following:

$$E_\phi = 24f\left(N_{ph}k_{w_1}/mp\right)^2 \left(\mu_0 Dl/h_g\right) I_\phi \tag{6.62}$$

Hence, the magnetizing reactance, $X_\phi = X_m$, in ohms per phase becomes

$$X_\phi = X_m = E_\phi/I_\phi = 24f\left(N_{ph}k_{w_1}/mp\right)^2 \left(\mu_0 Dl/h_g\right) \tag{6.63}$$

which can be rewritten as follows:

$$X_\phi = X_m = (24/\pi) f\left(N_{ph}k_{w_1}/mp\right)^2 \left(\mu_0 \pi Dl/h_g\right) \tag{6.64}$$

The last term in Eq. (6.64), $(\mu_0 \pi Dl/h_g)$, is the airgap cylindrical-shell permeance, $\mathbf{P}_{airgap}$, that is:

$$\mathbf{P}_{airgap} = \left(\mu_0 \pi Dl/h_g\right) \tag{6.65}$$

Hence, $X_\phi = X_m$ can be rewritten as follows:

$$X_\phi = X_m = (24/\pi) f\left(N_{ph}k_{w_1}/mp\right)^2 \mathbf{P}_{airgap} \tag{6.66}$$

Therefore, as the airgap permeance increases or decreases, so does the magnetizing reactance, $X_\phi = X_m$. Also, the magnetizing inductance $L_\phi = L_m = (X_\phi/2\pi f) = (X_m/2\pi f)$ in Henries becomes

$$L_\phi = L_m = \left(12/\pi^2\right)\left(N_{ph}k_{w_1}/mp\right)^2 \left(\mu_0 \pi Dl/h_g\right) \tag{6.67}$$

For further details on the magnetizing inductance/reactance and other inductance/reactance calculations, Ref. [1] should be consulted. Based on Eqs. (6.63)–(6.66), it should be pointed out that the larger the radial airgap height, h_g, the lower the magnitude of the magnetizing reactance, X_ϕ or X_m, the larger the magnetizing current, $\overline{I}_\phi$, in the circuit of Figure 6.17, the lower the overall power factor.

Now, we proceed to the calculation of the induction motor performance characteristics based on the simplified T-equivalent circuit, which is reshown in Figure 6.20 for convenience of the reader.

The input power into an induction motor, P_{in}, can be expressed as follows:

$$P_{in} = \mathrm{Re}\left(3\overline{I}_1^* \cdot \overline{V}_1\right) \tag{6.68}$$

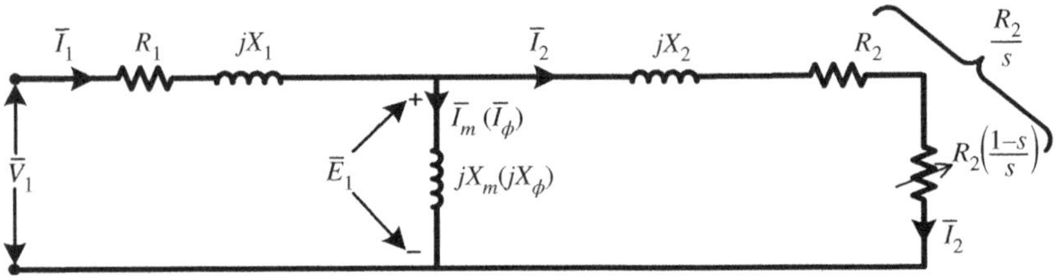

Figure 6.20 The simplified T-equivalent circuit of an induction motor per phase, lumping the core losses with the friction and windage losses, the so-called rotational losses.

The stator ohmic losses, $P_{stator\ ohmic}$, is

$$P_{stator\ ohmic} = 3I_1^2R_1 \tag{6.69}$$

and the airgap or electromagnetic power, P_{em}, is

$$P_{em} = 3I_2^2\left(R_2/s\right) \tag{6.70}$$

while the rotor ohmic losses, $P_{rotor\ ohmic}$, is

$$P_{rotor\ ohmic} = 3I_2^2R_2 \tag{6.71}$$

Accordingly, the gross-mechanical or developed power, $P_{dev.}$, can be expressed as follows:

$$P_{dev.} = P_{em} - P_{rotor\ ohmic} = 3I_2^2R_2\left(\left(1-s\right)/s\right) \tag{6.72}$$

Hence, the output power, P_{out}, can be expressed as follows:

$$P_{out} = P_{dev.} - P_{rotational} = 3I_2^2R_2\left(\left(1-s\right)/s\right) - P_{rotational} \tag{6.73}$$

Thus, the overall motor efficiency, η_{motor}, in per unit is as follows:

$$\eta_{motor} = \left(P_{out}/P_{in}\right) \tag{6.74}$$

or, in percent:

$$\eta_{motor} = \left(P_{out}/P_{in}\right) \times 100\% \tag{6.75}$$

Consequently, the power flow diagram, depicting the various power and loss components from the input stage to the shaft-output stage, in such an induction motor is shown in Figure 6.21. Recalling, I_2 expressed in Eq. (6.51), the developed power of Eq. (6.72), $P_{dev.}$, can be expressed as follows:

$$P_{dev.} = 3I_2^2R_2\left(\left(1-s\right)/s\right) = \frac{3V_{1a}^2R_2\left(\left(1-s\right)/s\right)}{\left[\left(R_{e1} + \left(R_2/s\right)\right)^2 + \left(X_{e1} + X_2\right)^2\right]} \tag{6.76}$$

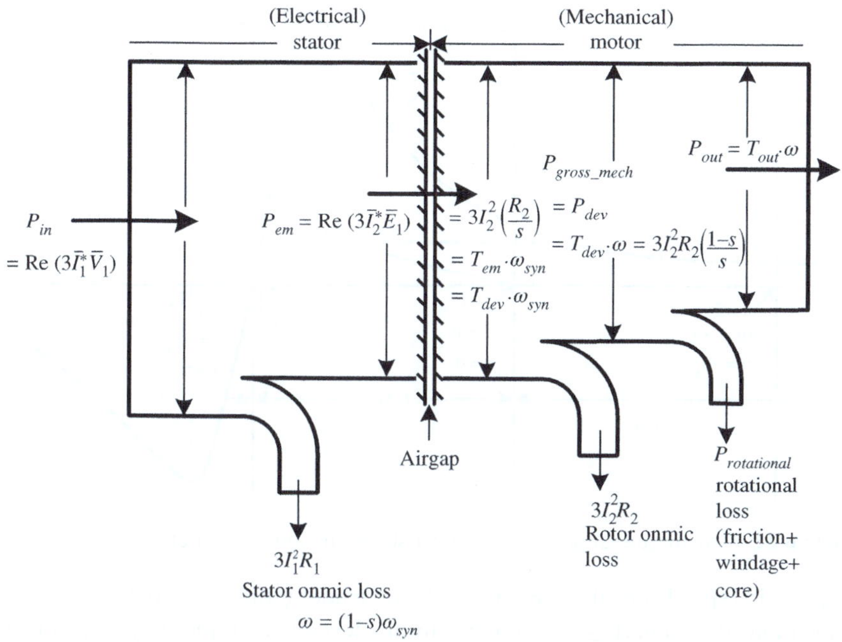

Figure 6.21 The input–output power (energy) flow in an induction motor using the simplified *T*-equivalent circuit.

which algebraically reduces to the following:

$$P_{dev.} = 3V_{1a}^2\,(R_2/s)\Big/\left[\left(R_{e1} + (R_2/s)\right)^2 + \left(X_{e1} + X_2\right)^2\right] \tag{6.77}$$

Therefore, using Eq. (6.77) for the developed power, Eq. (6.43) for the developed torque, $T_{dev.}$, and Eq. (6.44) for the angular speed, ω, the following can be written for the developed torque:

$$T_{dev.} = 3V_{1a}^2\,(R_2/s)\Big/\omega_{syn}\left[\left(R_{e1} + (R_2/s)\right)^2 + \left(X_{e1} + X_2\right)^2\right] \tag{6.78}$$

which reduces to the following for $T_{dev.}$:

$$T_{dev.} = 3V_{1a}^2R_2\,(s)\Big/\omega_{syn}\left[\left(sR_{e1} + R_2\right)^2 + s^2\left(X_{e1} + X_2\right)^2\right] \tag{6.79}$$

The plot of the developed torque, $T_{dev.}$, versus slip, s, and the speed is shown in Figure 6.22. The slip scale is for $(-1 < s \leq 2)$, thus covering the brake (or plugging) range $(2 \geq s > 1)$, and the motoring range $(1 \leq s < 0)$, and the generating range $(s < 0)$. It should be noticed that on the speed scale, $s = 2$ corresponds to a rotational speed of ($-n_{syn}$ in rev/min) or ($-\omega_{syn}$ in rad/sec), and, for $s = 1$, the rotational speed is "zero," and, for $s = 0$, the rotational speed is (n_{syn} in rev/min) or (ω_{syn} in rad/sec). Naturally, the torque $T_{dev.} = 0$ at $s = 0$. Furthermore, in this

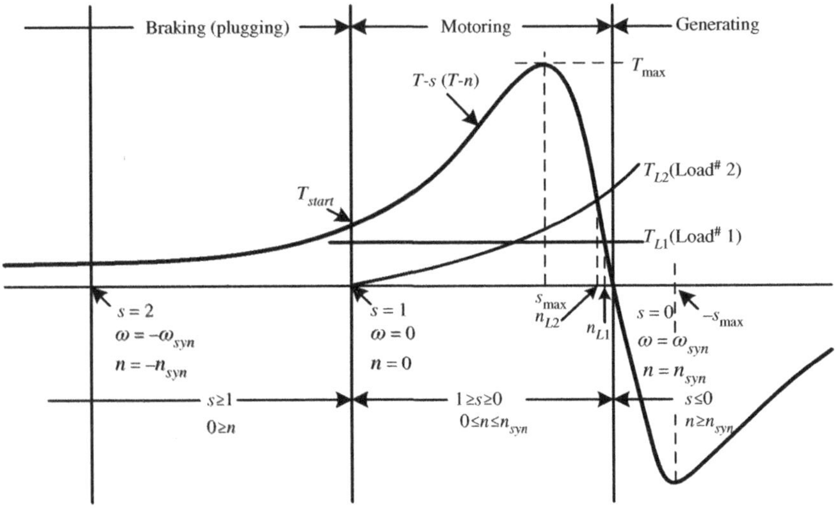

Figure 6.22 The torque–slip/speed characteristic of the induction motor.

torque–slip (speed) profile of Figure 6.22, the reader (student) should show that in the motoring range, $0 \leq s \leq 1$, the maximum developed torque, T_{max}, is given by

$$T_{max} = 3V_{1a}^2/2\omega_{syn}\left[R_{e1} + \sqrt{R_{e1}^2 + (X_{e1} + X_2)^2}\right] \tag{6.80}$$

and the slip at T_{max} and s_{max} is given by

$$s_{max} = R_2/\sqrt{R_{e1}^2 + (X_{e1} + X_2)^2} \tag{6.81}$$

Also, the starting torque, T_{start}, at $s = 1$, that is at a speed, $n = 0$ or $\omega_r = 0$, is given by

$$T_{start} = 3V_{1a}^2R_2/\omega_{syn}\left[(R_{e1} + R_2)^2 + (X_{e1} + X_2)^2\right] \tag{6.82}$$

6.5 On Torque–Slip (Speed) Simplification and Its Formulation

Furthermore, for very large induction motors, where the stator phase winding resistance, R_1, is negligible, hence, R_{e1} is negligible, the student should attempt to show that

$$s_{max} = R_2/(X_{e1} + X_2) \tag{6.83}$$

And, hence,

$$(X_{e1} + X_2) = R_2/s_{max} \tag{6.84}$$

and that for $T_{dev.}$, at any slip, s, the following is valid:

$$T_{dev.}/T_{\max} = 2/\left[\left(s_{\max}/s\right) + \left(s/s_{\max}\right)\right] \tag{6.85}$$

and for the starting torque, T_{start}, at $s = 1$, the following ratio is valid:

$$T_{start}/T_{\max} = 2/\left[s_{\max} + \left(1/s_{\max}\right)\right] \tag{6.86}$$

Equations (6.83)–(6.86) are very useful in obtaining torque–slip (speed) information on torque–slip (speed) characteristics of integral-horsepower induction motors with minimum effort of dynamometer testing of such induction motors. We next discuss induction motor starting transient using the simplified T-equivalent circuit formulation presented in Section 6.4; however, we follow first with some worked examples on induction motor performance and analysis computations.

Example 6.1

A 3-phase, 440 V (line to line), 60 Hz, Y-connected, 8-pole, 100 hp induction motor has the following equivalent circuit parameters expressed in ohms per phase referred to the stator:

$$R_1 = 0.06\,\Omega, R_2 = 0.048\,\Omega, X_1 = X_2 = 0.26\,\Omega, \text{and } X_\phi = 8.47\,\Omega.$$

The combined rotational and core losses are 3600 W at rated shaft output and can be taken constant over the normal operating range. The motor is directly connected to a 440 V, 3ϕ source.

(1.1) Derive the Thevenin equivalent circuit with $\overline{V}_{1a}$, R_{e1}, jX_{e1}, R_2, and X_2, for a slip, s, and draw a schematic of the circuit.

(1.2) For a slip of 2.363% ($s = 0.02363$ p.u.), find

- (1.2.1) The motor speed, in rev/min.
- (1.2.2) The rotor current, developed power, and developed torque.
- (1.2.3) The output shaft power and output shaft torque.
- (1.2.4) The input power, the line current and power factor, as well as overall motor efficiency.

(1.3) Find the starting slip and torque, maximum torque, and slip at maximum torque and draw a schematic of the torque–speed characteristic showing the speed at maximum torque.

Solution

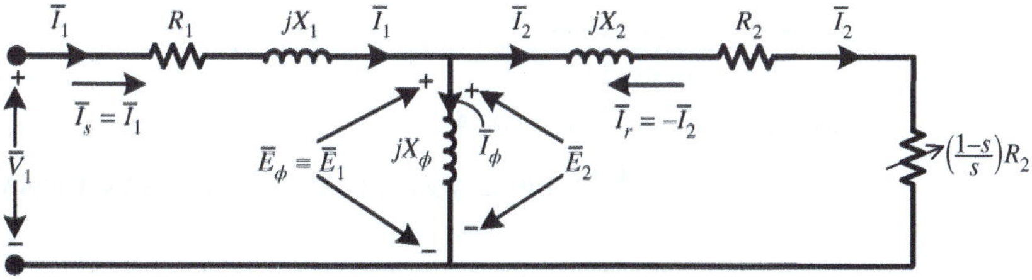

(1)

(1.1) $\overline{V}_{1a} = \frac{jX_\phi}{R_1+j(X_1+X_\phi)}\overline{V}_1, \overline{V}_1 = \frac{440}{\sqrt{3}}\angle 0^\circ \text{ V}, \overline{V}_1 = 254.03\angle 0^\circ \text{ V}$

$$\overline{V}_{1a} = \frac{j8.47\,(254.03\angle 0^\circ)}{0.06 + j\,(0.26 + 8.47)} = \frac{8.47\angle 90^\circ \times 254.03\angle 0^\circ}{8.73\angle 89.61^\circ}$$

$$\overline{V}_{1a} = 246.46\angle 0.39^\circ \text{ V}$$

$$R_{e1} = \frac{R_1X_\phi^2}{R_1^2 + (X_1 + X_\phi)^2} = \frac{0.06 \times (8.47)^2}{(0.06)^2 + (0.26 + 8.47)^2} = 0.0565\Omega$$

$$X_{e1} = \frac{X_\phi\left[R_1^2 + X_1\,(X_1 + X_\phi)\right]}{R_1^2 + (X_1 + X_\phi)^2} = \frac{8.47\left[(0.06)^2 + 0.26\,(0.26 + 8.47)\right]}{(0.06)^2 + (0.26 + 8.47)^2}$$

$$X_{e1} = \frac{8.47 \times 2.2734}{71.2165} = 0.2526\Omega$$

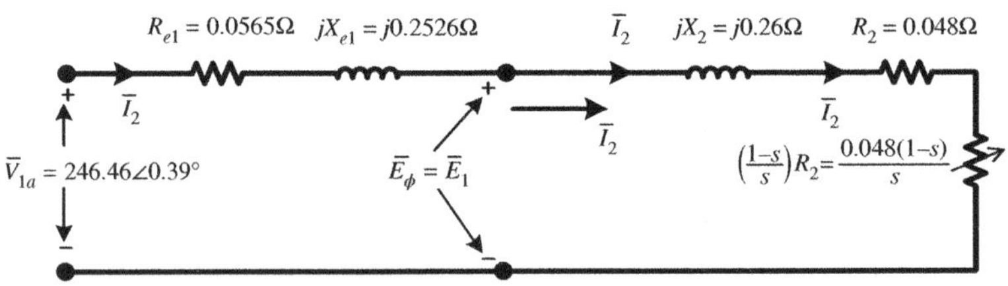

(1.2)

(1.2.1) $n_{syn} = \frac{120f}{p} = \frac{120 \times 60}{8} = 900 \text{ rev/min}$

$$n_r = (1 - s)\,n_{syn} = (1 - 0.02363) \times 900 = 878.7 \text{ rev/min}$$

$$\omega_r = 92.02 \text{ rad/s} \quad \left(\frac{2\pi}{60} \times n_r\right) = \omega_r$$

(1.2.2) $\overline{I}_2 = \frac{246.46\angle 0.39^\circ}{\left(0.0565 + \frac{0.048}{0.02363}\right) + j\,(0.2526 + 0.26)}$

$$= \frac{246.46\angle 0.39^\circ}{(0.056 + 2.0313) + j\,(0.5126)} = \frac{246.46\angle 0.39^\circ}{2.0878 + j0.5126}$$

$$\overline{I}_2 = \frac{246.46\angle 0.39^\circ}{2.1498\angle 13.79^\circ} = 114.64\angle - 13.40^\circ \text{ A}$$

hence, $\overline{I}_r = 114.64\angle - 193.40^\circ$ A

$$P_{dev} = 3I_2^2\left(\frac{1-s}{s}\right)R_2 = 3 \times (114.64)^2 \times 1.9833 = 78,195.5 \text{ W}$$

$$T_{dev} = \frac{P_{dev}}{\omega_r} = \frac{78,195.5}{92.02} = 849.77\ \text{Nm}$$

(1.2.3) $P_{out} = P_{dev} - P_{rotational} = 78,195.5 - 3600 = 74,595.5\ \text{W}$

$$T_{out} = \frac{74595.5}{92.02} = 810.64\ \text{Nm}$$

(1.2.4)

$$\overline{E}_\phi = (R_2/s + jX_2)\,\overline{I}_2 = \left(\frac{0.048}{0.02363} + j0.26\right) \times 114.64\angle - 13.40^\circ$$

$$= (2.0313 + j0.26) \times 114.64\angle - 13.40^\circ$$

$$= 2.0479\angle 7.29^\circ \times 114.64\angle - 13.40^\circ$$

$$= 234.77\angle - 6.11^\circ\ \text{V}$$

$$\overline{I}_\phi = \frac{\overline{E}_\phi}{jX_\phi} = \frac{234.77\angle - 6.11^\circ}{j8.47} = 27.72\angle - 96.11^\circ\ \text{A}$$

$$\overline{\Phi}_{sr} = \Phi_{sr}\angle - 96.11^\circ\ \text{Wb} \quad \overline{F}_{sr} = F_{sr}\angle - 96.11^\circ$$

$$\overline{I}_1 = \overline{I}_2 + \overline{I}_\phi = 114.64\angle - 13.40^\circ + 27.72\angle - 96.11^\circ$$

$$= (111.52 - 2.95) - j\,(26.57 + 27.56)$$

$$= 108.57 - j54.13 = 121.32\angle - 26.50^\circ\ \text{A}$$

hence, $\overline{I}_s = 121.32\angle - 26.50^\circ$ A

P.F. = 0.895 lagging ϕ = P.F.Angle = 26.50°

$$P_{in} = 3I_1V_1\cos\phi = 3 \times 121.32 \times 254.03 \times 0.895 = 82,743.0\ \text{W}$$

$$\eta = \frac{74,595.5}{82,743.0} \times 100\% = 90.15\%$$

(1.3) $s_{start} = 1.0$ p.u.

$$T_{start} = \frac{3I^2_{2(start)}R_2}{\omega_{syn}} = \frac{3I^2_{2(start)} \times 0.048}{94.248} = 1.5279 \times 10^{-3} I^2_{2(start)}$$

$$I_{2(start)} = \frac{(246.46)^2}{\sqrt{(0.0565 + 0.048)^2 + (0.2526 + 0.26)^2}} = 471.11\ \text{A}$$

$$I^2_{2(start)} = \frac{(246.46)^2}{0.27368} = 221,948\ \text{A}^2$$

$$T_{start} = 339.11\ \text{Nm}$$

$$s_{max} = \frac{R_2}{\sqrt{R_e^2 + (X_{e1} + X_2)^2}} = \frac{0.048}{\sqrt{(0.0565)^2 + (0.5126)^2}} = 0.0931 \text{ p.u.}$$

$$T_{max} = \frac{1.5V_{1a}^2}{\omega_{syn}\left[R_{e1} + \sqrt{R_e^2 + (X_{e1} + X_2)^2}\right]} = \frac{1.5(246.46)^2}{94.248\,[0.0565 + 0.5157]}$$

$$= 1689.51 \text{ Nm}$$

Example 6.2

A 3-phase, Y-connected, 480-V (line to line) 30 hp, 60-Hz, 6-pole induction motor has the following equivalent circuit constants in ohms per phase referred to the stator

$$R_1 = 0.24\,\Omega, R_2 = 0.20\,\Omega, X_1 = 1.3\,\Omega, X_2 = 1.2\,\Omega, X_\phi = 37\,\Omega$$

The total friction, windage, and core losses may be assumed constant at 1340 W. The motor is connected directly to a 480-V source.

(2.1) For a rated shaft output power of 30 hp, what is the internal (electromagnetic) power in Watts.

(2.2) Determine the slip and speed for the conditions in (a) using the simplified circuit with R_{e1}, jX_{e1}, jX_2, and R_2/s connected in series across $\overline{V}_{1a}$.

(2.3) Determine the internal and output torques in Newton-meters.

(2.4) Determine the phase current, terminal power factor, and input power, as well as efficiency.

(2) A 3-phase, Y-connected, 30-hp, 480-V, 60-Hz, 6-pole induction motor whose equivalent circuit is as follows:

Solution

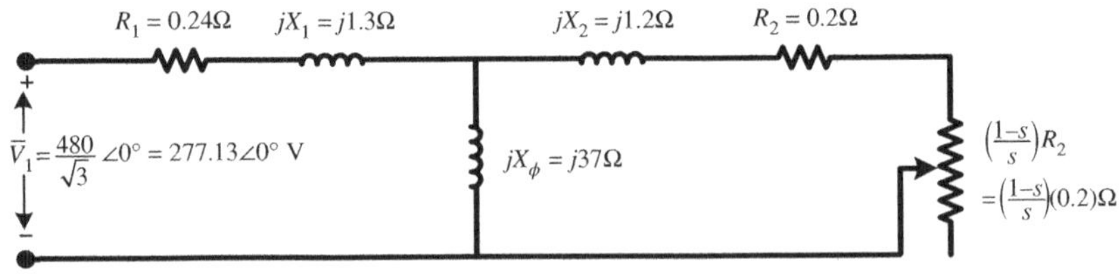

(2.1) This equivalent circuit can be reduced to the simpler series branch type as follows (the Thevenin equivalent):

$$\overline{V}_{1a} = \frac{jX_\phi \overline{V}_1}{R_1 + j(X_1 + X_\phi)} = \frac{j37 \times 277.13\angle 0^\circ}{0.24 + j(1.3 + 37)} = \frac{37\angle 90^\circ \times 277.13\angle 0^\circ}{38.3\angle 89.64^\circ}$$

hence, $\overline{V}_{1a} = 267.72\angle 0.26°$ V

$$R_{e1} = \frac{R_1 X_\phi^2}{R_1^2 + (X_1 + X_\phi)^2} = \frac{0.24 \times (37)^2}{(0.24)^2 + (38.3)^2} = 0.224\,\Omega$$

$$jX_{e1} = j\frac{X_\phi \left[R_1^2 + X_1 (X_1 + X_\phi)\right]}{R_1^2 + (X_1 + X_\phi)^2} = j\frac{37\left[(0.24)^2 + 1.3 \times 38.3\right]}{(0.24)^2 + (38.3)^2} = j1.257\,\Omega$$

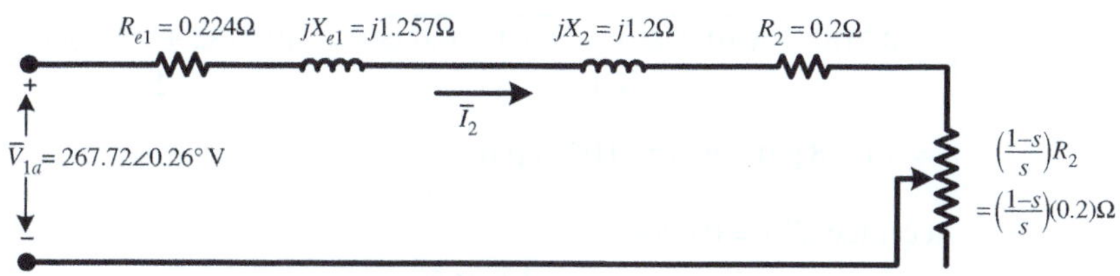

(2.2) $\overline{V}_{1a} = \frac{jX_\phi \overline{V}_1}{R_1 + j(X_1 + X_\phi)} = 267.72\angle 0.26°$ V

$$\overline{I}_2 = \frac{\overline{V}_{1a}}{(R_{e1} + R_2/s) + j(X_{e1} + X_2)} = \frac{267.72\angle 0.26°}{(0.224 + 0.2/s) + j(1.257 + 1.2)}$$

$$\overline{I}_2 = \frac{267.72}{\sqrt{(0.224 + 0.2/s)^2 + (2.457)^2}}\angle\left[0.26° - \tan^{-1}(2.457/(0.224 + 0.2/s))\right]$$

hence, $I_2 = \frac{267.72}{\sqrt{(0.224+0.2/s)^2+(2.457)^2}}, \theta_2 = 0.26° - \tan^{-1}\left(\frac{2.457}{0.224+0.2/s}\right)$

(2.3) $P_{Grossmech} = P_{\text{int}} = 3I_2^2\left(\frac{1-s}{s}\right)R_2$

$$P_{\text{int}} = \frac{3 \times (267.72)^2 \times 0.2\,(1-s)}{s\left[(0.224 + 0.2/s)^2 + (2.457)^2\right]}$$

$$P_{\text{int}} = \frac{43004.4s\,(1-s)}{(0.224s + 0.2)^2 + (2.457)^2 s^2} = \frac{43004.4s\,(1-s)}{6.08703s^2 + 0.0896s + 0.04}$$

$$P_{\text{int}} = \frac{43004.4s\,(1-s)}{6.08703s^2 + 0.0896s + 0.04}$$

(2.4) $P_{\text{int}} = P_{out} + P_{RotationalLosses}$
$P_{\text{int}} = 30 \times 746 + 1340 = 23720$ Watts

(2.5) From (c) and (d), at rated conditions, we must have

$$P_{\text{int}} = 23,720 = \frac{43,004.4s\,(1-s)}{6.08703s^2 + 0.08965s + 0.04}$$

$$\text{hence, } s(1-s) = \frac{23,720}{43,004.4}\left(6.08703s^2 + 0.08965s + 0.04\right)$$

$$s - s^2 = 3.3574s^2 + 0.04942s + 0.02206$$

$$4.3574s^2 - 0.95058s + 0.02206 = 0$$

$$s^2 - 0.2182s + 0.00506 = 0$$

$$s = \frac{-b \pm \sqrt{b^2 - 4ac}}{2a}$$

$$s = \frac{0.2182 \pm \sqrt{(0.2182)^2 - 4 \times 1 \times 0.00506}}{2 \times 1} = \frac{(0.2182 \pm 0.16544)}{2}$$

$$s = 0.1918\,\text{p.u.} \quad \text{or} \quad s = 0.0264\,\text{p.u.}$$

Accepted slip = 0.0264 p.u.

$$\text{hence, } n = (1-s)\,n_{syn} = (1 - 0.0264) \times \frac{120 \times 60}{6} = 1168.3\,\text{r.p.m}$$

$$\omega = \frac{2\pi}{60} \times n = \frac{2\pi}{60} \times 1168.3 = 122.344\,\text{rad/s.}$$

(2.6)

$$T_{\text{int}} = \frac{P_{\text{int}}}{\omega} = \frac{23,720}{122.344} = 193.88\,\text{Nm}$$

$$T_{out} = \frac{23,720 - 1340}{122.344} = 182.93\,\text{Nm}$$

$$\overline{I}_2 = \frac{267.72\angle 0.26^\circ}{\left(0.224 + \dfrac{0.2}{0.0264}\right) + j2.457} = \frac{267.72\angle 0.26^\circ}{8.178\angle 17.49^\circ} = 32.737\angle -17.23^\circ$$

$$\overline{I}_r = -\overline{I}_2$$

$$\text{hence, } \overline{I}_r = 32.737\angle 162.77^\circ\ \text{A}$$

$$\text{hence, } P_{in} = 3 \times 267.72 \times 32.737 \times 0.9538 = 25,078.31\ \text{W}$$

$$\text{hence, } \eta = \frac{P_{out}}{P_{in}} = \frac{30 \times 746}{25078.31} = 0.8924\,\text{p.u.} = 89.24\%$$

$$\overline{I}_2 = 32.737\angle -17.23^\circ\ \text{A}$$

$$\overline{E}_\phi = \left[(R_2/s) + jX_2\right]\overline{I}_2 = \left(\frac{0.2}{0.0264} + j1.2\right)\left(32.737\angle -17.27^\circ\right)\ \text{V}$$

$$\overline{E}_\phi = (7.57575 + j1.2)\left(32.737\angle -17.27^\circ\right) = \left(7.670\angle 9.0^\circ\right)\left(32.737\angle -17.27^\circ\right)$$

$$\overline{E}_\phi = 251.093\angle -8.27^\circ\ \text{V}$$

$$\overline{I}_\phi = \overline{E}_\phi / jX_\phi = 251.093\angle -8.27^\circ / j37 = 6.786\angle -98.27^\circ\ \text{A}$$

$$\overline{I}_1 = \overline{I}_\phi + \overline{I}_2 = 6.786\angle -98.27^\circ + 32.737\angle -17.27^\circ\ \text{A}$$

$$\bar{I}_1 = -0.976 - j6.715 + 31.261 - j9.719 \text{ A}$$

$$\bar{I}_1 = (31.261 - 0.976) - j(6.715 + 9.719) \text{ A}$$

$$\bar{I}_1 = 30.285 - j16.434 = 34.457\angle - 28.49° \text{ A}$$

$$P_{in} = \text{Re}\left(3\bar{I}_1^*\bar{V}_1\right) = \text{Re}(3(30.285 - j16.434)(277.13 + j0.0))$$

$$P_{in} = 3 \times 30.285 \times 277.13 = 25,178.65 \text{ W}$$

$$\text{P.F.} = \cos\left(-28.49°\right) = 0.879 \text{ lagging}$$

$$P_{out} = 30 \times 746 = 22,380 \text{ W}$$

$$\eta = \frac{P_{out}}{P_{in}} = \frac{22380}{25178.65} = 0.88885 \text{ p.u.}$$

$$\eta = 88.885\%$$

Hence, the phasor-space vector diagram is as follows:

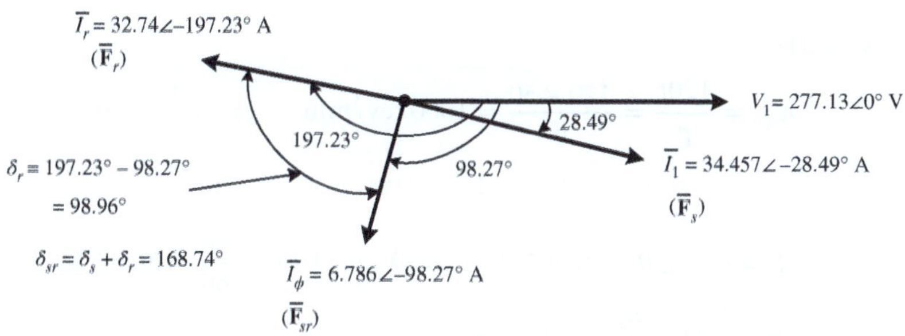

Example 6.3

(3) In an AC ASD, the motor was a 3-phase, 6-pole, squirrel-cage induction type, rated at 335 hp for an operating frequency of 60 Hz and terminal voltage of 2000 V (line to line). The motor is Y-connected and has the following equivalent circuit parameters per phase referred to the stator side at 60 Hz:

$$R_1 = 0.2\Omega, jX_1 = jX_2 = j0.707\Omega$$
$$R_2 = 0.203\Omega, jX_\phi = j77\Omega$$

and its total rotational losses (core, friction, and windage) can be taken constant and equal to 14,000 Watts. The motor was operating at a slightly overload condition at a slip, $s = 0.015$ p.u. at 60 Hz. It was decided to reduce its speed by reducing the supply frequency to 50 Hz while maintaining the same maximum (stalling) torque and operating at the same slip, $s = 0.015$ p.u.:

(3.1) What is the motor's operating speed, n in rev/min? What are its circuit parameters under this condition, including the terminal line-to-line and line-to-neutral voltages?
(3.2) Obtain the Thevenin equivalent circuit at a slip, $s = 0.015$ p.u.
(3.3) Under the new voltage and frequency conditions, for an $s = 0.015$ p.u., calculate the rotor current (referred to the stator), the developed mechanical power in Watts, and the developed torque in Nm.
(3.4) Calculate the output power in Watts and hp (1 hp = 764 W) and the output torque in Nm.
(3.5) Calculate the stator current and terminal power factor, as well as input power and motor efficiency.

Solution

(3) At 60 Hz: $R_1 = 0.2\,\Omega$, $R_2 = 0.203\,\Omega$, $jX_1 = jX_2 = j0.707\,\Omega$, $jX_\phi = j77\,\Omega$

$$p = 6 \text{ pole}, V_{L-L} = 2000 \text{ V}$$

(3.1) At 50 Hz:

$$n_{syn} = \frac{120f}{p} = \frac{120 \times 50}{6} = 1000 \text{ rev/min}$$

$$n = (1 - s)\, n_{syn} = (1 - 0.015) \times 1000 = 985 \text{ rev/min}$$

$$R_1 = 0.2\,\Omega, R_2 = 0.203\,\Omega, jX_1 = jX_2 = j0.707 \times \frac{50}{60} = j0.589\,\Omega$$

$$jX_\phi = j77 \times \frac{50}{60} = j64.167\,\Omega$$

$$V_{L-L} = \frac{50}{60} \times 2000 = 1666.67 \text{ V}$$

$$V_{ph} = \frac{1666.67}{\sqrt{3}} = 962.25 \text{ V}$$

(3.2) $$\overline{V}_{1a} = \frac{jX_{\phi 1}}{R_1 + j(X_1 + X_\phi)}\overline{V}_1, R_{e1} = \frac{R_1 X_\phi^2}{R_1^2 + (X_1 + X_\phi)^2}$$

$$jX_{e1} = j\frac{X_\phi \left[R_1^2 + X_1 (X_1 + X_\phi)\right]}{R_1^2 + (X_1 + X_\phi)^2}$$

$$\overline{V}_1 = 962.25\angle 0^\circ \text{ V}$$

$$\overline{V}_{1a} = \frac{j64.167 \times 962.25\angle 0^\circ}{0.2 + j(0.589 + 64.167)} = \frac{j61,744.69}{64.756\angle 89.82^\circ}$$

$$\overline{V}_{1a} = 953.5\angle 0.18^\circ \text{ V}$$

$$R_{e1} = \frac{0.2 \times (64.167)^2}{0.2^2 + (0.589 + 64.167)^2} = \frac{0.2 \times (64.167)^2}{4193.34} = 0.1964\,\Omega$$

$$jX_{e1} = j\frac{64.167\left[0.2^2 + 0.589 \times (0.589 + 64.167)\right]}{0.2^2 + (0.589 + 64.167)^2} = j\frac{64.167 \times 38.181}{4193.34}$$

$$jX_{e1} = j0.5843\,\Omega$$

$$\left(\frac{1-s}{s}\right) R_2 = \frac{0.985}{0.015} \times 0.203 = 13.330\,\Omega$$

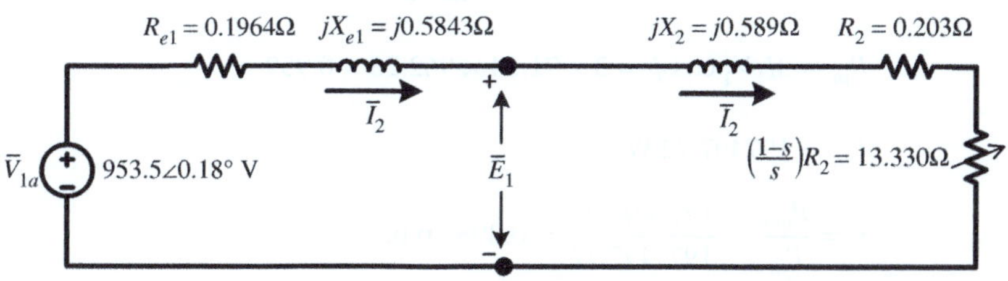

Thev. Equiv. AT $s = 0.015$ p.u.

(3.3) $$\overline{I}_2 = \frac{953.5\angle 0.18^\circ}{(0.1964 + 0.203 + 13.33) + j(0.5843 + 0.589)} = \frac{953.5\angle 0.18^\circ}{13.729 + j1.173}$$

$$\overline{I}_2 = \frac{953.5\angle 0.18^\circ}{13.779\angle 4.88^\circ} = 69.20\angle -4.70^\circ \text{ A}$$

$$P_{dev} = 3I_2^2\left(\frac{1-s}{s}\right) R_2 = 3 \times (69.20)^2 \times 13.33 = 191,497.71 \text{ W}$$

$$\omega_m = \frac{2\pi}{60} \times n = \frac{2\pi}{60} \times 985 = 103.149 \text{ rad/s}$$

$$T_{dev} = \frac{P_{dev}}{\omega_m} = \frac{191,497.71}{103.149} = 1865.5 \text{ Nm}$$

(3.4) $$P_{out} = P_{dev} - P_{rotational} = 191,497.71 - 14,000 = 177,497.71 \text{ W}$$

$$P_{out} = \frac{177,497.71}{746} = 237.93 \text{ hp}$$

$$T_{out} = \frac{177,497.71}{103.149} = 1720.8 \text{ Nm}$$

(3.5) $\overline{E}_1 = ((13.33 + 0.203) + j0.589) \times 69.20\angle - 4.70^\circ$

$$\overline{E}_1 = 13.546\angle 2.49^\circ \times 69.20\angle - 4.70^\circ = 937.38\angle - 2.21^\circ \text{ V}$$

$$\overline{I}_\phi = \frac{\overline{E}_1}{jX_\phi} = \frac{937.38\angle - 2.21^\circ}{j64.167} = 14.608\angle - 92.21^\circ \text{ A}$$

$$\overline{I}_1 = \overline{I}_2 + \overline{I}_\phi = 69.20\angle - 4.70^\circ + 14.608\angle - 92.21^\circ$$
$$= 68.967 - j5.670 - 0.563 - j14.597$$

$$\overline{I}_1 = 68.404 - j20.267 = 71.34\angle - 16.50^\circ \text{ A}$$

$$\text{PF} = \cos\left(-16.50^\circ\right) = 0.959 \text{ lagging}$$

$$P_{in} = 3I_1V_1\cos\phi = 3 \times 71.34 \times 962.25 \times 0.959$$

$$P_{in} = 197497.72 \text{ W}$$

$$\eta = \frac{P_{out}}{P_{in}} = \frac{177,497.71}{197,497.72} = 0.8987 \text{ p.u.}$$

$$\eta = 89.87\%$$

Hence, the phasor-space vector diagram is shown as follows:

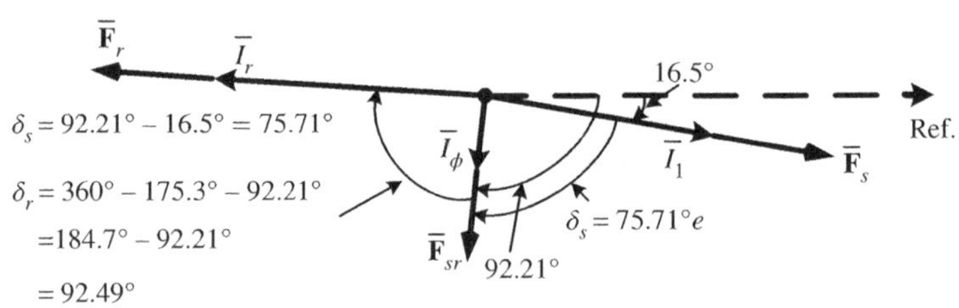

Example 6.4

(4) A 3-phase, 440 V (line to line), 60 Hz, Y-connected, 8-pole, 100 hp induction motor has the following equivalent circuit parameters expressed in ohms per phase referred to the stator:

$$R_1 = 0.06\,\Omega, R_2 = 0.048\,\Omega, X_1 = X_2 = 0.26\,\Omega, \text{and } X_\phi = 8.47\,\Omega$$

The combined rotational and core losses are 3600 W at rated shaft output and can be taken constant over the normal operating range of this motor. The motor is directly connected to a 440 V, 3-phase source.

(4.1) Drive the Thevenin equivalent circuit with V_{1a}, R_{e1}, jX_{e1}, R_2, and jX_2 for a slip, s, and draw a schematic of the circuit.

(4.2) For a slip, $s = 0.02363$ p.u., find: 1.2.1) the motor speed in rev/min; 1.2.2) the rotor current, developed power and developed torque; 1.2.3) the shaft output power and output shaft torque, and 1.2.4) the input power, the line current, and power factor, as well as motor efficiency.

Solution

(4.1) $\overline{V}_{1a} = \frac{jX_\phi}{R_1 + j(X_1 + X_\phi)} \overline{V}_1$

$$\overline{V}_1 = \left(440/\sqrt{3}\right) \angle 0° \text{ V} = 254.03\angle 0° \text{ V}$$

$$\overline{V}_{1a} = \frac{j8.47\,(254.03\angle 0°)}{0.06 + j\,(0.26 + 8.47)} = 246.46\angle 0.39° \text{ V}$$

$$R_{e1} = \frac{R_1 X_\phi^2}{R_1^2 + (X_1 + X_\phi)^2} = \frac{0.06 \times (8.47)^2}{(0.06)^2 + (0.26 + 8.47)^2} = 0.0565\ \Omega$$

$$X_{e1} = \frac{X_\phi\left[R_1^2 + X_1\,(X_1 + X_\phi)\right]}{R_1^2 + (X_1 + X_\phi)^2}$$

$$X_{e1} = \frac{8.47\left[(0.06)^2 + 0.26\,(0.26 + 8.47)\right]}{(0.06)^2 + (0.26 + 8.47)^2} = 0.2526\ \Omega$$

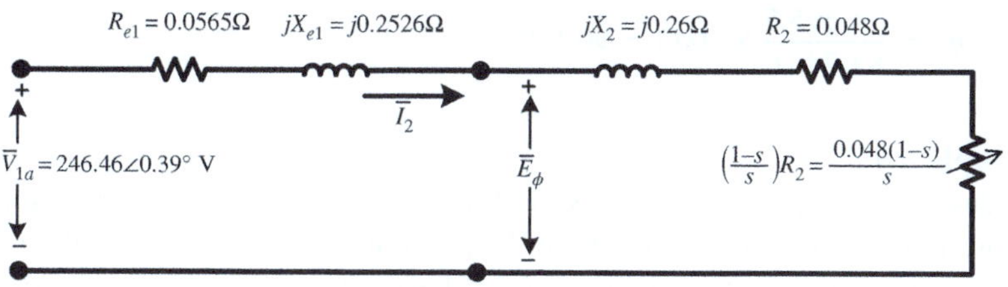

(4.2) $n_{syn} = \frac{120f}{p} = \frac{120 \times 60}{8} = 900$ rev/min

$$n_r = (1 - s)\,n_{syn} = (1 - 0.02363) \times 900 = 878.7 \text{ rev/min}$$

$$\overline{I}_2 = \frac{246.46\angle 0.39°}{(0.056 + 2.0313) + j\,(0.5126)}$$

$$\bar{I}_2 = \frac{246.46\angle 0.39^\circ}{2.14\,98\angle 13.79^\circ} = 114.64\angle -13.40^\circ \text{ A}$$

$$P_{dev} = 3I_2^2\left(\frac{1-s}{s}\right)R_2 = 3 \times (114.64)^2 \times 1.9833 = 78,195.5 \text{ W}$$

$$T_{dev} = \frac{P_{dev}}{\omega_r} = \frac{78,195.5}{92.02} = 849.77 \text{ Nm}$$

$$P_{out} = P_{dev} - P_{rotational} = 78,195.5 - 3600 = 74,595.5 \text{ W}$$

$$T_{out} = \frac{74,595.5}{92.02} = 810.64 \text{ Nm}$$

$$\bar{E}_\phi = \left(\frac{R_2}{s} + jX_2\right)\bar{I}_2 = \left(\frac{0.048}{0.02363} + j0.26\right) \times 114.64\angle -13.40^\circ$$
$$= 234.77\angle -6.11^\circ \text{ V}$$

$$\bar{I}_\phi = \frac{\bar{E}_\phi}{jX_\phi} = \frac{234.77\angle -6.11^\circ}{j8.47} = 27.72\angle -96.11^\circ \text{ A}$$

$$\bar{I}_1 = \bar{I}_2 + \bar{I}_\phi = 114.64\angle -13.40^\circ + 27.72\angle -96.11^\circ$$

$$\bar{I}_1 = 121.32\angle -26.50^\circ \text{ A}$$

$$\text{P.F. Angle} = -26.50^\circ = \theta$$

hence, P. F. $= \cos(-26.50^\circ) = 0.895$ lagging

$$P_{in} = 3I_1V_1\cos\theta = 3 \times 121.32 \times 254.03 \times 0.895$$

$$P_{in} = 82,743.0 \text{ W}$$

$$\eta_{\%} = \frac{74,595.5}{82,743.0} \times 100\% = 90.15\%$$

Example 6.5

The design of a 4-pole, 3-phase, 480 V (line to line), 60 Hz induction motor is to be based on a stator inner diameter of 9 in., for an airgap length of 0.012 in., and stator core length of 12 in. (1.0 in. = 0.0254 m). The stator winding distribution which has been chosen has a winding factor, $k_w = 0.92$. The stator winding is to be Y-connected. The designer must pick the number of armature turns per phase so that the peak fundamental flux density at mid-airgap is $1.2T$.

(5.1) Calculate the resultant fundamental flux per pole and the number of turns per phase. Also, calculate the peak value (amplitude) of resultant MMF, F_{sr}, on AT/pole.

(5.2) If under these conditions, the torque angle, $\delta_s = 120°$, and $\delta_{sr} = 150°$, find F_s and F_r in AT/pole ($F_s = F_{stator}$ and $F_r = F_{rotor}$) and calculate the stator phase current as well as the developed torque in Nm. Draw the MMF vector diagram showing δ_s, δ_r, and δ_{sr}, for a counterclockwise rotation. Calculate the approximate value of developed power in hp and in kW under such conditions.

Solution

$$L = 12 \text{ in.} = 12 \times 0.0254 = 0.3048 \text{ m}$$

$$D = 9 \text{ in.} = 9 \times 0.0254 = 0.2286 \text{ m}$$

$$\tau_p = \frac{\pi D}{p} = \frac{\pi \times 0.2286}{4} = 0.1795 \text{ m}$$

$$h_g = 0.012 \times 0.0254 = 3.048 \times 10^{-4} \text{ m}$$

$$E_{ph} = \frac{480}{\sqrt{3}} = 277.13 \text{ V}$$

(5.1) $\Phi_{\text{Res}} = \frac{2}{\pi}\tau_p L B_{\max} = \frac{2}{\pi} \times 0.1795 \times 0.3048 \times 1.2$

$$\Phi_{\text{Res}} = 0.0418 \text{ Wb}$$

$$E_{ph} = 4.44\, f N_{ph} k_w \Phi_{\text{Res}}$$

$$N_{ph} = \frac{E_{ph}}{4.44\, f k_w \Phi_{\text{Res}}} = \frac{277.13}{4.44 \times 60 \times 0.92 \times 0.0418}$$

$$N_{ph} = 27.05 \text{ Turns} \Rightarrow N_{ph} = 27 \text{ Turns/ph}$$

(5.2) Hence, the space vector diagram is shown as follows:

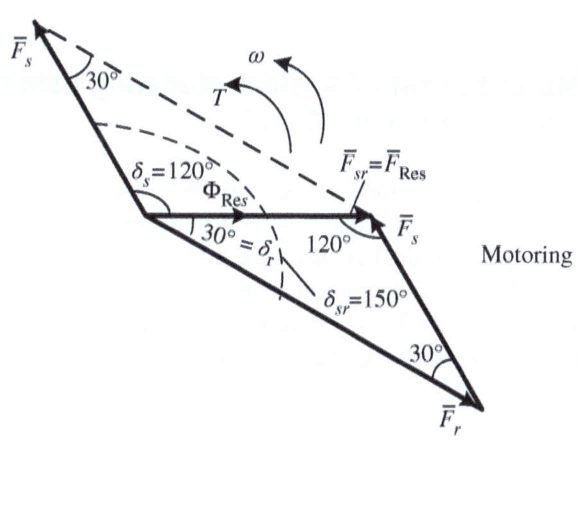

$$F_{sr} = F_{\text{Res}} = \frac{B_{\max}}{\mu_0} hg$$

$$F_{sr} = \frac{1.2 \times 3.048 \times 10^{-4}}{4\pi \times 10^{-7}} = 291.06\ \text{AT/pole}$$

$$F_r = \frac{F_{sr}}{\sin 30^\circ} \times \sin 120^\circ = \frac{F_{sr}}{0.5} \times \frac{\sqrt{3}}{2}$$

$$F_r = F_{sr} \times \sqrt{3} = 291.06 \times \sqrt{3} = 504.13\ \text{AT/pole}$$

$$F_r = \frac{F_{sr}}{\sin 30^\circ} \times \sin 30^\circ = F_{sr} = 291.06\ \text{AT/pole}$$

However,

$$F_s = \frac{3}{2}\left[\frac{4}{\pi}\left(\frac{N_{ph}}{p}\right) k_w \left(\sqrt{2} I_a\right)\right]$$

$$291.06 = \frac{3}{2} \times \frac{4}{\pi} \times \frac{27}{3} \times 0.92 \times \sqrt{2} I_a$$

hence, $I_a = \dfrac{2\pi \times 291.06}{81 \times 0.92\sqrt{2}} = 17.35\ \text{A}$

$$T = \frac{\pi}{2}(p/2)^2 \Phi_{\text{Res}} F_r \sin \delta_r$$

$$= \frac{\pi}{2}\left(\frac{4}{2}\right)^2 \times 0.0418 \times 504.13 \times \sin 30^\circ$$

$$T = 66.21\ \text{Nm}$$

$$P_{dev} \simeq \frac{T\omega_{syn}}{746} = \frac{66.21 \times \frac{2\pi}{60} \times 1800}{746}$$

$$\simeq 16.80\ \text{HP} \quad (12.53\ \text{kW})$$

6.6 The Motor Starting Transient Modeling Using the Simplified *T*-Equivalent Circuit

Based on earlier developments in Sections 6.4 and 6.5, for a wound-rotor induction machine, with access to the rotor phase windings through slip rings and brushes, the following poly-phase induction machine equivalent circuit with all circuit parameters referred to the stator side in ohms per phase, see Figure 6.23. Also, shown in Figure 6.23 is the Thevenin equivalent circuit including the possible presence of an external resistance, R_x, added in series in each phase of the rotor circuit.

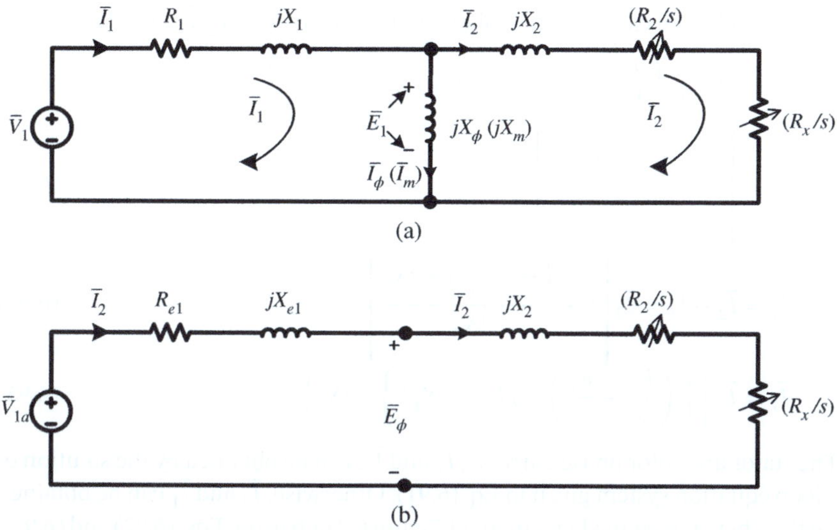

Figure 6.23 The induction motor equivalent circuit. (a) All parameters referred to stator side (R_x is an external resistance in a wound-rotor case). (b) Thevenin equivalent circuit.

Here, from Thevenin's theorem and Section 6.1, we have:

$$\bar{V}_{1a} = \frac{jX_\phi}{R_1 + j\left(X_1 + X_\phi\right)} \bar{V}_1 \tag{6.87}$$

$$R_{e1} = \left[\frac{R_1 X_\phi^2}{R_1^2 + \left(X_1 + X_\phi\right)^2} \right] \tag{6.88}$$

and

$$jX_{e1} = j \left[\frac{X_\phi \left[R_1^2 + X_1\left(X_1 + X_\phi\right)\right]}{R_1^2 + \left(X_1 + X_\phi\right)^2} \right] \tag{6.89}$$

Notice that in all the developments to follow, s is the slip in per unit:

$$s = \left(\omega_{syn} - \omega_{rm}\right)/\omega_{syn} = \left(n_{syn} - n_{rm}\right)/n_{syn} \tag{6.90}$$

Here, the two-loop equations of the circuit in part (a) of Figure 6.23 can be written in the matrix form as follows:

$$\begin{bmatrix} \left[R_1 + j\left(X_1 + X_\phi\right)\right] & \left[-jX_\phi\right] \\ \left[-jX_\phi\right] & \left[\left(R/s + R_x/s\right) + j\left(X_2 + X_\phi\right)\right] \end{bmatrix} \begin{bmatrix} \bar{I}_1 \\ \bar{I}_2 \end{bmatrix} = \begin{bmatrix} \bar{V}_1 \\ 0 \end{bmatrix} \tag{6.91}$$

From which, one can write the following for the current, $\overline{I}_2$, $\overline{I}_\phi$, and $\overline{I}_1$:

$$\overline{I}_2 = \left[\overline{V}_{1a}/\left\{\left(R_{e1} + \frac{R_2}{s} + \frac{R_x}{s}\right) + j\left(X_{e1} + X_2\right)\right\}\right] \tag{6.92}$$

$$\overline{I}_\phi = \left[\frac{\left(\frac{R_2}{s} + \frac{R_x}{s}\right) + jX_2}{jX_\phi}\right]\overline{I}_2 \tag{6.93}$$

$$\overline{I}_1 = \overline{I}_2 + \overline{I}_\phi = \overline{I}_2\left[1 + \frac{\left(\frac{R_2}{s} + \frac{R_x}{s}\right) + jX_2}{jX_\phi}\right] \tag{6.94}$$

$$\overline{I}_1 = \overline{I}_2\left[\left\{\left(\frac{R_2}{s} + \frac{R_x}{s}\right) + j\left(X_2 + X_\phi\right)\right\}/jX_\phi\right] \tag{6.95}$$

The stator and rotor phase currents, $\overline{I}_1$ and $\overline{I}_2$, can be obtained by the solution of the loop equation system given in Eq. (6.91). Otherwise, $\overline{I}_2$ and $\overline{I}_1$ can be obtained from the Thevenin equivalent circuit of Figure 6.23a,b using Eqs. (6.92) and (6.93), respectively. Notice that from Eqs. (6.93) and (6.94), one can write the following:

$$\begin{aligned}\overline{I}_1 = &\left[\left\{\left(\frac{R_2}{s} + \frac{R_x}{s}\right) + j\left(X_2 + X_\phi\right)\right\}\overline{V}_{1a}/\left(jX_\phi\right)\right.\\ &\left.\times\left\{\left(R_{e1} + \frac{R_2}{s} + \frac{R_x}{s}\right) + j\left(X_{e1} + X_2\right)\right\}\right]\end{aligned} \tag{6.96}$$

For the various power components in an induction motor, one should recall the following:

(1) The input electric power, P_{in}, in Watts, is given by

$$P_{in} = \text{Re}\left(3\overline{I}_1^*\overline{V}_1\right) = 3\left[I_1^2R_1 + I_2^2\left(\frac{R_2}{s} + \frac{R_x}{s}\right)\right] \tag{6.97}$$

(2) The stator ohmic power loss, P_{1ohmic}, in Watts, is

$$P_{1ohmic} = 3I_1^2R_1 \tag{6.98}$$

(3) The airgap (electromagnetic) power, P_{em}, flowing from the stator windings into the rotor windings, in Watts, is

$$P_{em} = 3I_2^2\left(\frac{R_2}{s} + \frac{R_x}{s}\right) \tag{6.99}$$

(4) The rotor ohmic power loss, P_{2ohmic}, in Watts, is

$$P_{2ohmic} = 3I_2^2 \left(R_2 + R_x\right) \tag{6.100}$$

(5) The developed mechanical power, P_{dev}, in Watts, is

$$P_{dev} = P_{em} - P_{2ohmic} = 3I_2^2 \left(\frac{R_2}{s} + \frac{R_x}{s}\right)(1 - s) \tag{6.101}$$

(6) The output mechanical power, P_{out}, in Watts, is

$$P_{out} = P_{dev} - P_{RL} = 3I_2^2 \left(\frac{R_2}{s} + \frac{R_x}{s}\right)(1 - s) - P_{RL} \tag{6.102}$$

where P_{RL} is the motor's rotational losses (friction plus windage plus core losses), which usually can be expressed as a second-order function of the rotor mechanical speed, ω_{rm} (mech. rad/s), as follows:

$$P_{RL} = P_{RL}\omega_{rm}^2 \tag{6.103}$$

where ω_{rm} is the rotor speed in mechanical radians per second. Hence,

$$P_{out} = 3I_2^2 \left(\frac{R_2}{s} + \frac{R_x}{s}\right)(1 - s) - K_{RL}\omega_{rm}^2 \tag{6.104}$$

where K_{RL} is a rotational loss constant in Watts/rad^2/s^2.

Therefore, the motor's efficiency, η_{motor}, is

$$\eta_{motor} = \frac{P_{out}}{P_{in}} \text{ pu} = \frac{3I_2^2\left(R_2 + R_x\right)\{(1-s)/s\} - K_{RL}\omega_{rm}^2}{\text{Re}\left(3\overline{I}_1^*\overline{V}_1\right)} \times 100\% \tag{6.105}$$

Now, for the torque components in an induction motor, one can write the following.

(1) The developed mechanical (electromechanical) torque, T_{dev}, in Nm, is

$$T_{dev} = \frac{P_{dev}}{\omega_{rm}} = \left[3I_2^2(1 - s)\left(\frac{R_2}{s} + \frac{R_x}{s}\right)\right] / \omega_{rm} \tag{6.106}$$

but

$$\omega_{rm} = (1 - s)\,\omega_{syn} \tag{6.107}$$

where ω_{rm} and ω_{syn} are in mech. rad/s, hence,

$$T_{dev} = \left[3I_2^2\left(\frac{R_2}{s} + \frac{R_x}{s}\right)\right] / \omega_{syn} \tag{6.108}$$

Substituting from Eq. (6.92) into Eq. (6.108), one can write the following for T_{dev}:

$$T_{dev} = \frac{3V_{1a}^2 \left(R_2 + R_x\right)/s}{\omega_{syn}\left[\left(R_{e1} + \left(\frac{R_2+R_x}{s}\right)\right)^2 + \left(X_{e1} + X_2\right)^2\right]} \tag{6.109}$$

which reduces to the following:

$$T_{dev} = \frac{3V_{1a}^2 s\left(R_2 + R_x\right)}{\omega_{syn}\left[\left(sR_{e1} + R_2 + R_x\right)^2 + s^2\left(X_{e1} + X_2\right)^2\right]} \tag{6.110}$$

Here, recall that for maximum developed torque value, $T_{dev\ max}$, one must have the corresponding slip, $s_{\max}$, be

$$s_{\max} = \frac{R_2 + R_x}{\sqrt{R_{e1}^2 + \left(X_{e1} + X_2\right)^2}} \tag{6.111}$$

and

$$T_{dev\ \max} = \frac{3V_{1a}^2}{2\omega_{syn}\left[R_{e1} + \sqrt{R_{e1}^2 + \left(X_{e1} + X_2\right)^2}\right]} \tag{6.112}$$

Also, the developed starting torque, at $s = 1$, is

$$T_{dev\ start} = \frac{3V_{1a}^2 \left(R_2 + R_x\right)}{\omega_{syn}\left[\left(R_{e1} + R_2 + R_x\right)^2 + \left(X_{e1} + X_2\right)^2\right]} \tag{6.113}$$

Meanwhile, from Eq. (6.96), the starting current, $\bar{I}_{1start}$, in amperes, becomes

$$\bar{I}_{1start} = \left[\left\{\left(R_2 + R_x\right) + j\left(X_2 + X_\phi\right)\right\}\bar{V}_{1a}/\left(jX_\phi\right)\left\{\left(R_{e1} + R_2 + R_x\right) + j\left(X_{e1} + X_2\right)\right\}\right] \tag{6.114}$$

That is,

$$I_{1start} = V_{1a}\sqrt{\left(R_2 + R_x\right)^2 + \left(X_2 + X_\phi\right)^2}/X_\phi\sqrt{\left(R_{e1} + R_2 + R_x\right)^2 + \left(X_{e1} + X_2\right)^2} \tag{6.115}$$

(2) The rotational loss torque, T_{RL}, in Nm, from Eq. (6.103) is

$$T_{RL} = P_{RL}/\omega_{rm} = K_{RL}\omega_{rm}^2/\omega_{rm} = K_{RL}\omega_{rm} = K_{RL}\left(1 - s\right)\omega_{syn} \tag{6.116}$$

(3) Accordingly, the output torque, T_{out}, can be written as follows:

$$T_{out} = T_{dev} - T_{RL} \tag{6.117}$$

And from Eqs. (6.110) and (6.116), T_{out}, in Nm, is

$$T_{out} = \frac{3V_{1a}^2 s\left(R_2 + R_x\right)}{\omega_{syn}\left[\left(sR_{e1} + R_2 + R_x\right)^2 + s^2\left(X_{e1} + X_2\right)^2\right]} - K_{RL}\left(1 - s\right)\omega_{syn} \tag{6.118}$$

Again, recall Eq. (6.107):

$$\omega_{rm} = (1 - s)\,\omega_{syn}$$

Here is a verification example to assist the readers' understanding of the aforementioned principles:

Consider a 460-V, 100-hp, Y-connected (stator and rotor windings), 60 Hz, 4-pole, wound rotor induction motor, see Reference [10], whose stator to rotor turns ratio, $(N_S/N_R) = 2.5$, and whose rotational loss torque in Nm, $T_{RL} = 0.06299\omega_{rm}$, where ω_{rm} is in mech. rad/s, and whose T-equivalent circuit parameters per phase, referred to the stator side are as follows:

$$R_1 = 0.05238\,\Omega, jX_1 = j0.23571\,\Omega, jX_\phi = j9.166\,\Omega$$

$$R_2 = 0.05238\,\Omega, jX_2 = j0.23571\,\Omega$$

The motor's moment of inertia for all its components' rotating masses, J_{motor} (J_m), in kg m^2, is 0.8807 kg m^2. For a rotor speed, $n_{rm} = 1760$ rev/min, and an external rotor circuit resistance in ohms per phase, $R_x = 0\,\Omega$.

Develop a computer algorithm and program in the language of your choice C++, or any other choice to verify that for $n_{rm} = 1760$ rev/min and $R_x = 0\Omega$.

The following results are essentially in agreement with your results within normal numerical round-off accuracy:

(1) $\omega_{rm} = 184.3$ mech.rad/s

$$s = 0.02222 \text{ p.u.}$$

$$\left(R_2 + R_x\right)/s = 2.357\,\Omega$$

(2) $\bar{I}_1 = 111.68\angle -24.71°$ A and $\bar{I}_2 = 105.61\angle -10.63°$ A

$$\text{pf} = 0.9085 \text{ lag}$$

(3) $P_{in} = 80,837$ W, $P_{1ohmic} = 1960$ W

$$P_{2ohmic} = 1753 \text{ W}, P_{RL} = 2140 \text{ W}$$

(4) $P_{dev} = 77,124$ W, $P_{out} = 74,984$ W

$$\eta_{motor} = 0.9276 \text{ p.u.} = 92.76\%$$

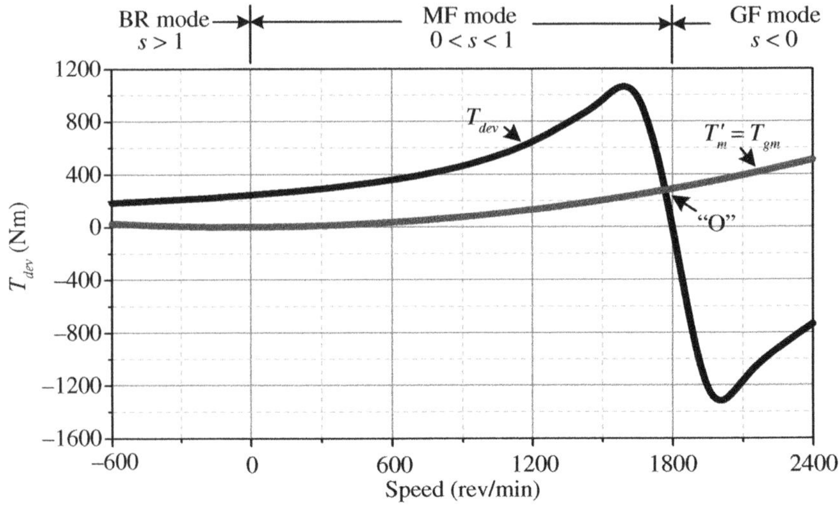

Figure 6.24 Torque–speed characteristic of the example motor.

(5) $T_{dev} = 418.45$ Nm, $T_{RL} = 11.61$ Nm, $T_{out} = 406.84$ Nm

(6) $s_{\max} = 0.1118$ p.u.

$$\omega_{rm}\left(\text{at } T_{dev\max}\right) = 167.42 \text{ mech.rad/s}$$
$$= 1599 \text{ rev/min}.$$

$$T_{dev\max} = 1029.43 \text{ Nm}$$

and $\bar{I}_2\left(\text{at } T_{dev\max}\right) = 371.59\angle -41.63^\circ$ A

(7) For a speed range ($-600 \le n_{rm} \le +2400$ rev/min), verify the following tabulation and torque–speed (slip) characteristic shown in Figure 6.24:

n_{rm} (rev/min)	I_1 (A)	P_{in} (kW)	P_{out} (kW)	T_{dev} (Nm)	pf (%)	η_{motor} (%)
−600.0	560.03	84.42	−11.96	186.38	18.92	******
−450.0	559.44	86.58	−9.49	198.39	19.42	******
−300.0	558.75	89.03	−6.72	212.03	20	******
−150.0	557.93	91.83	−3.59	227.67	20.66	******
0.0	556.94	95.07	0	245.77	21.42	0.00
150.0	555.72	98.85	4.18	266.94	22.32	4.23

n_{rm} (rev/min)	I_1 (A)	P_{in} (kW)	P_{out} (kW)	T_{dev} (Nm)	pf (%)	η_{motor} (%)
300.0	554.2	103.31	9.11	292.03	23.4	8.82
450.0	552.26	108.66	15.04	322.21	24.7	13.84
600.0	549.7	115.18	22.32	359.12	26.3	19.38
750.0	546.2	123.26	31.44	405.2	28.32	25.5
900.0	541.19	133.5	43.18	464.1	30.96	32.34
1050.0	533.6	146.79	58.77	541.37	34.53	40.03
1200.0	521.14	164.33	80.11	645.41	39.58	48.75
1350.0	498.3	187.29	109.94	786.6	47.17	58.7
1500.0	449.48	212.6	149.15	959.43	59.36	70.16
1650.0	323.47	203.14	169.26	990.45	78.82	83.32
1950.0	358.27	−208.87	−250.75	−1215.07	73.17	83.3
2100.0	496.08	−181.63	−260.06	−1168.72	45.95	69.84
2250.0	539.47	−128.05	−220.73	−921.95	29.79	58.01
2400.0	555.68	−89.8	−188.4	−733.8	20.28	47.66

(8) For the belt-pulley coupled motor-mechanical load system shown in Figure 6.25, see Reference [10], using the mechanical concepts of Chapter 4, verify the following calculations:

The 100-hp motor is driven from a balanced 3-ph, 460 V, 60 Hz source.
Motor pulley: radius = 15 cm = r_m

$$\omega_{rm} \cdot r_m = \omega_L \cdot r_L$$

hence,

$$\omega_{rm} = \left(\frac{r_L}{r_m}\right) \omega_L$$

where

$$\left(\frac{r_L}{r_m}\right) = \left(\frac{27}{15}\right)$$

Fan: 500 Nm @ 1000 rev/min
Fan pulley: radius = 27 cm = r_L

$$T_L = A_2 \omega_L^2$$

$$n_L = 1000 \text{ rev/min}, \omega_L = \frac{2\pi}{60} n_L \text{ mech.rad/s}$$

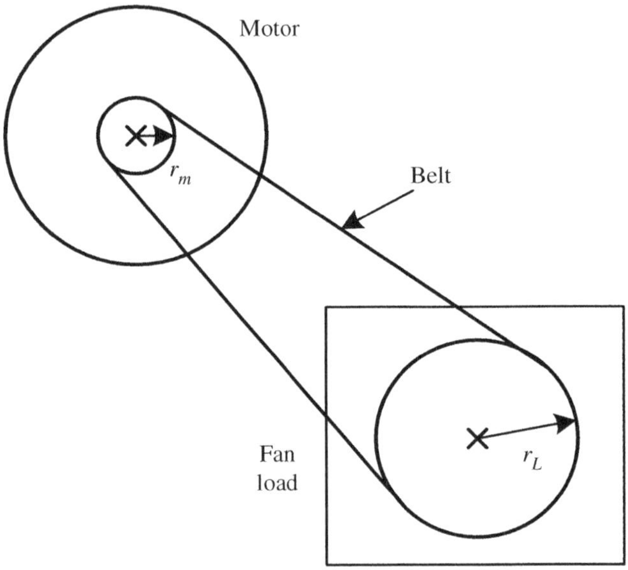

Figure 6.25 Motor-fan load system.

$$\omega_L = \frac{2\pi}{60} n_L = \frac{2\pi}{60} \times 1000 = 104.7 \text{ mech.rad/s}$$

$$\omega_{rm} = \left(\frac{r_L}{r_m}\right) \omega_L = \left(\frac{27}{15}\right) \times 104.7 = 188.5 \text{ mech.rad/s}$$

For $T_L = A_2\omega_L^2$, we have 500 Nm $= A_2(104.7)^2$
hence,

$$A_2 = 500/(104.7)^2 = 0.04559 \text{ Nm/mech.rad}^2/\text{sec}^2$$

hence,

$$T_L = 0.04559\omega_L^2 \text{ Nm}$$

But $T_m \cdot \omega_{rm} = T_L \cdot \omega_L$ (no belt slippage or deformation)
hence,

$$T_m \cdot \left(\frac{r_L}{r_m}\right) \omega_L = T_L \cdot \omega_L$$

hence,

$$T_m = \left(\frac{r_m}{r_L}\right) T_L = \frac{15}{27} T_L = \left(\frac{15}{27}\right) \times 0.04559\omega_L^2$$

$$T_m = 0.02533\, \omega_L^2 = 0.02533\left(\frac{r_m}{r_L}\right)^2 \omega_{rm}^2 \text{ Nm}$$

hence,

$$T_m = 0.02533\left(\frac{15}{27}\right)^2 \omega_{rm}^2 = 0.007818\omega_{rm}^2 \text{ Nm}$$

But recalling that $T_{RL} = 0.06299\omega_{rm}$ Nm
hence,

$$T_{gm} = T'_m = T_m + T_{RL} = 0.007818\omega_{rm}^2 + 0.06299\omega_{rm} \text{ Nm}$$

Recalling that the steady-state running speed for any motor-mechanical load system must be at the intersection of the developed motor torque, T_{dev}, and the gross-mechanical load torque, $T_{gm} = T'_m = T_m + T_{RL}$, that is, when

$$T_{dev} = T'_m$$

That is, when:

$$T_{dev} = \frac{3V_{1a}^2 s\,(R_2)}{\omega_{syn}\left[(sR_{e1} + R_2)^2 + s^2(X_{e1} + X_2)^2\right]} = T'_m \qquad (6.119)$$

where

$$T'_m = 0.007818\omega_{rm}^2 + 0.06299\omega_{rm} \text{ Nm} \qquad (6.120)$$

and

$s = \frac{(\omega_{syn} - \omega_{rm})}{\omega_{syn}}$, with $\omega_{syn} = \left(\frac{2\pi}{60}\right) \times 1800 = 60\pi$ mech.rad/s

$s = (60\pi - \omega_{rm})/60\pi = [1 - (\omega_{rm}/60\pi)]$ p.u.

Equations (6.119) and (6.120) must be solved numerically (graphically) for ω_{rm}, and the process leads to the following tabulation:

n_{rm} (rev/min)	$(T_{dev} - T_{RL})$ (Nm)	T_m (Nm)
1770.0	310.45	268.56
1773.0	280.27	269.47
1774.0	270.09	269.78
1774.1	269.06	269.81
1775.0	259.79	270.08
1780.0	207.78	271.61

$$T_{dev} - T_{RL} = T_m \text{ or } T_{dev} = T_{RL} + T_m = T'_m$$

Answer:

$n_{rm} = 1774.1$ rev/min. for motor operation
See point "O" in Figure 6.24.

(9) The reader is to verify that at $n_{rm} = 1774.1$ r/min, at 460-V and 60-Hz operation, see Reference [10], we have

- $\omega_{rm} = 185.78$ mech. rad/s and $s = 0.01439$ p. u.
- 460-V, 60 Hz operation
- $\bar{I}_1 = 76.57\angle - 28.04°$ A and $\bar{I}_2 = 69.61\angle - 6.88°$ A
- $T_{dev} = 280.77$ Nm, $T_{RL} = 11.70$ Nm, $T_{out} = 269.06$ Nm
- $P_{in} = 53.845$ kW, $P_{dev} = 52.162$ kW, $P_{out} = 49.988$ kW
- $P_{1ohmic} = 0.921$ kW, $P_{2ohmic} = 0.762$ kW, $P_{RL} = 2.174$ kW
- $\eta_{motor} = 92.84\,\%$ and pf = 0.8826 lagging

The dynamics of the starting transient of an induction motor using the steady-state T-equivalent circuit shown in Figure 6.23 as given below:

- Again, the motor-mechanical load rotation dynamics are governed by Eq. (6.121), namely, that is,

$$T_{dev} - T_{gm} = T_{dev} - T'_m = T_a = J\left(\frac{d\omega_{rm}}{dt}\right) \tag{6.121}$$

  where T_a is the acceleration torque.

- Hence, when $T_a > 0$, that is, $T_{dev} > T'_m$, the motor-load system accelerates.
- When $T_a < 0$, that is, $T_{dev} < T'_m$, the motor-load system decelerates.
- In order to conduct a complete analysis, Eq. (6.121) must be solved numerically because both T_{dev} and T'_m are complex nonlinear functions of ω_{rm}, and, hence, are both nonlinear functions of time, t.
- In the strictest sense, the motor steady-state T-equivalent circuit of Figure 6.23 is technically inapplicable because it is valid for balanced steady-state ac analysis only, with no time-varying slip, s, or speed, ω_{rm}.
- Accordingly, what is required is a more sophisticated and general model that includes enough rigor to model the time-varying electrical transient in the motor's windings (circuit), which will be covered in Chapter 10.
- However, experience shows that for a large number of practical applications, good results can be obtained by using a "quasi-ac" approach, by using the circuit of Figure 6.23 with the slip, s, varying from one time-step of integration, Δt, of Eq. (6.121) to the next. That is, from time, (t) to $(t + \Delta t)$.
- This "quasi-ac" approach can be used if the RMS voltage and speed do not change substantially within the selected integration time step, Δt, of Eq. (6.121) governing the dynamics of the motor-load system.
- For 60-Hz application, a plausible and appropriate time step, Δt, corresponds to approximately the period of a cycle, $T = (1/60)$ seconds, or approximately 0.016–0.02 second. In some applications, a smaller Δt may be necessary.

- The example motor-mechanical load system of Section 6.6, for instance, had a time constant of about 0.75 seconds. Thus, an integration time step $\Delta t = 0.015$ seconds, $\tau_{mech} = 0.75$ seconds. is 50 times larger than $\Delta t = 0.015$ seconds. Hence, "quasi-ac" analysis may be adequate.
- One must remember that starting the system is an important dynamic situation because of ohmic losses in the motor associated with the large magnitudes of starting currents and the concern for a motor's thermal overheating that might result in the windings and core.
- Hence, a starting transient will complicate a motor's over-current protection schemes.
- In a wound-rotor induction motor, external resistances connected in series with rotor windings present a practical way of limiting magnitudes of starting currents while increasing magnitudes of starting torque and, hence, shortening the duration of a starting transient.
- In a "quasi-ac" approach, at any time step sample, $n = 1, 2, 3, \ldots$ $t = [t_0 + (n-1)\Delta t]$, where Δt is the time step length per sample, and the motor speed is $\omega_{rm}(t)$. Hence, by solving Eq. (6.121), one determines $\omega_{rm}(t)$, at any time, t. Hence, it follows that at any time, t, we have

(1) $s(t) = (\omega_{syn} - \omega_{rm}(t))/\omega_{syn}$

(2) From Eq. (6.96), find I_1 at time, t:

$$I_1(t) = V_{1a}\sqrt{((R_2 + R_x)/s)^2 + (X_2 + X_\phi)^2}/X_\phi$$
$$\sqrt{(R_{e1} + (R_2 + R_x)/s)^2 + (X_{e1} + X_2)^2}$$

(3) From Eq. (6.110), find $T_{dev}(t)$ at time, t:

$$T_{dev}(t) = \frac{3V_{1a}^2 s(R_2 + R_x)}{\omega_{syn}\left[(sR_{e1} + R_2 + R_x)^2 + s^2(X_{e1} + X_2)^2\right]}$$

(4) Find $T'_m(t)$ at time, t, from the gross mechanical load torque function of the particular application:

$$T'_m(t) = fn\{\omega_{rm}(t)\}$$

(5) Find $T_a(t)$ at time, t:

$$T_a(t) = T_{dev}(t) - T'_m(t)$$

(6) Step time forward, $n = n + 1$ and $t = [t_0 + (n-1)\Delta t]$, that is, $t = t + \Delta t$, while numerically integrating forward Eq. (6.121), that is, forward integrating:

$$J(d\omega_{rm}/dt) = T_a(t) = T_{dev}(t) - T'_m(t)$$

(7) If speed, $\omega_{rm}(t)$, is still changing significantly, or time, t, is less than the expected duration of the starting transient, repeat steps (1) through (6).

(8) If speed, $\omega_{rm}(t)$, is not changing significantly, that is, steady-state running speed is achieved, or the starting transient, or any other transient is over and done, quit the forward time stepping and integration and output result.

On rotor equivalent circuit parameters for squirrel-cage induction machines.

- In wound-rotor induction machines, the current path in the rotor wires is well defined and not much affected by skin effects and other induced eddy current diffusion phenomena. Hence, the rotor circuit resistance and leakage inductance per phase, R_2 and L_2, are largely unaffected by the change in the rotor frequency, $f_r = sf$, and, hence, are largely independent of the rotor slip, s, or angular speed, ω_{rm}.
- In squirrel-cage rotor induction machines, depending on the shape (width and depth) of the squirrel-cage bars, both R_2 and L_2 are affected by the skin-effect phenomenon, hence, are the functions of the rotor frequency, $f_r = sf$, and consequently, both are functions of the slip, s, in per unit. Two general functions that can well approximate that effect are as follows:

$$R_2 = R_{20}\left(1 + as + bs^2\right) \tag{6.122}$$

and

$$L_2 = L_{20}\left(1 + cs + ds^2\right) \tag{6.123}$$

where R_{20} and L_{20} are the values of the rotor resistance and leakage inductance per phase, when the machine is running at $s = 0$, that is, at synchronous speed.
- The parameters, a, b, c, and d, in Eqs. (6.122) and (6.123) are dimensionless and their values depend on the type of squirrel cage that is in a particular induction machine (bar geometry/rotor slot shape).
- Squirrel-cage induction motors have been classified by NEMA (the National Electrical Manufacture's Association, USA) into five categories of design, namely, NEMA "A," NEMA "B," NEMA "C," NEMA "D," and NEMA "F," which are characterized by the shapes of their torque–slip (torque–speed) profiles and percentage (or per unit) torque, with 1.0 p.u. (100%) being the rated torque as shown in Figure 6.26.
- The shapes of the squirrel-cage bar cross-sections for the "A, B, C, D, and F" NEMA classes are shown in Figure 6.27.

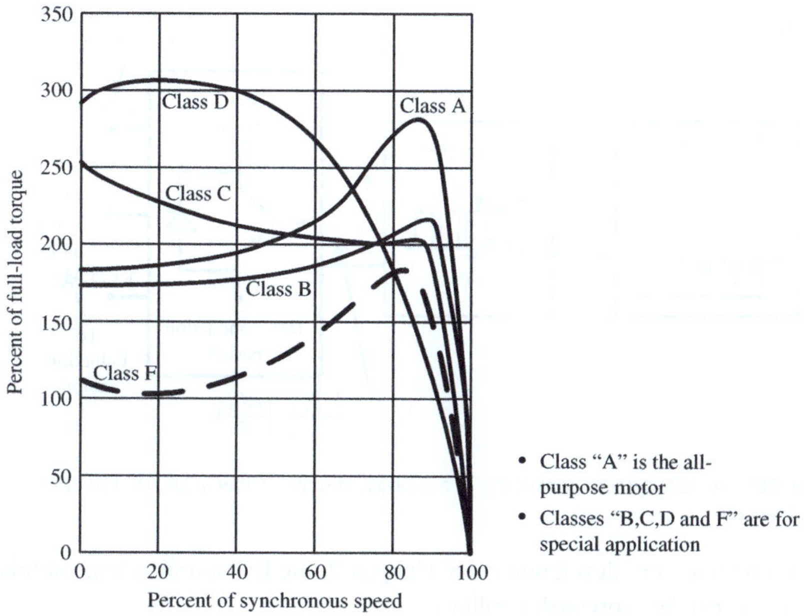

Figure 6.26 Torque–speed characteristics of NEMA classes "A, B, C, D, and F."

Figure 6.27 Approximate shapes of bars of NEMA classes A, B, C, D, and F squirrel-cage induction motors.

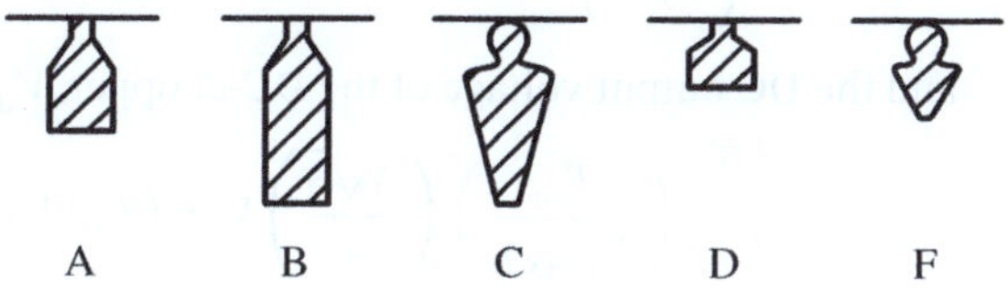

6.7 On Legacy and Modern Constant Volts Per Hertz Control of Torque and Speed of Induction Motors

Consider the three-phase full-wave rectifier–DC-chopper–six-step inverter system shown in Figure 6.28, which is energized from a three-phase supply whose balanced a, b, and c, phase-to-neutral voltages, v_a, v_b, and v_c, can be expressed in the time-domain as follows:

$$v_a = \sqrt{2}V\cos(\omega t) \tag{6.124}$$

$$v_b = \sqrt{2}V\cos(\omega t - 2\pi/3) \tag{6.125}$$

$$v_c = \sqrt{2}V\cos(\omega t - 4\pi/3) \tag{6.126}$$

where V is the magnitude of the RMS supply line-to-neutral voltage (voltage per phase), and $\omega = 2\pi f$ elec. rad/s

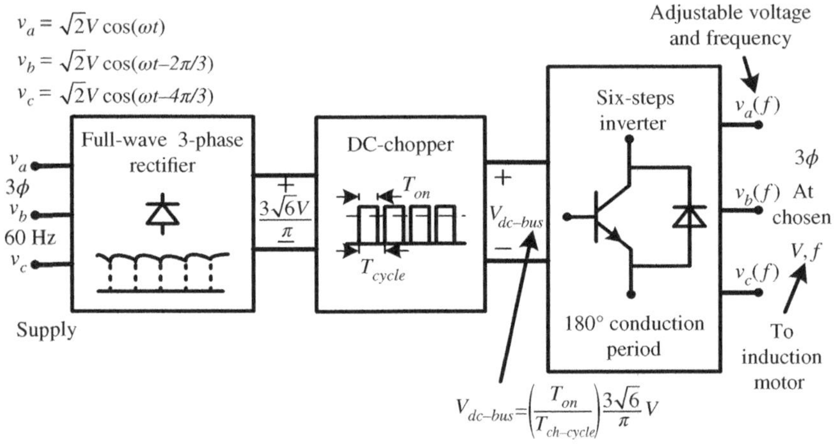

Figure 6.28 An ASD system consisting of a 3-phase diode rectifier, a DC/DC chopper, and an inverter.

According to earlier developments in Chapter 2, the DC output voltage of this rectifier, V_{dc}, can be expressed as follows:

$$V_{dc} = \left(\frac{3\sqrt{6}}{\pi}\right) V \tag{6.127}$$

and the DC output voltage of the DC-chopper, V_{dc-bus}, becomes

$$V_{dc-bus} = \left(\frac{T_{on}}{T_{ch-cycle}}\right)\left(\frac{3\sqrt{6}}{\pi}\right) V = \left(d_{ch}\right) V_{dc} \tag{6.128}$$

where

d_{ch} is the duty cycle of the DC-chopper, T_{on} is the "on" period of time of the DC-chopper, and $T_{ch-cycle}$ is the period of time of the cycle, that is, the "on and off" duration of time of the DC-chopper.

Naturally, see Figure 6.28, V_{dc-bus} is the input DC voltage into the six-step-inverter. Again, from the developments in Chapter (2), the fundamental RMS line-to-line voltage, $V_{1(L-L)}$, is

$$V_{1(L-L)} = \left(\frac{\sqrt{6}}{\pi}\right) V_{dc-bus} \tag{6.129}$$

while the line-to-neutral voltage, $V_{1(L-N)}$, is

$$V_{1(L-N)} = \left(\frac{\sqrt{2}}{\pi}\right) V_{dc-bus} \tag{6.130}$$

while the total line-to-line RMS voltage, $V_{(L-L)rms}$, is

$$V_{(L-L)rms} = \left(\sqrt{\frac{2}{3}}\right) V_{dc-bus} \tag{6.131}$$

and the total line-to-neutral RMS voltage,

$$V_{(L-N)rms} = \left(\frac{\sqrt{2}}{3}\right) V_{dc-bus} \tag{6.132}$$

The voltage and frequency control system of Figure 6.28 is also represented in a functional block diagram fashion in Figure 6.29, with a consequent effect on induction motor torque–speed (slip) characteristics shown in Figure 6.30 in which the concept of constant volts per Hertz is graphically summarized, where for frequencies, $f_1 < f_2 < f_3 < f_4$, and corresponding synchronous speeds, $n_{syn1} = \left(\frac{120f_1}{60}\right) < n_{syn2} = \left(\frac{120f_2}{60}\right) < n_{syn3} = \left(\frac{120f_3}{60}\right) < n_{syn4} = \left(\frac{120f_4}{60}\right)$ in rev/min, are also depicted.

At this stage, let us recall, from Section 6.4, that from Eqs. (6.80) and (6.81), we have

$$T_{\max} = 3V_{1a}^2/2\omega_{syn}\left[R_{e1} + \sqrt{R_{e1}^2 + (X_{e1} + X_2)^2}\right] \tag{6.80}$$

and,

$$s_{\max} = R_2/\sqrt{R_{e1}^2 + (X_{e1} + X_2)^2} \tag{6.81}$$

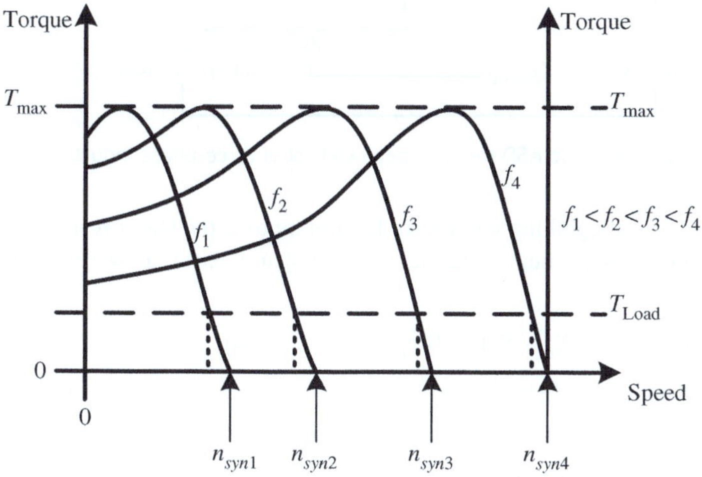

Figure 6.29 Effect of constant volts per Hertz (V/f) control on shaping an induction motor torque–speed (slip) characteristic.

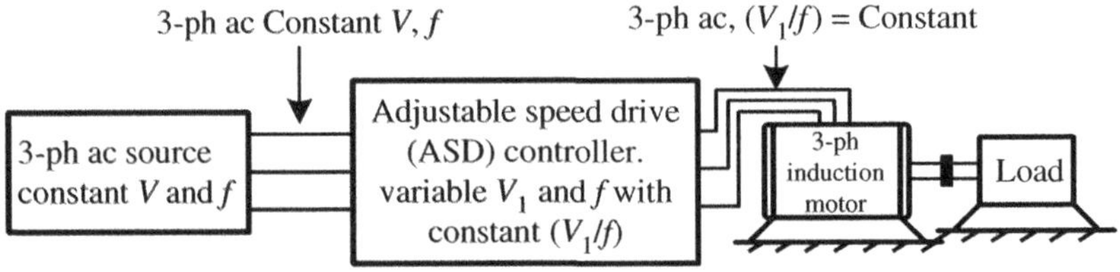

Figure 6.30 (V/f) an induction motor adjustable speed drive (IM-ASD).

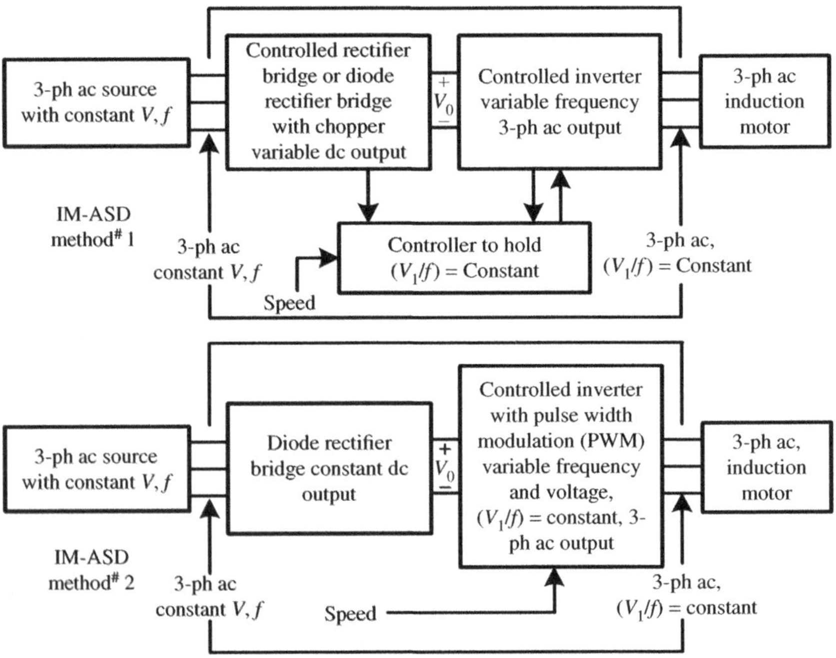

Figure 6.31 Two approaches to ASD torque–speed control of three-phase induction motors.

Therefore, for most integral horsepower induction motors, the stator resistance, R_1, is negligible, hence, R_{e1} is negligible, and Eqs. (6.80) and (6.81) can be rewritten and simplified as follows:

$$\begin{aligned} T_{\max} &\simeq 3V_{1a}^2/2\omega_{syn}\left(X_1+X_2\right) = 3V_{1a}^2/8\pi^2 f^2\left(L_1+L_2\right) \\ &= \left[3/8\pi^2\left(L_1+L_2\right)\right]\left(V_{1a}/f\right)^2 \end{aligned} \tag{6.133}$$

and,

$$s_{\max} \simeq R_2/\left(X_{e1}+X_2\right) = \left[R_2/2\pi\left(L_{e1}+L_2\right)\right]/f \tag{6.134}$$

This approach is also depicted in the functional block diagram of Method#1 in Figure 6.31 for six-step inverter-type systems and Method#2 in the same figure

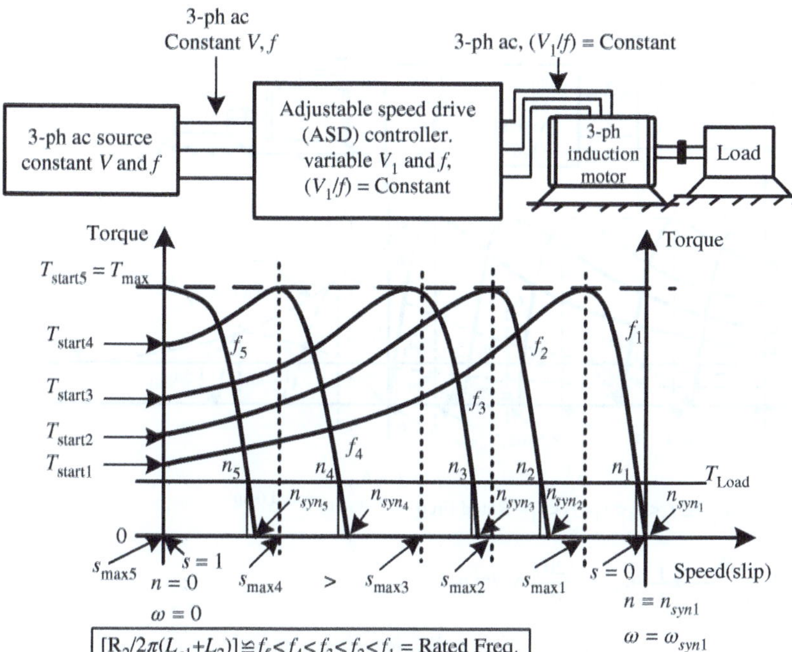

$[R_2/2\pi(L_{e1}+L_2)] \leqq f_5 < f_4 < f_3 < f_2 < f_1$ = Rated Freq.

$s_{max} \leqq \dfrac{R_2}{X_{e1}+X_2} \propto \dfrac{1}{f}, n_{syn} = \dfrac{120f}{p}$ rev/min

$T_{max} \leqq \dfrac{3V_{1a}^2}{2\omega_{syn}(X_{e1}+X_2)} \propto \left(\dfrac{V_1}{f}\right)^2$

$T_{start} \leqq \dfrac{3V_{1a}R_2}{\omega_{syn}(X_{e1}+X_2)^2} \propto \dfrac{1}{f}\left(\dfrac{V_1}{f}\right)^2$

Figure 6.32 Effects of adjustable voltage and frequency under the constant (V_1/f) = constant on the torque–speed (slip) characteristics of an induction motor.

for pulse-width-modulated (PWM) inverter-type systems. The resulting effects on torque–speed (slip) characteristics of induction motors are shown in Figure 6.32, in which T_{max} is held constant by the constant volts per Hertz, (V/f) = constant approach.

Furthermore, the extended torque–speed (slip) characteristics, in the flux weakening range, past the rated frequency, is shown in Figure 6.33, where the voltage is held constant in that range to the rated value, while the frequency is increased beyond the rated value. In this extended flux weakening range, T_{max} is progressively decreasing when the frequency is increased past the rated-frequency value, while the voltage is held constant at the rated value for the motor for purposes of protection of the dielectric integrity of the stator winding insulation, see

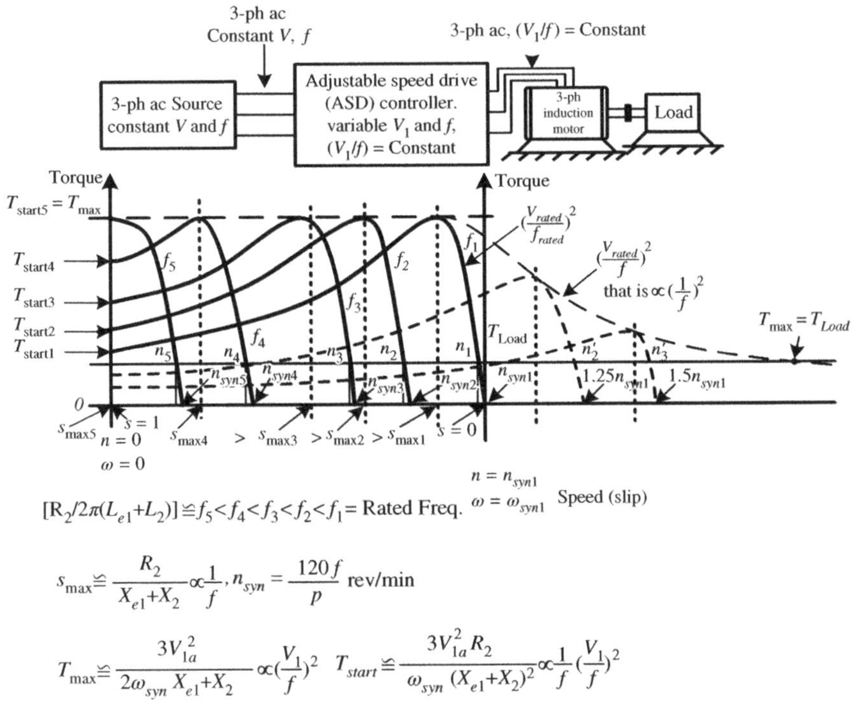

Figure 6.33 The flux-weakening-extended speed range.

Figure 6.33. For further in-depth details on time-stepping finite-element modeling of induction motors controlled by ASDs, Refs. [2–7] should be consulted.

6.8 Field-Oriented Control of Induction Motors

Field-oriented control, a widely used vector control method, is to control the space vectors of magnetic flux, current, and voltage. To achieve excellent dynamic performance, the field-oriented control of induction motors decouples the control of the stator magnetic field and the electromagnetic torque of the motor. The decoupling is implemented by transforming the stationary stator AC currents into two orthogonal DC current components in a rotating reference frame (i.e., the synchronous reference frame) that can be visualized with a rotating space vector. One DC current component controls the resultant magnetic flux of the motor, while the other controls the developed electromagnetic torque, see Figures 6.34 and 6.35. This is analogous to the torque–speed control of DC motor-drive systems, where the magnetic field is independently generated and controlled by

Figure 6.34 Stator current space vector and its components in the αβ stationary reference frame.

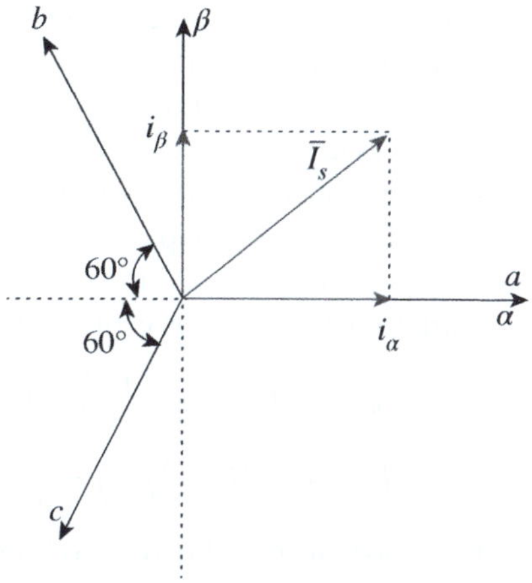

Figure 6.35 Stator current space vector and its components in the *dq* two-coordinate rotating reference frame.

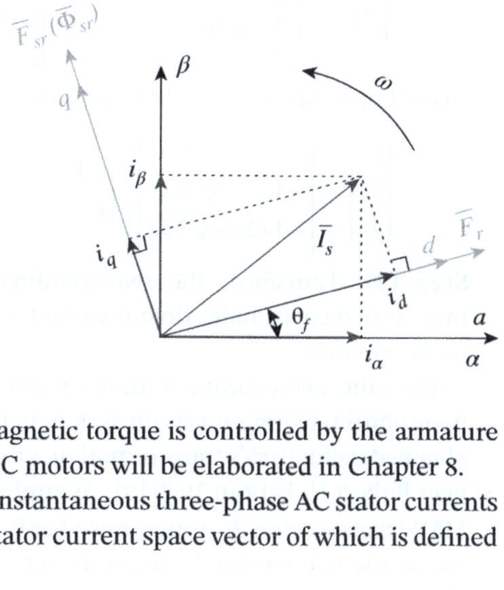

the field current, while the electromagnetic torque is controlled by the armature current. The operating principle of DC motors will be elaborated in Chapter 8.

Assuming that i_a, i_b, and i_c are the instantaneous three-phase AC stator currents of an induction motor, the complex stator current space vector of which is defined as follows:

$$\overline{I_s} = i_a + \alpha i_b + \alpha^2 i_c \tag{6.135}$$

where $\alpha = e^{j(2\pi/3)}$ and $\alpha^2 = e^{j(4\pi/3)}$ represent the rotation factors of the space-vector components.

The three-phase stator currents in a time-variant stationary reference frame are transformed into a time-invariant rotating reference frame through two steps:

Step (1): Transform the three-phase currents (i_a, i_b, and i_c) in the *abc* stationary reference frame into a two-coordinate time-variant system (i_α and i_β) based on Clark's transformation.

The current space vector can be transformed in a reference frame with two orthogonal axes (α, β), by using Clark's transformation, which was proposed in 1937–1938 [8] by Edith Clark, who was one of the earliest female electrical engineers in the United States. Assuming that the a-axis is in the same direction as the α-axis, the stator current space vector and the two orthogonal current components can be depicted, as shown in Figure 6.34.

By projecting i_b and i_c into the two-dimensional orthogonal system (α, β), the Clark transformation is expressed as follows:

$$\left.\begin{aligned} i_\alpha &= \tfrac{2}{3}\left(i_a - \tfrac{1}{2}i_b - \tfrac{1}{2}i_c\right) = i_a \\ i_\beta &= \tfrac{2}{3}\left(\tfrac{\sqrt{3}}{2}i_b - \tfrac{\sqrt{3}}{2}i_c\right) = \tfrac{1}{\sqrt{3}}\left(i_b - i_c\right) \end{aligned}\right\} \tag{6.136}$$

The matrix form of Eq. (6.136) is expressed as follows:

$$\begin{bmatrix} i_\alpha \\ i_\beta \end{bmatrix} = \frac{2}{3}\begin{bmatrix} 1 & -1/2 & -1/2 \\ 0 & \sqrt{3}/2 & -\sqrt{3}/2 \end{bmatrix} \cdot \begin{bmatrix} i_a \\ i_b \\ i_c \end{bmatrix} \tag{6.137}$$

Accordingly, the inverse Clark transformation is thereby provided as:

$$\begin{bmatrix} i_a \\ i_b \\ i_c \end{bmatrix} = \begin{bmatrix} 1 & 0 \\ -1/2 & \sqrt{3}/2 \\ -1/2 & -\sqrt{3}/2 \end{bmatrix} \cdot \begin{bmatrix} i_\alpha \\ i_\beta \end{bmatrix} \tag{6.138}$$

Step (2): Transform the two-coordinate time-variant system (i_α and i_β) into a two-coordinate time-invariant system (i_d and i_q) based on Park's transformation.

With the two-coordinate time-variant current components (i_α and i_β), the two-coordinate time-invariant system in the dq rotating reference frame can be obtained using Park's transformation, proposed by an American electrical engineer Robert H. Park in 1929 [9]. Assuming that the d-axis aligns with the rotor MMF space vector, $\overline{\mathcal{F}}_r$ (often mistakenly so-called rotor flux) of the induction motor, the relationship between the reference frames $\alpha\beta$ and dq is as shown in Figure 6.35.

Accordingly, the i_d (the rotor MMF, $\overline{\mathcal{F}}_r$) and the i_q torque component (the resultant MMF, $\overline{\mathcal{F}}_{sr}$) are determined as follows:

$$\left.\begin{aligned} i_d &= i_\alpha \cos\theta_f + i_\beta \sin\theta_f \\ i_q &= -i_\alpha \sin\theta_f + i_\beta \cos\theta_f \end{aligned}\right\} \tag{6.139}$$

here, i_d and i_q are used to control the rotor MMF $\overline{\mathcal{F}}_r$ (often mistakenly the so-called rotor flux) and developed electromagnetic torque component (the resultant MMF, $\overline{\mathcal{F}}_{sr}$) of the induction motor, respectively. The angle, θ_f, is the position of the rotor

MMF, $\overline{\mathcal{F}}_r$, or d-axis. Generally, there are two approaches to calculate the angle, θ_f. One approach, named **direct field-oriented control**, is to use magnetic flux sensing devices embedded in the motor or derived from the measured motor terminal voltages and currents. Another approach, named **indirect field-oriented control,** is to obtain the position of the rotor MMF, $\overline{\mathcal{F}}_r$, here designated by the angle, θ_f, given as follows:

$$\theta_f = \theta_r + \theta_{slip} \tag{6.140}$$

where θ_r is the rotor position angle which can be sensed by an encoder or resolver and θ_{slip} refers to the slip angle, defined in Chapter 10.

By utilizing Eqs. (6.137) and (6.139), a direct combined Clark–Park transformation from the abc reference frame to the dq reference frame is accordingly provided in the matrix form as follows:

$$\begin{bmatrix} i_d \\ i_q \end{bmatrix} = \frac{2}{3} \begin{bmatrix} cos\theta_f & cos\left(\theta_f - 2\pi/3\right) & cos\left(\theta_f - 4\pi/3\right) \\ -sin\theta_f & -sin\left(\theta_f - 2\pi/3\right) & -sin\left(\theta_f - 4\pi/3\right) \end{bmatrix} \cdot \begin{bmatrix} i_a \\ i_b \\ i_c \end{bmatrix} \tag{6.141}$$

It should be noted that the current variables in Eq. (6.141) can also refer to the voltage or MMF component of the induction motor.

Likewise, an inverse combined Clark–Park transformation from the dq reference frame to the abc reference frame can be derived as follows:

$$\begin{bmatrix} i_a \\ i_b \\ i_c \end{bmatrix} = \begin{bmatrix} cos\theta_f & -sin\theta_f \\ cos\left(\theta_f - 2\pi/3\right) & -sin\left(\theta_f - 2\pi/3\right) \\ cos\left(\theta_f - 4\pi/3\right) & -sin\left(\theta_f - 4\pi/3\right) \end{bmatrix} \cdot \begin{bmatrix} i_d \\ i_q \end{bmatrix} \tag{6.142}$$

Based on the Clark–Park transformations, a general closed-loop control block diagram is shown in Figure 6.36 to illustrate how the resultant flux and developed torque are controlled separately as in a DC motor. As can be seen, first of all, the three-phase stator currents are measured using current sensors. In practice, for a balanced three-phase motor-drive system, only two-phase currents such as (i_a and i_b) will be measured to save the cost of the current sensors and the associated signal conditioning interface circuits. The measured stator currents are provided as the input to the Clark transformation to obtain (i_α and i_β), which will be further provided along with the rotor MMF angle θ_f as the input to the Park transformation, in order to derive the two-coordinate time-invariant DC currents (i_d and i_q). The i_d and i_q components are compared to their reference values, i_d^* and i_q^*, namely, resultant flux reference and torque reference, respectively. Afterwards, the comparison errors will be provided as the input to the proportional–integral (PI) controllers, the outputs of which serve as an inverse Clark transformation to generate the voltage reference signals for the PWM block to control the power inverter in the ASD.

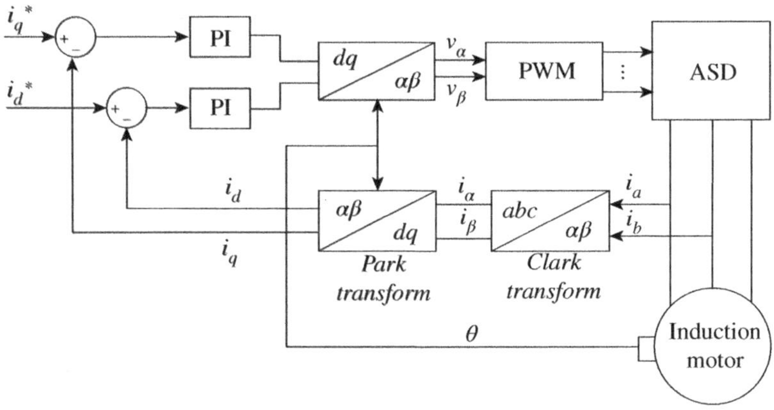

Figure 6.36 Block diagram of field-oriented control of an induction motor.

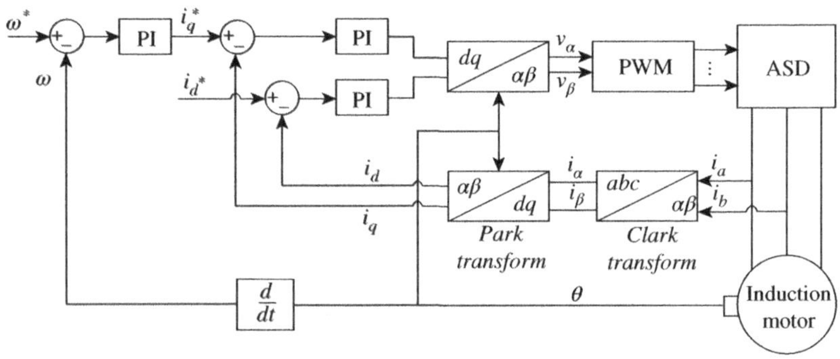

Figure 6.37 Block diagram of a field-oriented control system of an induction motor cascaded with an outer speed loop.

In industrial motor-drive applications, a speed controller can be cascaded with the torque controller, as shown in Figure 6.37. Specifically, a command speed is compared with the actual motor speed, and the speed error between the two is provided to a PI controller. Afterwards, the output of the PI controller will serve as the torque reference, namely, i_q^*. In a practical implementation of such a speed-torque cascaded closed-loop field-oriented control strategy, the dynamic response of the inner loop (i.e., torque or current loop) needs to be faster than the outer loop (i.e., the speed loop).

It should be pointed out that one advantage of the field-oriented control approach is the good portability for both induction motors and synchronous motors. To control permanent magnet synchronous motors, which will be covered in Chapter 7, the field-oriented control structure is almost identical to the approach presented here, except that the reference i_d^* will be set to zero in a normal constant torque operation mode since the rotor MMF, $\overline{\mathcal{F}}_r$, is fixed and determined by the permanent magnets mounted on the rotor. The field-oriented control of synchronous machines will be detailed in Chapter 7.

Homework Problems

6.1 A three-phase, two-pole, induction motor has the following physical characteristics: rotor diameter = 70 mm, stator axial conductor length = rotor length = 50 mm, an effective airgap height = 0.35 mm, an effective number of turns per phase, and $N_{ph}k_w = 420$ turns. The motor is to be operated from a 60-Hz, three-phase power supply. Stator resistance and leakage reactance may be neglected for the purposes of this analysis.

(a) Determine the appropriate value of the RMS stator voltage per phase if the peak airgap flux density is to be 0.80 T.

(b) Determine the magnetizing inductance and reactance per phase for this machine.

(c) Determine the magnetizing current for the stator voltage per phase computed in (b).

(d) Suppose this induction machine can support a stator armature loading around its mid-airgap circumference, A, of 8000 A/m without overheating. The motor is to be energized from a 60-Hz, 3-phase supply. Determine an appropriate value for its developed rated torque for a torque angle, $\delta_s \simeq 70°e$. What would be its developed power at a slip of 0.04 p.u. and estimate its power factor.

(e) For a torque angle, $\delta_r \approx 90°e$, find F_s, F_r, and F_{sr} in AT/pole and sketch the MMFs' space-vector diagram.

6.2 A 3-phase, 4-pole, 440-V (line-to-line), 60-Hz induction motor is supplied from a constant volts per Hertz ASD controller. The motor has a magnetizing inductance of 80 mH per phase, and an effective rotor resistance referred to the stator side of 0.4 Ω per phase. Leakage reactances and stator resistance may be neglected.

(a) Draw an approximate per phase equivalent circuit for this machine.

(b) When the ASD (constant volts per Hertz) controller is supplying the motor at its rated voltage and frequency of 440 V and 60 Hz, respectively, what will be the speed of the motor when it is driving a load of 45 Nm?

(c) What will be the rotor and supply (stator) currents and power factor for the condition in (b). Neatly draw the phasor diagram of the currents and deduce directions for the MMF space vectors, F_s, F_r, F_{sr}, as well as the values of the angles, δ_s, δ_r, and δ_{sr}, in electrical degrees for that operating condition.

(d) For a motor speed equal to half that in (b), what will be the supply frequency and line-to-line voltage for the same developed torque of 45 Nm, and what will be the phase current and power factor under this condition, and the corresponding values of the angles, δ_s, δ_r, and δ_{sr}. Compare your answer with those in (c) and comment.

6.3 A 3-phase, 6-pole, Y-connected, 335 hp, 60-Hz squirrel-cage induction motor has a rated line-to-line voltage of 2000 V. The motor has the following equivalent circuit parameters per phase referred to the stator side at 60 Hz:

$$R_1 = 0.201\ \Omega, R_2 = 0.204\ \Omega, jX_1 = jX_2 = 0.702\ \Omega, \text{and}\ jX_m = jX_\varphi = j76\ \Omega$$

The motor's total rotational losses (core, friction, and windage) can be taken constant and equal to 14,000 W throughout its normal operating range. At a slip within the normal operating range, $s = 0.014$ p.u., under rated voltage and frequency:

(a) Obtain the parameters and sketch a schematic circuit diagram of the Thevenin equivalent circuit of the motor at a slip, $s = 0.014$ p.u., under rated voltage and frequency conditions.

(b) Compute the speed of the stator MMF with respect to the stator frame, and the rotor MMF with respect to the rotor frame, as well as the rotor MMF with respect to the stator MMF, all in rev/min at this slip.

(c) Under the rated voltage and frequency conditions of (a), and a slip, $s = 0.014$ p.u., calculate the rotor current, $\bar{I}_2$, referred to the stator, and the developed mechanical power in Watts as well as developed torque in Newton meters.

(d) Calculate the output power in Watts and hp (1 hp = 746 W), and the output torque in Newton meters, as well as the input power, power factor, input line current, and overall motor efficiency.

6.4 A 3-phase, 6-pole, 60-Hz, 480-V (line-to-line), Y-connected 120-hp squirrel-cage induction motor has the following "T" equivalent circuit parameters in Ω/phase, all referred to the stator: $R_1 = 0.084\ \Omega$, $R_2 = 0.066\ \Omega$, $X_1 = 0.24\ \Omega$, $X_2 = 0.198\ \Omega$, and $X_\varphi = 8.28\ \Omega$. Its rotational losses are about 3770 W and can be taken constant throughout its normal operating range.

(a) Compute the parameters of the motor's Thevenin equivalent circuit at any slip, s, for a rated line-to-line voltage of 480 V. Sketch the circuit and state which of the stator and rotor windings are armature or field windings, and why?

(b) At a slip, $s = 0.035$ p.u., determine the rotor speed in rev/min and mechanical radians/second. Also, compute the rotor current $\overline{I}_2$, the developed power in watts and hp, as well as the developed torque in Nm.
(c) For the same slip, $s = 0.035$ p.u., compute the output power in watts and hp, as well as the output torque in Nm.
(d) Compute the magnetizing current, $\overline{I}_{\phi}$, and stator current, $\overline{I}_1$, and the motor power factor.
(e) Find the total input power and motor efficiency, at the same slip, $s = 0.035$ p.u.

6.5 Consider the motor-belt-driven fan load given in Section 6.6, see Figure 6.25. The motor is the 100-hp, 4-pole, 460-V, 60-Hz, Y-connected wound rotor induction motor given in the previous section as a verification example. If the motor's inertia, $J_{motor} = 0.8807$ kg m^2, and the mechanical load (fan) inertia, $J_{load} = 6.5$ kg m^2, with a load-to-motor pulley ratio, $(r_L/r_m) = (27/15)$, then the total inertia at the motor, J, becomes

$$J = \left[J_{motor} + \left(r_m/r_L\right)^2 J_{load}\right] = \left[0.8807 + (0.5556)^2 \times 6.5\right] \text{ kg m}^2$$

$$J = 0.8807 + 2.0065 = 2.8872 \text{ kg m}^2$$

And a gross mechanical load torque at the motor, T'_m, is expressed as follows:

$$T'_m = \underbrace{0.007818\omega_{rm}^2}_{T_m} + \underbrace{0.06299\omega_{rm}}_{T_{RL}} \text{ Nm}$$

(a) Obtain the starting transient by integrating:

$$J\left(d\omega_{rm}/dt\right) = T_{dev}(t) - T'_m(t) = T_a(t)$$

where $T_{dev}(t)$ is obtained from Eq. (6.110), for $R_x = 0\ \Omega$, with an initial condition, $\omega_{rm}(t = 0) = 0$ mech. rad/s. Perform the solution over a period of at least 1.5 seconds, using a step size, Δt, of your choice (try $\Delta t = 0.005$ seconds as a start). Defining, W_{loss}, the cumulative energy losses in the motor in Joules, as follows:

$$W_{loss} = \int_0^t \left(3I_1^2R_1 + 3I_2^2R_2 + P_{RL}\right) dt = \sum_0^t \left(3I_1^2R_1 + 3I_2^2R_2 + P_{RL}\right) \Delta t$$

Verify the validity of your results versus the following time tabulation of: $n_{rm} = \left(\frac{60}{2\pi}\omega_{syn}\right)$, current, I_1, developed torque, T_{dev}, mechanical load torque, T_m, acceleration torque, T_a, and cumulative motor loss, W_{loss}, given in Table P5.1 (plot all the starting transient's variables as functions of time showing all or most of the tabulations on them).

Table P5.1

Time, t (seconds)	n_{rm} (rev/min)	I_1 (A)	T_{dev} (Nm)	T_m (Nm)	T_a (Nm)	$V_1\%V_{total}$ (%)	W_{loss} (kJ)
0.00	0	446.9	453	0	453	100.0	0
0.12	182	427.6	434	3	430	100.0	13.2
0.24	349	410.5	421	10	409	100.0	24.4
0.36	507	395.1	414	22	391	100.0	34.3
0.48	659	380.8	413	37	374	100.0	43.2
0.60	804	367.4	418	55	360	100.0	51.2
0.72	945	354.2	429	76	350	100.0	58.3
0.83	1071	341.8	447	97	345	100.0	64.2
0.94	1197	328.2	474	122	347	100.0	69.5
1.04	1313	313.1	509	146	356	100.0	73.9
…							
1.32	1661	195.9	584	235	341	100.0	82.3

(b) Obtain the starting transient by integrating Eq. (6.121) over a period of at least 1.0 seconds, with an initial condition of $\omega_{rm}\,(t = 0) = 0$ mech.rad/s, if, R_x, the external rotor resistance referred to the stator side is made to vary from the starting condition according to the following regime:

$$R_x = 0.2727\ \Omega \text{ for } 0.00 \le \omega_{rm} < 0.40\omega_{syn}\ (\text{mech})$$

$$R_x = 0.1605\ \Omega \text{ for } 0.40\omega_{syn} \le \omega_{rm} < 0.80\omega_{syn}$$

$$R_x = 0.0\ \Omega \text{ for } 0.80\omega_{syn} \le \omega_{rm} < \omega_{syn}$$

Verify the validity of your results versus the following time tabulation of the results on n_{rm}, I_1, T_{dev}, T_m, T_a, and W_{loss}, given in Table P5.3 (do repeat similar plots and of all these variables with comparison to this tabulation on them).

(c) If the motor in (1) and (2) was replaced by a NEMA Class "B" motor with the following characteristic parameters and rating:
Squirrel-cage Class "B", 100-hp, 460-V, 60-Hz, 4-pole induction motor whose equivalent circuit values at rated stator frequency referred to the stator side are as follows:

$$R_1 = 0.04643\ \Omega, X_1 = 0.33856\ \Omega, X_\phi = 8.464\ \Omega$$
$$R_{20} = 0.08210\ \Omega, X_{20} = 0.50784\ \Omega, \text{with skin effect}$$

Table P5.2 Results of motor performance during a starting transient.

Time, t (seconds)	n_{rm} (rev/min)	I_1 (A)	T_{dev} (Nm)	T_m (Nm)	T_a (Nm)	R_x (Ω)	W_{loss} (kJ)
0.00	0	556.9	246	0	246	0	0
0.19	165	555.6	269	2	265	0	18.5
0.37	329	553.9	296	9	285	0	35.5
0.53	485	551.8	329	20	306	0	50.5
0.68	643	548.9	370	35	331	0	64.5
0.81	791	545.2	418	53	360	0	76.4
0.92	929	540.3	474	73	395	0	86.4
1.02	1067	533	547	96	444	0	95.3
1.10	1193	522.8	633	120	505	0	102.2
1.16	1300	509.3	725	143	574	0	107.2
...						0	
1.33	1697	268.6	895	244	641	0	117.6

Table P5.3

Time, t (seconds)	n_{rm} (rev/min)	I_1 (A)	T_{dev} (Nm)	T_m (Nm)	T_a (Nm)	R_x (Ω)	W_{loss} (kJ)
0.00	0	444.4	970	0	970	0.2727	0
0.05	179	429.6	996	2	993	0.2727	3.2
0.10	344	411.7	1017	9	1006	0.2727	5.9
0.16	524	457.6	940	22	914	0.1605	9.8
0.21	678	439	981	38	939	0.1605	12.9
0.26	765	545.9	408	49	354	0	17
0.32	837	543.7	435	59	370	0	22.5
0.37	900	541.5	461	68	387	0	27
0.43	979	538	499	81	411	0	32.4
0.49	1064	533.2	545	96	443	0	37.7
...							
0.67	1376	495	803	160	634	0	52.8

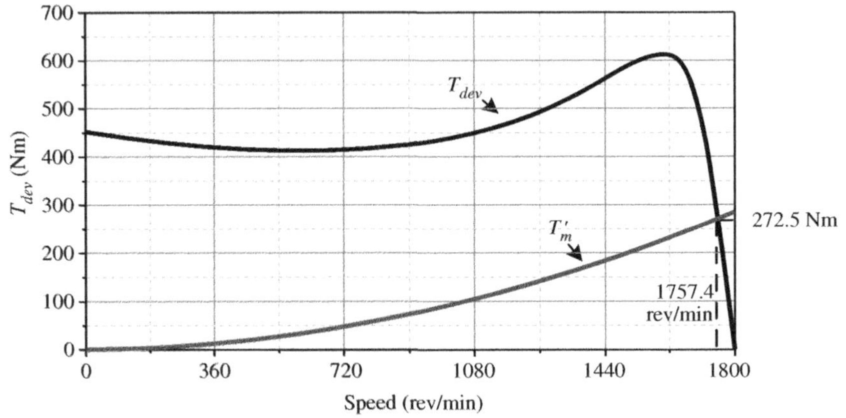

Figure 6.38 Torque–speed characteristics.

Parameters, $a = 0.71$, $b = 0.12$, $c = -0.55$, $d = 0.00$, while its moment of inertia, $J_m = 0.8807$ kg m^2, and its rotational loss torque, T_{RL}, is given by

$$T_{RL} = 0.042\omega_{rm} \text{ Nm}$$

where ω_{rm} is the rotor mechanical speed in rad/s.

(3.1) Verify the motor's, T_{dev}, and gross mechanical load, T'_m, torque–speed characteristics shown in Figure 6.38, if the load is the same as the fan load of (1) and (2), and, hence, the gross mechanical load torque, T'_m, is as follows:

$$T'_m = 0.007818\omega_{rm}^2 + 0.042\omega_{rm} \text{ Nm}$$

$$\text{and } R_2 = R_{20}\left(1 + as + bs^2\right) = 0.0821\left(1 + 0.71s + 0.12s^2\right)\,\Omega$$

$$\text{also } L_2 = L_{20}\left(1 + cs + ds^2\right) = 1.3471 \times 10^{-3}\,(1 - 0.55s) \text{ H}$$

$$\text{and total inertia, } J = 2.8872 \text{ kg m}^2$$

(3.2) Also, verify the motor-mechanical load starting transient data given in Table P5.3 and plot all variables of the starting transient for at least 1.5 seconds (compare all of this tabulated data to your plots of the transient for n_{rm}, I_1, T_{dev}, T_m, T_a, and W_{loss}).

References

1 Say, M.G. (1976). *Alternating Current Machines*. Pitman.

2 Demerdash, N.A., Bangura, J.F., and Arkadan, A.A. (1999). A time-stepping coupled finite element – state space model for induction motor drives – Part I: model formulation and machine parameter computation. *IEEE Transactions on Energy Conversion* 14 (4): 1465–1471.

3 Bangura, F.J., Isaac, F.N., Demerdash, N.A., and Arkadan, A.A. (1999). A time-stepping coupled finite element – state space model for induction motor drives – Part II: machine performance computation and verifications. *IEEE Transactions on Energy Conversion* 14 (4): 1472–1478.

4 Demerdash, N.A. and Bangura, J.F. (1999). Characterization of induction motors in adjustable-speed drives using a time-stepping coupled finite-element state space method including experimental validation. *IEEE Transactions on Industry Applications* 35 (4): 790–802.

5 Bangura, F.J. and Demerdash, N.A. (1999). Simulation of inverter-fed induction motor drives with pulse-width modulation by a time-stepping coupled finite element-flux linkage-based state space model. *IEEE Transactions on Energy Conversion* 14 (3): 518–525.

6 Bangura, F.J. and Demerdash, N.A. (1999). Performance and torque-ripple characterization in induction motor in adjustable-speed drives using a time-stepping coupled finite-element state space techniques. *IEEE Transactions on Industry Applications* 35 (5): 982–990.

7 Bangura, F.J. and Demerdash, N.A. (1999). Diagnosis and characterization of effects of broken bars and connectors in squirrel-cage induction motors by a time-stepping coupled finite element-state space modeling approach. *IEEE Transactions on Energy Conversion* 14 (4): 1167–1176.

8 Clark, E. (1938). Problems Solved by Modified Symmetrical Components. *General Electric Review (Schenectady, New York)* 41 (11 and 12): 488-94–545-49.

9 Park, R.H. (1929). Two-reaction theory of synchronous machines generalized method of analysis-part I. *Transactions of the American Institute of Electrical Engineers* 48 (3): 716–727.

10 Gross, C.A. (2006). *Electric Machines*, 1e. CRC Press.

7

The Poly-phase Synchronous Machine

Poly-phase synchronous machines, and specifically three-phase synchronous machines, particularly those of the wound-field variety, are mainly used for power generation. These machines are of the cylindrical rotor, or salient-pole rotor, variety where both rotor types are shown in Figures 7.1 and 7.2, respectively. Those machines with cylindrical rotors are associated with higher speed turbine-driven generators, of 3600 r/min (2-poles) or 1800 r/min (4-poles) for 60-Hz rated frequency, in power systems, or even higher speeds corresponding to 400-Hz frequency or higher, in aerospace power generation applications. The salient-pole rotor variety is associated with hydro-turbine prime mover in hydro-electric power-plant generators, and their speeds are much lower than their cylindrical rotor counterparts. The number of poles in such machines can well exceed six-poles. In this chapter, the emphasis will be on studying the synchronous machines of the cylindrical rotor variety, which are simpler to model and analyze. A more complete treatment of the salient-pole rotor variety will be given in Chapter 9, including the dynamic/transient modeling of such machines, which applies to both the salient-pole and cylindrical rotor types.

The stators of such machines are typical three-phase windings, of the double-layer layouts, such as shown in Figure 3.28 and Figures 5.1–5.6. Meanwhile, Figure 7.3 shows a photograph of the stator of a typical steam or gas turbine-driven generator, while Figure 7.4 shows a photograph of the stator of a typical hydro-turbine-driven generator.

7.1 Main Constructional Features of the Salient-Pole and Cylindrical Rotor Varieties

The stators of both the salient-pole and cylindrical rotor varieties of synchronous machines consist of slotted and laminated ferrous cylindrical-shell cores, made

Electric Machinery and Drives: An Electromagnetics Perspective, First Edition.
Nabeel A. O. Demerdash, JiangBiao He, and Hao Chen.

Companion website: www.wiley.com/go/demerdash/electricmachineryanddrives

Figure 7.1 A typical cylindrical rotor of a wound-field synchronous machine.

Figure 7.2 A typical salient-pole rotor of a wound-field synchronous machine.

of silicon-steel laminations of various grades and thicknesses, depending on the rated frequency and application in which a machine is used. Usually, the higher the rated operating frequency, the thinner the thicknesses of such laminations. Figure 7.5 shows such a slotted cylindrical-shell of a laminated stator core of a synchronous machine. Figure 7.6 shows such a stator core, with the three-phase stator winding embedded in the stator slots, including a view of the stator winding's end-turn region.

The rotors of salient-pole synchronous machines are built with laminated ferrous material, with field coils surrounding the salient-poles, and damper bars imbedded in the pole-shoes (pole-tips) region, to form a shorted cage with end-rings on both ends of a rotor, to form the damper-winding/cage, such as shown in Figures 7.7 and 7.8.

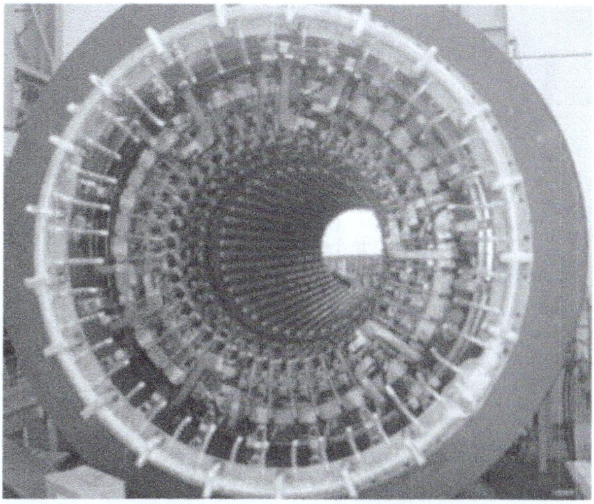

Figure 7.3 The stator of a typical turbogenerator. Source: Yang Zhao et.al., 2014 / IEEE.

Figure 7.4 The stator of a typical hydro-turbine-driven generator. Source: Clarion Events North America / https://www.hydroreview.com/technology-and-equipment/generators-and-electrical-components/hydro-review-stator-core-and-winding-repair-alternatives-after-a-fault-with-significant-damage// last accessed January 29, 2024

Figure 7.5 The slotted cylindrical-shell of a laminated stator core of a synchronous machine. Source: EC21.

Figure 7.6 A stator core with the three-phase winding embedded in the stator slots.

Meanwhile, rotors of turbo-type synchronous generators (turbogenerators) are made of solid magnetic steel forgings, see Figure 7.9, which are slotted as shown in this figure to house the field excitation coils/windings, as shown in the finished and completely assembled rotor of Figure 7.10, where the damper winding bars are buried in the rotor slots on top of the field winding coils, shorted by the retaining end-rings at both ends of the rotor.

Figure 7.7 The laminated rotor structure of a salient-pole synchronous machine. Source: John Wiley & Sons, Inc.

Figure 7.8 The complete rotor structure of a salient-pole synchronous machine including field winding coils and damper bar cage.

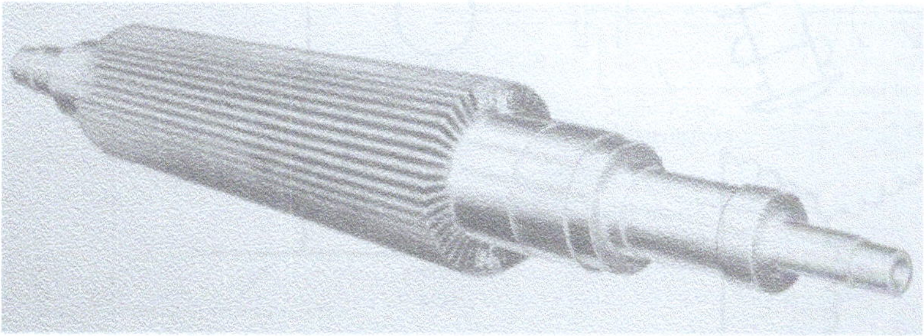

Figure 7.9 The rotor slotted forging of a synchronous turbogenerator.

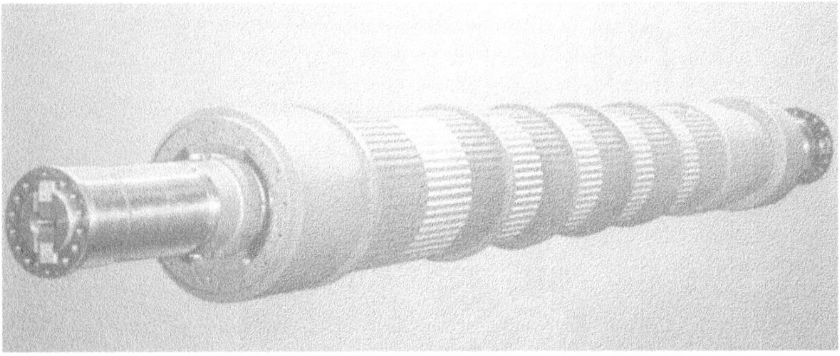

Figure 7.10 The complete assembly of a cylindrical rotor of a synchronous turbogenerator.

7.2 The Types of Synchronous Machine Exciter-Conventional and Modern Brushless Excitation Systems

In a synchronous machine, the field winding mounted on the rotor must be energized with a DC current from a source referred to as the "exciter," and its associated system is referred to as the "excitation system." Such an excitation system can be one of two types. One type is the conventional brush-slipring excitation system, see Figure 7.11. In such a system, a conventional brush-commutator DC machine functions as a generator supplying DC excitation current through a

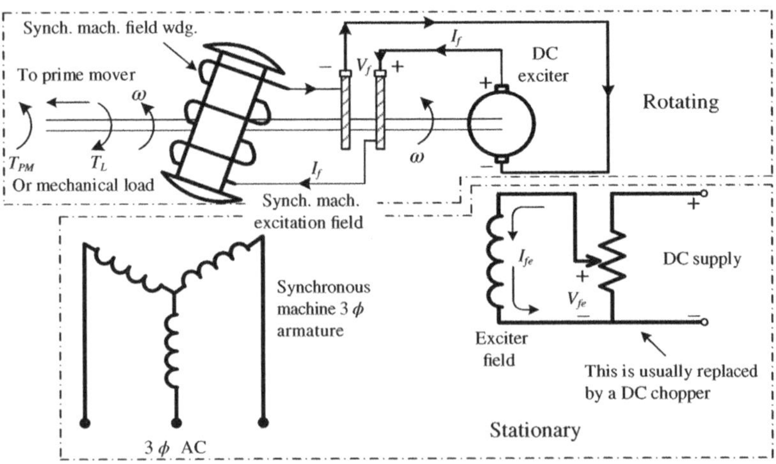

Figure 7.11 A conventional brush-slipring excitation system for wound-field synchronous machines.

system of sliprings and brushes to the main field excitation winding of the synchronous machine. In this system, the DC exciter, sliprings, and rotor-mounted field winding of the synchronous machine are rotating, while the field winding of the DC exciter and the three-phase stator (armature) winding of the synchronous machine are stationary; see the schematic diagram of the entire synchronous machine-exciter system in Figure 7.11.

The other type of an excitation system for a synchronous machine is the more modern brushless (rotating diode-bridge) system shown in Figure 7.12. This excitation system consists of a pilot exciter and a main exciter, as depicted in Figure 7.12. The stationary portions of this system are clearly delineated in Figure 7.12 within a rectangular frame designated "STATIONARY," while the rotating portions of this system are delineated in the same figure within a rectangular frame designated "ROTATING." The "STATIONARY" portion consists of the three-phase winding of the pilot-exciter, its associated three-phase full-wave bridge rectifier with its associated DC chopper feeding the field winding of the main-exciter, and the three-phase stator (armature) winding of the synchronous machine. Meanwhile, the "ROTATING" portion consists of the permanent magnet (PM) rotor of the pilot-exciter, the three-phase AC armature winding of the main-exciter which is feeding into a three-phase full-wave diode-bridge rectifier, and the DC output terminals of which are directly supplying the DC excitation current, I_f, into the rotating field-excitation winding of the synchronous machine

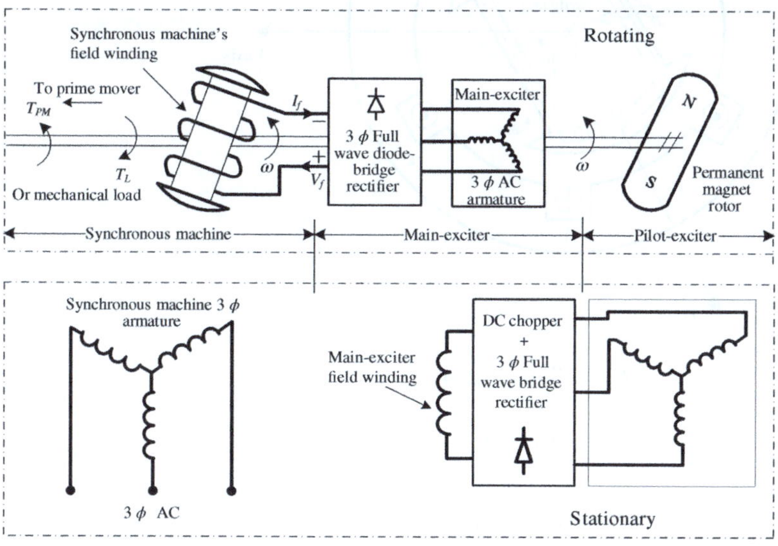

Figure 7.12 A modern brushless (rotating diode-bridge) excitation system for wound-field synchronous machines.

exciting its three-phase stator (armature) winding. Next, we develop the elements of the model and the equivalent circuit of the cylindrical rotor synchronous machine, without the effects of rotor magnetic circuit saliency, which will be modeled and fully treated in Chapter 9.

7.3 The Equivalent Circuit Model of the Idealized Cylindrical Rotor Synchronous Machine

A careful examination of the nature of the elementary three-phase stator winding in the elementary synchronous machine of Figure 7.13 reveals the resemblance to that of the circuit schematic of the stator armature of the induction machine of Figure 6.8, which led to the equivalent circuit model discussed in Section 6.4, and

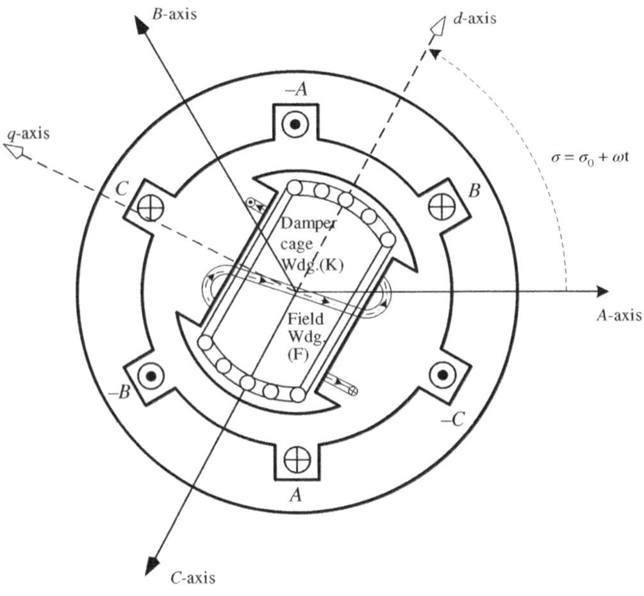

$$\bar{E}_{af} = \frac{\omega}{\sqrt{2}} L_{afm} I_f \angle{-\delta}$$

$$s = \frac{\omega_{syn} - \omega}{\omega_{syn}}$$

$$\omega = (1-s)\omega_{syn}$$

Figure 7.13 Schematic of an elementary three-phase synchronous machine including armature, field excitation, and damper windings.

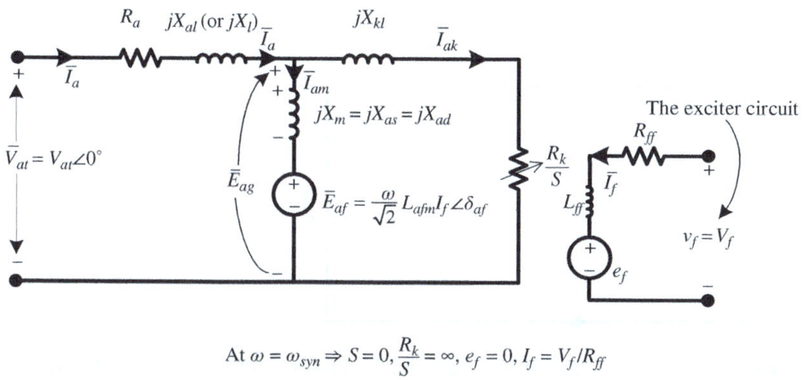

Figure 7.14 Effect of a field-excitation winding on equivalent circuits of three-phase stator windings.

specifically to the equivalent circuit model in Figure 6.17. This leads directly to the present equivalent circuit model of the synchronous machine given in Figure 7.14, in which the presence of the DC excitation field winding introduces an induced EMF, $\overline{E}_{af}$, in series with the magnetizing reactance, jX_m, where

$$E_{af} = (\omega L_{afm})I_f/\sqrt{2} \tag{7.1}$$

in which I_f is the DC field winding excitation current and L_{afm} is the maximum value of the mutual inductance between a phase winding and the field excitation winding, when the magnetic axes of a phase winding and the field excitation winding, that is the d-axis, see Figure 7.13, are congruent. Also, recalling Eq. (6.63) in Section 6.4 of Chapter 6, one can write for $X_m = X_\phi$ the following:

$$X_m = 24f(N_{ph}k_{w_1}/mp)^2(\mu_0 Dl/h_g) \tag{7.2}$$

All terms in Eq. (7.2) are identical to those defined earlier in Section 6.4, Eqs. (6.63) and (6.64). Here, this magnetizing reactance is referred to as the armature synchronous reactance, X_{as}, or direct-axis armature synchronous reactance, X_{ad}, thus one can write the following:

$$X_{as} = X_{ad} = X_m = 24f(N_{ph}k_{w_1}/mp)^2(\mu_0 Dl/h_g) \tag{7.3}$$

where X_m is as given earlier in Eq. (7.2). Here, the damper-bars cage is very similar to the squirrel-cage of an induction machine, and hence the damper winding/cage branch $(R_k/s + jX_{kl})$ is analogous to the $(R_2/s + jX_2)$ branch of the induction motor's equivalent circuit of Figure 6.20 in Section 6.4 of Chapter 6. Here, s is the slip and is as defined earlier in Eq. (6.3) of Section 6.2 of Chapter 6. However, when the synchronous machine is rotating at the synchronous speed, $n_{syn} = (120f/p)$ rev/min, that is, at ω_{syn}, the slip $s = 0$. Hence, the damper winding

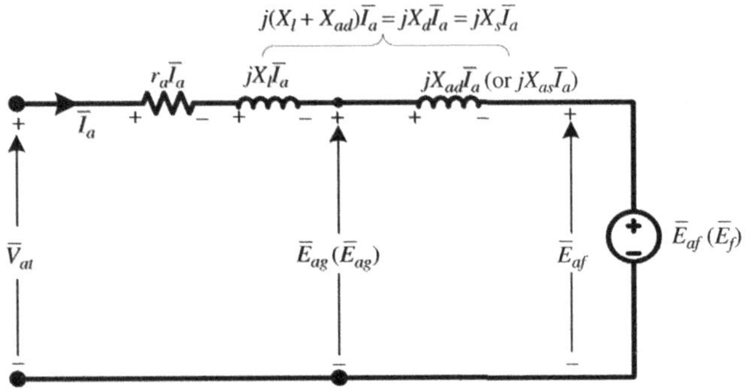

Figure 7.15 The equivalent circuit of an idealized cylindrical rotor synchronous machine.

branch with (R_k/s) is tending to infinity, or open circuit; the equivalent circuit of the synchronous machine, rotating at synchronous speed, ω_{syn} or n_{syn}, is rendered as shown in Figure 7.15. In Figure 7.15, we define the following based on a Y-connection configuration:

$\overline{V}_{at}$ is the armature (stator winding) terminal voltage per phase (line-to-neutral) in volts.

$\overline{I}_a$ is the armature phase current in amperes.

r_a is the armature resistance in ohms per phase.

X_l is the armature leakage reactance in ohms per phase; details on this reactance can be found in Ref. [1].

$X_{ad} = X_{as}$ is the direct-axis (or armature component) of the synchronous reactance in ohms per phase.

$X_d = X_s = (X_l + X_{ad}) = (X_l + X_{as})$ is the direct-axis synchronous reactance in ohms per phase.

$\overline{E}_{af} = \overline{E}_f$ is the induced back EMF due to the field winding excitation in volts per phase, where its magnitude E_{af} was defined earlier in Eq. (7.1); for further details on the mutual inductance, L_{afm}, Ref. [2–11] should be consulted on the computation of this parameter using finite element (FE) techniques.

$\overline{E}_{ag}$ is the so-called airgap EMF induced in the armature winding due to the resultant airgap flux, again in volts per phase.

Here, in RMS phasor consumer system notation, the following voltage–current relationships can be written, based on the equivalent circuit of Figure 7.15 for a synchronous machine per phase:

$$\overline{V}_{at} = (r_a + jX_l)\overline{I}_a + \overline{E}_{ag} \tag{7.4}$$

However, the airgap EMF, $\overline{E}_{ag}$, can be expressed as follows:

$$\overline{E}_{ag} = (jX_{ad})\overline{I}_a + \overline{E}_{af} = (jX_{as})\overline{I}_a + \overline{E}_{af} \tag{7.5}$$

Accordingly, from Eqs. (7.4) and (7.5), one can write the following:

$$\overline{V}_{at} = (r_a + j(X_l + X_{ad}))\overline{I}_a + \overline{E}_{af} \tag{7.6}$$

or

$$\overline{V}_{at} = (r_a + j(X_l + X_{as}))\overline{I}_a + \overline{E}_{af} \tag{7.7}$$

Here, the direct-axis synchronous reactance, X_d, which is simply the synchronous reactance, X_s, for idealized cylindrical rotor synchronous machine, means that,

$$X_d = X_s = (X_l + X_{ad}) = (X_l + X_{as}) \tag{7.8}$$

Therefore, Eqs. (7.7) and (7.8) can be rewritten as follows:

$$\overline{V}_{at} = (r_a + jX_d)\overline{I}_a + \overline{E}_{af} \tag{7.9}$$

or

$$\overline{V}_{at} = (r_a + jX_s)\overline{I}_a + \overline{E}_{af} \tag{7.10}$$

Next, we discuss the phasor diagrams of both the overexcited and underexcited synchronous motors and generators and the associated real and reactive power relationships to the main voltage components, $\overline{V}_{at}$ and $\overline{E}_{af}$, as well as the armature (stator winding) current, $\overline{I}_a$, all quantities per phase.

7.4 Phasor-Vector Diagrams of Synchronous Motors and Generators

Here, we discuss the so-called overexcited and underexcited motors and generators, in Cases 7.1–7.4, respectively. In all these cases, the terminal voltage of the synchronous machine per phase, $\overline{V}_{at}$, will be taken as the reference of the phasor diagrams. That is, in all these Cases 7.1–7.4, we can write the following:

$$\overline{V}_{at} = V_{at}\angle 0^{\circ} = V_{at} + j0.0 \tag{7.11}$$

in volts per phase, that is also line-to-neutral.

Case 7.1 The Over-Excited Synchronous Motor

In this case of an over-excited motor, the armature (stator) phase current, $\overline{I}_a$, can be written, see the phasor-space-vector diagram of Figure 7.16, as follows:

$$\overline{I}_a = I_a\angle\theta = (I_a \cos\theta) + j(I_a \sin\theta) \tag{7.12}$$

(Continued)

Case 7.1 (Continued)

Hence,

$$\overline{I}_a^* = I_a\angle -\theta = (I_a\cos(-\theta)) + j(I_a\sin(-\theta)) = (I_a\cos\theta) - j(I_a\sin\theta). \quad (7.13)$$

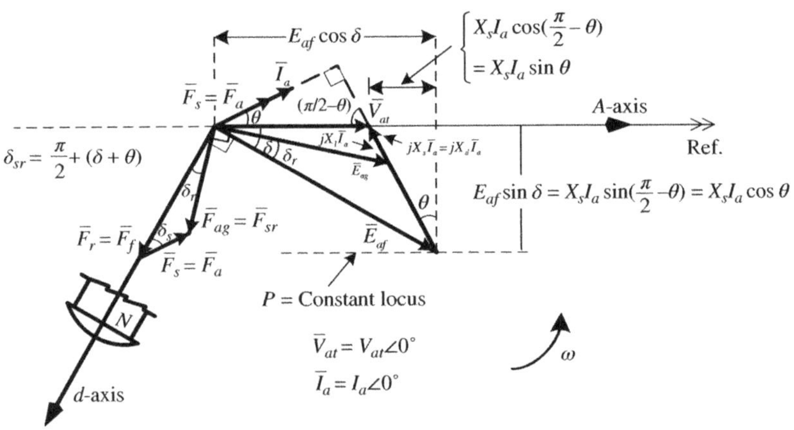

Figure 7.16 The phasor-space-vector diagram of an over-excited synchronous motor.

Accordingly, the complex power, $\overline{S}$, consumed by this motor is given, based on Eqs. (7.11) and (7.13), as follows:

$$\overline{S} = 3\overline{I}_a^* \cdot \overline{V}_{at} = 3V_{at}[(I_a\cos\theta) - j(I_a\sin\theta)] \quad (7.14)$$

That is,

$$\overline{S} = (3V_{at}I_a\cos\theta) + j(-3V_{at}I_a\sin\theta) = P + jQ \quad (7.15)$$

which means that the real power, P, is as follows:

$$P = 3V_{at}I_a\cos\theta \quad (7.16)$$

That is, real power is positive and is representing consumed watts as in normal motoring operating modes in synchronous machines. Meanwhile, the reactive power, Q, is as follows:

$$Q = -3V_{at}I_a\sin\theta \quad (7.17)$$

That is, reactive power is negative and is representing the production (generation) of vars (voltamperes reactive) and hence is a source of improving the power factors of resistive-inductive loads and supplying vars that need not be transmitted over transmission lines from power plants.

It should further be noticed, upon careful examination of the associated magnetomotive force (MMF) space-vector, $\overline{F}_a = \overline{F}_s$, the stator armature MMF, and the field excitation winding MMF, $\overline{F}_f = \overline{F}_r$, as well as the resultant air-gap MMF, $\overline{F}_{ag} = \overline{F}_{sr}$, that $\overline{F}_a = \overline{F}_s$ has a space orientation along the armature current phasor, $\overline{I}_a$, as pointed out earlier in Chapter 5 in discussions on the theory of rating MMFs in polyphase AC machines, and the $\overline{F}_f = \overline{F}_r$ is along the direct-axis (*d*-axis) of the rotor (the location of north "N" pole on the rotor) and is in quadrature to the EMF phasor, $\overline{E}_{af}$, in a lagging direction with respect to the phasor-space-vector diagram's direction of synchronous rotational orientation, ω, indicated in Figure 7.16. Furthermore, $\overline{F}_{ag} = \overline{F}_{sr}$ is in quadrature to the EMF phasor, $\overline{E}_{ag}$, also in a lagging direction with respect to the phasor-space-vector diagram's direction of synchronous rotational orientation, ω, see Figure 7.16.

It is important to observe, in this overexcited synchronous motor case, that the projection of the field MMF, $\overline{F}_f$, along the direction of orientation of the resultant MMF, $\overline{F}_{sr} = \overline{F}_{ag}$, indicates that $\overline{F}_f$ is a magnetizing MMF, and the projection of the armature (stator) MMF, $\overline{F}_a = \overline{F}_s$, along the direction of orientation of MMF, $\overline{F}_{sr} = \overline{F}_{ag}$, indicates that $\overline{F}_a = \overline{F}_s$ is a demagnetizing MMF, hence the need for an ample magnitude of the MMF, $\overline{F}_f$, which is referred to here as overexcitation.

It should be pointed out that in the phasor-space-vector diagram of Figure 7.16, the power angle, δ, is the phase angle between the armature terminal voltage per phase, $\overline{V}_{at}$, and the armature-induced EMF due to the field winding excitation in volts per phase, $\overline{E}_{af}$, while δ_r is the angle between the phasors, $\overline{E}_{ag}$ and $\overline{E}_{af}$, which is the same angle between $\overline{F}_f$ and $\overline{F}_{sr}$ in the space-vector diagram. Here, the angle, δ_{sr}, between the space-vectors, $\overline{F}_f$ and $\overline{F}_a$, is given by the following expression:

$$\delta_{sr} = (\pi/2) + (\delta + \theta) \tag{7.18}$$

and the angle, δ_s, is accordingly given by the following:

$$\delta_s = \pi - (\delta_{sr} + \delta_r) = \pi - [(\pi/2) + (\delta + \theta) + \delta_r]$$

that is,

$$\delta_s = (\pi/2) - \delta - \theta - \delta_r \tag{7.19}$$

Obviously, from Eq. (7.18), the angle, δ_{sr}, is an obtuse angle, indicating the demagnetizing effect of the armature (stator) MMF, $\overline{F}_a$.

Now, we examine the real-power, *P*, computation in terms of the various voltages and angles in the phasor diagram of Figure 7.16, and the developed

(Continued)

Case 7.1 (Continued)

electromechanical torque, T_{dev}, computation in terms of the field excitation MMF, $\overline{F}_f$, and armature (stator) MMF, $\overline{F}_a$, space-vectors and the angles, δ_{sr}, δ_s, and δ_r, in the space-vector diagram in Figure 7.16. We start here with the developed torque computation.

Based on the torque expression developed earlier in Chapter 5, Section 5.6, one can write the following for the developed torque, T_{dev}, using Eq. (5.143):

$$T_{dev} = -(p/2)(\mu_0 \pi Dl/2h_g)F_a F_f \sin(\delta_{sr}) \tag{7.20}$$

Substituting for δ_{sr} from Eq. (7.18) into Eq. (7.19) yields the following:

$$T_{dev} = -(p/2)(\mu_0 \pi Dl/2h_g)F_a F_f \cos(\delta + \theta) \tag{7.21}$$

where all the terms, D, l, and h_g, were defined earlier in Chapter 5, on the basis of the assumption that most of the magnetic field's energy is stored in a uniform cylindrical airgap shell, neglecting any saliency effects. Here, the MMF amplitudes of $\overline{F}_a$ and $\overline{F}_f$ can be computed on the basis of earlier developments in Chapter 5 as follows:

$$F_a = (3/2)[(4/\pi)\sqrt{2}k_{w_1}(N_{ph}/p)(I_a/m)] \tag{7.22}$$

where F_a is expressed in Eq. (7.22) in ampere-turns per pole, and all other terms of this equation were defined earlier in Chapter 5, and from Eq. (5.57), one can write the following for F_f:

$$F_f = (4/\pi)k_{f1}(N_f/p)I_f \tag{7.23}$$

where I_f is the excitation field winding's DC excitation current, N_f is the total number of field winding turns, and all other terms are as defined earlier in Chapter 5.

Again, in this overexcited synchronous motor case, from Eqs. (7.16) and (7.17), we have the following for the three-phase powers P and Q:

$$P = 3V_{at}I_a \cos\theta \tag{7.24}$$

$$Q = -3V_{at}I_a \sin\theta \tag{7.25}$$

However, from the geometry of the phasor diagram of Figure 7.16, one can write the following:

$$X_s I_a \cos\theta = X_d I_a \cos\theta = E_{af} \sin\delta \tag{7.26}$$

Hence,

$$I_a \cos\theta = (E_{af}/X_s)\sin\delta = (E_{af}/X_d)\sin\delta \tag{7.27}$$

Therefore, substituting for ($I_a \cos \theta$) from Eq. (7.27) into Eq. (7.16), one can write the following for the total real-power, P, in watts:

$$P = (3V_{at}E_{af}/X_s)\sin\delta = (3V_{at}E_{af}/X_d)\sin\delta \tag{7.28}$$

Also, from the geometry of the phasor diagram of Figure 7.16, one can write the following:

$$X_sI_a\sin\theta = X_dI_a\sin\theta = (E_{af}\cos\delta - V_{at}) \tag{7.29}$$

Hence,

$$I_a\sin\theta = (E_{af}\cos\delta - V_{at})/X_s = (E_{af}\cos\delta - V_{at})/X_d \tag{7.30}$$

Accordingly, substituting for ($I_a \sin \theta$) from Eq. (7.30) into Eq. (7.17), one can write the following for the total reactive-power, Q, in vars:

$$Q = -\left[(3V_{at}E_{af}/X_s)\cos\delta - \left(3V_{at}^2/X_s\right)\right] \tag{7.31}$$

or

$$Q = -\left[(3V_{at}E_{af}/X_d)\cos\delta - \left(3V_{at}^2/X_d\right)\right] \tag{7.32}$$

Accordingly, one can write the following for the developed torque, T_{dev}, relationship to torque angle, δ:

$$T_{dev} = P/\omega_{syn} = (3V_{at}E_{af}/\omega_sX_s)\sin\delta = (3V_{at}E_{af}/\omega_sX_d)\sin\delta \tag{7.33}$$

That is,

$$T_{dev} = T_{\max}\sin\delta \tag{7.34}$$

where

$$T_{\max} = (3V_{at}E_{af}/\omega_sX_s) = (3V_{at}E_{af}/\omega_sX_d) \tag{7.35}$$

Substituting E_{af} as $E_{af} = (\omega L_{afm}I_f/\sqrt{2})$ gives the following for $T_{\max}$:

$$T_{\max} = [3V_{at}(2\pi f)L_{afm}I_f/\sqrt{2}\omega_sX_s] = [3V_{at}(2\pi f)L_{afm}I_f/\sqrt{2}\omega_sX_d] \tag{7.36}$$

However, here, ω_s is in mechanical radians per second, which means that

$$\omega_s = 2\pi f/(p/2) = (4\pi f/p) \tag{7.37}$$

Accordingly, $T_{\max}$ becomes

$$T_{\max} = 3V_{at}(2\pi f)L_{afm}I_f/(4\pi f/p)\sqrt{2}X_d \tag{7.38}$$

(Continued)

Case 7.1 (Continued)

or

$$T_{\max} = 3(V_{at}/X_d)(L_{afm}I_f/\sqrt{2})(p/2) \tag{7.39}$$

Here, $X_d = \omega L_d = (2\pi f)L_d$. This yields the torque, T, to power angle, δ, relationship depicted graphically in Figure 7.17, where

$$T = T_{\max}\sin\delta \tag{7.40}$$

Here, δ is the power or torque angle in the phasor diagram of Figure 7.16. Also, in this figure, one can write as follows:

$$T_{\max} = p(3L_{afm}/4\pi\sqrt{2}L_d)I_f(V_{at}/f) \tag{7.41}$$

Hence, for constant volts per Hertz torque–speed control, if (V_{at}/f) is kept constant and the field excitation current, I_f, is kept constant, $T_{\max}$ in Eq. (7.41) will be kept constant, in a similar fashion to what was presented for induction motors in Chapter 6, Section 6.7.

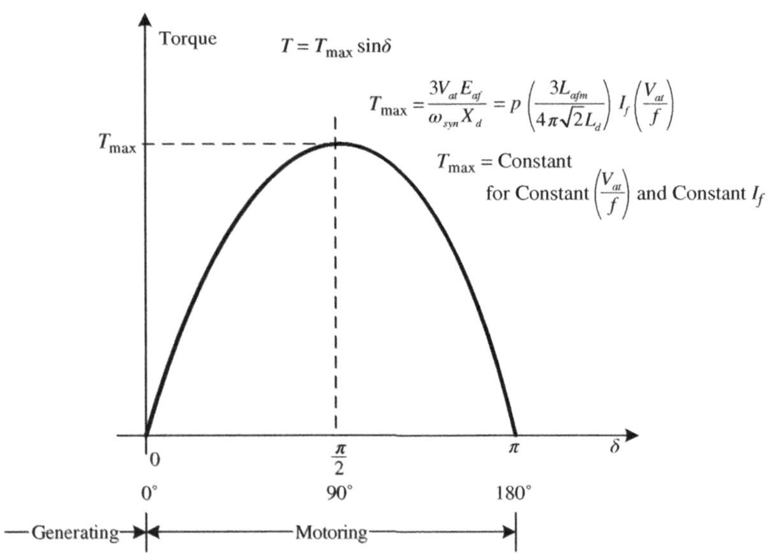

Figure 7.17 The torque-power (torque) angle, δ, characteristic for motoring modes in synchronous machines.

Case 7.2 The Under-Excited Synchronous Motor

In this case of an underexcited motor, the armature (stator) phase current, $\overline{I}_a$, can be written, see the phasor-space-vector diagram of Figure 7.18, as follows:

$$\overline{I}_a = I_a\angle - \theta = (I_a \cos\theta) - j(I_a \sin\theta) \tag{7.42}$$

Hence,

$$\overline{I}_a^* = I_a\angle\theta = (I_a \cos\theta) + j(I_a \sin\theta) \tag{7.43}$$

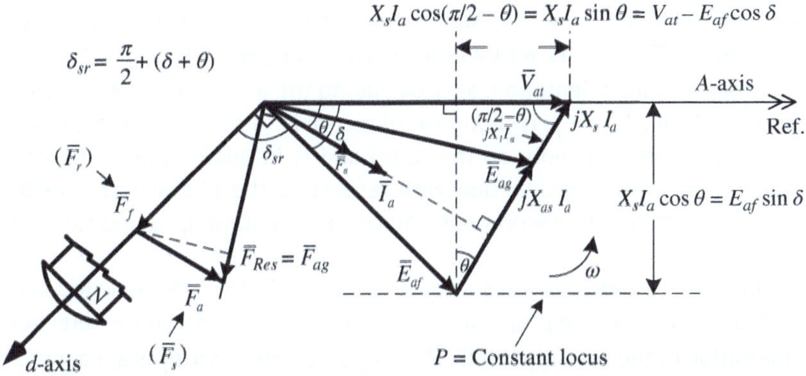

Figure 7.18 The phase-space-vector diagram of an under-excited synchronous motor.

Accordingly, the complex power, $\overline{S}$, consumed by this motor is given, based on Eqs. (7.11) and (7.43), as follows:

$$\overline{S} = 3\overline{I}_a^* \cdot \overline{V}_{at} = 3V_{at}[(I_a\cos\theta) + j(I_a\sin\theta)] \tag{7.44}$$

That is,

$$\overline{S} = (3V_{at}I_a\cos\theta) + j(3V_{at}I_a\sin\theta) = P + jQ \tag{7.45}$$

which means that the real power, P, is as follows:

$$P = 3V_{at}I_a\cos\theta \tag{7.46}$$

That is, real power is positive and is representing consumed watts as in normal motoring operating modes in synchronous machines. Meanwhile, the reactive power, Q, is as follows:

$$Q = 3V_{at}I_a\sin\theta \tag{7.47}$$

(Continued)

Case 7.2 (Continued)

That is, reactive power is positive and is representing the consumption of, such as static resistive-inductive loads, thus behaving in a similar manner to an induction motor. This is, in general, an undesirable way in which one would want to use a synchronous motor because of its more expensive cost caused by the need for an exciter, which is a costly item of extra hardware. Therefore, this mode of operation occurs very rarely in the utilization of synchronous motors and is usually a temporary/momentary mode of operation.

It should further be noticed, upon careful examination of the associated MMF space-vectors, $\overline{F}_a = \overline{F}_s$, the stator armature MMF, and the field excitation winding MMF, $\overline{F}_f = \overline{F}_r$, as well as the resultant airgap MMF, $\overline{F}_{ag} = \overline{F}_{sr}$, that $\overline{F}_a = \overline{F}_s$ has a space orientation, as usual, along the armature current phasor, $\overline{I}_a$, and the MMF, $\overline{F}_f = \overline{F}_r$, is along the direct-axis (d-axis) of the rotor (the location of a north "N" pole on the rotor) and is in quadrature to the EMF phasor, $\overline{E}_{af}$, in a lagging direction with respect to the phasor-space-vector diagram's direction of synchronous rotational orientation, ω, indicated in Figure 7.18.

Meanwhile, it is important to also observe that in this underexcited synchronous motor case, the projection of the field MMF, $\overline{F}_f$, along the direction of orientation of the resultant MMF, $\overline{F}_{sr} = \overline{F}_{ag}$, indicates that $\overline{F}_f$ is a magnetizing MMF, and the projection of the armature (stator) MMF, $\overline{F}_a = \overline{F}_s$, along the direction of orientation of the resultant MMF, $\overline{F}_{sr} = \overline{F}_{ag}$ indicates that $\overline{F}_a = \overline{F}_s$ is also a magnetizing MMF aiding/strengthening the magnetizing effect of the field MMF, $\overline{F}_f$, because the motor has insufficient field excitation current and consequently insufficient associated MMF, $\overline{F}_f$, hence this state is referred to as the underexcited synchronous motor operating mode. Here again, in this case, the angle $\delta_{sr} = (\pi/2 + \delta - \theta)$, which once more is an obtuse angle as pointed out earlier in the previous case of the overexcited synchronous motor.

The "over-excited" and "under-excited" synchronous motor operating modes lead to a relationship between the magnitudes of the armature currents, I_a, and field excitation currents, I_f, in synchronous motors, the so-called V-curves, which are depicted here graphically in the plots of the magnitudes of I_a versus I_f in Figure 7.19. We next briefly discuss the cases of the "over-excited" and "under-excited" synchronous generators.

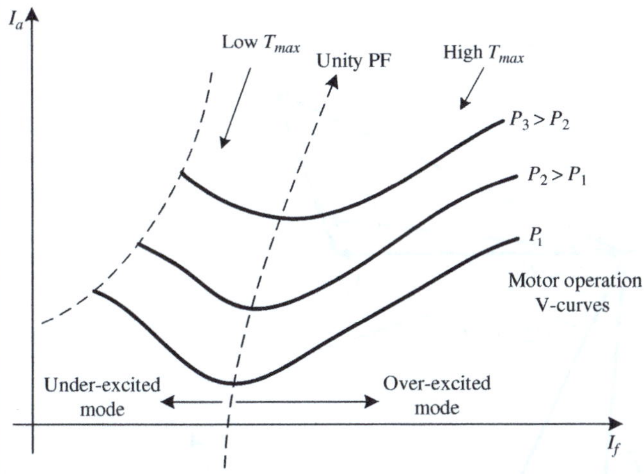

Figure 7.19 The V-curves of a synchronous motor modes of operation from the under-excited to unity power factor to the overexcited modes.

Case 7.3 The Over-Excited Synchronous Generator

In this case, the stator (armature) current, $\overline{I}_a$, see the phasor-space-vector diagram of Figure 7.20, can be expressed as follows:

$$\overline{I}_a = I_a \angle(\pi - \theta) = I_a \cos(\pi - \theta) + jI_a \sin(\pi - \theta) = -I_a \cos\theta + jI_a \sin\theta \quad (7.48)$$

That is,

$$\overline{I}_a^* = -I_a \cos\theta - jI_a \sin\theta \quad (7.49)$$

Hence, the complex power, $\overline{S}$, associated with this generator case, can be expressed as follows:

$$\overline{S} = 3\overline{I}_a^* \cdot \overline{V}_{at} = 3V_{at}[-I_a \cos\theta - jI_a \sin\theta] \quad (7.50)$$

Hence, one can write the following for the complex power, $\overline{S} = P + jQ$:

$$\overline{S} = P + jQ = (-3I_a V_{at} \cos\theta) + j(-3I_a V_{at} \sin\theta) \quad (7.51)$$

Hence, in this over-excited generator case, we have the real power, P, as follows:

$$P = -3I_a V_{at} \cos\theta \quad (7.52)$$

(Continued)

Case 7.3 (Continued)

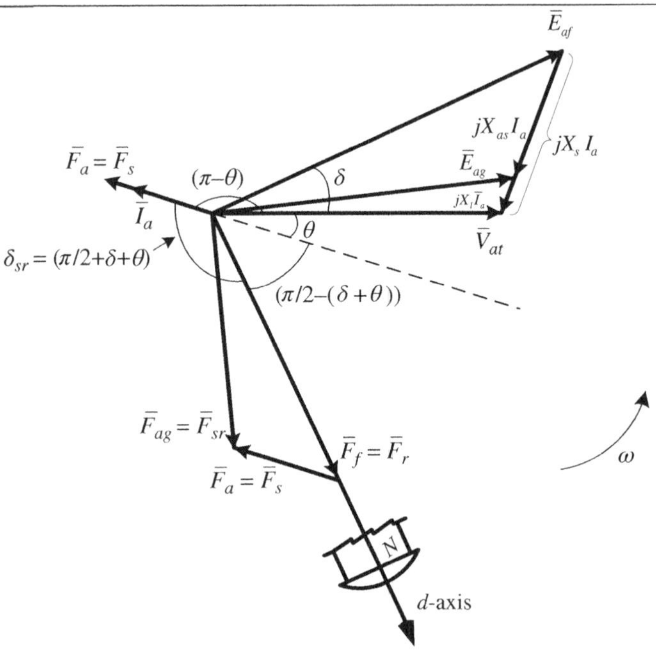

Figure 7.20 The phasor-space-vector diagram of an over-excited generator.

This negative real power means that watts is being generated, and for the reactive power, Q, we have the following:

$$Q = -3I_a V_{at} \sin\theta \tag{7.53}$$

This negative reactive power, Q, means that vars is also being generated by this over-excited synchronous generator. This is typically the case for most power plant generators.

It should further be observed upon close examination of the space-vector portion of the diagram in Figure 7.20 that the armature (stator) MMF space vector, $\overline{F}_a = \overline{F}_s$, is a demagnetizing factor on the resultant MMF, $\overline{F}_{ag} = \overline{F}_{sr}$, hence the need for a large field MMF, $\overline{F}_f = \overline{F}_r$, and a corresponding large amount of field excitation current, I_f, hence the term "overexcited" for this mode of operation of a synchronous generator. We next examine the "under-excited" synchronous generator case. Once again, the reader should notice the obtuse nature of the angle, $\delta_{sr} = (\pi/2 + \delta + \theta)$.

Case 7.4 The Under-Excited Synchronous Generator

In this case, the stator (armature) current, $\overline{I}_a$, as shown in the phasor-space-vector diagram of Figure 7.21, can be expressed as follows:

$$\overline{I}_a = I_a \angle(\pi + \theta) = I_a \cos(\pi + \theta) + jI_a \sin(\pi + \theta) = -I_a \cos\theta - jI_a \sin\theta \quad (7.54)$$

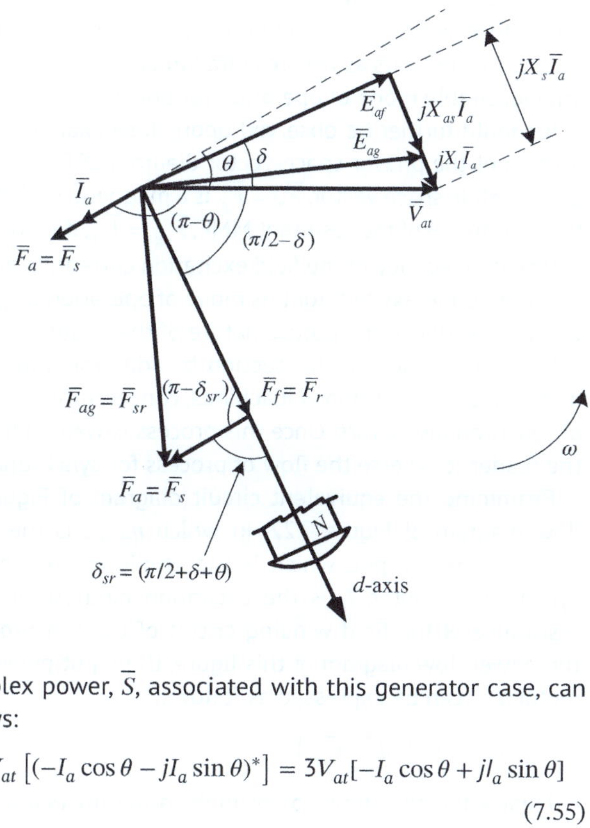

Figure 7.21 The phasor-space-vector diagram of an under-excited generator.

Therefore, the complex power, $\overline{S}$, associated with this generator case, can be expressed as follows:

$$\overline{S} = 3\overline{I}_a^* \cdot \overline{V}_{at} = 3V_{at}\left[(-I_a \cos\theta - jI_a \sin\theta)^*\right] = 3V_{at}[-I_a \cos\theta + jI_a \sin\theta] \quad (7.55)$$

Hence, one can write the following for the complex power, $\overline{S} = P + jQ$, as follows:

$$\overline{S} = P + jQ = (-3I_a V_{at} \cos\theta) + j(3I_a V_{at} \sin\theta) \quad (7.56)$$

Accordingly, in this underexcited generator case, we have the real power, P, as follows:

$$P = -3I_a V_{at} \cos\theta \quad (7.57)$$

(Continued)

Case 7.4 (Continued)

Again, this negative real power means that watts is being generated, and for the reactive power, Q, we have the following:

$$Q = 3I_a V_{at} \sin\theta \tag{7.58}$$

This positive reactive power, Q, means that vars is being consumed by this underexcited synchronous generator. This is atypical for most power plant generators, which are usually operated in the overexcited mode generating both watts and vars as we presented in Case 7.3 earlier. This is a very rare and an undesirable mode of operation for power plant generators.

It should further be observed upon close examination of the space-vector portion of the phasor-space-vector diagram of Figure 7.21 that the armature (stator) MMF space vector, $\overline{F}_a = \overline{F}_s$, is a magnetizing factor aiding in increasing the magnitude of the resultant MMF, $\overline{F}_{ag} = \overline{F}_{sr}$, due to the weakness, or lack of sufficient adequacy, of the field excitation current, I_f, and MMF, $\overline{F}_f = \overline{F}_r$, hence the term "underexcited" for this mode of operation of synchronous generators. Once more, notice the obtuse nature of the angle, $\delta_{sr} = (\pi/2 + \delta + \theta)$.

We finally conclude this section by addressing input, output real powers, and losses in synchronous machines, concentrating our attention on the case of synchronous motors. Once this process is well understood, it will be left to the reader to reverse the flow of process for synchronous generators.

Examining the equivalent circuit diagram of Figure 7.15 and the power flow diagram of Figure 7.22, in which $\eta_{exciter}$ is the exciter efficiency, $I_f^2 R_f$ is the exciter output, which is the ohmic loss in the field winding of the synchronous motor, I_f is the excitation field current, and R_f is the ohmic resistance of the field winding circuit of the synchronous machine. Also, in the power flow diagram of this figure, the input power into the synchronous motor, P_{in}, can be expressed as follows:

$$P_{in} = Re\left(3\overline{I}_a^* \cdot \overline{V}_{at}\right) \tag{7.59}$$

Meanwhile, the ohmic losses in the armature (stator) winding, P_{ohmic}, are as follows:

$$P_{ohmic} = 3I_a^2 R_a \tag{7.60}$$

and the airgap or electromagnetic-airgap power, P_{em}, can be expressed in two ways as follows:

$$P_{em} = P_{in} - P_{ohmic} = Re\left(3\overline{I}_a^* \cdot \overline{V}_{at}\right) - 3I_a^2 R_a \tag{7.61}$$

or

$$P_{em} = 3I_a E_{af} \cos(\angle\overline{I}_a \& \overline{E}_{af}) \tag{7.62}$$

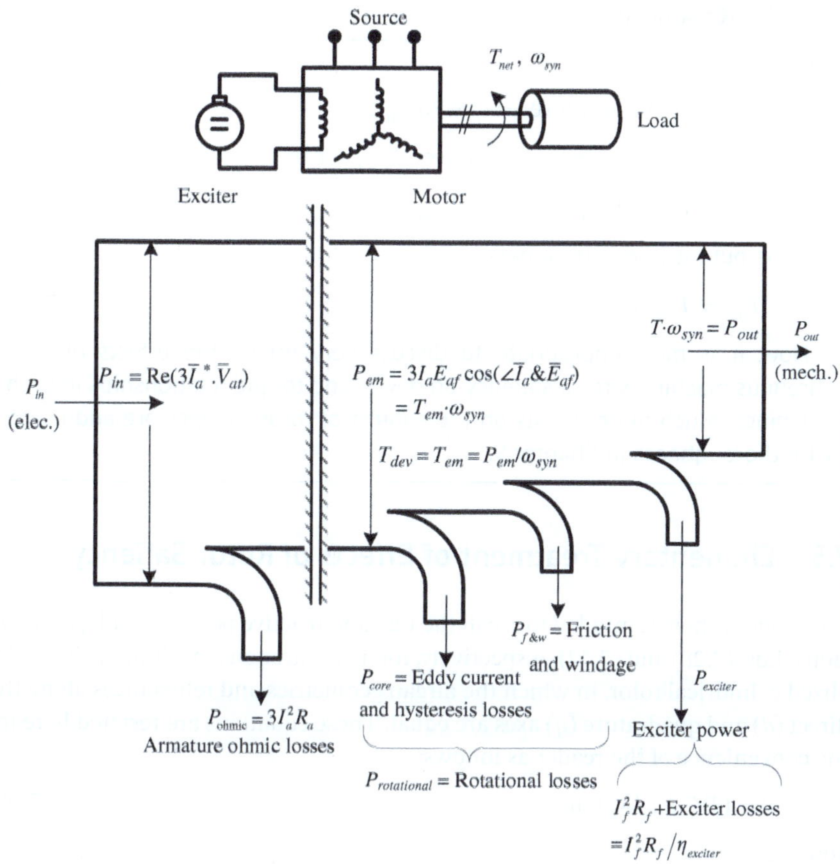

Figure 7.22 The input–output powers and losses and efficiency in a synchronous motor.

Hence, the developed eletromechanical torque,

$$T_{dev} = T_{em} = P_{em}/\omega_{syn} \tag{7.63}$$

where ω_{syn} is the synchronous speed in mechanical radians per second. Other losses are as follows:

$$P_{core} = \text{Eddy current and hysteresis losses in the laminated core of the stator} \tag{7.64}$$

$$P_{f\&w} = \text{Friction and windage losses} \tag{7.65}$$

(Continued)

Case 7.4 (Continued)

and

$$P_{exciter} = \text{Exciter Power} = I_f^2 R_f / \eta_{exciter} \quad (7.66)$$

Hence, the output mechanical power, P_{out}, becomes

$$P_{out} = P_{in} - (P_{ohmic} + P_{core} + P_{f\&w} + P_{exciter}) \quad (7.67)$$

and the output torque becomes

$$T_{out} = P_{out} / \omega_{syn} \quad (7.68)$$

Now, it is most appropriate to discuss very briefly the effects of synchronous machinery rotor saliency on power and torque relationships in such machines. Much more details on the subject of rotor saliency are addressed with thoroughness in Chapter 9.

7.5 Elementary Treatment of Effects of Rotor Saliency

The reader is now reminded to recall the real and reactive power, P and Q, expressions, Eqs. (7.28) and (7.32), respectively, for a synchronous machine with an idealized cylindrical rotor, in which the airgap geometries and reluctances along the direct (d) and quadrature (q) axes are equal. These equations are restated here for the convenience of the reader as follows:

$$P = (3V_{at}E_{af}/X_d)\sin\delta \quad (7.69)$$

and

$$Q = \left[\left(3V_{at}^2/X_d\right) - (3V_{at}E_{af}/X_d)\cos\delta\right] \quad (7.70)$$

where, again, X_d is the d-axis synchronous reactance, where $X_d = X_q$, and δ is the so-called power or torque angle in electrical measure between the terminal voltage phasor, $\overline{V}_{at}$, and the internal field-induced EMF phasor, $\overline{E}_{af}$.

Now, if the rotor saliency manifested in the salient-pole structure, shown in the elementary schematic of Figure 7.13, is taken into account, there will be a d-axis synchronous reactance, X_d, and a q-axis synchronous reactance, X_q, the detail on which will be given in Chapter 9. Meanwhile, X_d and X_q arise mainly from the differences in the magnetic permeances/reluctances along the d- and q-axes. This difference in the magnitudes of X_d and X_q leads to a different set of equations for the real power, P, and reactive power, Q, which can be expressed, saving the details for Chapter 9, as follows:

$$P = \left[(3V_{at}E_{af}/X_d)\sin(\delta) + \left(3V_{at}^2(X_d - X_q)/2X_dX_q\right)\sin(2\delta)\right] \quad (7.71)$$

and

$$Q = -\left[(3V_{at}E_{af}/X_d)\cos\delta - 3V_{at}^2(\cos^2\delta/X_d + \sin^2\delta/X_q)\right] \tag{7.72}$$

The developed torque, T_{dev}, in such a salient-pole synchronous machine can be expressed as follows:

$$T_{dev} = P/\omega_{syn} \tag{7.73}$$

where ω_{syn} in mechanical radians per second is given by

$$\omega_{syn} = (2/p)2\pi f = 4\pi f/p \tag{7.74}$$

Here,

$$X_d = (2\pi f)L_d \tag{7.75}$$

and

$$X_q = (2\pi f)L_q \tag{7.76}$$

where L_d and L_q are the d-axis and q-axis synchronous inductances, respectively, see Refs. [5–11], and recalling that E_{af} is given by

$$E_{af} = \omega L_{afm}I_f/\sqrt{2} = (2\pi f/\sqrt{2})L_{afm}I_f \tag{7.77}$$

The developed torque given in Eq. (7.73) can therefore be expressed for the salient-pole synchronous machine as follows:

$$\begin{aligned} T_{dev} &= [(3pL_{afm}/4\pi\sqrt{2}L_d)I_f(V_{at}/f)]\sin(\delta) \\ &\quad + \left[(3P(L_d - L_q)/16\pi^2 L_d L_q)(V_{at}/f)^2\right]\sin(2\delta) \end{aligned} \tag{7.78}$$

The first term in the torque Eq. (7.78) is the so-called synchronous torque term, while the second term in Eq. (7.78) is the so-called synchronous reluctance torque, because this term will vanish to zero value if and only if $X_d = X_q$, such as in an idealized cylindrical rotor synchronous machine. Of course, in the real world, all synchronous machines, even those with cylindrical rotor forgings, have an inherent degree of saliency because of the existence of rotor slottings to house the field winding, giving rise to different degrees of rotor magnetic reluctances along the d- and q-axes of such cylindrical rotors.

7.6 Synchronous Motor Under Constant Volts Per Hertz Control

Figure 7.23 depicts a functional block diagram of an adjustable speed drive (ASD)-synchronous motor system, of the constant volts-per-hertz control. From

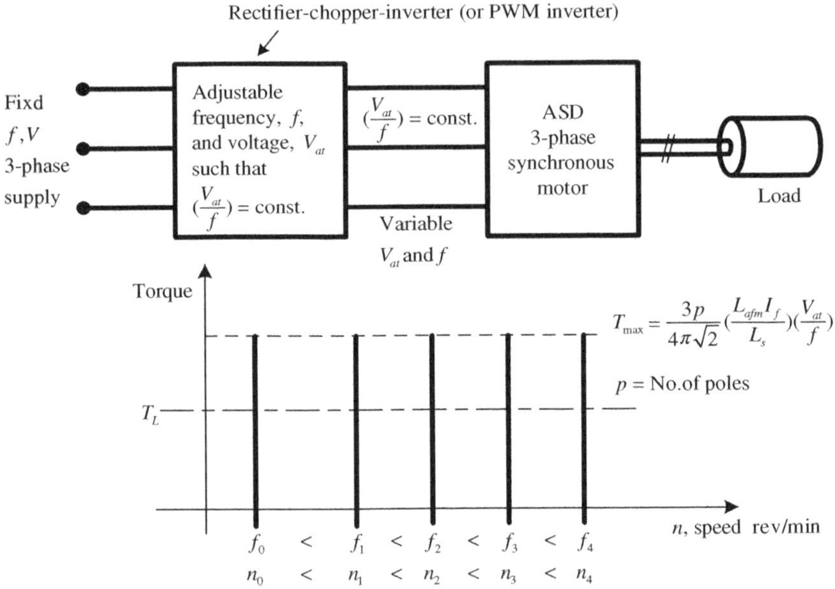

Figure 7.23 A synchronous motor torque–speed characteristic in a constant volts per Hertz (V/f) adjustable speed drive (ASD).

previous developments in Section 7.4, Eq. (7.41), the developed maximum torque, T_{max}, is given by

$$T_{max} = (3/4\pi\sqrt{2})(pL_{afm}/L_d)(I_f)(V_{at}/f) \tag{7.79}$$

where p is the number of poles, I_f is the excitation DC current in the field winding of the synchronous motor, and all other variable symbols are as defined earlier in this and previous chapters.

Also, for a given value of the field current, I_f, and a constant (V_{at}/f) ASD operation, we show the sweep of the motor's torque–speed characteristic as the terminal voltage, V_{at}, and frequency are kept such that

$$(V_{at}/f) = \text{constant} \tag{7.80}$$

For various frequencies, $f_4 > f_3 > f_2 > f_1 > f_0$, the speed, n, in rev/min is given by

$$n_4 = 120f_4/p > n_3 = 120f_3/p > n_2 = 120f_2/p > n_1 = 120f_1/p > n_0 = 120f_0/p \tag{7.81}$$

Here, the sweep of the speed versus torque, while T_{max} is kept constant as long as (V_{at}/f) and I_f are kept constant, is shown in the torque–speed characteristic as shown in Figure 7.23.

We now offer some solved examples on synchronous motor performance calculations and other motor characteristics.

Example 7.1 *A Synchronous Motor in an ASD System*

A 6-pole synchronous motor, in an adjustable speed drive system, draws 50 kW when the motor is energized at 220 V, 60 Hz (3-phase) at a power factor (PF) of 0.85 leading (over-excited). The motor is Y-connected and has a synchronous reactance, $X_s = 0.7\ \Omega$/phase, at 60 Hz. The armature resistance is negligible. Without any changes beyond the level of field excitation under these conditions, determine the outer envelope of the maximum torque in the torque–speed plane for a constant volts/Hertz power conditioning up to 60 Hz, then as voltage is held constant at its rated value of 220 V (line-to-line), while increasing the frequency up to a value of 120 Hz. Specify the maximum torque and speed at 90 and 120 Hz.

Solution

$$V_{L-L} = 220\ \text{V}, V_{ph} = \frac{220}{\sqrt{3}} = 127.02\ \text{V}$$

6-pole, 60 Hz, Y-connected synchronous motor with $X_s = 0.7\Omega$, 0.85 PF (leading over-excited)

$$n_{syn}(\text{at } 60\ \text{Hz}) = \frac{120f}{p} = \frac{120 \times 60}{6} = 1200\ \text{r/min}$$

$$\omega_{syn} = \frac{2\pi}{60} \times 1200 = 40\pi\ \ \text{rad/sec}$$

Hence, $\overline{V}_{at} = 127.02\angle 0^\circ$ V

$$\text{P.F.} = \cos\theta = 0.85 \Rightarrow \theta = \cos^{-1}(0.85) = 31.79^\circ$$

$$P = 3I_a V_{ta} \cos\theta = 50 \times 10^3\ \text{W}$$

Hence, $3 \times I_a \times 127.02 \times 0.85 = 5 \times 10^4$ W

Hence, $I_a = \dfrac{5 \times 10^4}{3 \times 127.02 \times 0.85} = 154.37$ A

Hence, $\overline{I}_a = 154.37\angle 31.79^\circ$

Because $\overline{E}_{af} + jX_s\overline{I}_a = \overline{V}_{at}$

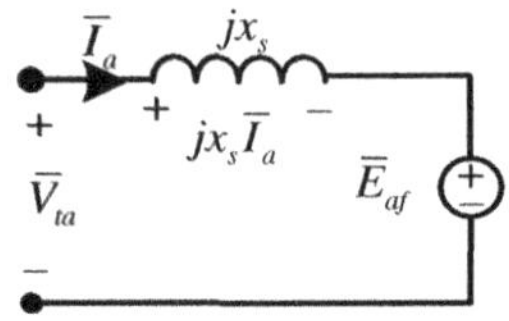

Hence, $\overline{E}_{af} = \overline{V}_{at} - jX_s\overline{I}_a = 127.02\angle 0° - j0.7 \times 154.37\angle 31.79°$

$$\overline{E}_{af} = 127.02\angle 0° + 108.06\angle - 58.21°$$

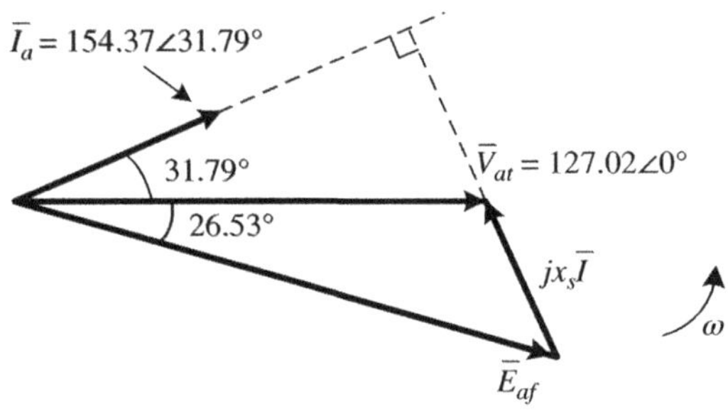

$$\overline{E}_{af} = 183.95 - j91.85 = 205.61\angle - 26.53° \text{ V}$$

Check on E_{af} and δ

$$P = \frac{3V_{at}E_{af}}{X_s}\sin\delta = \frac{3 \times 127.02 \times 205.61}{0.7}\sin(26.53°)$$

$P = 49950$ watts checks with 50,000 W

Because $T_{\max} = \dfrac{3V_{at}E_{af}}{\omega_{syn}X_s} = \dfrac{3E_{af}V_{at}}{\frac{2\pi}{60} \times \frac{120f}{p} \times X_s} = \dfrac{3pE_{af}}{4\pi X_s}\left(\dfrac{V_{at}}{f}\right)$

Hence, for $0 < f \leq 60$, $\left(\dfrac{V_{ta}}{f}\right) = \text{Constant}$

$$\left(\frac{V_{at}}{f}\right) = \frac{127.02}{60} = 2.117 \text{ V/Hz}$$

$$T_{\max} = \frac{3 \times 6 \times 205.61}{4\pi \times 0.7}\left(\frac{V_{at}}{f}\right) = 420.74\left(\frac{V_{at}}{f}\right) \text{ Nm}$$

$$T_{\max} = 420.74 \times 2.117 = 890.71 \text{ Nm}$$

Hence, for $60 < f < \infty$, $\frac{V_{at}}{f} = \frac{127.02}{f}$

Hence, $T_{max} = 420.74 \times \frac{127.02}{f} = \frac{53442.4}{f}$ Nm

For $f = 60$ Hz $\Rightarrow T_{max} = 890.71$ Nm

$$f = 90 \text{ Hz} \Rightarrow \quad T_{max} = \frac{53,442.4}{90} = 593.8 \text{ Nm}$$

$$f = 120 \text{ Hz} \Rightarrow \quad T_{max} = \frac{53,442.4}{120} = 445.36 \text{ Nm}$$

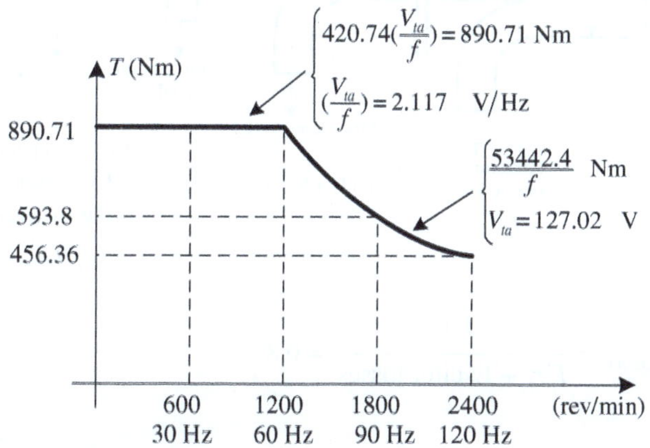

Example 7.2 ***Motor Losses and Efficiency***

A 150-hp, 12-pole, 0.8 p.f., 2300 V, 38 A, 60 Hz, 3-phase, Y-connected synchronous motor has a direct-connected exciter (on the same shaft) to supply its field current. For the purposes, of this problem, the efficiency of the exciter may be assumed constant at a value of 80%. The synchronous motor is run at no-load from a 2300 V, 60 Hz supply, with its field current supplied by its exciter, and the following readings were taken:

Armature voltage between terminals = 2300 V, armature current = 38 A, 3-phase power input = 13.7 kW, field current = 20 A, and voltage applied to field winding = 300 V.

When the synchronous motor is loaded so that its input is 38 A at 0.8 p.f. and 2300 V between terminals, its field current is found to be 17.3 A. Under these conditions, what is the efficiency of the synchronous motor (7.82) exclusive of exciter losses and (7.82) inclusive of exciter losses? What is the useful mechanical power output in kilo watts and horsepower, and what is the net output torque in Newton meters.

Solution

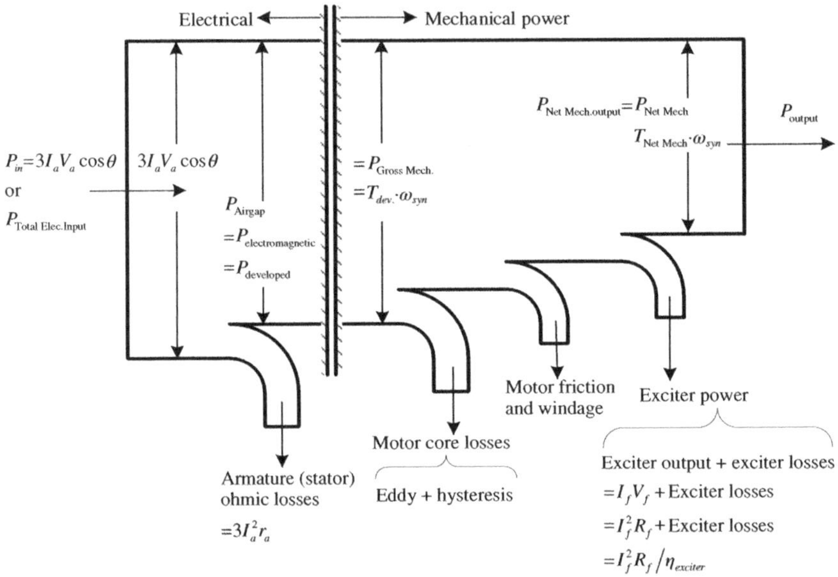

Because $\eta_{exciter} = \dfrac{I_f^2 r_f}{I_f^2 r_f + \text{Exciter losses}} = 0.8$

Hence,

$$\text{Exciter losses} = 0.25 I_f^2 r_f \tag{7.82}$$

Under any operating condition, we have the following:

$$P_{\text{input}} = P_{\text{output}} + \sum \text{Losses} \tag{7.83}$$

At no-load, we have the following:

$$P_{\text{output}} = 0.0$$

Hence, from (7.82), one can write the following:

$$P_{\text{input}} = \sum \text{Losses} = 13.7 \times 10^3\ \text{W}$$

Hence, $13.7 \times 10^3 = 3I_a^2 r_a + \text{Core losses} + \text{Friction and windage} + (I_f^2 r_f + \text{Exciter losses})$

But from the no-load test data, we have the following:

$$V_t = 2300\ \text{V}, I_a = 38.0\ \text{A}, V_f = r_f I_f = 300\ \text{V}$$

Hence, $r_f = \dfrac{300}{I_f} = \dfrac{300}{20} = 15\ \Omega$

Hence, $13{,}700 = 3I_a^2 r_a + \text{Core losses} + \text{Friction and windage} + 1.25 \times 20^2 \times 15$

Hence,

$$3I_a^2 r_a + \text{Core losses} + \text{Friction and windage} = 6200\ \text{W} \tag{7.84}$$

Under load, we have the following:
$I_a = 38.0$ A unchanged from no-load condition.
Hence, $3I_a^2 r_a$ losses component is unchanged from no-load condition.
$V_t = 2300$ V, unchanged from no-load condition.
Also, frequency is the same.
Hence, flux ($\Phi \cong V_t/4.44fN_{ph}k_w$) is unchanged from no-load condition.
Hence, core losses ($\propto\Phi^2 f^2$) are unchanged from no-load condition.
Also, friction and windage losses ($\propto$ speed) will also remain unchanged under load condition.
Hence, under load, we have, from (7.82):

$$3I_a^2 r_a + \text{Core losses} + \text{Friction and windage} = 6200\ \text{W}$$

However, $I_f^2 r_f = 17^2 \times 15 = 4500$ W

$$\text{Exciter losses} = 0.25 I_f^2 r_f = 1125\ \text{W}$$

Because $P_{\text{input}} = 3I_a V_t \cos\theta$

Hence, $P_{\text{input}} = 3 \times 38 \times \dfrac{2300}{\sqrt{3}} \times 0.8 = 121{,}000$ W

Without Exciter Losses, using energy balance diagram in Figure 7.22, we have the following:

$$P_{\text{output}} = 121{,}000 - 6200 - 4500 = 147.9\ \text{HP} = 110{,}300\ \text{W}$$

$$\eta_{motor} = \frac{110{,}300}{121{,}000} = 0.9116\ \text{p.u} = 91.16\%$$

Because $n_{syn} = \dfrac{120f}{p} = \dfrac{120 \times 60}{12} = 600\ \text{r/min}$

Hence, $T_{output} = T_{NetMech.} = \dfrac{P_{out}}{\frac{2\pi}{60}(n_{syn})} = \dfrac{60 \times 110{,}300}{2\pi \times 60} = 1755.5\ \text{Nm}$

With Exciter Losses, using energy balance diagram in Figure 7.22, we have the following:

$$P_{output} = 121{,}000 - 6200 - 4500 - 1125 = 146.3\ \text{HP} = 109{,}175\ \text{W}$$

$$\eta_{motor} = \frac{109{,}175}{121{,}000} = 0.9023\ \text{p.u} = 90.23\%$$

Hence, $T_{output} = \dfrac{60 \times 109{,}175}{2\pi \times 60} = 1737.6\ \text{Nm}$

7.7 Vector Control of Synchronous Motors in ASDs

Similar to the vector control of induction motors covered in Chapter 6, the vector control of synchronous motors is essentially to decouple the developed electromagnetic torque and rotor MMF (also often mistakenly called rotor flux) producing components from the stator AC currents, by applying the Clark and Park transformations. Such a decoupling is to achieve the control convenience of complex three-phase synchronous motors in the same manner as DC motors with separate excitations. Considering the broad and increasing applications of permanent magnet synchronous motors (PMSM) in industry, the vector control to be explained here in this section will focus on three-phase PMSMs.

Unlike induction motors that require a positive *d*-axis current reference (i_d^*) to generate the magnetizing current, the magnets in the rotor of PMSM produce fixed rotor flux linkages, which means that *d*-axis current reference (i_d^*) should be set to zero in the normal operation region. The *q*-axis current reference (i_q^*) will be generally the output of the speed controller, which is the same as in vector control systems of induction motors, shown in Figure 7.24.

During motor operation, as the rotor speed increases, the associated back EMF voltage increases proportionally, which reduces the applied voltage across the motor stator windings fed by the ASD. The rated speed of the motor (also named base speed) is determined by the maximum output voltage of an ASD, and such maximum output voltage is generally constrained by the rated DC-bus voltage of the inverter in the ASD. When the required applied voltage is greater than the maximum voltage available from an ASD, the drive no longer supplies current to the motor stator windings and the motor will not generate any torque. If the rotor is externally forced to rotate faster, the back EMF voltage will exceed the power

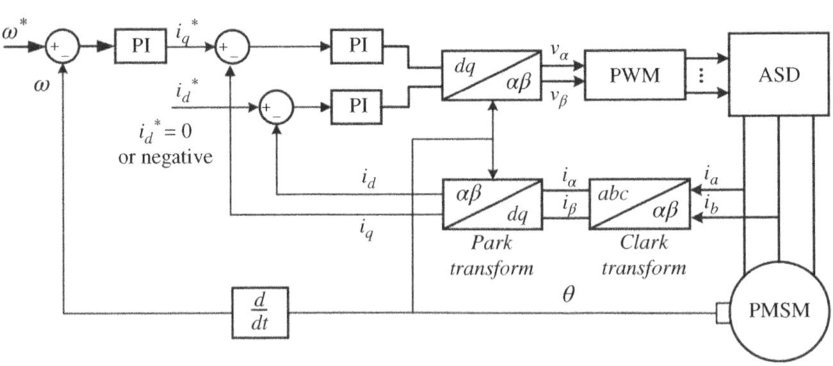

Figure 7.24 Vector control of a three-phase permanent magnet synchronous motor.

supply voltage and the current will attempt to flow from the stator windings to the ASD, resulting in a negative torque counter to the rotation direction.

A wide speed range is preferred in many applications. For instance, for traction motors used in electric vehicles, constant power speed ratio (CPSR), which is a ratio of the maximum speed to the rated speed, is used as a critical parameter to evaluate the performance of traction motors. To achieve a wide speed range beyond the rated speed, an approach is needed to reduce the back EMF at high speed. Since the back EMF is a function of the magnetic flux linkage between the rotor magnets and the stator windings, reducing this magnetic flux linkage will reduce the back EMF voltage. Eventually, a higher rotor speed can be achieved, although this would be at the expense of a reduced torque. The technology of reducing magnetic field strength by applying an opposing magnetic field on the motor windings out of phase (180°e) with respect to the rotor magnetic field is widely called field-weakening control. Field-weakening control is generally implemented in the vector control by setting the i_d^* as a negative current. The variations of the stator voltage and q-axis current during the constant torque region and constant power region of the PMSM are depicted in Figure 7.25. As can be seen, during the constant torque region, the rotor speed is proportional to the increase of the stator voltage, until reaching the rated base speed, while the d-axis current stays at zero during this region. Meanwhile, in the constant power region (i.e., above the rated speed), the stator voltage reaches the maximum value due to the limited voltage rating of the DC bus in the ASD, and the q-axis current i_q is controlled as a negative value to weaken the magnetic flux. Specifically, the higher the rotor speed, a greater negative i_q needs to be controlled, as shown in Figure 7.25. Further details on field weakening are given in Chapter 8, where PM brushless DC motor operation is discussed.

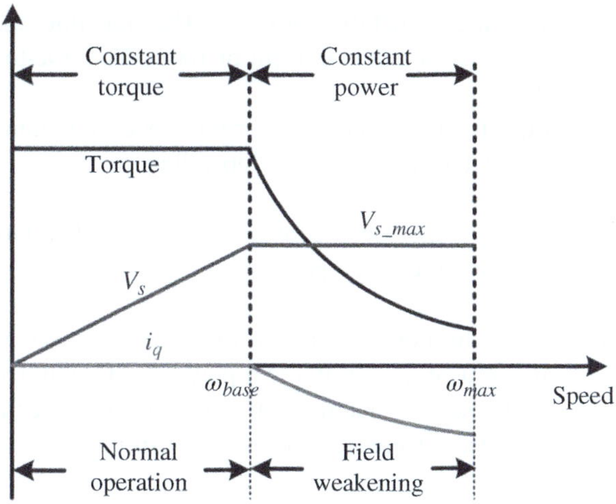

Figure 7.25 Variations of the stator voltage and q-axis current during the constant torque region and constant power region of PMSMs.

Homework Problems

7.1 A 4-pole, 60-Hz, Y-connected, 3-phase synchronous generator has a regulated terminal line-to-neutral voltage of $\overline{V_{as}} = 260V_{rms}\angle 0^{\circ}$, a synchronous reactance of 0.06 Ω, a stator resistance of 0.003 Ω, and a rotor field winding inductance of 0.02 H. The balanced 3-phase load draws 2 MW at a lagging 0.8 p.f.

(a) Compute the three-phase complex power of the load (magnitude and angle).
(b) Compute the generator phase current (magnitude and angle; hint: the angle of the current should be negative for the load to have a lagging power factor).
(c) Find the required excitation voltage.
(d) Calculate the required field current.
(e) Calculate the required speed of the prime mover.

7.2 Given the same machine as in Problem 7.1, if the excitation voltage is $E_a = 360V_{rms}\angle 14^{\circ}$ and the field current is adjusted to maintain the same terminal voltage as in Problem 1,

(a) Find the new generator phase current.
(b) Calculate the three-phase real power consumed by the load and identify the new power factor.
(c) Calculate the generator efficiency (consider stator winding losses as the only losses).

7.3 A 3-phase, Y-connected, 6-pole, 60 Hz synchronous generator has a synchronous reactance of 1 Ω, negligible stator resistance, and a rotor field winding inductance of 0.2 H. The machine operates at a power angle of 31.48° with a field current of 100 A. The line-to-neutral terminal voltage is $2400V_{\text{rms.}}$

(a) Find the speed of the prime mover in radians per seconds.
(b) Determine the excitation voltage.
(c) Find the line current.
(d) Find the power consumed by the load and its power factor.
(e) Find the developed torque.

7.4 A synchronous machine with a synchronous reactance of 1.28 p.u. is operating as a generator at a real power loading of 0.6 p.u. connected to a system with a series reactance of 0.07 p.u. An increase in its field current is observed to cause a decrease in armature current.

(a) Before the increase, was the generator supplying or absorbing reactive power from the power system?
(b) As a result of this increase in excitation, did the generator terminal voltage increase or decrease?
(c) Repeat parts (a) and (b) if the synchronous machine is operating as a motor.

References

1 Say, M.G. (1976). *Alternating Current Machines*. Pitman.
2 Nehl, T.W., Fouad, F.A., and Demerdash, N.A. (1982). Determination of saturated values of rotating machinery incremental and apparent inductances by an energy perturbation method. *IEEE Transactions on Power Apparatus and Systems* PAS-101: 4441–4451.
3 Demerdash, N.A., Fouad, F.A., and Nehl, T.W. Determination of winding inductances in ferrite type permanent magnet electric machinery by finite elements. *IEEE Transactions on Magnetics* MAG-18: 1052–1054.
4 Demerdash, N.A., Hijazi, T.M., and Arkadan, A.A. (1988). Computation of winding inductances of permanent magnet brushless DC motors with damper windings by energy perturbation. *IEEE Transactions on Energy Conversion* EC-3 (3): 705–713.
5 Chaudrhry, S.R., Ahmed-Zaid, S., and Demerdash, N.A. (1995). Coupled finite-element/state-space modeling of turbogenerators in the ABC frame of reference – the no-load case. *IEEE Transactions on Energy Conversion* 10 (1): 56–62.
6 Chaudrhry, S.R., Ahmed-Zaid, S., and Demerdash, N.A. (1995). Coupled finite/state-space modeling of turbogenerators in the ABC frame of reference – the short-circuit and load cases including saturated parameters. *IEEE Transactions on Energy Conversion* 10 (1): 63–70.
7 Deng, F. and Demerdash, N.A. (1996). A coupled finite element – state space approach for synchronous generators – Part I: model development. *IEEE Transactions on Aerospace and Electronic Systems* 32 (2): 775–784.
8 Deng, F., Demerdash, N.A., Vaidya, J.G., and Shah, J.J. (1996). A coupled finite element – state space approach for synchronous generators – Part II: applications. *IEEE Transactions on Aerospace and Electronic Systems* 32 (2): 785–794.
9 Deng, F. and Demerdash, N.A. (1997). CFE-SS approach for synchronous generators under unbalances – Part I: the unbalanced loads. *IEEE Transactions on Aerospace and Electronic Systems* 33 (1): 142–152.

10 Demerdash, N.A. and Deng, F. (1997). CFS-SS approach for synchronous generators under unbalances – Part II: the balanced and unbalanced rectifier loads. *IEEE Transactions on Aerospace and Electronic Systems* 33 (1): 152–162.

11 Deng, F. and Demerdash, N.A. (1998). Comprehensive salient-pole synchronous machine parametric design analysis using time-step finite element – state space modeling techniques. *IEEE Transactions on Energy Conversion* 13 (3): 221–229.

8

Brush-Commutator and Brushless DC Machines

In this chapter, we will discuss and analyze legacy brush-commutator-type DC machines and modern brushless DC machines. We will discuss their main constructional features, stator and rotor configurations and windings, their equivalent circuits, and performance computation and modeling. This is in addition to means of torque production, torque–speed characteristics and control, the input–output powers and losses, and other ancillary performance features and characteristics. Here, we start with the main constructional features of brush-commutator-type and brushless DC machines.

8.1 Main Constructional Features of Brush-Commutator and Brushless DC Machines

An appropriate start to the discussion of the main constructional features of the legacy brush-commutator DC machines and brushless DC machines is presented in the geometric schematics (a), (b), and (c) in Figure 8.1. In this figure, schematic (a) shows the cross-section of a two-pole wound-field excited brush-commutator-type DC machine, with the slotted rotor housing the armature winding and the associated commutator segments with the positive and negative brushes pressing against the outer cylindrical surface of the segmented commutator. Meanwhile, schematic (b) shows the cross-section of a two-pole permanent magnet (PM) excited brush-commutator-type DC machine, with a rotor similar to that of the wound-field excited case of schematic (a). While in schematic (c) of Figure 8.1, a brushless DC machine cross-section is shown, in which the stator is housing a three-phase winding with phases (*a*), (*b*), and (*c*), that connects to a three-phase inverter (not shown in the figure), and the rotor has two surface-mounted PMs to form a two-pole rotor, the position of which is

Electric Machinery and Drives: An Electromagnetics Perspective, First Edition.
Nabeel A. O. Demerdash, JiangBiao He, and Hao Chen.

Companion website: www.wiley.com/go/demerdash/electricmachineryanddrives

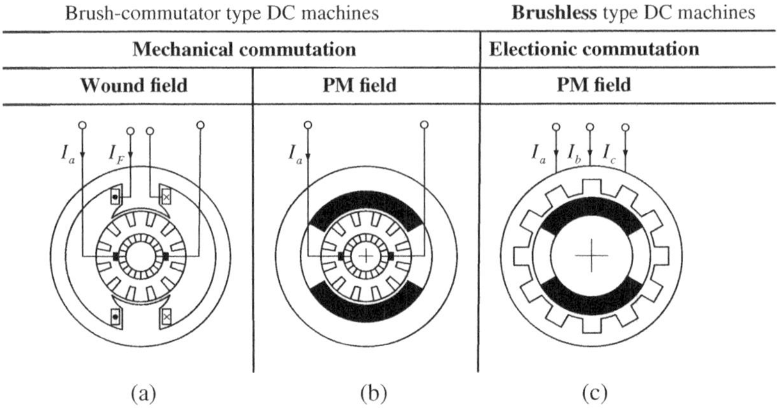

Figure 8.1 Brush-commutator and brushless DC machines. (a) Wound field brushed DC motor (b) PM field brushed DC motor (c) PM field brushless DC motor.

detected by a resolver (not shown in the figure) that initiates the switching in the inverter as will be detailed later on in this chapter.

In Figure 8.2, we show more schematic details of the constructional features of a two-pole brush-commutator DC machine, with shunt and series-field excitation coils (windings) wrapped around the two main poles. Also, two interpoles (or commutation poles) are shown in this figure, which are used to mitigate the adverse effects of the DC armature magnetomotive force (MMF) on the commutation process. Here, the DC armature is housed on the rotor, with proper slotting arrangements, while the brushes are depicted to indicate that they are in electrical contact with conductors (or coil sides) that are located midway in the interpolar regions of the machine. All these windings constituting a brush-commutator DC machine are shown in a schematic representation in Figure 8.3. That is, this figure schematically depicts the DC armature, the interpolar, the series, and shunt field excitation windings.

Figure 8.4 shows the main constructional features of an elementary two-pole, rotor surface-mounted, permanent magnet (SPM) brushless DC machine. The corresponding elementary three-phase A, B, and C phase windings are housed in a six-slotted stator, with the proper slotting sequence: (A), ($-C$), (B), ($-A$), (C), and ($-B$), to give rise to an elementary three-phase winding housed in a six-slots laminated stator assembly. Meanwhile, Figure 8.5 depicts the main elementary constructional features of two types of two-pole inner rotor-mounted bar-shaped and V-shaped, inner permanent magnet brushless DC machines, and their elementary three-phase six slots: (A), ($-C$), (B), ($-A$), (C), and ($-B$) stators. The switching sequence of the associated inverter will be explained in detail later in Section 8.9.

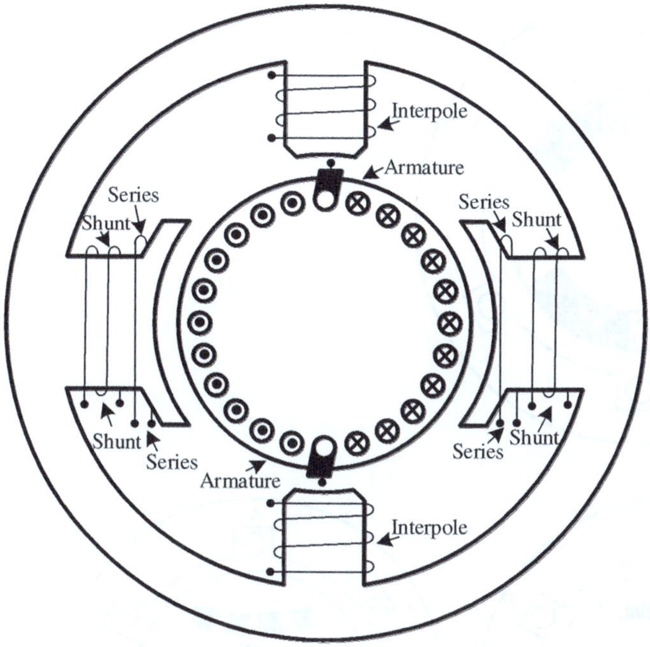

Figure 8.2 Main constructional features of brush-commutator DC machines.

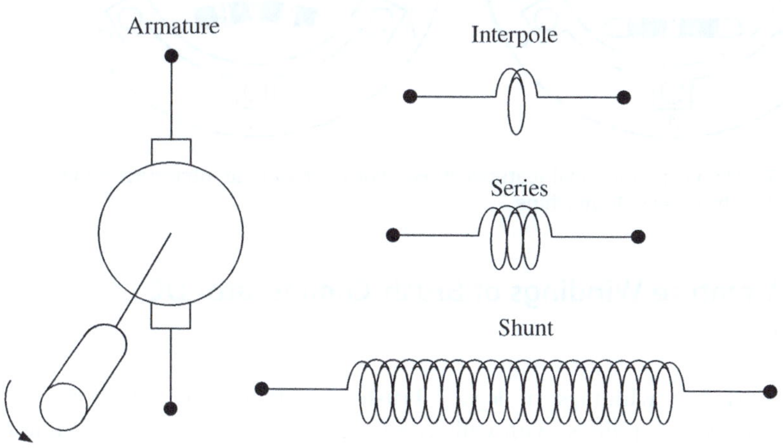

Figure 8.3 A schematic representation of windings in a brush-commutator DC machine.

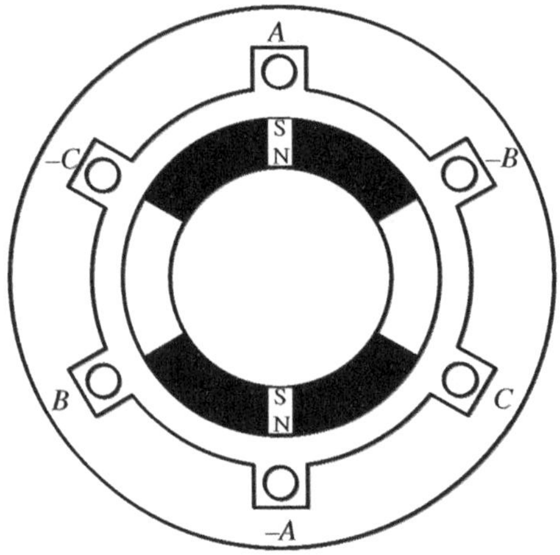

Figure 8.4 Main constructional features of a two-pole SPM brushless DC machine.

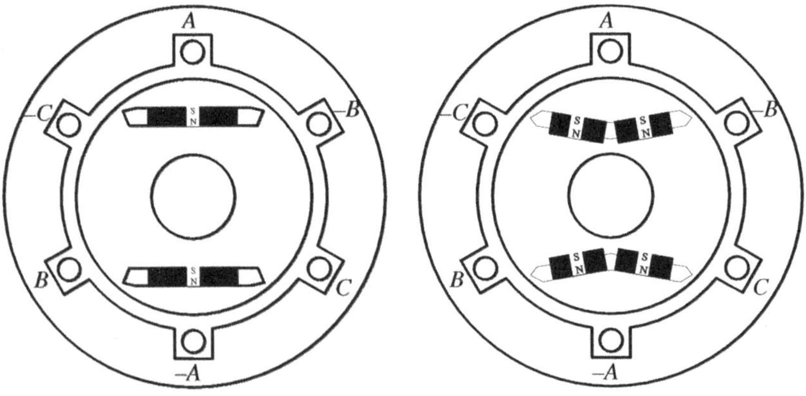

Figure 8.5 Main constructional features of two types of two-pole interiorpermanent magnet (IPM) brushless DC machine.

8.2 Armature Windings of Brush-Commutator DC Machines

Armature (rotor) windings of legacy brush-commutator DC machines are double-layer-type windings, with coils connected to a segmented commutator, as shown in Figure 8.6 for a lap winding and in Figure 8.7 for a wave winding. In Figures 8.6 and 8.7, the coil span (or pitch), τ_c, and pole pitch, τ_p, are shown.

In electrical measure, τ_p = 180° electrical. However, the coil pitch, τ_c, is usually much less than 180° electrical. The positive and negative brushes are always 180° electrical apart, that is, a pole pitch, τ_p, apart. In Figures 8.6 and 8.7, the double-layer nature of the windings is shown by the cross-sectional drawing of the slots, and the solid as well as dotted lines representing the coil sides. It should further be pointed out that for simple-lap windings, Figure 8.6, the number of parallel paths, m, is equal to the number of poles, p, in such a machine, that is, $m = p$. Meanwhile, for simple-wave winding, Figure 8.7, the number of parallel paths, m, is equal to two, that is, $m = 2$. Meanwhile, Figure 8.8 shows a coil in the armature of a DC machine as its position advances in time from position #①, to position #②, to position #③, over an arc of a full pole pitch, τ_p. In position #①, the induced voltage in both coil sides is "zero" value because of the presence of both coil sides in the middle of the interpolar region with "zero" radial flux densities. That induced voltage reaches its maximum value when this coil is at position #② because of the exposure of the coil sides to the highest flux densities, and this induced voltage collapses back to "zero" value in position #③ as in position #①, again because of the presence of both coil sides in the middle of another interpolar region. It should be pointed out that this coil experiences being shorted by the brushes when it is located at positions #① and #③, when the

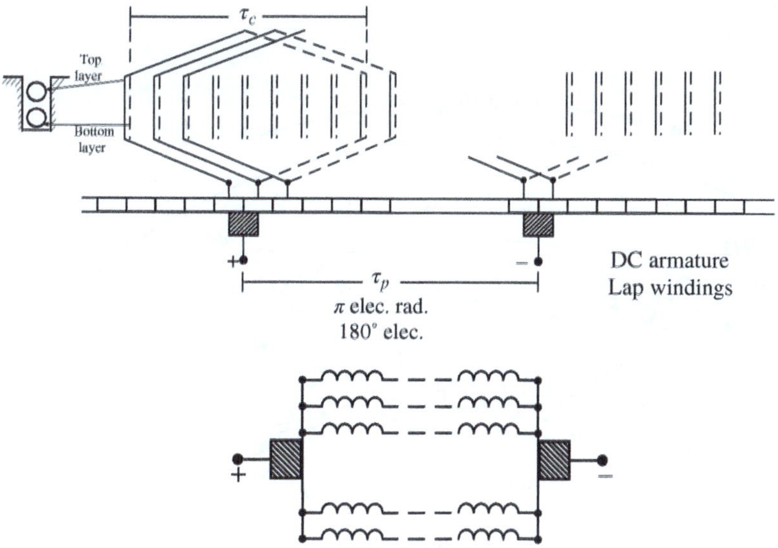

Figure 8.6 Simple lap winding schematic for brush-commutator DC machines, $m = p$.

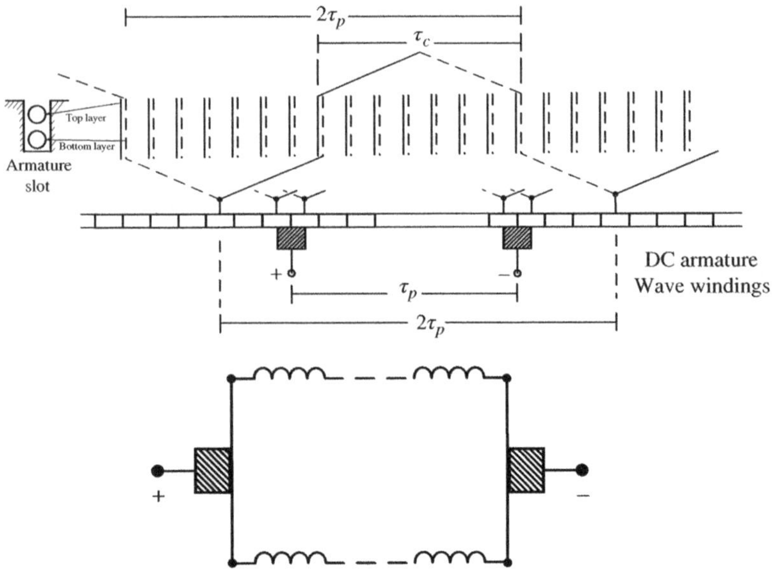

Figure 8.7 Simple wave winding schematic for brush-commutator DC machines, $m = 2$.

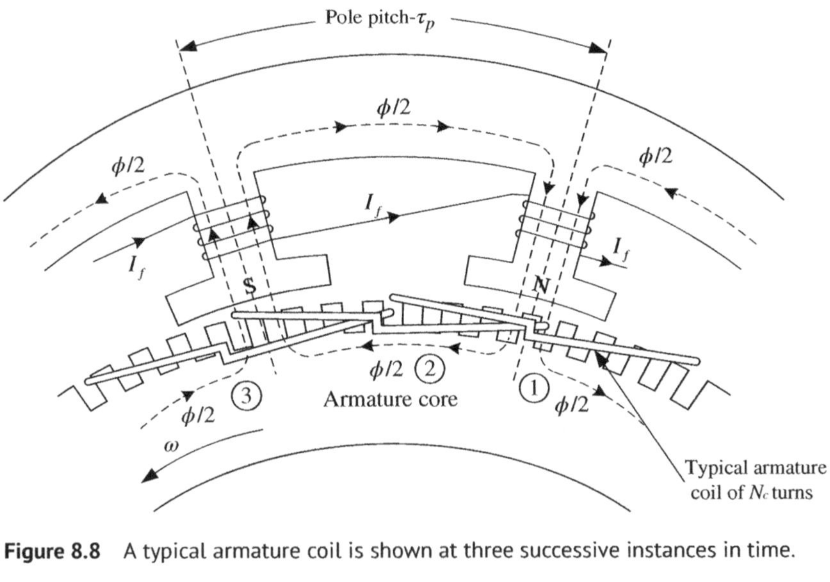

Figure 8.8 A typical armature coil is shown at three successive instances in time.

commutation process, that is, the mechanical rectification of AC voltage induced in the coil, is rectified and is appearing through the brushes as a rectified DC voltage.

Meanwhile, the amplitude of the armature MMF, $\overline{\mathbf{F}}_a$, as a space vector, for which from earlier developments in Chapter 5, Eq. (5.71), one can write the following:

$$F_a = (8/\pi^2)(N_a I_a/mp) \tag{8.1}$$

where F_a is the amplitude of the armature MMF in ampere-turns per pole, N_a is the number of DC armature turns, I_a is the DC armature current, m is the number of parallel paths in the armature winding, and p is the number of poles.

Meanwhile, the field winding MMF, $\overline{\mathbf{F}}_f$, as a space vector, where its magnitude from earlier developments in Chapter 5, Eq. (5.57), one can write the following:

$$F_f = (4/\pi)k_{f_1}(N_f/p)I_f \tag{8.2}$$

where F_f is the amplitude of the field winding's MMF in ampere-turns per pole, N_f is the number of turns in the field winding, k_{f_1} is the fundamental harmonic form factor, and I_f is the DC current flowing in the field winding.

Furthermore, it is of paramount importance to point out, by close examination of the elementary nature of the two-pole DC machine of Figure 8.9, the inherent electrical angular quadrature nature of the two space vectors, $\overline{F}_a$ and $\overline{F}_f$ (or MMFs $\overline{\mathbf{F}}_a$ and $\overline{\mathbf{F}}_f$, regardless of the number of poles, p, in any DC machine). Hence, from torque derivation developments presented earlier in Chapter 5, one can write the following for the developed electromechanical torque, T_{dev}, in Nm in any DC machine, see Figure 8.9:

$$T_{dev} = -(p/4)(\mu_0 \pi Dl/h_g)F_f F_a \sin \delta_{fa} \tag{8.3}$$

where p is the number of poles, D is the diameter at mid-airgap, l is the active conductor/stack length of the armature laminations, h_g is the radial height of the airgap between the pole face and the armature laminations, and F_f as well as F_a are as given in Eqs. (8.1) and (8.2). However, since, $\delta_{fa} = \delta_{sr} = \pi/2$ elec. rad. $= 90\,°$e, it follows that T_{dev} in magnitude is given by

$$T_{dev} = (p/4)(\mu_0 \pi Dl/h_g)F_f F_a \tag{8.4}$$

Now, we are in a position to develop further the equivalent electrical circuits and performance characteristics of the separately excited DC motor. However, before proceeding further, one should observe the locations of induced electromagnetic "N" and "S" poles on the stator and rotor in the machine cross-section of Figure 8.9 to realize that the attraction forces between the "N" and "S" polarities on the rotor

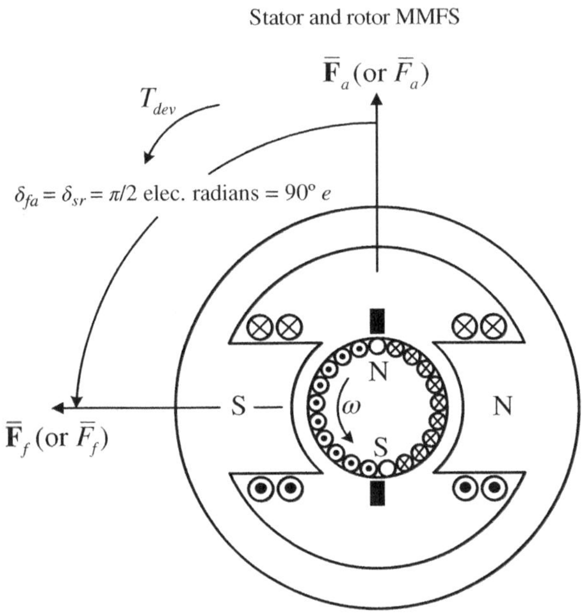

Figure 8.9 The MMFs of the armature and field windings, $\overline{\mathbf{F}}_a$ and $\overline{\mathbf{F}}_f$, in a brush-commutator DC machine.

and stator clearly indicate a torque, T_{dev}, orientation on the rotor (armature) of the DC machine in the counterclockwise direction.

8.3 Separately Excited DC Motors, Modeling, and Formulations

Figure 8.10 shows the field and armature winding connections as well as the equivalent electric circuits of a separately excited DC motor, where E_a is the back electromotive force (EMF) induced in the armature due to the field flux, see Chapter 5, Eqs. (5.39) and (5.40) for details, R_a is the armature circuit resistance, I_a is the armature current, and V_a is the armature terminal voltage. Meanwhile, in the same figure, I_f is the field winding current, R_f is the field winding resistance, and V_f is the applied voltage at the terminals of the field winding.

Meanwhile, Figure 8.11 depicts a separately excited DC motor with its field winding being supplied from a variable voltage source through a potentiometer/variable contacts on rheostat. The equivalent circuit model of both the armature and the field winding circuits is also shown in Figure 8.11.

The reader should recall, from earlier developments in Chapter 5, that the back EMF, E_a, in volts is given by the following in a DC machine:

$$E_a = [pC_a/(2\pi m)]\omega_m \Phi = k_a \omega_m \Phi \tag{8.5}$$

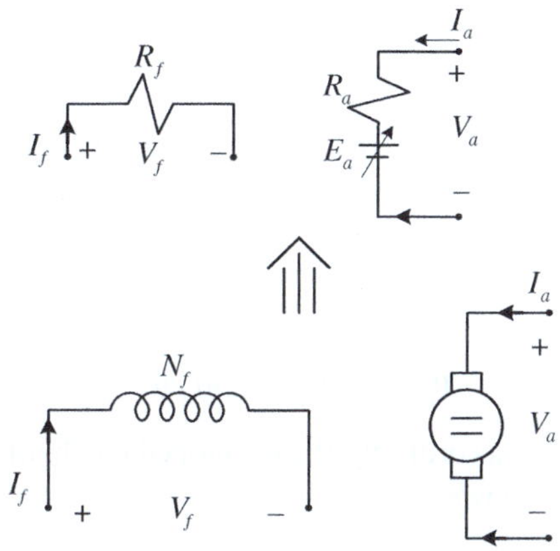

Figure 8.10 Schematic winding representation and equivalent circuit model of separately excited DC machine.

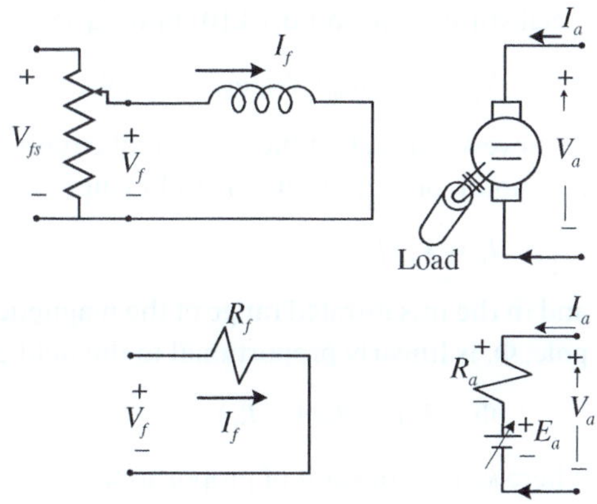

Figure 8.11 Equivalent circuit model of separately excited DC motors.

where C_a is the number of conductors in a DC machine's armature winding, m is the number of parallel paths in such an armature winding, and ω_m is the angular speed of rotation of the armature winding in mechanical radians per second, while Φ is the flux per pole in Webers.

Here also,

$$\omega_m = (2\pi/60)n \tag{8.6}$$

where n is the angular speed of rotation of the armature in r/min, hence, one can express the back EMF, E_a, as follows:

$$E_a = (pC_a/60m)n\Phi = k'_a n\Phi \tag{8.7}$$

It should be pointed out that in Eq. (8.5),

$$k_a = (pC_a/2\pi m) \tag{8.8}$$

while, in Eq. (8.7)

$$k'_a = (pC_a/60m) \tag{8.9}$$

In such a DC machine, the developed power, P_{dev}, is given by the product of the armature current, I_a, times the back EMF, E_a, that is,

$$P_{dev} = I_a E_a = k_a \omega_m I_a \Phi \tag{8.10}$$

or

$$P_{dev} = I_a E_a = k'_a n I_a \Phi \tag{8.11}$$

Accordingly, the developed mechanical torque, T_{dev}, in Nm, can be expressed as follows:

$$T_{dev} = P_{dev}/\omega_m = I_a E_a/\omega_m \tag{8.12}$$

Substituting from Eq. (8.10) into Eq. (8.12), it follows that

$$T_{dev} = (k_a \omega_m I_a \Phi)/\omega_m = k_a I_a \Phi \tag{8.13}$$

However, based on the equivalent circuit of Figure 8.10, for a separately excited DC motor, one can write the following:

$$I_f = V_f/R_f \tag{8.14}$$

and in the unsaturated range of the magnetic circuit in a DC machine, the flux per pole, Φ, is linearly proportional to the field current, I_f, that is,

$$\Phi = k_f I_f = k_f(V_f/R_f) \tag{8.15}$$

where k_f is a constant of proportionality.

Also, from the armature's equivalent circuit, Figures 8.10 and 8.11, we write

$$V_a = R_a I_a + E_a \tag{8.16}$$

and substituting for E_a from Eq. (8.5) into Eq. (8.16), one writes the following:

$$V_a = R_a I_a + k_a \Phi \omega_m \tag{8.17}$$

However, from Eq. (8.13), the armature current $I_a = T_{dev}/(k_a\Phi)$; therefore, Eq. (8.17) can be written in terms of the developed torque, T_{dev}, as follows:

$$V_a = R_a[T_{dev}/(k_a\Phi)] + k_a \Phi \omega_m \tag{8.18}$$

Accordingly,

$$\omega_m = [V_a/(k_a\Phi)] - \left[R_a/\left(k_a^2\Phi^2\right)\right] T_{dev} \tag{8.19}$$

which results in the drooping speed–torque characteristic depicted in Figure 8.12. In the motor's magnetic circuit linear unsaturated range, where the flux, Φ, and

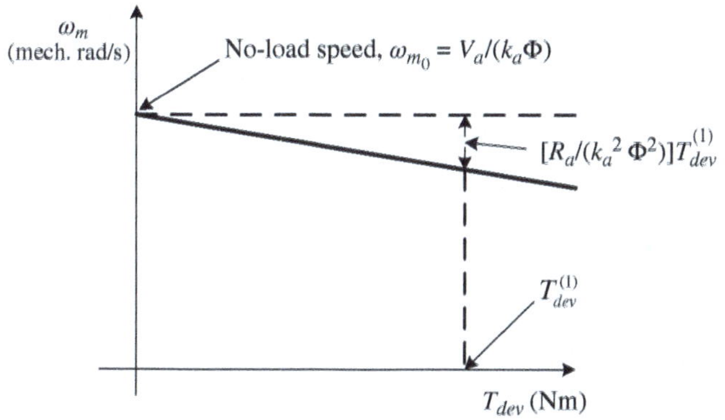

Figure 8.12 Torque–speed characteristic of a separately excited DC motor.

field current, I_f, and voltage, V_f, are linearly proportional to the flux, Eq. (8.19) of the torque–speed characteristic can be rewritten in two ways in terms of I_f or V_f, as follows:

$$\omega_m = [V_a/(K_a K_f I_f)] - \left[R_a/\left(k_a^2 k_f^2 I_f^2\right)\right] T_{dev} \tag{8.20}$$

or

$$\omega_m = [V_a R_f/(K_a K_f I_f)] - \left[R_a R_f^2/\left(k_a^2 k_f^2 V_f^2\right)\right] T_{dev} \tag{8.21}$$

Now, it should be observed that the droop-characteristic slope in Eqs. (8.19), (8.20), and (8.21) are $[R_a/(k_a^2\Phi^2)]$, $[R_a/\left(k_a^2 k_f^2 I_f^2\right)]$, and $[R_a R_f^2/\left(k_a^2 k_f^2 V_f^2\right)]$, respectively. Hence, based on the earlier discussion, it becomes obvious that the natural control variables in such a separately excited DC motor, from the external motor terminals point-of-view, are the armature terminals voltage, V_a, and field winding terminal voltage, V_f. The simplest way to control these voltages is through a power-electronic DC-DC converter (i.e., chopper). An example of such a one-quadrant chopper is shown in Figure 8.13.

In this one-quadrant chopper of Figure 8.13, the duty cycles of the electronic solid-state switch, T_1, and diode, D_2, are also shown in this figure, as well as a representation of the chopper's instantaneous output voltage, v_{out}, and instantaneous output current, i_{out}, where the average output voltage, V_{avg}, is expressed as

$$V_{avg} = (T_{on}/T_{cycle})V_{in} \tag{8.22}$$

or

$$V_{avg} = d_{cycle} V_{in} \tag{8.23}$$

where the chopper's duty cycle, d_{cycle}, is defined as

$$d_{cycle} = (T_{on}/T_{cycle}) \tag{8.24}$$

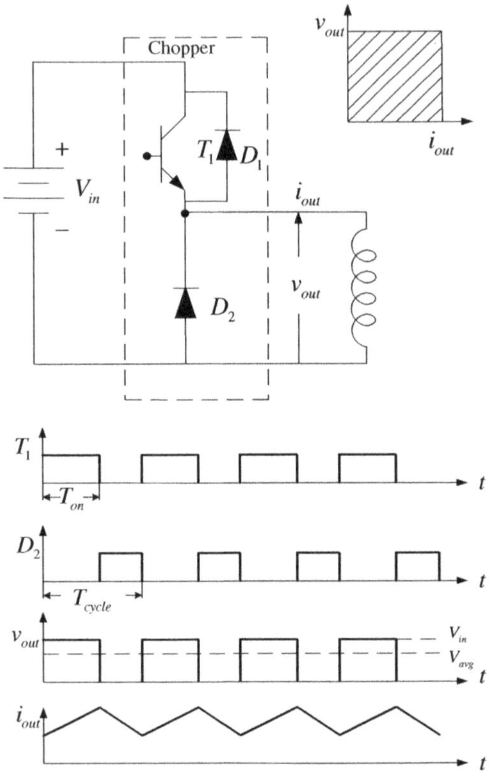

Figure 8.13 A one-quadrant chopper circuit topology, PWM control and output signals.

See Figure 8.13 for details, and see Chapter 2 for a complete treatment of various chopper types and operations.

Utilizing such choppers for control of field winding voltages, V_f, in separately excited DC motors is schematically shown in Figure 8.14. In this approach, the control variable V_f affects both the droop of the torque–speed characteristic, $\left[R_aR_f^2/\left(k_a^2k_f^2V_f^2\right)\right]$, and the no-load speed intercept, $[V_aR_f/(k_ak_fV_f)]$. Meanwhile, utilizing such choppers for control of armature winding voltage, V_a, in separately excited DC motors is schematically shown in Figure 8.15. In this approach, the control variable is V_a, which does not affect the droop of the torque–speed characteristic, $\left[R_aR_f^2/\left(k_a^2k_f^2V_f^2\right)\right]$, but directly affects the no-load intercept, $[V_aR_f/k_ak_fV_f)]$.

It is of paramount importance to point out that a DC motor with permanent magnet (PM) field excitation of the brush-commutator variety behaves in a very similar fashion to a separately excited DC motor, in which the field winding is energized from a constant-current source, that is, as if the field excitation current, I_f, is a constant.

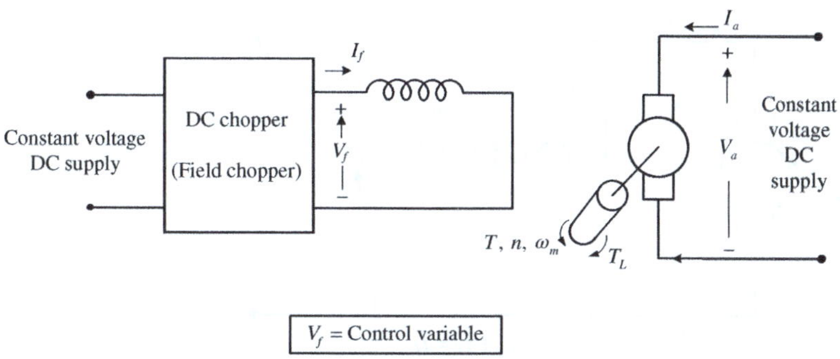

Figure 8.14 Field voltage control of separately excited DC motors by DC choppers.

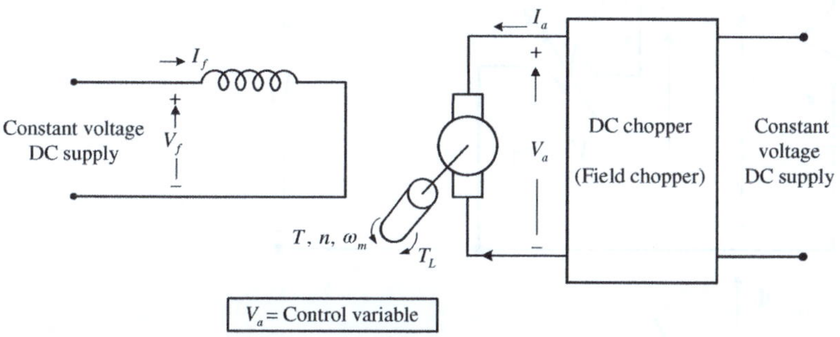

Figure 8.15 Armature voltage control of separately excited DC motors by DC choppers.

8.4 Shunt-Excited DC Motors Modeling and Formulations

In this section, we discuss the torque–speed characteristics of the shunt-field connected DC motors, the circuit connection of which is depicted in Figure 8.16. The equivalent circuit model of such a shunt-field connected DC motor of the brush-commutator variety is depicted in Figure 8.16. Examining this circuit, one can write the following for the field current, I_f:

$$I_f = (V_f/R_f) \tag{8.25}$$

where V_f is the terminal voltage of the filed winding, and R_f is the field winding's ohmic resistance.

In the linear unsaturated machine magnetic circuit range, the flux per pole, Φ, is linearly proportional to the field excitation current, I_f. Therefore, one can write the following:

$$\Phi = k_{f_1} I_f \tag{8.26}$$

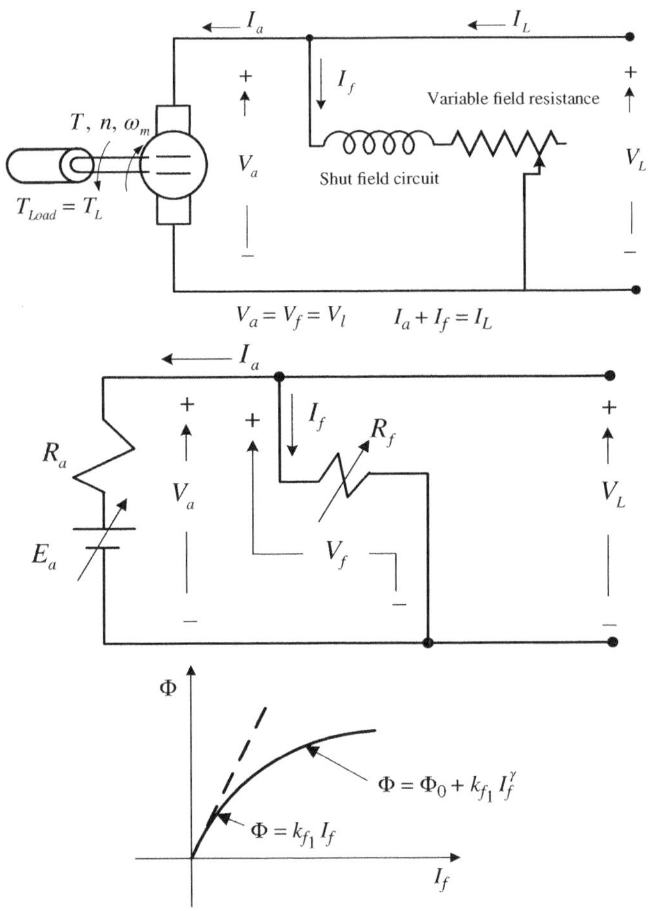

Figure 8.16 A shunt field control connected DC motor and its equivalent circuit.

Once the magnetic circuit begins to experience magnetic saturation, the flux relationship to the field excitation current, I_f, will take the following form:

$$\Phi = \Phi_0 + k_{f_2} I_f^\gamma \tag{8.27}$$

where γ is a fractional exponent depending on the magnetic material(s) used in the motor construction. See the $\Phi - I_f$ nonlinear curve depicted in Figure 8.16.

In the linear unsaturated range, using Eqs. (8.25) and (8.26), one can write the following for the flux per pole, Φ:

$$\Phi = k_{f_1} I_f = (k_{f_1}/R_f)V_f \tag{8.28}$$

Now, upon examining the armature branch of the equivalent circuit of Figure 8.16, one can write the following for the line voltage, V_L, and armature voltage, V_a:

$$V_L = V_a = R_a I_a + E_a \tag{8.29}$$

Substituting for E_a from Eq. (8.5) into Eq. (8.29) yields the following:

$$V_L = V_a = R_a I_a + k_a \omega_m \Phi \tag{8.30}$$

where k_a is as defined earlier in Eq. (8.8). However, from Eq. (8.13), one can express the armature current, I_a, in terms of the developed torque, T_{dev}, as follows:

$$I_a = T_{dev}/(k_a \Phi) \tag{8.31}$$

Therefore, substituting for the armature current, I_a, from Eq. (8.31) into Eq. (8.30) gives

$$V_L = V_a = [R_a/(k_a \Phi)]T_{dev} + k_a \omega_m \Phi \tag{8.32}$$

From Eq. (8.32), with proper algebraic manipulations, the relationship between the mechanical rotational speed, ω_m, in mach. rad/s, and the developed torque, T_{dev}, can be written as follows:

$$\omega_m = [V_a/(k_a \Phi)] - \left[R_a/\left(k_a{}^2 \Phi^2\right)\right]T_{dev} \tag{8.33}$$

or,

$$\omega_m = [V_L/(k_a \Phi)] - \left[R_a/\left(k_a{}^2 \Phi^2\right)\right]T_{dev} \tag{8.34}$$

The speed–torque (or torque–speed) characteristics of Eqs. (8.33) and (8.34) are graphically depicted in Figure 8.17, in which the no-load speed is $\omega_{m_0} = V_a/(k_a \Phi)$ or $\omega_{m_0} = V_L/(k_a \Phi)$, and the droop slope is $R_a/(k_a{}^2 \Phi^2)$, which bears great similarity to the torque–speed characteristic of the separately excited DC motor of Figure 8.12. This is why the reference to a "separately excited" or "shunt-excited" torque–speed characteristics is mutually exchangeable in the discussions of torque–speed characteristics of DC motors, as well as other AC motors.

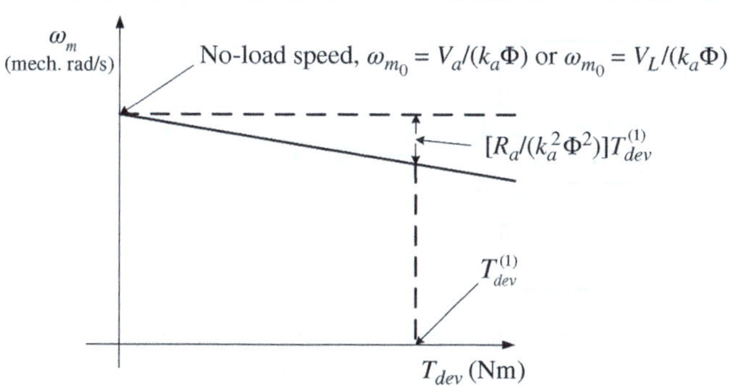

Figure 8.17 Torque–speed characteristic of a shunt-field-excited DC motor.

Now, we will discuss two types of torque–speed control of shunt-field-excited DC motors using DC choppers such as the ones discussed in Chapter 2. One approach utilizes a DC chopper connected to the field-excitation winding, see Figure 8.18, while the second approach utilizes a DC chopper connected to the armature terminals, see Figure 8.19.

Let us first discuss the case of torque–speed control using a DC chopper connected to the field-excitation winding, Figure 8.18. In this case, the field current, I_f, and voltage, V_f, are increased or decreased depending on the duty cycle of the chopper, and, hence, the flux, Φ, in the torque–speed governing Eqs. (8.33) and (8.34). This affects the no-load speed, $\omega_{m_0} = V_a/(k_a\Phi)$ or $\omega_{m_0} = V_L/(k_a\Phi)$, decreasing when Φ, and, hence, V_f and I_f are increasing, and ω_{m_0} increases when Φ and, hence, V_f and I_f are decreasing. This also affects the droop slope,

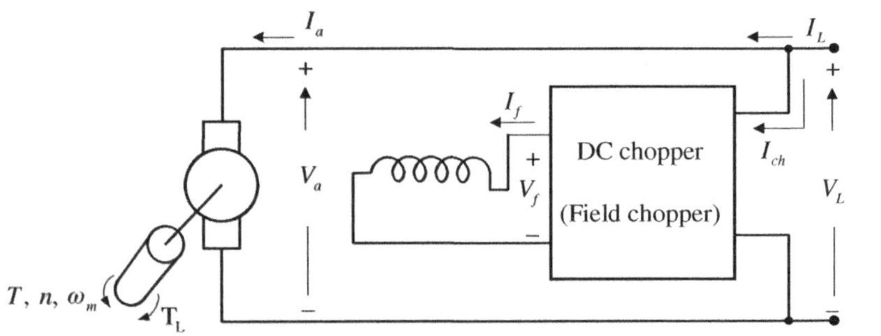

$I_L = I_a + I_{ch}$, I_f = function (I_{ch})

$V_a = V_L$ | V_f = Control variable

Figure 8.18 Field chopper control of torque–speed characteristics of shunt motors.

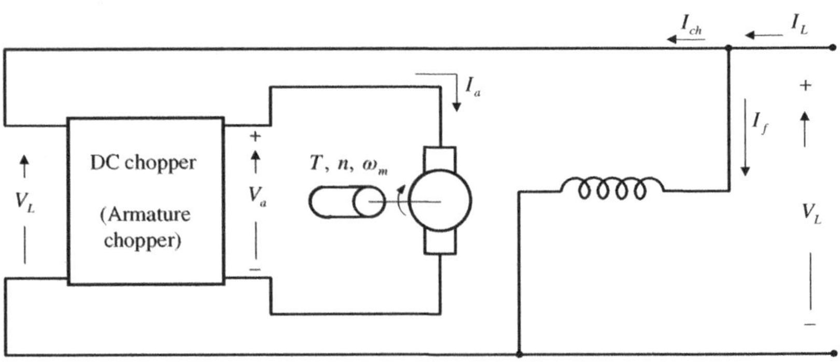

$V_f = V_L$ & $I_L = I_f + I_{ch}$, I_a = function (I_{ch})

V_a = Control variable

Figure 8.19 Armature chopper control of torque–speed characteristics of shunt motors.

$[R_a/(k_a{}^2\Phi^2)]$, which decreases this droop slope when Φ, and, hence, V_f and I_f are increased by the action of the field chopper, and vice versa increases this droop slope when Φ and, hence, V_f and I_f are decreased, see Figure 8.20.

Next, let us discuss the case of torque–speed control using a DC chopper connected across the terminals of the armature winding, Figure 8.19. In this case, the field current, I_f, voltage, V_f, and, hence, the flux, Φ, are fixed (constant) and are directly linked to the line voltage, V_L Meanwhile, the armature voltage, V_a, is different from the line voltage, depending on the settings of the armature chopper, see the schematic of Figure 8.19. In this torque–speed control method, the droop slope, $[R_a/(k_a{}^2\Phi^2)]$, remains constant (fixed), while the no-load speed, $\omega_{m_0} = V_a/(k_a\Phi)$, decreases or increases, depending on whether the armature terminal voltage, V_a, decreases or increases with the setting of the armature chopper, see Figure 8.21.

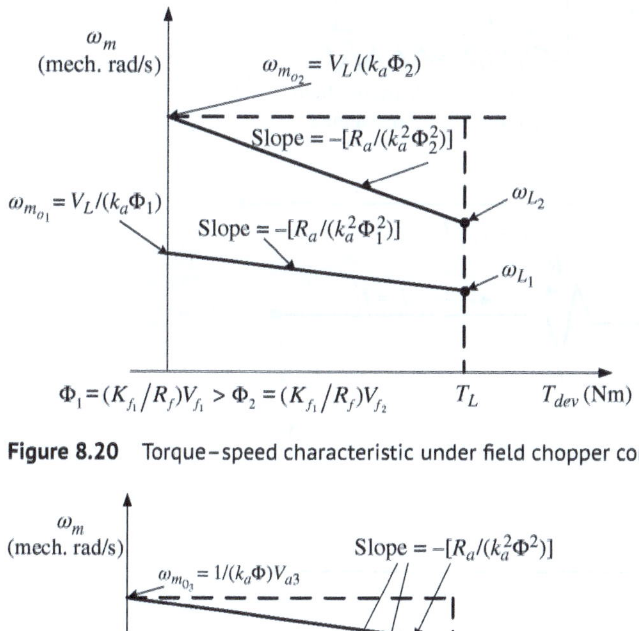

Figure 8.20 Torque–speed characteristic under field chopper control.

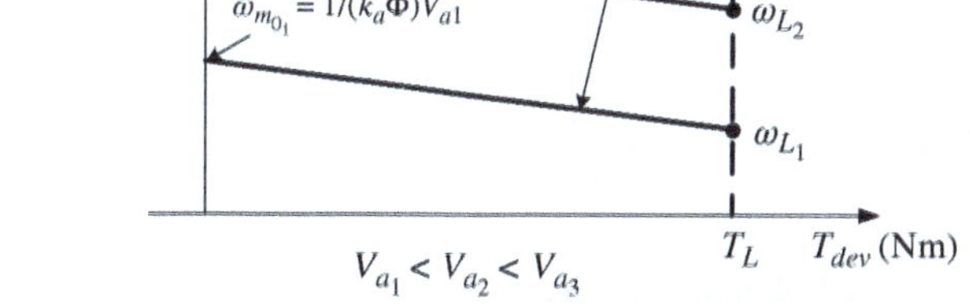

Figure 8.21 Torque–speed characteristic under armature chopper control.

8.5 Series-Excited DC Motors Modeling and Formulations

Figure 8.22 shows the schematic of the winding connections of a series-excited DC motor, and its corresponding equivalent electric circuit. In this class of DC motors, the series-connected field excitation winding current, I_s, is the same as the armature winding's current, I_a, and is the same as the line current, I_L. That is:

$$I_s = I_a = I_L \tag{8.35}$$

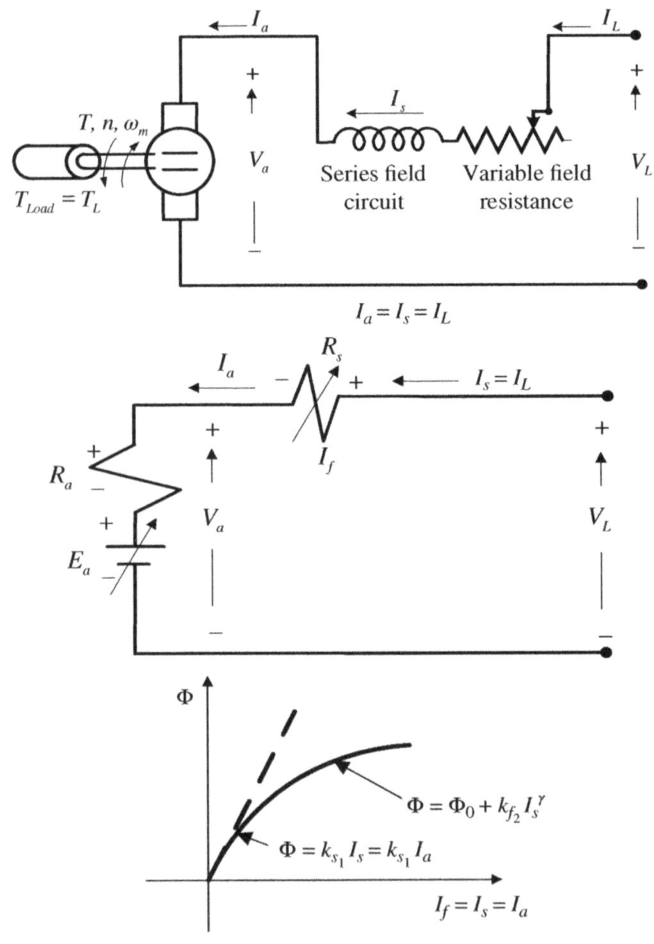

Figure 8.22 A series-field connected DC motor and its equivalent circuit.

Accordingly, in the magnetically unsaturated linear range in a series-excited DC motor's magnetic circuit, the flux per-pole, Φ, is linearly proportional to $I_a = I_s$. Therefore, one can write the following for Φ:

$$\Phi = k_{s_1} I_a = k_{s_1} I_s \tag{8.36}$$

where k_{s_1} is the linear proportionality constant. Meanwhile, in the nonlinear magnetic saturation range, one can express the flux per pole, Φ, as follows:

$$\Phi = \Phi_0 + k_{s_2} I_a^{\gamma} = \Phi_0 + k_{s_2} I_s^{\gamma} \tag{8.37}$$

where Φ_0 is a nonlinear zero-excitation interception point on the flux co-ordinate, k_{s_2} is a proportionality constant, and γ is a fractional exponent resulting from the curve-fitting algebra. See the bottom schematic of the $\Phi - I_f$ relationship in Figure 8.22.

Now, upon examination of the equivalent circuit in Figure 8.22, one can write the following relationships among the armature voltage, V_a, the line voltage, V_L, and the armature back EMF, E_a:

$$V_a = R_a I_a + E_a \tag{8.38}$$

and

$$V_L = V_a + R_s I_a = (R_a + R_s) I_a + E_a \tag{8.39}$$

where R_a is the armature winding/circuit resistance, R_s is the series-field excitation winding resistance, and E_a is the induced back EMF in the armature winding. In Eq. (8.39), one should recall that the back EMF, E_a, can be expressed as follows:

$$E_a = K_a \omega_m \Phi \tag{8.40}$$

where ω_m is the angular rotational speed of the rotating armature winding in mechanical radians per second (mach. rad/s).

Substituting for E_a from Eq. (8.40) into Eq. (8.39), one can express the terminal line voltage, V_L, as follows:

$$V_L = (R_a + R_s) I_a + k_a \omega_m \Phi \tag{8.41}$$

However, the developed torque, T_{dev}, based on earlier developments is given by

$$T_{dev} = k_a \Phi I_a \tag{8.42}$$

Hence,

$$I_a = T_{dev}/(k_a \Phi) \tag{8.43}$$

Substituting for I_a from Eq. (8.43) into Eq. (8.41) gives the following:

$$V_L = [(R_a + R_s)/(k_a \Phi)] T_{dev} + K_a \omega_m \Phi \tag{8.44}$$

Accordingly, solving for the angular rotational speed, ω_m, one obtains from Eq. (8.44) the following:

$$\omega_m = [V_L/(k_a\Phi)] - \left[(R_a + R_s)/\left(k_a{}^2\Phi^2\right)\right] T_{dev} \tag{8.45}$$

Meanwhile, in the motor magnetic circuit's unsaturated linear range, the flux, Φ, according to Eq. (8.36), is given by, $\Phi = k_{s_1} I_a$, and, hence, the developed torque, T_{dev}, Eq. (8.42), becomes,

$$T_{dev} = k_a(k_{s_1} I_a)I_a \tag{8.46}$$

Hence,

$$I_a^2 = T_{dev}/(k_a k_{s_1}) \tag{8.47}$$

That is,

$$I_a = \sqrt{T_{dev}}/\sqrt{k_a k_{s_1}} \tag{8.48}$$

and

$$\Phi = k_{s_1} I_a = (\sqrt{k_{s_1}/k_a})(\sqrt{T_{dev}}) \tag{8.49}$$

Substituting for (Φ) from Eq. (8.49) into Eq. (8.45) yields the following for the angular speed, ω_m:

$$\omega_m = \left\{ V_L/[k_a(\sqrt{k_{s_1} T_{dev}/k_a})] - (R_a + R_s)/\left[k_a{}^2(k_{s_1}/k_a)T_{dev}\right] \right\} T_{dev}$$

That is,

$$\omega_m = [V_L/\sqrt{k_a k_{s_1} T_{dev}}] - [(R_a + R_s)/(k_a k_{s_1})] \tag{8.50}$$

The torque–speed characteristic given by Eq. (8.50) is depicted in the plot of Figure 8.23.

One simple way of torque–speed control of such series-field excited DC motors is through the increase of or decrease of R_s through a variable resistance/rheostat

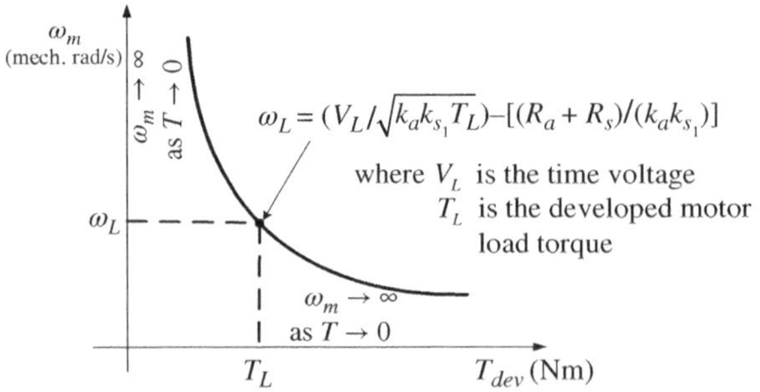

Figure 8.23 Torque–speed characteristic of a series-field-excited DC motor.

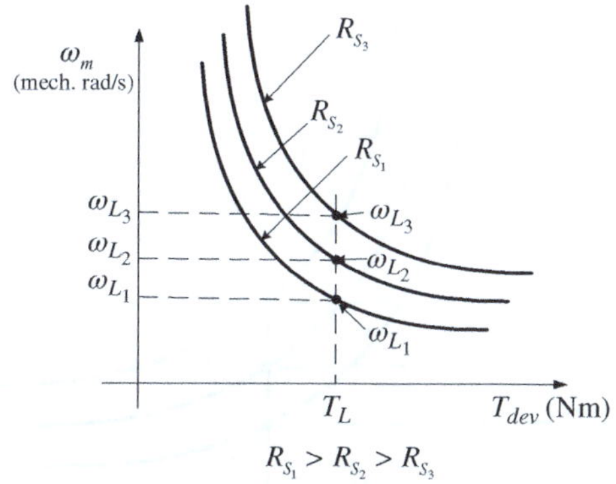

Figure 8.24 The torque–speed characteristic of a series-field-excited DC motor under variable armature circuit resistance.

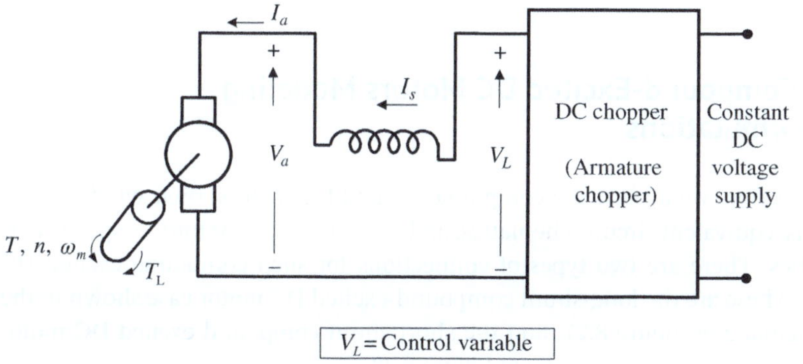

Figure 8.25 DC chopper line-voltage control of torque–speed characteristics of DC series-field motors.

connected in series with the field winding. The result of this method of torque–speed control is shown in Figure 8.24, for resistances, $R_{S_1} > R_{S_2} > R_{S_3}$, at load torque, T_L, leading to the corresponding three angular speeds, ω_{L_1}, ω_{L_2}, and ω_{L_3}, respectively.

Another more modern approach to torque–speed control of series-field-excited DC motors is through the use of a power-electronic DC chopper, Figure 8.25, to control the line voltage, V_L. More details on such DC choppers are included in Chapter 2. Upon close examination of the torque–speed governing equation, Eq. (8.50), one realizes that increasing or decreasing, V_L, the torque–speed characteristic is moved upward for $V_{L_3} > V_{L_2} > V_{L_1}$, vice versa as shown in Figure 8.26, with corresponding speeds, at a developed load torque, T_L, of $\omega_{L_3} > \omega_{L_2} > \omega_{L_1}$, etc.

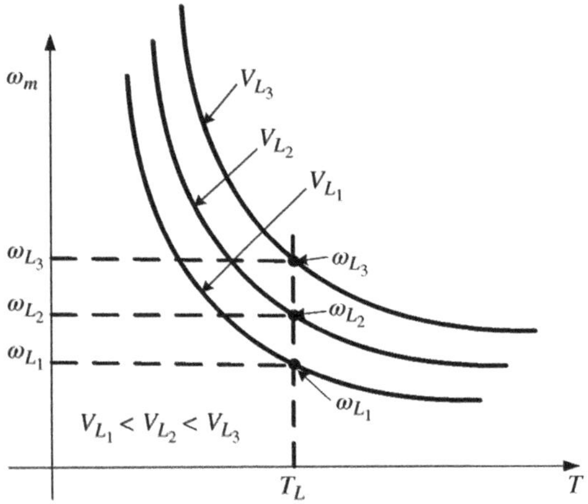

Figure 8.26 Torque–speed control of a series-field-excited DC motor by a DC chopper.

8.6 Compound-Excited DC Motors Modeling and Formulations

In this section, we discuss the compound-excited DC motors, their winding connections, equivalent circuit schematics, and the nature of their torque–speed characteristics. There are two types of connections for such compound-excited DC motors. These are the long-shunt compound-excited DC motor case shown in the top schematic of Figure 8.27, and the short-shunt compound-excited DC motor case shown in the top schematic of Figure 8.28.

In the case of the long-shunt compound-excited DC motor, see the top and bottom schematics in Figure 8.27, one can write the following equivalent circuit relationships:

$$I_L = I_a + I_f \tag{8.51}$$

where I_L is the line current, I_a is the armature current, and I_f is the shunt field current.

Here also,

$$V_L = R_f I_f \tag{8.52}$$

that is,

$$I_f = V_L / R_f \tag{8.53}$$

Furthermore,

$$V_L = (R_s + R_a)I_a + E_a \tag{8.54}$$

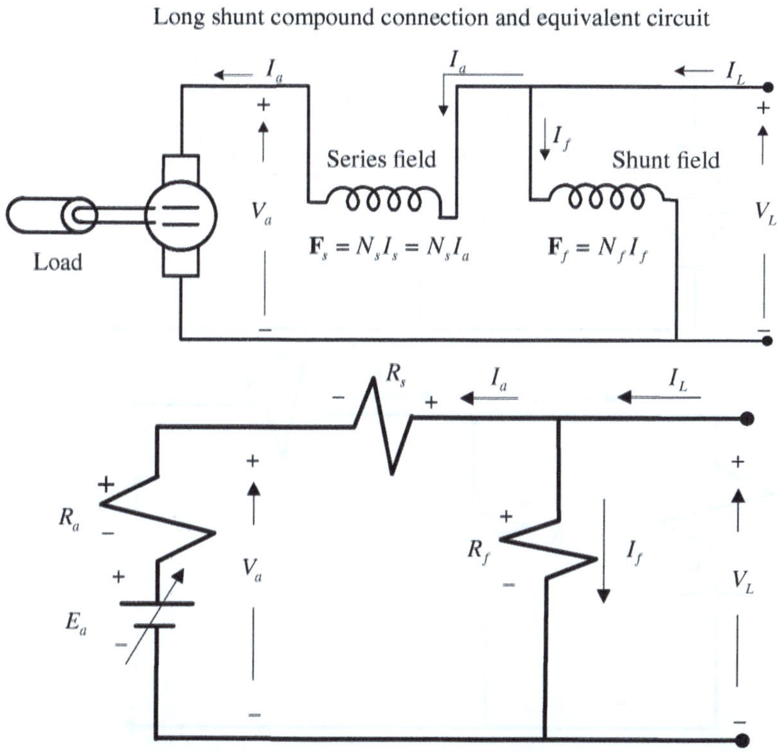

Figure 8.27 A long-shunt field-connected-compound DC motor and its equivalent circuit.

and

$$V_a = R_a I_a + E_a \tag{8.55}$$

where V_L is the line voltage, and V_a is the armature terminal voltage.

Meanwhile, in the case of the short-shunt compound-excited DC motor, see the top and bottom schematics of Figure 8.28, one can write the following equivalent circuit relationships:

$$I_L = I_a + I_f \tag{8.56}$$

however, here

$$V_L = R_s I_L + R_f I_f \tag{8.57}$$

or

$$V_L = R_s I_L + V_a \tag{8.58}$$

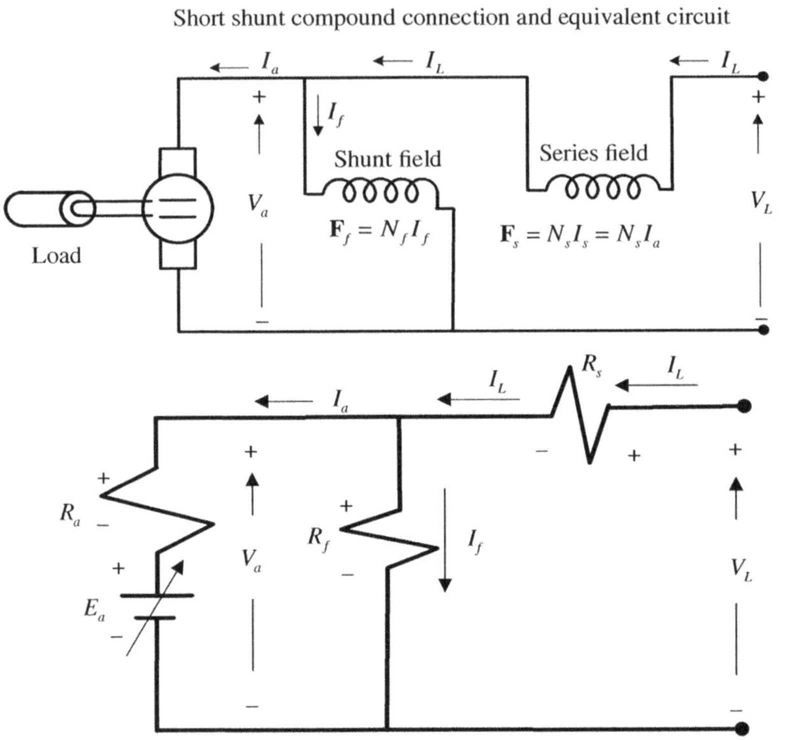

Figure 8.28 A short-shunt-field connected-compound DC motor and its equivalent circuit.

because

$$V_a = R_f I_f \tag{8.59}$$

Also,

$$V_a = R_a I_a + E_a \tag{8.60}$$

In addition, the developed torque, T_{dev}, in both the long- and short-shunt compound-excited DC motors is given by the following:

$$T_{dev} = I_a E_a / \omega_m \tag{8.61}$$

where ω_m is the angular rotational motor speed in mechanical radians/second (mach. rad/s).

Now, the resulting torque–speed characteristic of a shunt compound-excited motor is a mix between the pure-shunt and pure-series characteristics, as shown in Figure 8.29, depending on the relative strength of the MMFs of the shunt-field

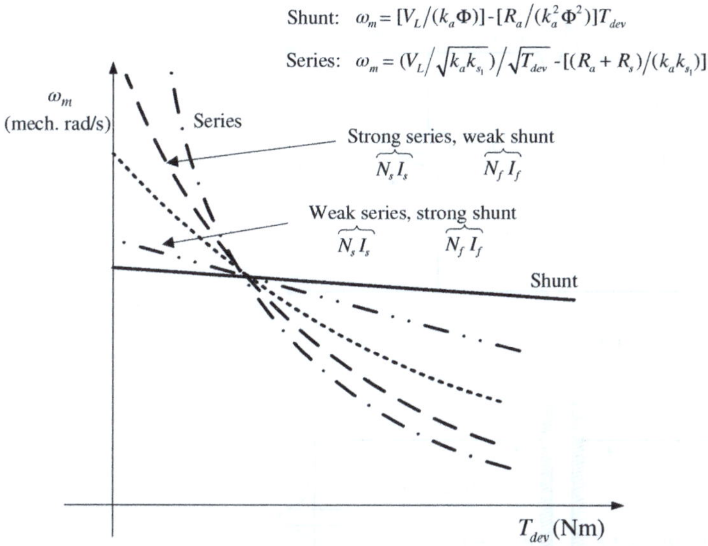

Figure 8.29 Torque–speed characteristic of a compound-excited DC motor.

and series-field windings, respectively. These are indicated in Figure 8.29 as strong shunt, weak shunt, strong series, and weak series, respectively.

8.7 DC Motors Input–Output Powers, Losses, and Efficiency

The input–output powers and losses will be based on the long-shunt compound-excited DC motor, the equivalent circuit of which is given in the top schematic of Figure 8.30. Here, the input, output powers, and losses can be formulated as follows:

$$P_{in} = I_L V_L \tag{8.62}$$

The shunt-field ohmic losses, $P_{sh-ohmic}$, is

$$P_{sh-ohmic} = I_f^2 R_f \tag{8.63}$$

The armature circuit (armature and series-field) ohmic losses, $P_{arm-ohmic}$, is

$$P_{arm-ohmic} = I_a^{\ 2}(R_a + R_s) \tag{8.64}$$

The developed airgap/electromagnetic power, $P_{em} = P_{airgap} = P_{dev}$, is

$$P_{dev} = I_a E_a \tag{8.65}$$

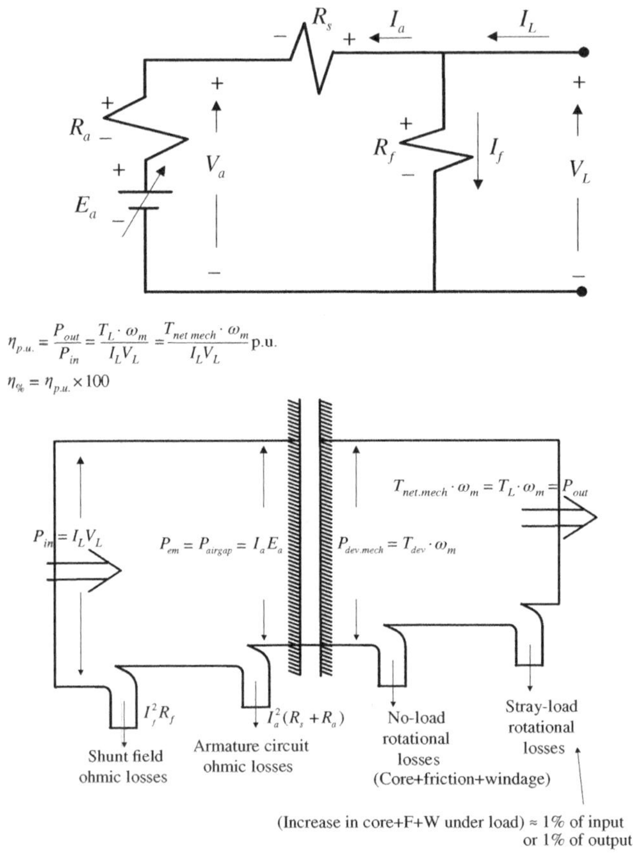

Figure 8.30 Input–output and losses in a compound-excited DC motor and its efficiency, on the basis of the general case of a long-shunt compound motor.

The developed torque, T_{dev}, is

$$T_{dev} = P_{dev}/\omega_m = I_a E_a/\omega_m \tag{8.66}$$

Here, $P_{rotational}$ is the sum of no-load core plus friction and windage losses, and P_{stray} load is the stray rotational losses under-load, which can be taken to equal 1.0% of either the input or the output power of the motor.

It should be noted that all these input, loss, and output power components are well delineated in the power flow diagram of Figure 8.30. Accordingly, the output power, P_{out}, can be expressed as follows:

$$P_{out} = P_{in} - \sum P_{losses} \tag{8.67}$$

That is,

$$P_{out} = I_L V_L - I_f^2 R_f - I_a^2 (R_a + R_s) - P_{rotational} - P_{stray\ load} \tag{8.68}$$

and, consequently, the output torque, $T_{out} = T_{net} = T_{load}$, is,

$$T_{out} = P_{out}/\omega_m \tag{8.69}$$

Hence, the overall motor efficiency in p.u., η_{motor}, becomes

$$\eta_{motor} = (P_{out}/P_{in}) \tag{8.70}$$

or, in percent,

$$\eta_{motor} = (P_{out}/P_{in}) \times 100\% \tag{8.71}$$

Now, we next introduce a sample of worked examples.

8.8 Worked Examples on DC Motors

(1) A 15-ph, 230-V, DC-shunt motor has an armature-circuit resistance of 0.20 Ω and a field resistance of 140 Ω. At no-load and rated voltage, the speed is 1150 r/min, and the armature current is 9 A. At full load and rated voltage, the line current is 58.50 A. Neglecting armature reaction and magnetic saturation effects (i.e., flux is linearly proportional to field current), compute the speed in r/min at full load and find the developed power in watts and hp as well as the developed torque in Nm.

(2) A chopper-controlled DC series motor runs at 900 r/min, when the chopper is supplied from a 230-V supply, and it is fully on. That is, its duty cycle, $T_{on}/T_{cycle} = 1.0$. The chopper supplies the entire armature and field circuits which are connected in series. The total armature and series-field resistance is 0.20 Ω. The armature current was found to be 75 A under this condition.

(2.1) Compute the developed torque in Nm under the earlier condition.

(2.2) With the chopper duty cycle $T_{on}/T_{cycle} = 1.0$, the load torque was reduced so that the armature current reading was found to be 30 A. Assuming that effects of magnetic saturation and armature reaction (MMF) are to be neglected, that is, flux is linearly proportional to the series-field current (in this case, it is also the armature current), fine the new motor speed in r/min and developed torque in Nm.

(2.3) For a load torque equal to that of (6.2.1), the chopper duty cycle was reduced to a value, $T_{on}/T_{cycle} = 0.7$, what would be the new motor speed in r/min.

Example 8.1

Solution

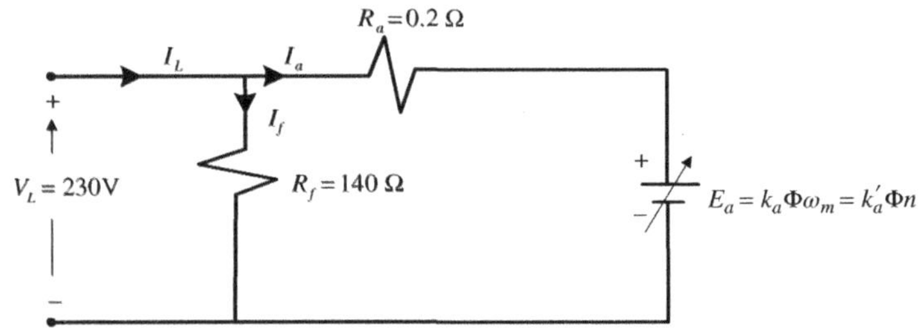

At no-load:

$$V_L = 230\,\text{V}, I_a = 9\,\text{A}, n = 1150\,\text{r/min},$$

$$I_f = \frac{230}{140} = 1.643\,\text{A}$$

$$E_a = 230 - 0.2 \times 9 = 230 - 1.8 = 228.2\,\text{V}$$

At full-load:

$$I_f = 1.643\,\text{A because } V_L = \text{Const} = 230\,\text{V}$$

$$V_L = 230\,\text{V}, I_L = 58.50\,\text{A} \rightarrow I_a = 58.50 - 1.643 = 56.857\,\text{A}$$

$$E_a = 230 - 0.2 \times 56.857 = 218.63\,\text{V}$$

hence,

$$\frac{E_a(\text{FL})}{E_a(\text{NL})} = \frac{218.63}{228.20} = \frac{k_a' \times \Phi \times n_{FL}}{k_a' \times \Phi \times n_{NL}} = \frac{n_{FL}}{1150}$$

hence,

$$n_{FL} = 1150 \times \frac{218.63}{228.20} = 1101.77\,\text{r/min}$$

$$P_{dev} = I_a E_a = 56.857 \times 218.63 = 12,430.65\,\text{W}$$

$$P_{dev} = 12,430.65/746 = 16.66\,\text{hp}$$

$$\omega_{FL} = \frac{2\pi}{60} \times 1101.77$$

$$T_{dev} = P_{dev}/\omega_{FL} = 12430.65 \Big/ \left(\frac{2\pi}{60} \times 1101.77\right)$$

$$T_{dev} = 107.74\,\text{Nm}$$

Example 8.2

Solution

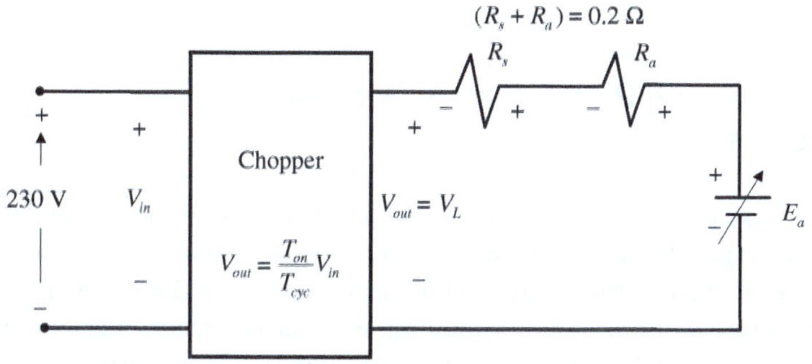

(2.1) $I_{a_1} = 75\text{ A}, T_{on}/T_{cycle} = 1.0, n_1 = 900\text{ r/min}$

$$V_L = V_{out} = 1.0 \times 230 = 230\text{ V}$$

$$E_{a_1} = 230 - 0.2 \times 75 = 230 - 15 = 215\text{ V}$$

$$\omega_1 = \frac{2\pi}{60} \times 900 = 30\pi\text{ rad/s}$$

$$T_{dev} = 16{,}125/30\pi = 171.09\text{ Nm}$$

(2.2) $I_{a_2} = 30\text{ A}, T_{on}/T_{cycle} = 1.0 \rightarrow V_L = V_{out} = 230\text{ V}$ hence,

$$E_{a_2} = 230 - 0.2 \times 30 = 230 - 6 = 224\text{ V}$$

$$\frac{E_{a_2}}{E_{a_1}} = \frac{k'_a \times \Phi_2 \times n_2}{k'_a \times \Phi_1 \times n_1} = \frac{n_2}{900} \times \frac{30}{75} = \frac{224}{215}\text{ V}$$

$$n_2 = 900 \times \frac{75}{30} \times \frac{224}{215} = 2344.19\text{ r/min}$$

$$\omega_2 = \frac{2\pi}{60} \times 2344.19$$

$$T_{dev} = \frac{I_{a_2} E_{a_2}}{\omega_2} = \frac{30 \times 224}{2\pi \times 2344.19} \times 60 = 27.375\text{ Nm}$$

(2.3) $V_L = \frac{T_{on}}{T_{cycle}} \times V_{in} = 0.7 \times 230 = 161\text{ V}$

$$E_{a_3} = 161 - 75 \times 2 = 161 - 15 = 146\text{ V}$$

$\Phi_3 = \Phi_1$ because $I_{a_3} = I_{a_1} = 75\text{ A}$ (series motor)

$$\text{but } \frac{E_{a_3}}{E_{a_1}} = \frac{k'_a \times \Phi_3 \times n_3}{k'_a \times \Phi_1 \times n_1} = \frac{n_3}{n_1} = \frac{146}{215}$$

$$n_3 = n_1 \times \frac{146}{215} = 900 \times \frac{146}{215} = 611.2 \text{ r/min}$$

8.9 Operating Principles and Modeling of Brushless DC Motors

PM brushless DC motors, see Figures 8.4 and 8.5, pose several advantages over their counterpart legacy brush-commutator-type DC motors. First, among these advantages is the fact that their armatures are located on the stator side. Hence, these armature windings are not rotating, with far less mechanical stresses and with more areas to house conductors with higher current-carrying capacities, for the same overall motor outside diameter/volume. Also, the stationary nature of such an armature winding means that it has improved heat transfer/cooling qualities. Second, because of the rotor-mounted PM field excitation, there are less rotor-associated losses. Also, this leads to smaller rotor volumes and lower inertia for rotor dynamics, and, consequently, higher motor speeds are possible. Third, there are no mechanical commutators, and, hence, there are no commutator segments, or brushes, and/or slip rings. This reduces the overall motor/shaft length and leads to increased motor reliability, elimination of electrical arcing, and reduction in electromagnetic interference (EMI). Fourth, the aforementioned factors lead to excellent brushless DC motor performance in high-performance adjustable speed drives (ASDs). The cross-sections of two sample 15 hp, high speed (>8000 r/min), brushless DC motors are shown in Figures 8.31 and 8.32. Figure 8.31 shows a samarium–cobalt-based PM, 6-pole 18 stator slots, brushless DC motor, with *A*, *B*, and *C* winding labels. In both motors, the PMs are mechanically retained on the rotor by nonmagnetic retaining stainless steel sleeves, as shown in the cross-sections of Figures 8.31 and 8.32. See Ref. [1] for further details on both motors.

A functional block diagram of the components of a PM brushless DC motor in an ASD system is shown in Figure 8.33. A small but of paramount importance component in this ASD system, is a rotorpositionsensor (resolver or encoder), the main function of which is to sense the positions of the north "N" poles of the rotor-mounted PMs with respect to the *A*, *B*, and *C* magnetic axes of these three-phase windings. This "N" pole positions signal is the "key signal feed-back" to the controller of the power-electronic switches in a three-phase inverter to set the flow of the excitation currents in the stator-mounted three-phase stator (armature) winding of such a PM brushless DC motor, as will be explained later on in detail next in Section 8.9.

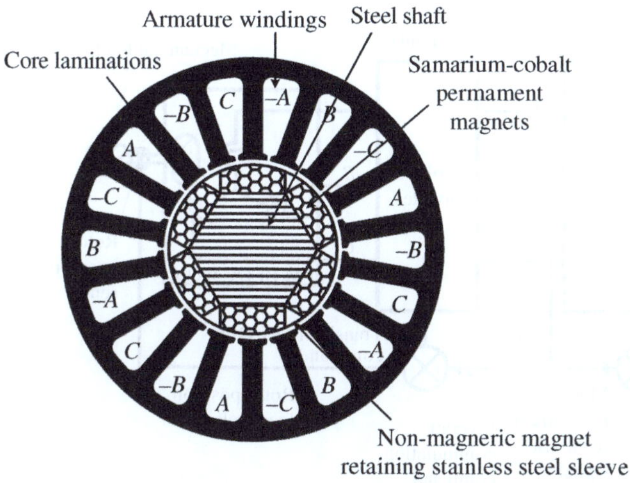

Figure 8.31 Six-pole samarium–cobalt PM Brushless DC motor. Source: Adapted from Demerdash et al. [1].

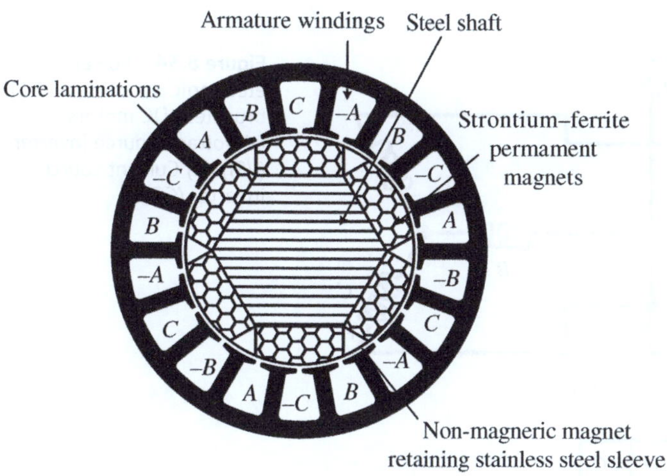

Figure 8.32 Six-pole strontium–ferrite PM brushless DC motor. Source: Adapted from Demerdash et al. [1].

Here, we introduce the simplified circuit schematics of a six-switch voltage source inverter in part (a) of Figure 8.34 and a six-switch current source inverter in part (b) of Figure 8.34. Here, in both cases, we will only discuss the 120 °*e* conduction period per switch, Q_1 through Q_6, and the corresponding six states, #1 through #6, see Figure 8.35 for the 120 °*e* conduction period of each switch over the 360 °*e* electrical cycle.

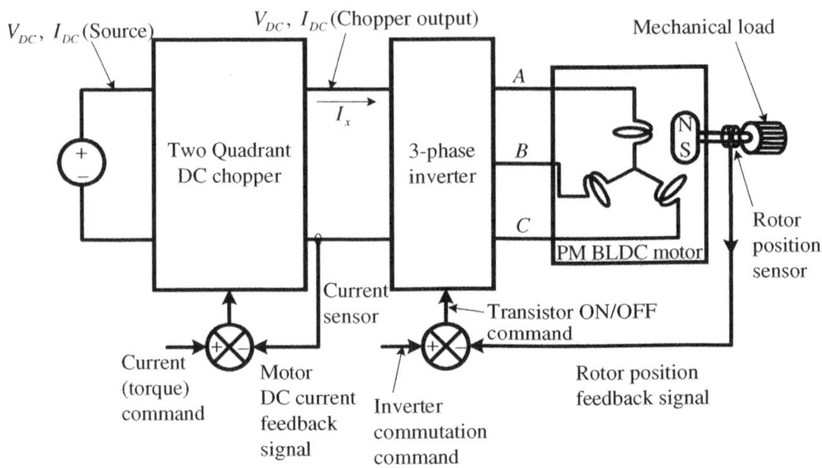

Figure 8.33 Functional block diagram of the components of a PM brushless DC motor (BLDC) ASD system.

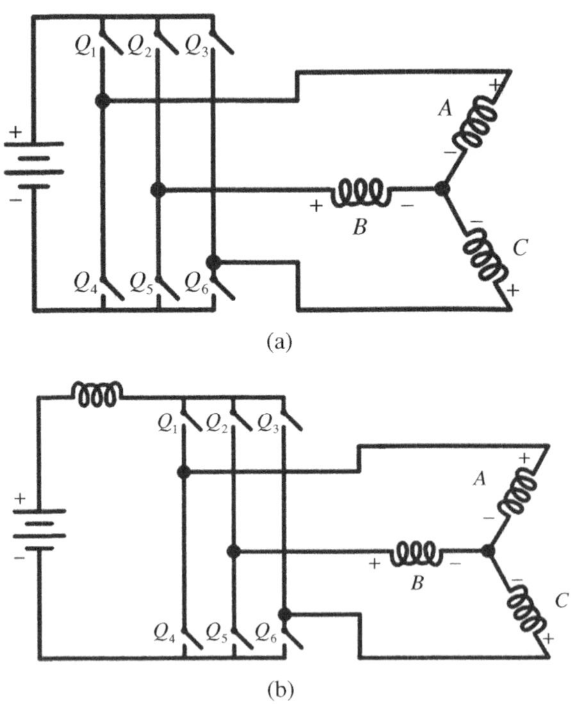

Figure 8.34 Power electronic inverters for PM brushless DC motors. (a) Voltage source inverter (VSI). (b) Current source inverter (CSI).

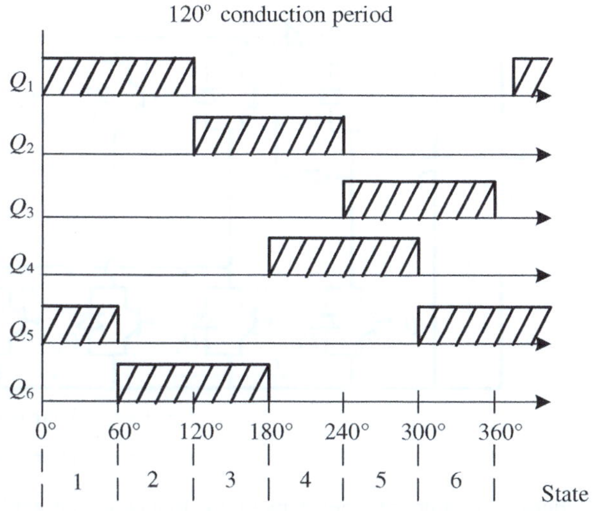

Figure 8.35 120° Conduction cycles of switches for voltage-source and current-source inverters for brushless DC motors.

The power-electronic switches and corresponding circuit schematic of a voltage source six-switch inverter feeding a three-phase load is shown in Figure 8.36. Also, the power-electronic switches of a two-quadrant DC chopper connected to a six-switch inverter to form a current–source inverter feeding a three-phase load (motor) is shown in Figure 8.37. Also, the path of an increasing phase current and a decreasing phase current, during the electronic commutation of current from phase (a) to phase (b) in the circuit diagram, is shown in Figure 8.37. The reader is advised to further consult Refs. [2, 3] and the material covered earlier in Chapter 2. The reader should notice that there are six states, states #1 through #6; each state covers 60 °e to form the complete cycle of 360 °e; the pair of switches, Q_1 and Q_4,

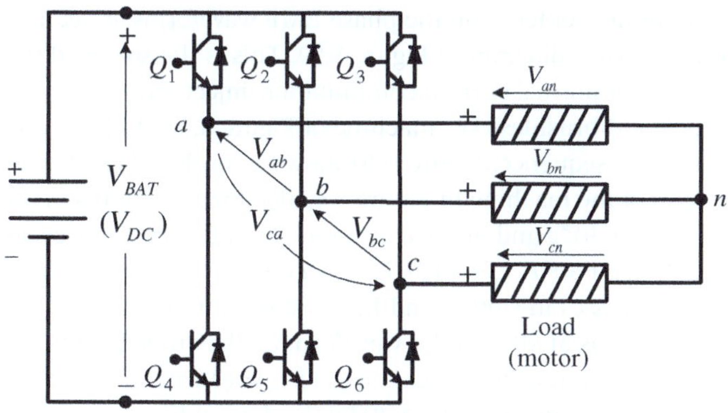

Figure 8.36 Three-phase voltage source feeding a 3-phase load.

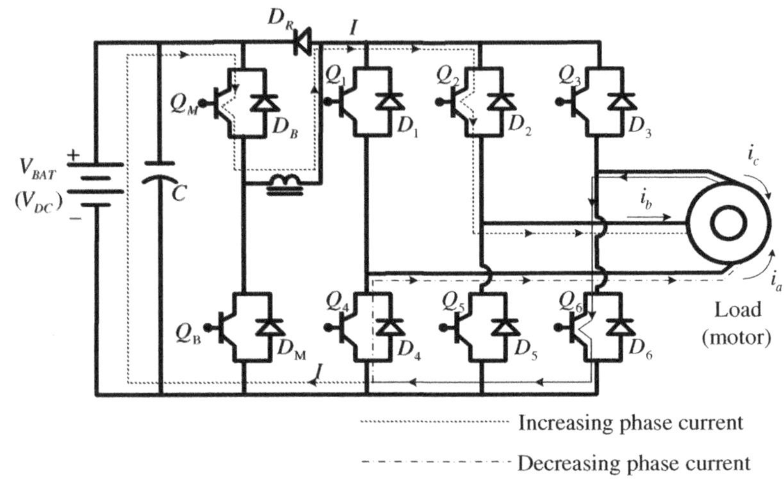

Figure 8.37 Current–source inverter with a two-quadrant DC chopper with current patterns during the commutation of current from phase (*a*) to phase (*b*).

are in the "on" state from (0 °*e* to 120 °*e*) for Q_1 and from (180 °*e* to 300 °*e*) for Q_4, respectively, see Refs. [2–4]; the pair of switches, Q_2 and Q_5, are in the "on" state from (120 °*e* to 240 °*e*) for Q_2, and (300 °*e* to 420 °*e* [i.e., 60 °*e*]) for Q_5, respectively; and the pair of switches, Q_3 and Q_6, are in the "on" state from (240 °*e* to 360 °*e*) for Q_3, and (60 °*e* to 180 °*e*) for Q_6, respectively, see Refs. [2–4].

These conduction periods for switches, Q_1 through Q_6, can be seen clearly in the "on" duty cycles of these switches illustrated in the current–source-fed *A*, *B*, and *C* phase armature windings of a motor depicted in Figure 8.38.

It should be noticed that the rotor position sensor is adjusted, under normal commutation, to inject the currents, i_a, i_b, and i_c, in such a manner that their 120 °*e* conduction periods are centered on the phase EMF waveforms, e_a, e_b, and e_c, as shown in the waveforms' diagram of Figure 8.39. This is the so-called 0 °*e* advance-of-commutation, or normal current commutation/injection.

Consider the two-pole PM brushless DC machine of Figure 8.4 with its *A*, −*C*, *B*, −*A*, *C*, and −*B* stator slot sequence. Figure 8.40 shows a developed (flattened) redraw of the stator slots of the machine in Figure 8.4, albeit with currents in the slots at two-time instants at 30 °*e* and at 90 °*e* in the i_a, i_b, and i_c currents cycle, also shown in Figure 8.40. At 30 °*e*, $i_a = (+I)$, $i_b = (-I)$, and $i_c = 0$, as shown in the slots by the dots and crosses in slots *A* and *B*, and the resulting stator MMF, $\overline{F}_S$, will have its magnetic axis at 90 °*e* and a waveform as shown below the top stator slots. Now, at 90 °*e* instant in the cycle, $i_a = (+I)$, $i_b = (0)$, and $i_c = (-I)$, as shown below the lower stator slots of Figure 8.40 by the dots and crosses in slots *A* and *C*, and the resulting stator MMF, $\overline{F}_S$, will have its magnetic axis at 150 °*e*,

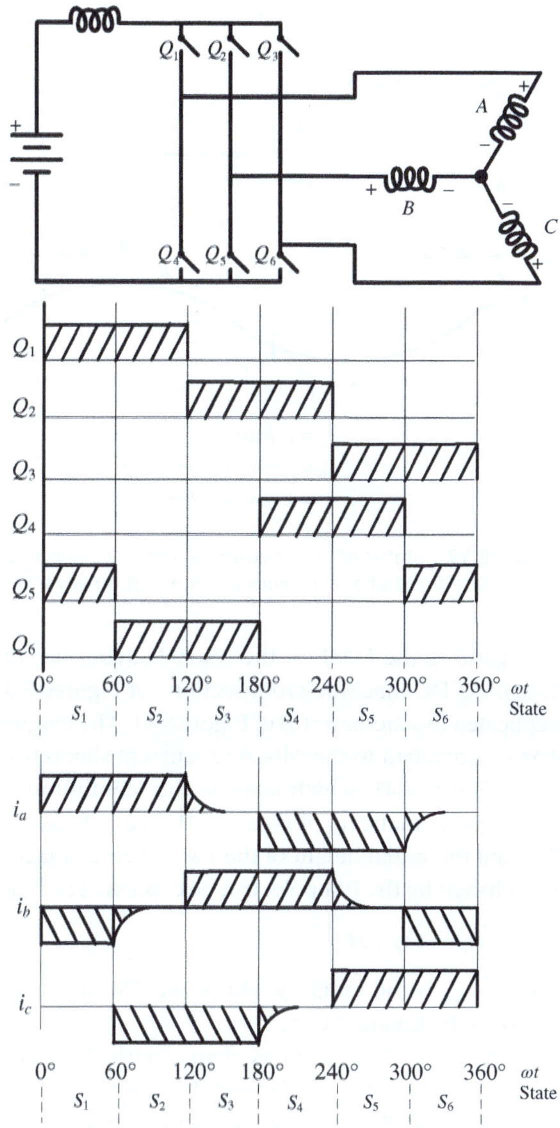

Figure 8.38 Duty cycles of switches, Q_1 through Q_6, and the idealized phase current, i_a, i_b, and i_c, from a current-source inverter 120 °e "on" cycle per switch.

and a waveform as shown below the lower stator slots. This constitutes a forward "jump" in the location of the magnetic axis by 60 °e from a location at 90 °e to a location at 150 °e. This "jumping or hopping" nature of the stator MMF, $\overline{F}_S$, is the result of the rectangular cyclic nature of the time-domain waveforms of the phase currents, i_a, i_b, and i_c, for this type of 120 °e conduction period for the six-switches in the inverter-bridge of Figure 8.37.

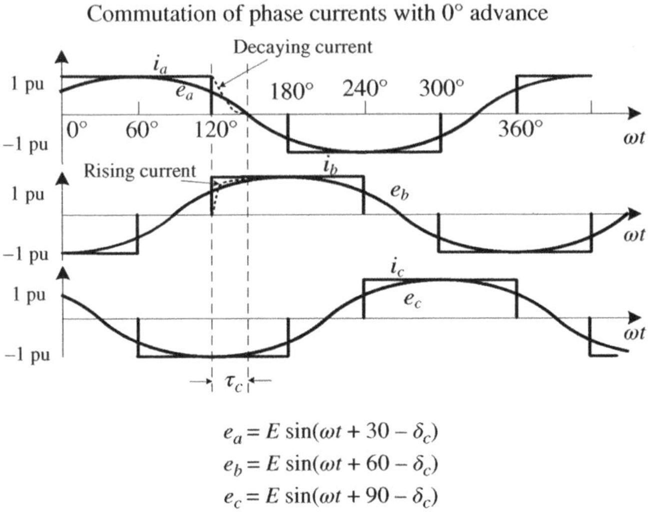

Figure 8.39 Status of phase currents and EMFs during commutation of phase current from phase (*a*) to phase (*b*), with a commutation advance angle, $\delta_c = 0°$.

Regarding the MMF of the rotor, $\overline{F}_R$, consider the PM shapes in the example brushless DC machine cross-sections of Figures 8.31 and 8.32. This PM shape is replicated in schematic (a) of Figure 8.41. The magnetomotive (MMF) effect of this PM is equivalent to the "discrete" current sheets (or excitation coils) of schematic (c) of Figure 8.41, which leads to the "continuous" current sheets (or excitation coils) of schematic (c) of Figure 8.41, in all of which the coercivity of PM material, H_c, and the radial height of the PM are crucial factors, where the rotor MMF, $\overline{F}_R$, established by the PM pole, can be expressed as follows:

$$\overline{F}_R = (h_{PM})H_c \tag{8.72}$$

where h_{PM} is the radial height of the PM in the direction of magnetization, see Figure 8.41, schematic (a).

Here, if the PM pole-arc is of an electrical angular spread, ρ, electrical radians, the amplitude of the fundamental component, F_{R1}, of the rotor MMF, $\overline{F}_R$, can be expressed, based on developments in Chapter 5, as follows:

$$F_{R1} = (4/\pi)(\sin(\rho/2))(h_{PM}H_c) \tag{8.73}$$

Again, based on the developments given earlier in Chapter 5, the fundamental component amplitude, F_{S1}, of the stator three-phase winding's MMF, $\overline{F}_S$, can be expressed as follows:

$$F_{S1} = (3/2)(4/\pi)k_W(N_{ph}/p)(I/m) \tag{8.74}$$

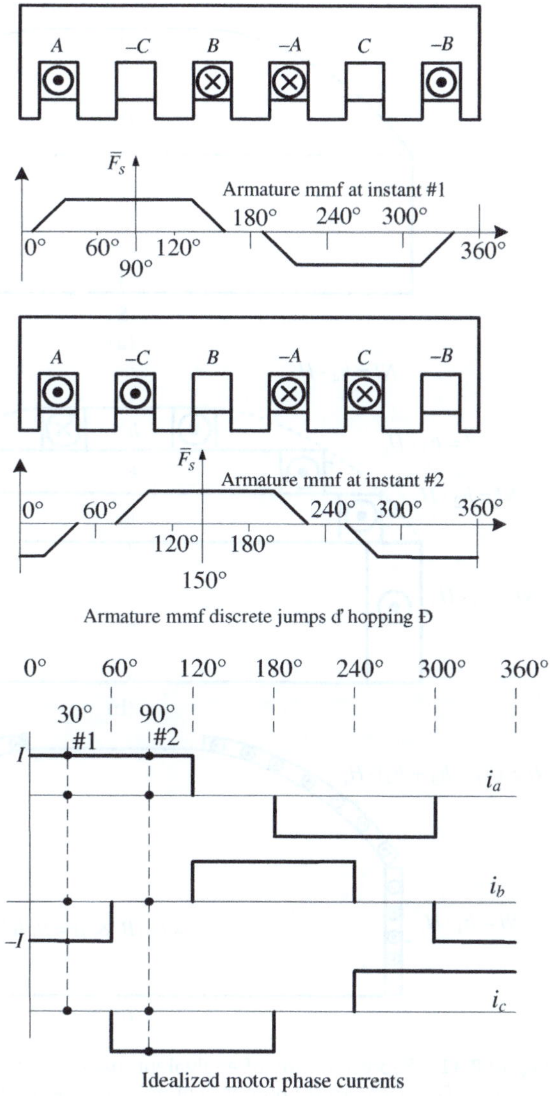

Figure 8.40 The discrete hopping nature of the stator MMF, F_S.

That is,

$$F_{S1} = (6/\pi)k_W(N_{ph}/p)(I/m) \tag{8.75}$$

where k_W is the winding factor defined earlier in Chapter 5, N_{ph} is the number of turns per phase, p is the number of poles, I is the DC current supplied to the six-switch inverter bridge, see Figure 8.37, and m is the number of parallel paths per phase.

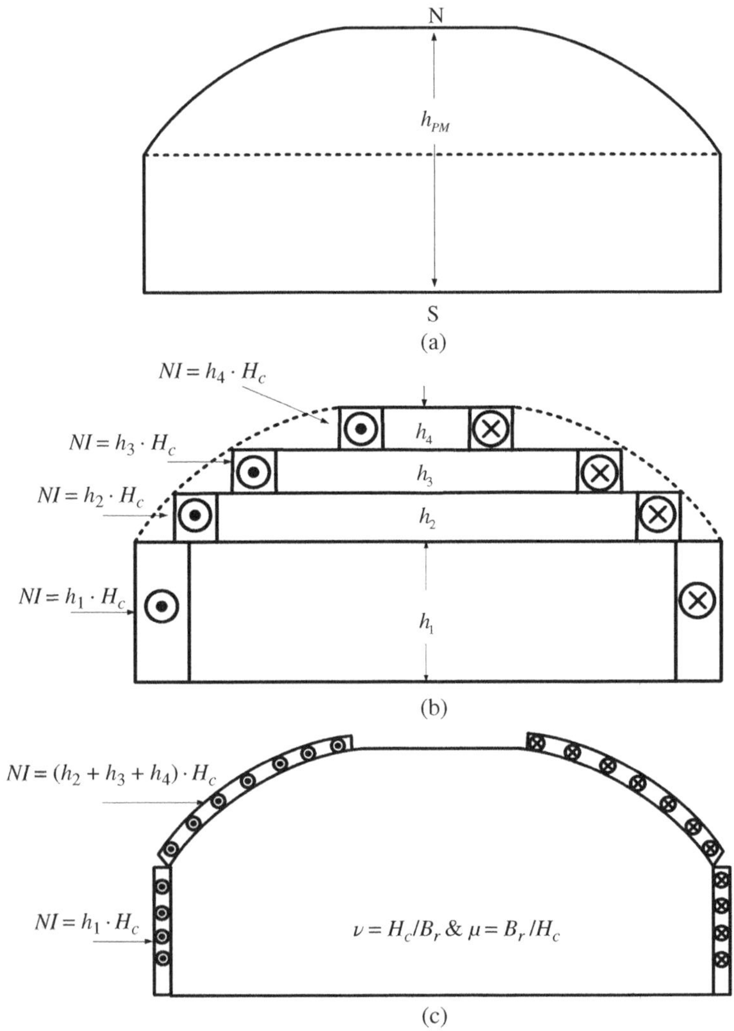

Figure 8.41 Representation of equivalent magnetic effect for the magnet shapes of the brushless DC machines for purposes of finite-element field analysis and PM MMF equivalence.

Now, the magnitude of the developed electromagnetic/electromechanical torque, T_{dev}, see Eq. (5.145), can be expressed as follows:

$$T_{dev} = (\mu_0/2)(p/2)(\pi Dl/h_g)F_{R1}F_{S1}\sin\delta_{sr} \tag{8.76}$$

that is,

$$T_{dev} = (\pi/4)(\mu_0 pDl/h_g)F_{R1}F_{S1}\sin\delta_{sr} \tag{8.77}$$

where D is the mid-airgap diameter, l is the length of the stator lamination stack, h_g is the radial height of the airgap opposite the PM pole arc, and δ_{sr} is the angle in electrical measure between the magnetic axes of the MMFs, $\overline{F}_S$ and $\overline{F}_R$.

Here, because of the "discretely jumping or hopping" nature of $\overline{F}_S$ explained earlier, it follows that the torque angle, δ_{sr}, varies between 120 °*e* and 60 °*e*, see the MMF waveform schematics of Figure 8.42, that is, $60\,°e \leq \delta_{sr} \leq 120\,°e$. Included in this figure is the developed torque, T_{dev}, profile over a complete electrical cycle of 360 °*e*, both idealized according to Eq. (8.77) and realistic with sharper peaks and deeper dips, see Refs. [2, 3] for further details. Here, the average torque, T_{dev-av}, can be obtained as follows:

$$T_{dev-av} = (3/\pi)\int_{\pi/3}^{2\pi/3} T_{max}\sin\delta_{sr}d(\delta_{sr}) = (3/\pi)T_{max} \tag{8.78}$$

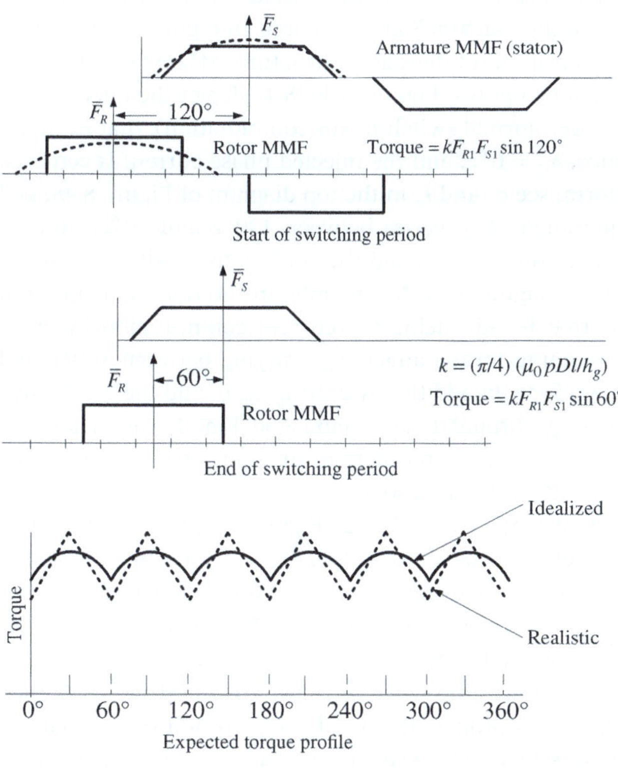

Figure 8.42 The process of torque production in brushless DC motors.

where,

$$T_{\max} = (\pi/4)(\mu_0 pDl/h_g)F_{R1}F_{S1} \tag{8.79}$$

Here, Eq. (8.78) yields the following:

$$T_{dev-av} = (3/\pi)T_{\max} \tag{8.80}$$

Hence, this yields from Eqs. (8.79) and (8.80) the following:

$$T_{dev-av} = (3/4)(\mu_0 pDl/h_g)F_{R1}F_{S1} \tag{8.81}$$

where F_{R1} is defined in Eq. (8.73), and F_{S1} is defined in Eq. (8.75).

The torque variation with time during any one of the six states covering a cycle of 360 °e is further illustrated in Figure 8.43.

The process of torque production in a brushless DC motor, and the motor–inverter interaction at the beginning, in the middle, and at the end of state #1, S_1, and the beginning of state #2, S_2, can further be understood by observing the "on" status of the six switches of the inverter, Q_1 through Q_6, and the corresponding direction of current flows in the coil sides in the armature slots, A, $-C$, B, $-A$, C, and $-B$, in the machine's cross-sections in Figures 8.44–8.47, as well as the corresponding positions of the stator (armature) MMFs, $\overline{F}_S$, and rotor (PM) MMFs, $\overline{F}_R$, in these four figures. Figures 8.44–8.47 depict the operation of the inverter and motor under normal switching (or commutation), that is, under 0 °e commutation advance, $\delta_C = 0$ °e, and the injected phase current is centered on the back EMF waveform, see e_a and i_a in the top diagram of Figure 8.48, and the corresponding torque angle, δ_{SR}, varies between 120 °e and 60 °e. That is, 60 °$e \leq \delta_{SR} \leq 120$ °e. Also, see the i_a, i_b, i_c, and the "on" status of the six inverter switches, Q_1 through Q_6, in Figure 8.49. Meanwhile, in the middle diagram of Figure 8.48, $\delta_C = 30$ °e, that is, advancing the current commutation by 30 °e, which leads to a corresponding torque angle, δ_{SR}, varying between 150 °e and 90 °e. That is, 90 °$e \leq \delta_{SR} \leq 150$ °e. In addition, see the i_a, i_b, i_c, and the "on" status of the six inverter switches, Q_1 through Q_6, in Figure 8.50. Finally, for an advance of commutation with, $\delta_C = 60$ °e, see the bottom diagram of Figure 8.48, the torque angle, δ_{SR}, becomes 120 °$e \leq \delta_{SR} \leq 180$ °e.

It should be observed that when $\delta_C = 0$ °e, see Figure 8.48, the average value of the torque angle, δ_{SR}, between the stator MMF, $\overline{F}_s$, and the rotor MMF, $\overline{F}_R$, is 90 °e. Therefore, on average $\overline{F}_s$ does not magnetize or demagnetize the rotor MMF, $\overline{F}_R$. However, when $\delta_C = 30$ °e, on average δ_{SR}=120 °e. Therefore, on average, the stator MMF, $\overline{F}_s$, demagnetizes the rotor MMF, $\overline{F}_R$. Accordingly, this condition of advanced commutation/firing when $\delta_C = 30$ °e is in effect, a so-called "flux-weakening" condition. A stronger "flux-weakening" condition is produced when $\delta_C = 60$ °e, see Figure 8.49, in which case on average $\delta_{SR} = 150$ °e, and, consequently, $\overline{F}_S$ has an even stronger demagnetizing effect on the rotor MMF, $\overline{F}_R$.

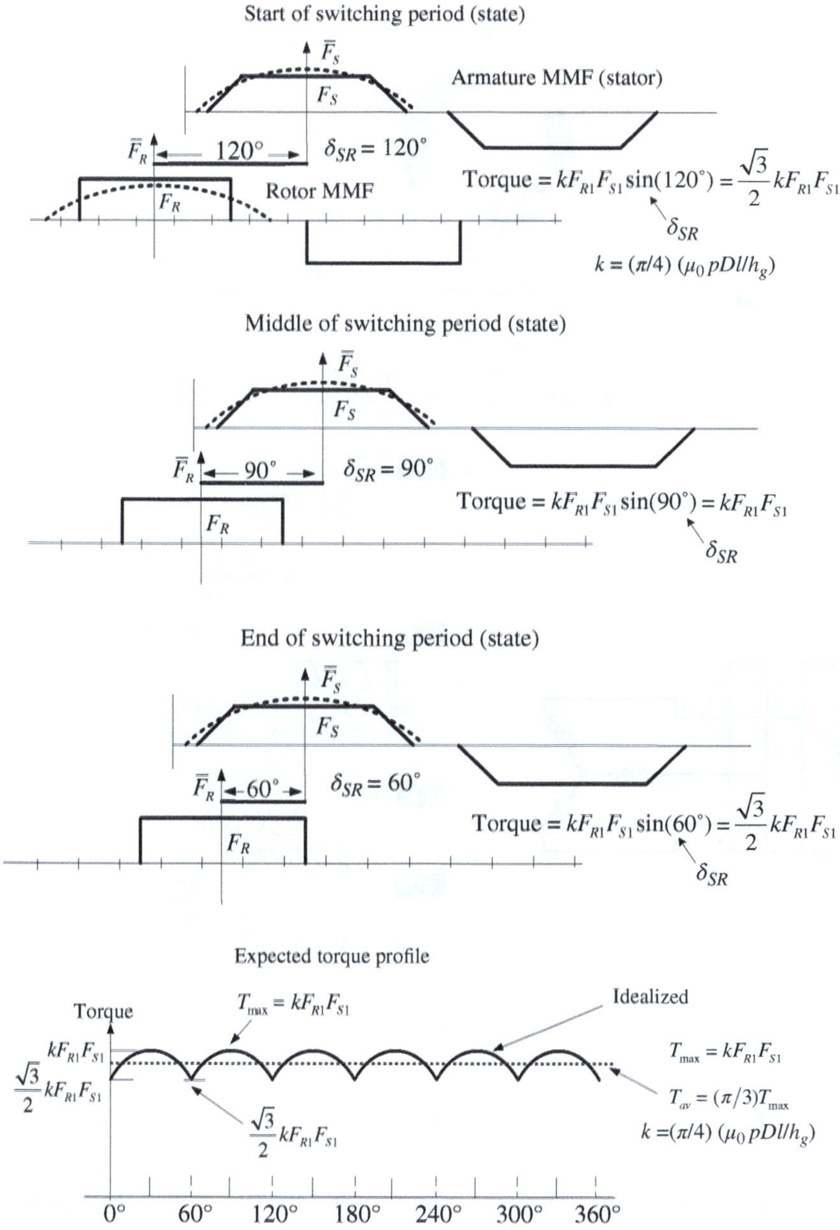

Figure 8.43 Torque variation during one of the six states (periods) of switching in a brushless DC motor.

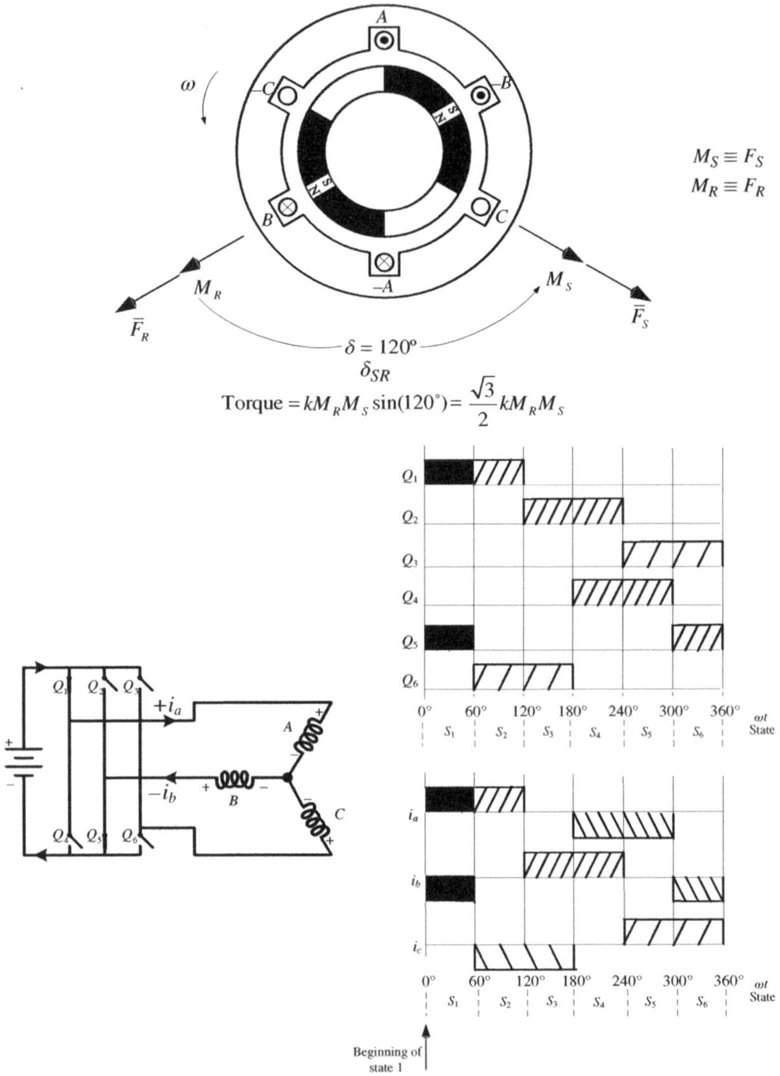

Figure 8.44 The MMFs $\overline{F}_S$ and $\overline{F}_R$ at the beginning of state #1.

In general, whenever the average value of $\delta_{SR} > 90\,^\circ e$, a "flux weakening" condition prevails. Therefore, an advanced commutation/firing angle, $\delta_C > 0\,^\circ e$, leads to the so-called flux weakening condition.

However, it should be pointed out that despite the reduction in the rotor MMF, $\overline{F_R}$, this results in a significant increase in the magnitudes of the stator

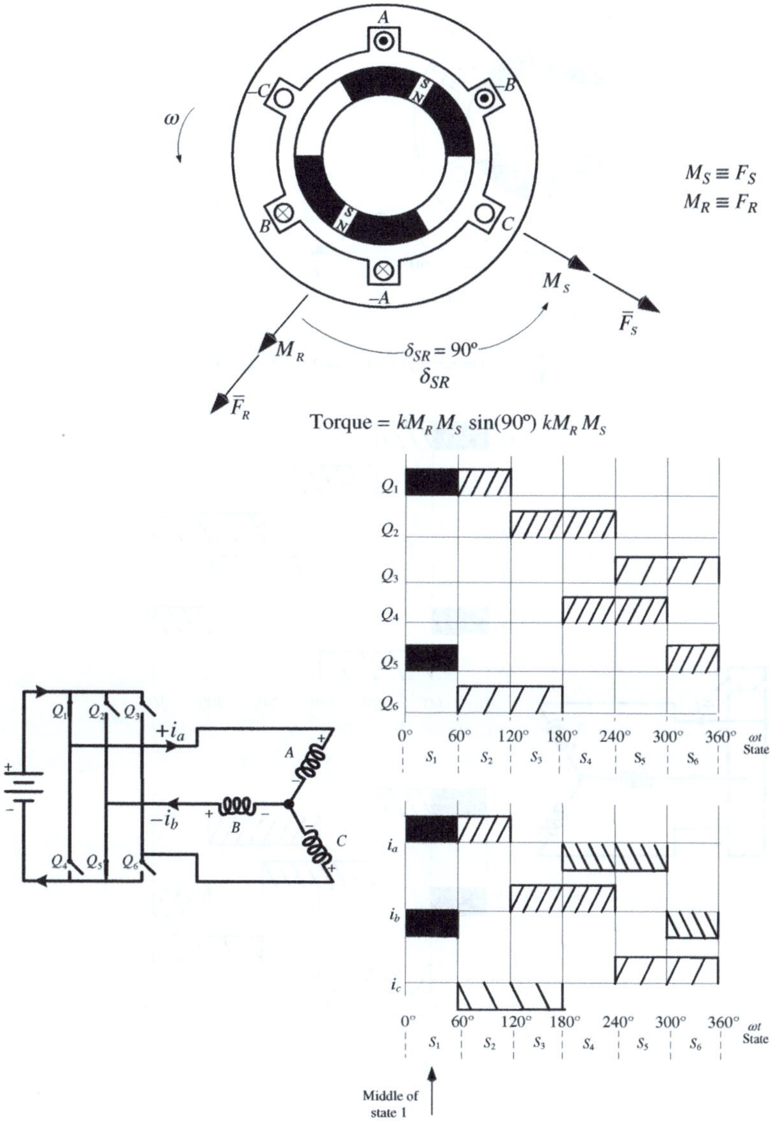

Figure 8.45 The MMFs $\overline{F}_S$ and $\overline{F}_R$ at the middle of state #1.

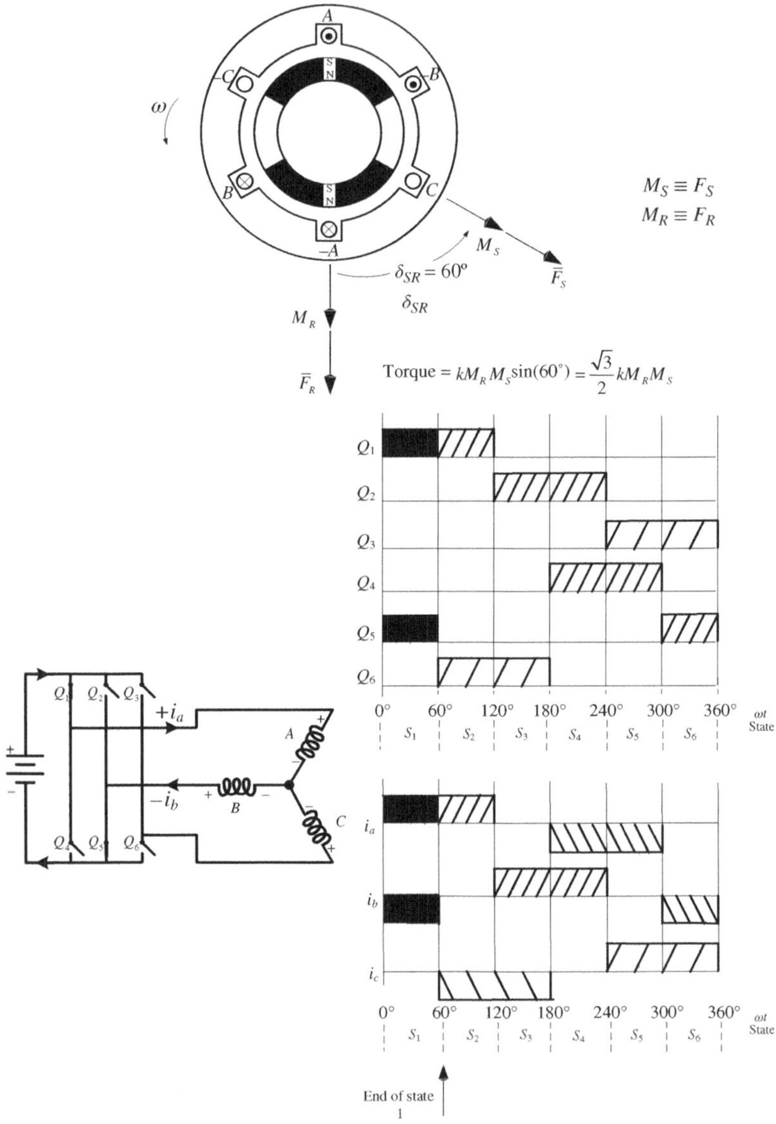

Figure 8.46 The MMFs $\overline{F}_S$ and $\overline{F}_R$ at the end of state #1.

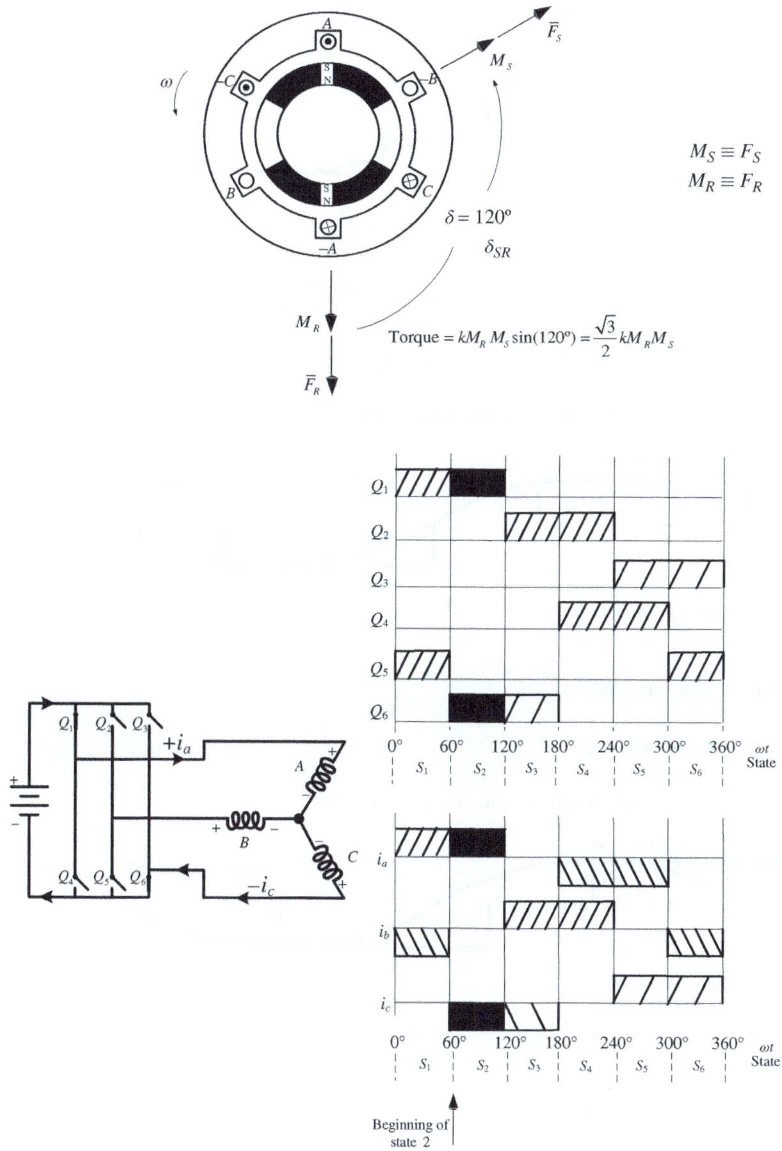

Figure 8.47 The MMFs $\overline{F}_S$ and $\overline{F}_R$ at the beginning of state #2.

0° Commutation advance ($\delta_c = 0°$)

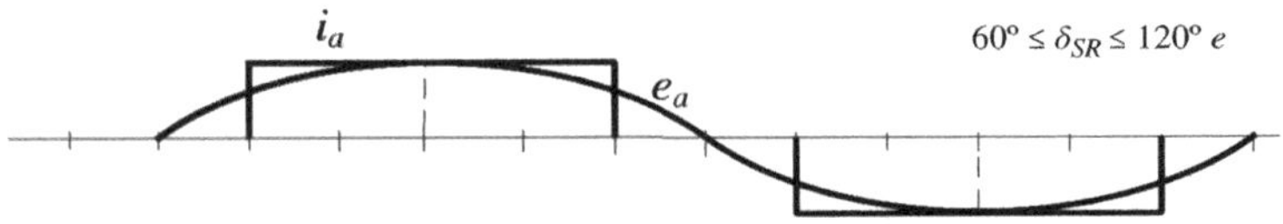

Start of state: $T = K_T \sin(120°) = \frac{\sqrt{3}}{2} K_T$

Middle of state: $T = K_T \sin(90°) = K_T$

End of state: $T = K_T \sin(60°) = \frac{\sqrt{3}}{2} K_T$

$k = \frac{p}{4}\left[\frac{\mu_0 \pi D l}{g}\right] \quad K_T = kF_S F_R$

$K_T = \frac{p}{4}\left[\frac{\mu_0 \pi D l}{g}\right] F_S F_R$

$F_S \equiv M_S \quad F_R \equiv M_R$

30° Commutation advance ($\delta_c = 30°$)

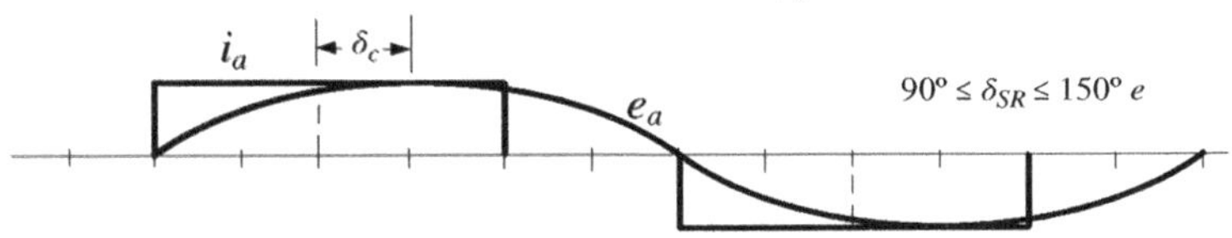

Start of state: $T = K_T \sin(150°) = \frac{1}{2} K_T$

Middle of state: $T = K_T \sin(120°) = \frac{\sqrt{3}}{2} K_T$

End of state: $T = K_T \sin(90°) = K_T$

60° Commutation advance ($\delta_c = 60°$)

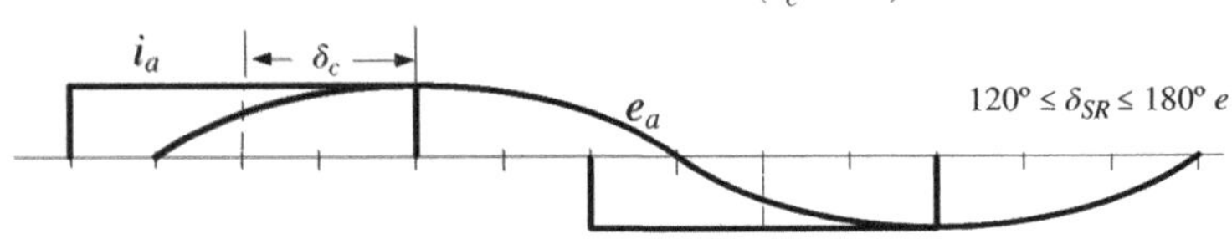

Start of state: $T = K_T \sin(180°) = 0$

Middle of state: $T = K_T \sin(150°) = \frac{1}{2} K_T$

End of state: $T = K_T \sin(90°) = \frac{\sqrt{3}}{2} K_T$

NOTE: Torque relationships consider only fundamental MMFs.

Figure 8.48 Effect of commutation angle, δ_c, on range of the torque angle, δ_{SR}.

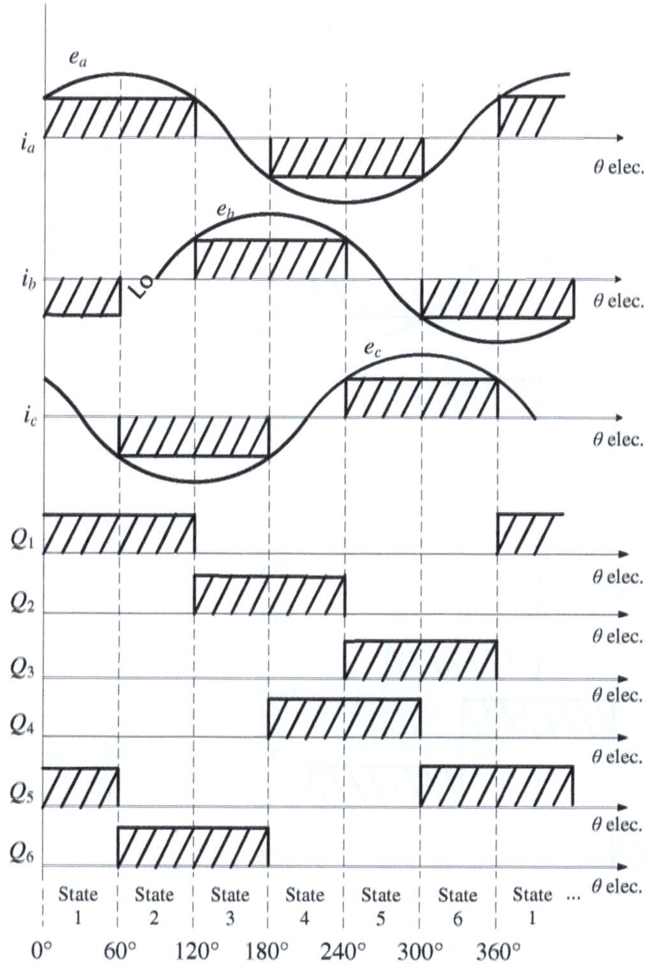

Figure 8.49 Inverter switching sequence during motoring, 0 ° commutation advance.

phase currents, with a corresponding considerable increase in the stator MMF, $\overline{\mathcal{F}}_s$, and consequently, an overall increase in the developed torque and power of such PM brushless DC motors. This has been shown to be the case through simulation/modeling and experimental test results of such motors as documented in the literature in Refs. [1, 3, 5, 6].

The effect of commutation advance angle, δ_C, on the developed electromechanical/electromagnetic torque profile, for $\delta_C = 0\,°e$, $\delta_C = 30\,°e$, and $\delta_C = 60\,°e$, is shown in Figure 8.51. In Figure 8.51, the effect of the commutation angle, δ_C, on the

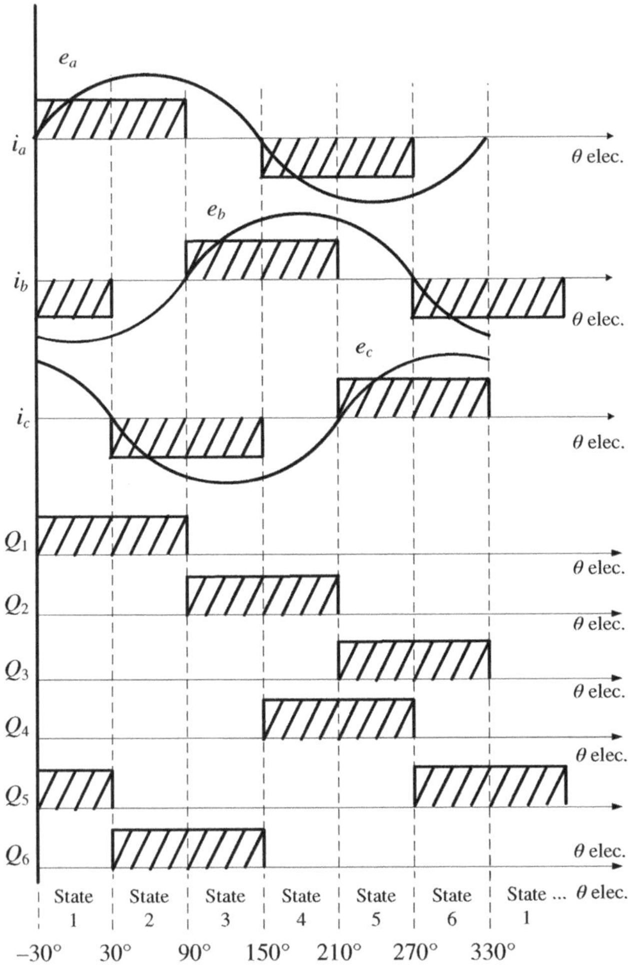

Figure 8.50 Inverter switching sequence during motoring, 30 ° commutation advance.

developed average torque, T_{dev-av}, is also shown where the $T_{dev-av} = 0.955$ p.u. for $\delta_C = 0\,°e$, $T_{dev-av} = 0.860$ p.u. for $\delta_C = 30\,°e$, and $T_{dev-av} = 0.477$ p.u. for $\delta_C = 60\,°e$. Here, $T_{dev-\max} = 1.0$ p.u. We next discuss the modeling for torque–speed control of PM brushless DC motors. Once again, the schematics in Figures 8.48, 8.49, and 8.50 depict normal and advanced commutation/firing for $\delta_C = 0\,°e$, $\delta_C = 30\,°e$, and $\delta_C = 60\,°e$, where for $\delta_C = 30\,°e$, and $\delta_C = 60\,°e$, also the so-called flux-weakening conditions prevail.

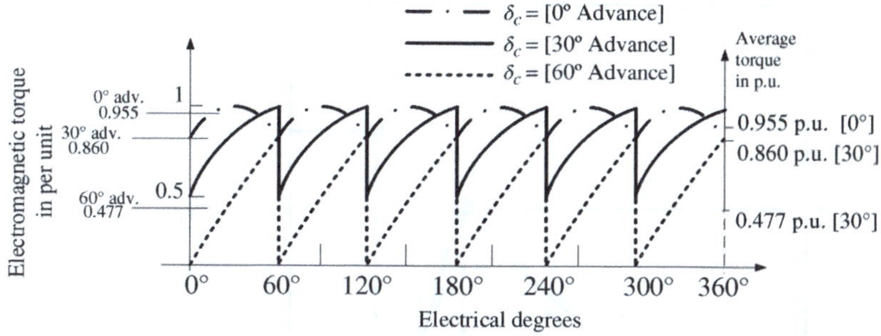

Figure 8.51 Effect of the commutation angle, δ_c, on the torque profile for $\delta_c = 0\,°e$, $\delta_c = 30\,°e$, and $\delta_c = 60\,°e$, see Ref. [2, 3].

8.10 Brushless DC Motors Torque–Speed Control

Here, upon examination of the DC chopper–inverter–PM brushless DC machine schematic of Figure 8.37, the machine's equivalent circuit model appears from the DC bus (source) as an ohmic resistance, R, in series with a DC back-EMF, E, in volts, which can be analytically expressed as follows:

$$E = k_E\omega \tag{8.82}$$

where k_E is a back EMF constant in volts per mechanical radians per second, which is a function of the strength of the machine's PMs and the machine winding design particulars, and ω is the motor shaft speed in mechanical radians per second, see Figure 8.52.

Here, in the electric circuit of Figure 8.52, the voltage, V, is the DC-bus/chopper output voltage, and I is the DC-bus current. Also, I_{max} is the maximum DC current output of the chopper, which is the steady current-carrying limit of the chopper switches, and the inverter switches, and V_{max} is the maximum output voltage when the upper chopper switch is continuously in an "on" state at all times. Hence, when the back EMF, E, reaches a value equal to V_{max}, the current $I = 0$, and, hence, that point defines the no-load speed, ω_{NL}, that is, when

$$E = k_E\omega_{NL} = V_{max} \tag{8.83}$$

At all other speed, $0 \le \omega \le \omega_0$, see Figure 8.52, we have the following:

$$V - RI = E = k_E\omega \tag{8.84}$$

and the developed power, P, is,

$$P = EI = k_E\omega I \tag{8.85}$$

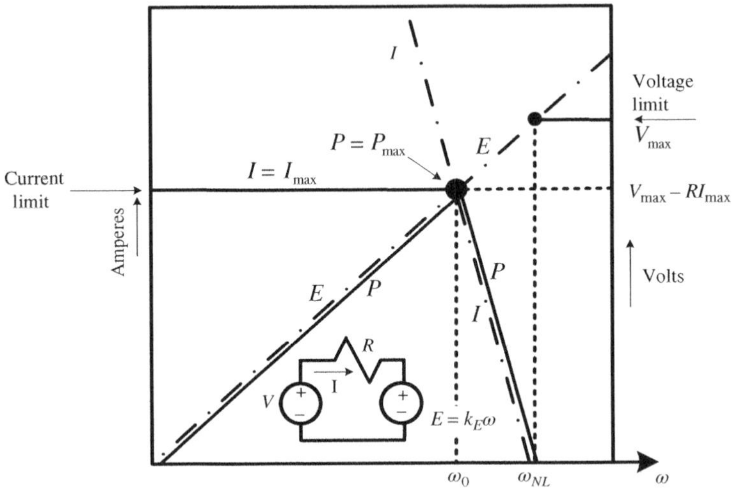

$V - RI = E = k_E\omega$

$E = k_E\omega,\ P = EI = k_E\omega I,\ T = \frac{P}{\omega} = k_E I$

Hence,

$\omega = \frac{1}{k_E}(V - RI) = \frac{V}{k_E} - \frac{R}{k_E}\left(\frac{T}{k_E}\right) = \frac{V}{k_E} - \frac{RT}{k_E^2}$

Inverter limits: V_{max}, I_{max}

Figure 8.52 Idealized characteristics of PM brushless DC motors.

Hence, from Eq. (8.85), one can express the developed torque, T_{dev}, as follows:

$$T_{dev} = (P/\omega) = (k_E\omega I/\omega) = k_E I \tag{8.86}$$

Hence, the current, I, can be expressed in terms of the developed torque, T_{dev}, as follows:

$$I = T_{dev}/k_E \tag{8.87}$$

Therefore, Eq. (8.84) can be rewritten as follows:

$$k_E\omega = V - R(T_{dev}/k_E) \tag{8.88}$$

That is the basis for "the torque–speed equation" in a brushless DC machine, which can be deduced from Eq. (8.88) to yield the following relationship:

$$\omega = (V/k_E) - \left(R/k_E^2\right) T_{dev} \tag{8.89}$$

Equation (8.89) leads to the torque–speed family of plots, Figure 8.53, for various levels of DC output voltages of the DC chopper of Figure 8.37, for output voltages,

$V = V_{max} > V = V_1 > V = V_2 > V = V_3$, etc., for a developed torque,

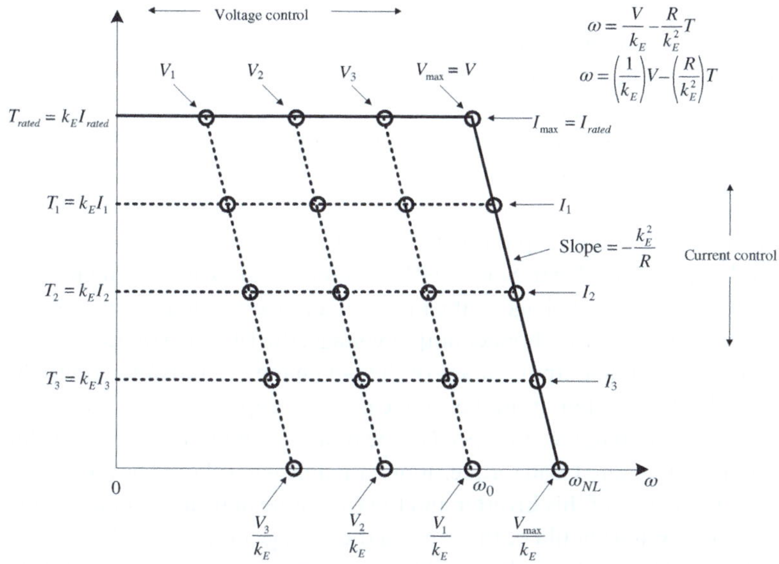

Figure 8.53 Idealized torque–speed characteristics of PM brushless DC motors.

$$T_{dev} = T_{max} = T_{rated} = k_E I_{rated} > T_{dev} = k_E I_1 > T_{dev} = k_E I_2 > T_{dev} = k_E I_3, \text{etc.}$$

Here, in this torque–speed plane, the corresponding no-load speeds, ω_{NL}, are,

$$\omega_{NL} = V_{max}/k_E > V_1/k_E > V_2/k_E > V_3/k_E, \text{etc.}$$

The reader should notice the similarity between the sweep of the torque–speed plane through control of the output DC voltage of the chopper, $V_{max} > V_1 > V_2 > V_3$, etc., and the sweep of the torque–speed plane for the induction motor under constant volts per hertz control as shown in Chapter 6, Figure 6.42. For further details on more advanced coupled finite-element-power conditioner modeling of PM brushless DC machines, the interested reader should consult Refs. [4–13].

Homework Problems

8.1 A constant volts per hertz power electronic converter (i.e., power conditioner) feeds a 3-phase, Y-connected, 460 V (line-to-line, RMS), 30-hp, 60 Hz, 4-pole induction motor in an ASD system. The converter consists of a three-phase full-wave diode rectifier whose dc output is fed into a chopper for DC link voltage control. The DC link voltage is fed into a six-switch three-phase inverter whose switching is based on a 180 °*e* duty cycle per switch, without pulse-width modulation. In other words, the voltage control is achieved through the chopper's duty cycle, and the frequency control is achieved through the inverter's operating (switching) frequency. The rectifier is supplied by the power company through a suitable transformer with tap-changing capability.

Draw a functional block diagram of the complete ASD system from the electric utility outlet to the motor.

(a) When the chopper is fully on, that is, $(T_{on}/T_{cycle}) = 1$, the fundamental component of the output line-to-line voltage of the inverter is supposed to have an Root Mean Square (RMS) value equal to 460 V, which is the rated line-to-line voltage of the motor at an inverter frequency of 60 Hz. What should be the chopper output voltage, the DC link voltage, and the rms line-to-line output voltage of the tap-changing transformer supplying the ASD system from the source power company?

(b) It is desired to reduce the speed of the motor to about half its rated value for the same rated load torque, approximately what should be the operating frequency of this inverter and the rms value of its line-to-line output voltage? What should be the value of the average dc output voltage of the chopper? What should be the chopper's duty cycle $(T_{\text{on}}/T_{\text{cycle}})$ per unit? Assume that the DC output voltage of the rectifier and the position of the taps on the transformer windings are unchanged (held constant).

(c) Repeat (b) for a motor speed equal to one-tenth its rated value for the same rated load torque as in (a) and (b).

(Hint: Assume ideal rectifier, chopper, and inverter in all the questions.)

8.2 A three-phase, six-switch, voltage source inverter, whose circuit topology is identical to the inverter shown in Figure 8.36, is feeding a 3-phase Y-connected balanced load from a 116 volts DC busbar. The load circuit per phase consists of a resistance, $R = 0.3\,\Omega$, connected in series with an inductance, $L = 1.4494 \times 10^{-3}$ H, and an EMF, $e(t)$ per phase. The EMF terms for the a, b, and c, phases are as follows:

$e_a(t) = 46.111\,\text{Sin}(958.186t)$ V

$e_b(t) = 46.111\,\text{Sin}\,(958.186t - 2\pi/3)$ V

and $e_c(t) = 46.111\,\text{Sin}\,(958.186t - 4\pi/3)$ V

(a) The voltage source inverter switches, Q_1 through Q_6, are timed to "switch-on" at times, $(958.186t) = 0, 2\pi/3, 4\pi/3, \pi, 5\pi/3$, and $7\pi/3$, respectively, and sequentially in a $180\,°e$ ($\pi\ e$ – radians) conduction cycle. Find the harmonic Fourier series of the line-to-neutral (phase) voltage of the inverter output voltage for phase (a), $v_{an}(t)$.

(b) Using superposition, find the RMS values of the fundamental, 5th, 7th, and 11th harmonics of the load phase (a) current, and write the complete harmonic Fourier series of the phase (a) current, $i_a\,(t)$. Plot the current waveform (versus time) over two cycles.

(c) Obtain the real power consumption of the load including the harmonic effects up to and including the 11th harmonic. Also, obtain an estimate of the dc line current assuming zero inverter losses.

Problem for Graduate Students

8.3 Simulate the system described in Problem 8.2 using MATLAB-Simulink (Power Tool Box) or other software of your choosing, plot the load's terminal voltage to neutral and phase current waveforms (versus time), and compare this current waveform you obtained in (b) with one of the phase currents of the simulation and comment. Compute the average real power consumption of the load and compare your answer with that you obtained in (c) and comment.

8.4 A two-pole laboratory-size conventional brush-commutator type dc motor has an inner stator bore (pole arc) radius of 2.524 in., an airgap height of 0.012 in., and an active armature conductor length of 6.9 in. The field winding has 480 turns and a fundamental space harmonic winding factor, $k_f = 0.93$. The field winding is separately excited with a DC field current, $I_f = 0.383$ A.

(a) Calculate the peak of the fundamental airgap flux density in Tesla and the flux per pole in Webers.

(b) For an armature speed of 1800 r/min, it is desired to generate an armature EMF of 120 V. Obtain a suitable number of armature turns, assuming a lap-connected winding.

(c) Compute the peak of the fundamental component of the armature MMF in ampere-turns (AT)/pole for an armature current of 4.0 A.

(d) Compute the torque in Nm under these conditions using the armature winding and field winding fundamental MMF peaks.

(e) Repeat the computation of the torque using the field flux per pole and the peak of the fundamental component of the armature MMF.

(f) Compute the torque in Nm using the armature EMF, the armature current, and motor speed. Also, compare this answer with those you obtained in (d) and (e) and comment succinctly.

8.5 A single-phase full-wave-controlled bridge rectifier is used to control the speed of a 115-V, 30-A, 1200 r/min permanent magnet brush-commutator-type DC motor. The armature resistance is 0.4 Ω. The available utility supply is 60-Hz, 120-V source. Rotational losses may be neglected.

(a) Determine the firing angle to give a motor speed of 1000 r/min when the motor is operating at the rated armature current.
(b) Determine the supply power factor for the condition in (a).
(c) If the developed torque is 16 Nm and the rectifier firing angle is 60 °*e*, what is the motor speed?
(d) What is the RMS current in each switch in the rectifier for the condition in (c).
(e) If the above rectifier is replaced with a combination of an uncontrolled rectifier feeding the motor through a two-quadrant chopper, what would be the chopper's duty cycle for a motor-operating condition identical to that in (a)?

8.6 In a 6-pole, 155-V, 9500 r/min permanent magnet brushless dc motor, each of the rotor's permanent magnet poles can be approximated by an equivalent sinusoidally distributed MMF, with a peak value, $F_r = 10{,}000$ AT/pole. Under the rated normal inverter commutation conditions, the motor develops 30-hp at 9500 r/min.
(a) Find the average value of the developed torque and find its peak value if the torque angle, δ_{sr}, varies such that $60\,°e \le \delta_{sr} \le 120\,°e$ during each of the six states of the operating ac cycles.
(b) Given that the motor's mid-airgap diameter, D = 0.09 m, the effective armature conductor length, ℓ = 0.12 m, and the airgap height, $h_g = 0.025$ m, find the stator MMF, F_s, in the AT pole.
(Hint: $T_{dev} = [(P/4)\,(\mu o\pi D\ell/h_g)\,F_sF_r\,\sin\delta_{sr}])$
(c) If the inverter and chopper losses, as well as the motor losses are neglected, estimate this motor's dc-line (link) current for a DC-link voltage of 155-V and estimate its inverter switching frequency at 9500 r/min.
(d) Using the approximating assumptions of (c), thus E = 155 V at 9500 r/min, express the average developed torque in terms of the dc-line (link) current.
(e) Now, with $E = 155$ V, and dc-link current as computed in (c), estimate the dc-link supply voltage, V, for a total motor–controller circuit resistance of 0.04 Ω and obtain the motor's torque–speed characteristic equation in mech. rad/s versus Nm and in r/min versus Nm. Also, express the speed–torque characteristics as a parametric equation in terms of a general dc-line voltage, V. Sketch the characteristic for this computed voltage in the speed-torque plane.

8.7 A 15-hp, 230-V, dc-shunt motor has an armature–circuit resistance of 0.19 Ω and a field resistance of 145 Ω. At no-load and rated voltage, the speed is 1150 r/min, and the armature current is 9 A. At full load and rated voltage, the line current is 57.45 A.

(a) Neglecting armature reaction and magnetic saturation effects (i.e., flux is linearly proportional to field current), compute the speed in r/min at full load.
(b) Find the developed power in Watts and hp as well as the developed torque in Nm.
(c) Would this motor be a suitable choice for traction load applications? And why? (Traction load requires high torque at low speeds and low torque at high speeds.)
(d) Would this motor be a suitable choice for a constant torque load application? And why?

8.8 A dc series motor runs at 900 r/min when it is supplied from a 230-V supply. The total armature and series-field resistance is 0.22 Ω. The armature current was found to be 75 A under this condition.

(a) Compute the developed torque in Nm under the earlier condition.
(b) The load torque was reduced so that the armature current reading was found to be 25 A. Assuming that the effects of magnetic saturation are to be neglected, that is, the flux is linearly proportional to the series-field current (in this case, it is also the armature current), find the new motor speed in r/min and developed torque in Nm.
(c) Would this motor be a suitable choice for a load with constant torque demand at all speeds? And why?
(d) Would this motor be a suitable choice for traction load applications? And why? (See traction load as defined in (c).)

8.9 In a 6-pole, 125-V, 8700 r/min, permanent magnet brushless DC motor, designed for propulsion of subcompact-size commuter-type electric vehicles, each of the rotor's permanent magnet poles has an equivalent sinusoidally distributed MMF, with a peak value, $F_r = 8800$ AT/pole. Under the rated normal inverter commutation conditions, the stator's peak MMF, $F_s = 750$ AT/pole. Calculate the highest and lowest values of the developed torque in Nm for an inverter with a 120 °e "on" period per switch per AC cycle, without PWM. Then:

(a) Estimate an average developed torque in Nm, and the average value of the developed power in Watts and hp at a rotor speed of 8700 r/min. The motor's mid-airgap diameter, $D = 0.079$ m, effective armature conductor length, $l = 0.102$ m, and airgap height, g $= 0.019$ m. (Hint: $T_{dev} = [(p/4)(\mu_0 \pi D l F_s F_r \sin \delta_{sr})/g]$)

(b) If the inverter and chopper losses, as well as the motor losses are neglected, for a DC-link voltage of 125-V, estimate this motor's DC-line (link) current and inverter switching frequency under the earlier rated conditions.

(c) Neglecting the ohmic resistive voltage drop in the motor-controller circuit, thus $E = 125$ V at 8700 r/min, calculate the DC-line current assuming the developed power is equal to the input DC link power and sketch the idealized waveforms of one of the motor's phase currents for a 120 °e conduction period per AC cycle per inverter switch. Also, calculate the EMF constant, k_E, in volts/mech. rad/s.

(d) Now, with $E = 125$ V and a DC-link current as calculated in (c), estimate the DC-link supply voltage, V, for a total motor-controller circuit resistance of 0.05 Ω and obtain the torque–speed characteristic equation in Nm versus mech. rad/s. And in Nm versus r/min. Also, express the torque–speed characteristic as a parametric equation in terms of a general DC-link voltage, V.

Note: Assume an inverter without PWM in all the above, and a "fully-on" chopper.

8.10 In a DC ASD, a DC shunt motor is used, which has independent chopper control systems applied to both its armature circuit and field circuit (i.e., both field and armature chopper control are present and are considered to be ideal choppers with zero losses). The motor is rated at 50 hp, and it draws an armature current of 160 A and a field current of 4 A at 1000 r/min when both the armature chopper and the field chopper are operating at duty cycles of 100% (choppers fully-on), and the armature as well as field terminal voltages are 250 V. The motor has an armature circuit resistance of 0.03 Ω. The motor is driving a constant torque load. Neglecting effects of magnetic saturation and armature reaction:

(a) Calculate the line current, input power, developed power, output power, and output torque, if the rotational losses are considered negligible throughout.

(b) If the speed of the motor is to be reduced to 750 r/min, while the field chopper is to remain on a 100% duty cycle, determine the necessary terminal voltage of the armature circuit and the corresponding duty cycle of the armature chopper, if the DC line voltage is to remain at 250 V.

(c) If after the conditions in (b) are achieved, the field chopper duty cycle is changed from 100% to 90%, what happens to the field and armature currents, speed, input, and output power, assuming the load torque to remain constant and the rotational losses to be negligible.

8.11 When armature chopper torque–speed control is applied to a DC series motor whose armature circuit resistance is 0.4 Ω, it was found that the chopper input voltage was 550 V and its output voltage applied to the armature terminals was 500 V, when the motor was drawing an armature current of 75 A at an operating speed of 300 r/min. (For this problem, consider the chopper to be an ideal device with zero losses.)

(a) What is the chopper's duty cycle in percent? And what are the developed power in Watts and torque in Nm?

(b) What would be the new armature current and speed if the torque is to remain constant, while the duty cycle of the chopper is increased to 100%?

References

1 Demerdash, N.A., Miller, R.H., Nehl, T.W. et al. (1983). Comparison between features and performance characteristics of fifteen hp samarium cobalt and ferrite based brushless dc motors operated by same power conditioner. *IEEE Transactions on Power Apparatus and Systems* PAS-102: 104–112.

2 Demerdash, N.A. and Nehl, T.W. (1980). Dynamic modeling of brushless DC motors for aerospace actuation. *IEEE Transactions on Aerospace and Electronic Systems* AES-16 (6): 811–821.

3 Nehl, T.W., Fouad, F.A., Demerdash, N.A., and Maslowski, E. (1982). Dynamic simulation of radially oriented permanent magnet type electronically operated synchronous machine with parameters obtained from finite element field solutions. *IEEE Transactions on Industrial Applications* Vol. IA-I 8: 172–182.

4 Nehl, T.W., Fouad, F.A., and Demerdash, N.A. (1981). Digital simulations of power conditioner – machine interaction for electronically commutated DC permanent magnet machines. *IEEE Transactions on Magnetics* MAG-17: 3284–3286.

5 Nehl, T.W., Demerdash, N.A., and Fouad, F.A. (1985). Impact of winding inductances and other parameters on the design and performance of brushless DC motors. *IEEE Transactions on Power Apparatus and Systems* PAS-104: 2206–2213.

6 Hijazi, T.M. and Demerdash, N.A. (1988). Computer-aided modeling and experimental verification of power conditioner operated permanent magnet brushless DC motors including rotor damping effects. *IEEE Transactions on Energy Conversion* EC-3 (3): 714–721.

7 Fouad, F.A., Nehl, T.W., and Demerdash, N.A. (1981). Magnetic field modeling of permanent magnet type electronically operated synchronous machines using

finite elements. *IEEE Transactions on Power Apparatus and Systems* PAS-100: 4125–4135.

8 Fouad, F.A., Nehl, T.W., and Demerdash, N.A. (1981). Permanent magnet modeling for use in vector potential finite element analysis in electrical machinery. *IEEE Transactions on Magnetics* MAG-17: 3002–3004.

9 Nehl, T.W., Fouad, F.A., and Demerdash, N.A. (1982). Determination of saturated values of rotating machinery incremental and apparent inductances by an energy perturbation method. *IEEE Transactions on Power Apparatus and Systems* PAS-101: 4441–4451.

10 Demerdash, N.A., Fouad, F.A., and Nehl, T.W. Determination of winding inductances in ferrite type permanent magnet electric machinery by finite elements. *IEEE Transactions on Magnetics* MAG-18: 1052–1054.

11 Demerdash, N.A., Nyamusa, T.A., and Nehl, T.W. (1985). Comparison of effects of overload on parameters and performance of samarium-cobalt and strontium-ferrite radially oriented permanent magnet brushless DC motors. *IEEE Transactions on Power Apparatus and Systems* PAS-104: 2223–2232.

12 Demerdash, N.A., Hijazi, T.M., and Arkadan, A.A. (1988). Computation of winding inductances of permanent magnet brushless DC motors with damper windings by energy perturbation. *IEEE Transactions on Energy Conversion* EC-3 (3): 705–713.

13 Alhamadi, M.A. and Demerdash, N.A. (1991). Modeling of effects of skewing of rotor mounted permanent magnets on the performance of brushless DC motors. *IEEE Transactions on Energy Conversion* EC-6 (4): 721–729.

9

State-Space Modeling of Synchronous Machines Including Full Effects of Rotor Saliency

As discussed in Chapter 7, three-phase synchronous machines, particularly those of the wound-field variety, are mainly used for power generation. The constructional features of the cylindrical-rotor and salient-pole variety were detailed in the introduction and Section 7.1 of Chapter 7, and, specifically, Figures 7.1, 7.2, and Figures 7.5–7.6. This includes their wound-field excitation systems in Figures 7.11 and 7.12, as well as the synchronous machine's steady-state equivalent circuit models in Figures 7.13–7.15. Here, in this chapter, we develop the dynamic time-domain state-space model of synchronous machines, including the effects of rotor-saliency, as well as the dynamic time-domain state-space model of permanent magnet synchronous machines (PMSM). At this stage, in Section 9.1, we will discuss the discrete representation of windings in a synchronous machine, including the effects of rotor saliency, in the natural *abc* frame of reference.

9.1 Discrete Representation of the Windings of a Synchronous Machine in the Natural Frame of Reference

With reference to Figure 9.1, we show the schematic diagram of the (a,$-a$), (b,$-b$), and (c,$-c$) of the phase-belts, in the three-phase stator of a two-pole synchronous machine with a salient-pole rotor. Also, shown in the cross-section is the rotor-mounted field winding, as well as the pole-head-inserted damper bars. Meanwhile, in Figure 9.2, we show the schematic coil representation of the phases, a, b, and c, as well as the field winding coil representation, f, and the equivalent damper winding representations by the shorted coils, kd and kq. In Figures 9.1 and (9.2), the magnetic axes, the a-Axis, b-Axis, and c-Axis are shown in an *abc* positive sequence for counterclockwise rotation of the rotor. Also, shown in Figures 9.1 and 9.2 are the direct-axis, the d-axis, and the quadrature-axis, the q-axis, which are referred to throughout the remaining

Electric Machinery and Drives: An Electromagnetics Perspective, First Edition.
Nabeel A. O. Demerdash, JiangBiao He, and Hao Chen.

Companion website: www.wiley.com/go/demerdash/electricmachineryanddrives

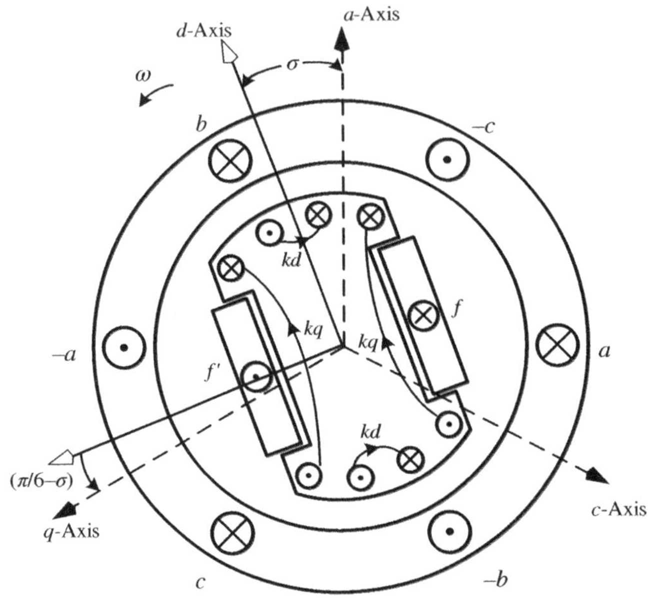

Figure 9.1 Schematic of the windings in a wound-field salient-pole synchronous machine including the damper bars and the direct (*d*) and quadrature (*q*) axes.

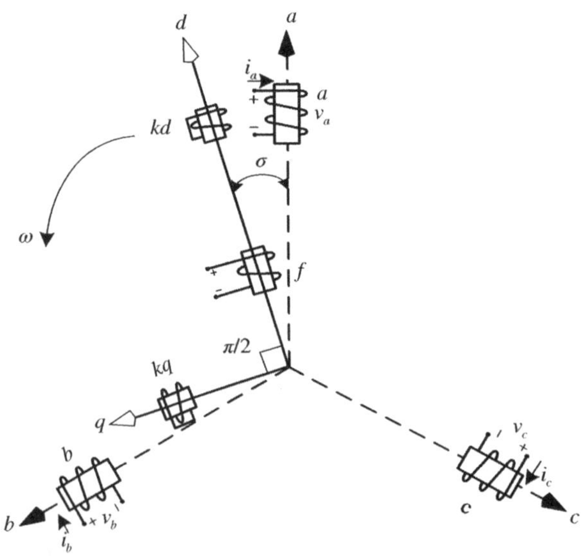

Figure 9.2 The discrete winding (coil) representation of the *a*, *b*, *c* phases, the excitation-field, *f*, and the direct (*d*) and quadrature (*q*) damper coils, *kd* and *kq*.

chapters of this book as the d-axis and q-axis, respectively. It is very important to point out that, henceforth, the q-axis is assumed to lead the d-axis by $(\pi/2)$ electrical radians, that is, $90°e$ in the counterclockwise direction, which is always assumed to be the positive direction of rotation of rotors, and the direction of rotation of the rotating magnetic field and magnetomotive forces (MMFs) in poly-phase and three-phase machines. Notice the angle, σ, in electrical measure (electrical radians or degrees), which is the angle of location of the d-Axis (or d-axis) with respect to a fixed stator space reference, which is here always taken to be the axis of phase (a) on the stator, that is, the a-Axis (or a-axis). Hence, the location of the q-Axis (or q-axis) is at an angle, $(\sigma + \pi/2)$ in electrical radians or $(\sigma + 90^{\circ})$ in electrical degrees.

According to Figures 9.1 and 9.2, the a-axis is at an angle, $(0°e)$, the b-axis is at an angle, $(2\pi/3)$ e-radians or $(120°e)$, and the c-axis is at an angle, $(4\pi/3)$ e-radians or $(240°e)$. All other angles between the various axes can be easily computed/derived by the reader from the geometries of Figures 9.1 and 9.2. Figure 9.3 shows the circuit schematic coils of a synchronous machine with two wound-field coils, with terminal voltages, v_{f_1} and v_{f_2}, and field excitation currents, i_{f_1} and i_{f_2}, which is encountered in cases of doubly excited synchronous machines.

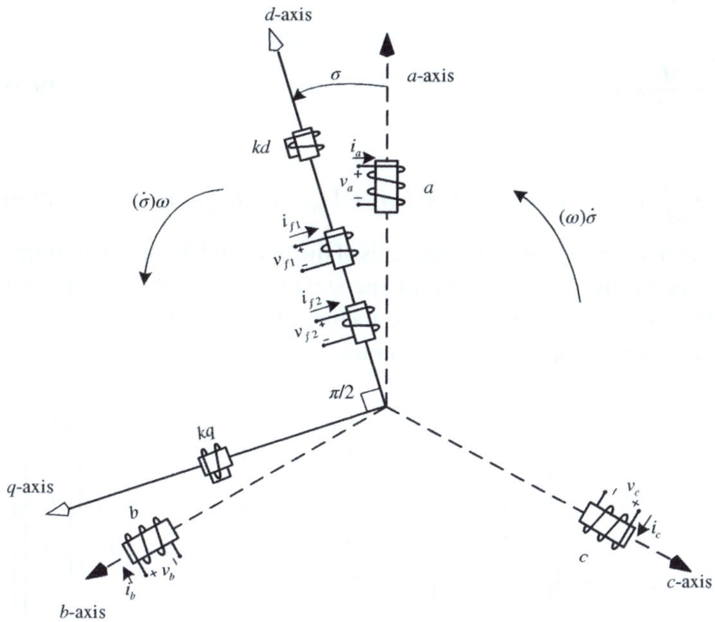

Figure 9.3 Schematic of armature and two-field windings synchronous machine.

9.2 State-Space Modeling Derivations of a Synchronous Machine in the Natural ABC Frame of Reference of the Poly-Phase Stator Windings

Consider phase a, Figure 9.2. This phase is represented schematically and analytically by a coil. The terminal voltage, v_a, can be expressed in terms of the phase ohmic resistance, r_a, phase current, i_a, and phase flux-linkage, λ_a, from Faraday's law as follows:

$$v_a = r_a i_a + \frac{d}{dt}(\lambda_a) \tag{9.1}$$

Hence, in terms of inductances and current, one can write the following:

$$v_a = r_a i_a + \frac{d}{dt}(L_{aa} i_a + L_{ab} i_b + L_{ac} i_c + L_{af} i_f + L_{akd} i_{kd} + L_{akq} i_{kq}) \tag{9.2}$$

For phases b and c, similar expressions are written as follows:

$$v_b = r_b i_b + \frac{d}{dt}(\lambda_b) \tag{9.3}$$

and, hence,

$$v_b = r_b i_b + \frac{d}{dt}(L_{ba} i_a + L_{bb} i_b + L_{bc} i_c + L_{bf} i_f + L_{bkd} i_{kd} + L_{bkq} i_{kq}) \tag{9.4}$$

Also,

$$v_c = r_c i_c + \frac{d}{dt}(\lambda_c) \tag{9.5}$$

and, hence,

$$v_c = r_c i_c + \frac{d}{dt}(L_{ca} i_a + L_{cb} i_b + L_{cc} i_c + L_{cf} i_f + L_{ckd} i_{kd} + L_{ckq} i_{kq}) \tag{9.6}$$

Similar expressions can be written for the coils of the field and damper windings, f, kd and kq, respectively. The equation for the six windings, a, b, c, f, kd, and kq, can be written, with the principle of the reciprocity of mutual inductances, $L_{ab} = L_{ab}$, $L_{bc} = L_{cb}$, etc., in matrix form as follows:

$$\begin{bmatrix} v_a \\ v_b \\ v_c \\ v_f \\ 0 \\ 0 \end{bmatrix} = \begin{bmatrix} r_a & 0 & 0 & & & \\ 0 & r_b & 0 & & \underline{0} & \\ 0 & 0 & r_c & & & \\ & & & r_f & 0 & 0 \\ & \underline{0} & & 0 & r_{kd} & 0 \\ & & & 0 & 0 & r_{kq} \end{bmatrix} \begin{bmatrix} i_a \\ i_b \\ i_c \\ i_f \\ i_{kd} \\ i_{kq} \end{bmatrix} + \frac{d}{dt}\left\{ \begin{bmatrix} L_{aa} & L_{ab} & L_{ac} & L_{af} & L_{akd} & L_{akq} \\ L_{ab} & L_{bb} & L_{bc} & L_{bf} & L_{bkd} & L_{bkq} \\ L_{ac} & L_{bc} & L_{cc} & L_{cf} & L_{ckd} & L_{ckq} \\ L_{af} & L_{bf} & L_{cf} & L_{ff} & L_{fkd} & L_{fkq} \\ L_{akd} & L_{bkd} & L_{ckd} & L_{fkd} & L_{kdkd} & L_{kdkq} \\ L_{akq} & L_{bkq} & L_{ckq} & L_{fkq} & L_{kdkq} & L_{kqkq} \end{bmatrix} \begin{bmatrix} i_a \\ i_b \\ i_c \\ i_f \\ i_{kd} \\ i_{kq} \end{bmatrix} \right\} \tag{9.7}$$

where $r_a = r_b = r_c = r_s$ is the stator winding resistance per phase, r_f is the filed winding resistance, and r_{kd} and r_{kq} are the direct and quadrature damper winding resistances, respectively.

All of the inductance coefficients are dependent on the angular position of the rotor, σ. Hence, all the inductance coefficients are functions of time. Therefore, the inductance matrix cannot be taken outside the derivative operator (d/dt). Let us find the expressions for these inductance coefficients as discussed below.

The mutual inductance between the field and armature phase windings is as follows:

$$L_{af} = L_{afm} \cos(\sigma) \tag{9.8}$$

where

$$\sigma = (\omega t + \sigma_0) \tag{9.9}$$

if the synchronous machine is operating at steady state (constant rotor speed), ω, namely the synchronous speed in electrical radians per second (Figure 9.4).

Hence, one can write the following:

$$L_{af} = L_{afm} \cos(\omega t + \sigma_0) \tag{9.10}$$

Also, from Figures 9.1 and 9.2, similar expressions for L_{bf} and L_{cf} can be written as follows:

$$L_{bf} = L_{afm} \cos(\sigma - 2\pi/3) \tag{9.11}$$

or

$$L_{bf} = L_{afm} \cos(\omega t + \sigma_0 - 2\pi/3) \tag{9.12}$$

if the angular rotor speed, ω, is constant, namely again the synchronous speed. Also,

$$L_{cf} = L_{afm} \cos(\sigma - 4\pi/3) \tag{9.13}$$

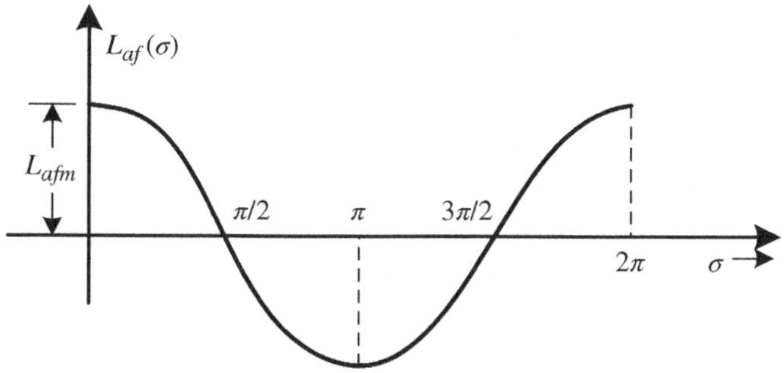

Figure 9.4 Phase (*a*) winding to field winding mutual inductance.

or

$$L_{cf} = L_{afm} \cos(\omega t + \sigma_0 - 4\pi/3) \tag{9.14}$$

again, if the angular speed of the rotor, ω, can be considered constant, where σ_0 is the initial rotor position at $t = 0$.

The mutual inductance between the d-axis damper circuit, k_d, and the armature phases, a, b, and c, is L_{akd}, L_{bkd}, and L_{ckd}. These inductances are obtainable by expressions similar to Eqs. (9.8)–(9.14) as follows:

$$L_{akd} = L_{akdm} \cos(\sigma) = L_{akdm} \cos(\omega t + \sigma_0) \tag{9.15}$$

$$L_{bkd} = L_{akdm} \cos(\sigma - 2\pi/3) = L_{akdm} \cos(\omega t + \sigma_0 - 2\pi/3) \tag{9.16}$$

$$L_{ckd} = L_{akdm} \cos(\sigma - 4\pi/3) = L_{akdm} \cos(\omega t + \sigma_0 - 4\pi/3) \tag{9.17}$$

The mutual inductances between the quadrative axis damper circuit, k_q, and armature phases are obtained from expressions similar to Eqs. (9.15)–(9.17), except for the amplitude, which becomes L_{akqm}, and a space angle shift of $(\frac{\pi}{2})$. Hence, one can write the following:

$$L_{akq} = L_{akqm} \cos(\sigma + \pi/2) = -L_{akqm} \sin(\omega t + \sigma_0) \tag{9.18}$$

$$L_{bkq} = L_{akqm} \cos(\sigma + \pi/2 - 2\pi/3) = -L_{akqm} \sin(\omega t + \sigma_0 - 2\pi/3) \tag{9.19}$$

$$L_{bkq} = L_{akqm} \cos(\sigma + \pi/2 - 4\pi/3) = -L_{akqm} \sin(\omega t + \sigma_0 - 4\pi/3) \tag{9.20}$$

Again, the last expressions in Eqs. (9.18)–(9.20) are for the constant rotor angular speed case.

The self-inductances of the rotor circuit (i.e., the field circuit, the d-axis damper circuit, and q-axis damper circuit) are independent of the rotor position, σ, due to the symmetry of the stator structure all around. These inductances are represented by constant, L_{ff}, L_{kdkd}, and L_{kqkq}.

The mutual inductance between the field circuit and the d-axis damper circuit is independent of rotor position, and, hence, can be considered as a constant, L_{fkd}. Meanwhile, the mutual inductances between the field and q-axis damper circuit and between the d-axis damper circuit and q-axis damper circuit are zeros. This is obvious by inspection of Figures 9.1 and 9.2. Hence, $L_{fkq} = 0$ and $L_{kdkq} = 0$.

From Figure 9.1, the self-inductance of phase (a) is maximum when the d-axis coincides with the axis of phase (a), that is, when $\sigma = 0$, and is also maximum when $\sigma = (\pi)$. It is also minimum when $\sigma = (\pi/2)$ and $\sigma = (3\pi/2)$. Thus, assuming a sinusoidal flux density distribution in the airgap, a plot of such an inductance would be as shown in Figure 9.5.

Hence, one can write the following based on Figure 9.5:

$$L_{aa}(\sigma) = L_{sa} + L_{sv} \cos(2\sigma) = L_{sa} + L_{sv} \cos(2\omega t + 2\sigma_0) \tag{9.21}$$

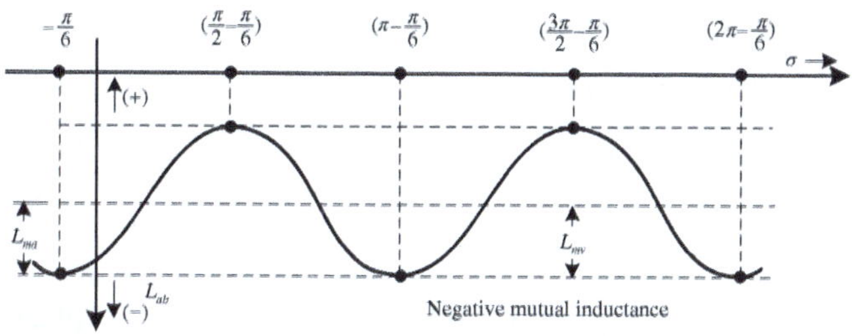

Figure 9.5 The self-inductance of phase (a), $L_{aa}(\sigma)$.

The last term in Eq. (9.21) is for the constant rotor speed case, ω.

The self-inductances of phases b and c have similar expressions with a delay (lag) angle of $(2\pi/3)$ for phase b and $(4\pi/3)$ for phase c, hence, the following can be written:

$$L_{bb}(\sigma) = L_{sa} + L_{sv}\cos(2\sigma - 2(2\pi/3)) = L_{sa} + L_{sv}\cos(2\omega t + 2\sigma_0 - 4\pi/3) \tag{9.22}$$

and

$$L_{cc}(\sigma) = L_{sa} + L_{sv}\cos(2\sigma - 2(4\pi/3)) = L_{sa} + L_{sv}\cos(2\omega t + 2\sigma_0 - 2\pi/3) \tag{9.23}$$

Again, the last terms in Eqs. (9.22) and (9.23) are for the constant rotor speed case.

The armature phase mutual inductances are the hardest to visualize and write. From Figure 9.1, it can be seen, after careful inspection, that the mutual inductance, L_{ab}, peaks (negatively) when $\sigma = (-\pi/6)$ and $\sigma = (\pi - \pi/6)$. Also, this inductance is minimum (negatively) when $\sigma = (\pi/2 - \pi/6)$ and $\sigma = (3\pi/2 - \pi/6)$. Thus, the profile representing this inductance variation is shown in Figure 9.6.

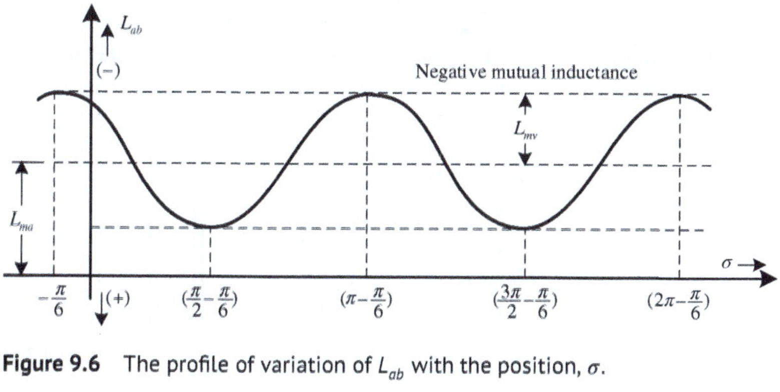

Figure 9.6 The profile of variation of L_{ab} with the position, σ.

Hence,

$$L_{ab} = -\{L_{ma} + L_{mv}\cos[2\sigma + 2(\pi/6)]\} \tag{9.24}$$

$$L_{ab} = -L_{ma} - L_{mv}\cos(2\omega t + 2\sigma_0 + \pi/3) = -L_{ma} + L_{mv}\cos(2\omega t + 2\sigma_0 - 2\pi/3) \tag{9.25}$$

and, consequently,

$$L_{bc} = -L_{ma} - L_{mv}\cos(2\omega t + 2\sigma_0 + \pi/3 - 2(2\pi/3)) \tag{9.26}$$

$$L_{bc} = -L_{ma} - L_{mv}\cos(2\omega t + 2\sigma_0 - \pi) \tag{9.27}$$

$$L_{bc} = -L_{ma} + L_{mv}\cos(2\omega t + 2\sigma_0) \tag{9.28}$$

Similarly,

$$L_{ca} = -L_{ma} + L_{mv}\cos(2\omega t + 2\sigma_0 - 4\pi/3) \tag{9.29}$$

Again, the expression of Eqs. (9.25)–(9.29) is given for the constant rotor speed case.

All these inductance coefficients, given by Eqs. (9.8)–(9.29), when substituted into Eq. (9.7), give the synchronous machine model. Upon such a substitution, the following is obtained:

$$\begin{bmatrix} v_a \\ v_b \\ v_c \\ \hdashline v_f \\ 0 \\ 0 \end{bmatrix} = \left[\begin{array}{ccc:ccc} r_a & 0 & 0 & & & \\ 0 & r_b & 0 & & \underline{0} & \\ 0 & 0 & r_c & & & \\ \hdashline & & & r_f & 0 & 0 \\ & \underline{0} & & 0 & r_{kd} & 0 \\ & & & 0 & 0 & r_{kq} \end{array}\right] \begin{bmatrix} i_a \\ i_b \\ i_c \\ \hdashline i_f \\ i_{kd} \\ i_{kq} \end{bmatrix} +$$

$$\frac{d}{dt}\left\{\left[\begin{array}{ccc:ccc} (L_{sa} + L_{sv}\cos(2\sigma)) & (-L_{ma} + L_{mv}\cos(2\sigma - 2\pi/3)) & (-L_{ma} + L_{mv}\cos(2\sigma - 4\pi/3)) & (L_{afm}\cos(\sigma)) & (L_{akdm}\cos(\sigma)) & (-L_{akqm}\sin(\sigma)) \\ (-L_{ma} + L_{mv}\cos(2\sigma - 2\pi/3)) & (L_{sa} + L_{sv}\cos(2\sigma - 4\pi/3)) & (-L_{ma} + L_{mv}\cos(2\sigma)) & (L_{afm}\cos(\sigma - 2\pi/3)) & (L_{akdm}\cos(\sigma - 2\pi/3)) & (-L_{akqm}\sin(\sigma - 2\pi/3)) \\ (-L_{ma} + L_{mv}\cos(2\sigma - 4\pi/3)) & (-L_{ma} + L_{mv}\cos(2\sigma)) & (L_{sa} + L_{sv}\cos(2\sigma - 2\pi/3)) & (L_{afm}\cos(\sigma - 4\pi/3)) & (L_{akdm}\cos(\sigma - 4\pi/3)) & (-L_{akqm}\sin(\sigma - 4\pi/3)) \\ \hdashline (L_{afm}\cos(\sigma)) & (L_{afm}\cos(\sigma - 2\pi/3)) & (L_{afm}\cos(\sigma - 4\pi/3)) & L_{ff} & L_{fkd} & 0 \\ (L_{akdm}\cos(\sigma)) & (L_{akdm}\cos(\sigma - 2\pi/3)) & (L_{akdm}\cos(\sigma - 4\pi/3)) & L_{fkd} & L_{kdkd} & 0 \\ (-L_{akqm}\sin(\sigma)) & (-L_{akqm}\sin(\sigma - 2\pi/3)) & (-L_{akqm}\sin(\sigma - 4\pi/3)) & 0 & 0 & L_{kqkq} \end{array}\right] \begin{bmatrix} i_a \\ i_b \\ i_c \\ \hdashline i_f \\ i_{kd} \\ i_{kq} \end{bmatrix}\right\} \tag{9.30}$$

or for constant rotor speed, $\sigma = \omega t + \sigma_0$, one can write the following:

$$
\begin{bmatrix} v_a \\ v_b \\ v_c \\ \hdashline v_f \\ 0 \\ 0 \end{bmatrix}
=
\left[\begin{array}{ccc:ccc}
r_s & 0 & 0 & & & \\
0 & r_s & 0 & & \underline{0} & \\
0 & 0 & r_s & & & \\
\hdashline
 & & & r_f & 0 & 0 \\
 & \underline{0} & & 0 & r_{kd} & 0 \\
 & & & 0 & 0 & r_{kq}
\end{array}\right]
\begin{bmatrix} i_a \\ i_b \\ i_c \\ \hdashline i_f \\ i_{kd} \\ i_{kq} \end{bmatrix}
+
$$

$$
+\frac{d}{dt}\left\{\left[\begin{array}{ccc:ccc}
(L_{sa}+L_{sv}\cos(2\omega t+2\sigma_0)) & (-L_{ma}+L_{mv}\cos(2\omega t+2\sigma_0-2\pi/3)) & (-L_{ma}+L_{mv}\cos(2\omega t+2\sigma_0-4\pi/3)) & (L_{afm}\cos(\omega t+\sigma_0)) & (L_{akdm}\cos(\omega t+\sigma_0)) & (-L_{akqm}\sin(\omega t+\sigma_0)) \\
(-L_{ma}+L_{mv}\cos(2\omega t+2\sigma_0-2\pi/3)) & (L_{sa}+L_{sv}\cos(2\omega t+2\sigma_0-4\pi/3)) & (-L_{ma}+L_{mv}\cos(2\omega t+2\sigma_0)) & (L_{afm}\cos(\omega t+\sigma_0-2\pi/3)) & (L_{akdm}\cos(\omega t+\sigma_0-2\pi/3)) & (-L_{akqm}\sin(\omega t+\sigma_0-2\pi/3)) \\
(-L_{ma}+L_{mv}\cos(2\omega t+2\sigma_0-4\pi/3)) & (-L_{ma}+L_{mv}\cos(2\omega t+2\sigma_0)) & (L_{sa}+L_{sv}\cos(2\omega t+2\sigma_0-2\pi/3)) & (L_{afm}\cos(\omega t+\sigma_0-4\pi/3)) & (L_{akdm}\cos(\omega t+\sigma_0-4\pi/3)) & (-L_{akqm}\sin(\omega t+\sigma_0-4\pi/3)) \\
\hdashline
(L_{afm}\cos(\omega t+\sigma_0)) & (L_{afm}\cos(\omega t+\sigma_0-2\pi/3)) & (L_{afm}\cos(\omega t+\sigma_0-4\pi/3)) & L_{ff} & L_{fkd} & 0 \\
(L_{akdm}\cos(\omega t+\sigma_0)) & (L_{akdm}\cos(\omega t+\sigma_0-2\pi/3)) & (L_{akdm}\cos(\omega t+\sigma_0-4\pi/3)) & L_{fkd} & L_{kdkd} & 0 \\
(-L_{akqm}\sin(\omega t+\sigma_0)) & (-L_{akqm}\sin(\omega t+\sigma_0-2\pi/3)) & (-L_{akqm}\sin(\omega t+\sigma_0-4\pi/3)) & 0 & 0 & L_{kqkq}
\end{array}\right]
\begin{bmatrix} i_a \\ i_b \\ i_c \\ \hdashline i_f \\ i_{kd} \\ i_{kq} \end{bmatrix}\right\}
\tag{9.31}
$$

Equations (9.30) and (9.31), with the partitioning patterns shown by the dotted lines above, can be written in short-hand matrix notation as follows:

$$
\overbrace{\begin{bmatrix} \underline{V}_{abc} \\ \underline{V}_{fkdkq} \end{bmatrix}}^{\underline{V}}
=
\overbrace{\begin{bmatrix} \underline{R}_{ss} & \underline{0} \\ \underline{0} & \underline{R}_{rr} \end{bmatrix}}^{(\underline{R})}
\overbrace{\begin{bmatrix} \underline{I}_{abc} \\ \underline{I}_{fkdkq} \end{bmatrix}}^{\underline{I}}
+ \frac{d}{dt}\overbrace{\left\{
\underbrace{\begin{bmatrix} \underline{L}_{ss} & \underline{L}_{sr} \\ \underline{L}_{sr}^{t} & \underline{L}_{rr} \end{bmatrix}}_{(\underline{L})}
\underbrace{\begin{bmatrix} \underline{I}_{abc} \\ \underline{I}_{fkdkq} \end{bmatrix}}_{\underline{I}}
\right\}}^{(\underline{\Lambda})}
\tag{9.32}
$$

That is,

$$\underline{V} = (\underline{R}) \cdot \underline{I} + \frac{d}{dt}(\underline{\Lambda})$$

where

$$\underline{\Lambda} = (\underline{L}) \cdot \underline{I}$$

Since the inductance matrix is time dependent, upon differentiation, Eq. (9.32) can be rewritten as follows:

$$\begin{bmatrix} \underline{V}_{abc} \\ \underline{V}_{fkdkq} \end{bmatrix} = \begin{bmatrix} \underline{R}_{ss} & \underline{0} \\ \underline{0} & \underline{R}_{rr} \end{bmatrix} \begin{bmatrix} \underline{I}_{abc} \\ \underline{I}_{fkdkq} \end{bmatrix} + \underbrace{\overbrace{\begin{bmatrix} \dot{\underline{L}}_{ss} & \dot{\underline{L}}_{sr} \\ \dot{\underline{L}}^{t}_{sr} & \dot{\underline{L}}_{rr} \end{bmatrix}}^{(\dot{\underline{L}})} \begin{bmatrix} \underline{I}_{abc} \\ \underline{I}_{fkdkq} \end{bmatrix}}_{\text{Rotational voltage}} + \underbrace{\begin{bmatrix} \underline{L}_{ss} & \underline{L}_{sr} \\ \underline{L}^{t}_{sr} & \underline{L}_{rr} \end{bmatrix} \begin{bmatrix} \dot{\underline{I}}_{abc} \\ \dot{\underline{I}}_{fkdkq} \end{bmatrix}}_{\text{Transformer voltage}} \tag{9.33}$$

The vector represented by the first term in the differential expansion is referred to as the rotational voltage, because it is caused by the rotational motion in the machine. The induced rotational voltage in the a, b, and c windings is the cause for the energy conversion (power production) process. This induced rotational voltage vector, $\underline{E}_{r_{abc}}$, is therefore given as follows:

$$\underline{E}_{r_{abc}} = \dot{\underline{L}}_{ss} \cdot \underline{I}_{abc} + \dot{\underline{L}}_{sr} \cdot \underline{I}_{fkdkq} \tag{9.34}$$

The power, P, produced by the rotational voltage is given as follows:

$$P = \underline{I}^{t}{}_{abc} \cdot \underline{E}_{r_{abc}} \tag{9.35}$$

Here, P, is a real instantaneous power, and, hence, in a torque calculation, it can be divided by the speed to attain the torque.

The transformer voltage induced in the windings, a, b, and c, is $\underline{E}_{t_{abc}}$ and can be written from Eq. (9.33) as follows:

$$\underline{E}_{t_{abc}} = \dot{\underline{L}}_{ss} \cdot \dot{\underline{I}}_{abc} + \dot{\underline{L}}_{sr} \cdot \dot{\underline{I}}_{fkdkq} \tag{9.36}$$

The transformer voltage does not play any direct role in the energy conversion (power production) process but is proportional to energy that shuttles back and forth from the synchronous machine to its associated external circuits.

Equation (9.33) can be further abbreviated and rewritten as follows:

$$\underline{V} = \underline{R} \cdot \underline{I} + \dot{\underline{L}} \cdot \underline{I} + \underline{L} \cdot \dot{\underline{I}} \tag{9.37}$$

or

$$\dot{\underline{I}} = -\underline{L}^{-1} \cdot (\dot{\underline{L}} + \underline{R}) \cdot \underline{I} + \underline{L}^{-1} \cdot \underline{V} \tag{9.38}$$

This is the current frame of reference state-space model of a synchronous machine, which is useful if saturation effects can be neglected.

However, one can clearly see that $(\underline{L}^{-1})$ is time dependent, as in the case with $(\underline{L})$. Therefore, Eq. (9.38) does not quite fit the standard state equation $(\dot{\underline{X}} = \underline{A} \cdot \underline{X} + \underline{B} \cdot \underline{U})$, in which $\underline{A}$ and $\underline{B}$ are time-independent matrices.

Therefore, further reformulation has to be done to the model of Eq. (9.32) to render the inductance matrices time independent. This is accomplished by a change of the frame of reference (*abc*) to another frame of reference through linear transformations.

Since linear transformations use in rotating electrical machines (electromechanical devices) is very common, its use will be studied in Sections 9.4 and 9.5. It will then be shown that by proper choice of a transformation, the time (angular-position) dependence of the inductance matrix of a machine can be eliminated.

The state-space model of a synchronous machine given in Eq. (9.32) can be rewritten in matrix notation in the flux linkage frame of reference as follows:

$$\underline{V} = (\underline{R}) \cdot \underline{I} + \frac{d}{dt}(\underline{\Lambda}) \tag{9.39}$$

or

$$\underline{V} = (\underline{R}) \cdot \underline{I} + \underline{\dot{\Lambda}} \tag{9.40}$$

where

$$\underline{\Lambda} = \underline{L} \cdot \underline{I} \tag{9.41}$$

Hence,

$$\underline{I} = \underline{L}^{-1} \cdot \underline{\Lambda} \tag{9.42}$$

Substituting Eq. (9.42) into Eq. (9.40) gives the following:

$$\underline{V} = (\underline{R} \cdot \underline{L}^{-1}) \cdot \underline{\Lambda} + \underline{\dot{\Lambda}} \tag{9.43}$$

Rearranging Eq. (9.43) yields the synchronous machine's state-space model in the flux linkage frame of reference as follows:

$$\underline{\dot{\Lambda}} = (-\underline{R} \cdot \underline{L}^{-1}) \cdot \underline{\Lambda} + \underline{V} \tag{9.44}$$

in which the state coefficients matrix $\underline{A} = (-\underline{R} \cdot \underline{L}^{-1})$ and the control (input) variables matrix $\underline{B}$ = identity matrix with a dimension (6×6) for a six-state variables model of a synchronous machine. This formulation can be used to account for saturation if the inductance matrix is obtained from a nonlinear magnetic field solution, see Refs. [1–14]. In Eq. (9.44), we can define the following:

$$\underline{\Lambda} = \begin{bmatrix} \underline{\Lambda}_{abc} \\ \underline{\Lambda}_{fkdkq} \end{bmatrix} = \underline{L} \begin{bmatrix} \underline{I}_{abc} \\ \underline{I}_{fkdkq} \end{bmatrix} \tag{9.45}$$

$$\underline{V} = \begin{bmatrix} \underline{V}_{abc} \\ \underline{V}_{fkdkq} \end{bmatrix} \tag{9.46}$$

$$\underline{R} = \begin{bmatrix} \underline{R}_{ss} & \underline{0} \\ \underline{0} & \underline{R}_{rr} \end{bmatrix} \tag{9.47}$$

and

$$\underline{L} = \begin{bmatrix} \underline{L}_{ss} & \underline{L}_{sr} \\ \underline{L}_{sr}^t & \underline{L}_{rr} \end{bmatrix} \tag{9.48}$$

also,

$$\begin{bmatrix} \underline{I}_{abc} \\ \underline{I}_{fkdkq} \end{bmatrix} = \begin{bmatrix} \underline{L}_{ss} & \underline{L}_{sr} \\ \underline{L}_{sr}^t & \underline{L}_{rr} \end{bmatrix}^{-1} \begin{bmatrix} \underline{\Lambda}_{abc} \\ \underline{\Lambda}_{fkdkq} \end{bmatrix} \tag{9.49}$$

Next, we discuss synchronous machinery modeling from a magnetic field distribution point of view, for both salient-pole and cylindrical-rotor machines.

9.3 A Magnetic Field Flux-Map Finite-Element Computed Point of View of Salient-Pole and Cylindrical Rotor Machines

As previously shown in the phasor-space-vector diagram of Figure 7.20 for the overexcited synchronous generator and in Figure 7.16 for the overexcited synchronous motor in Chapter 7, one can develop from Figure 7.20 the corresponding relative positions of the stator MMF waveform, $\mathbf{F}_s$, and rotor MMF waveform, $\mathbf{F}_r$, for a synchronous generator which is as shown in Figure 9.7. Also, shown in Figure 9.7 are the relative angular positions of the corresponding space vectors, $\overline{F}_s$ and $\overline{F}_r$, in a counterclockwise rotation of the rotor and the rotating

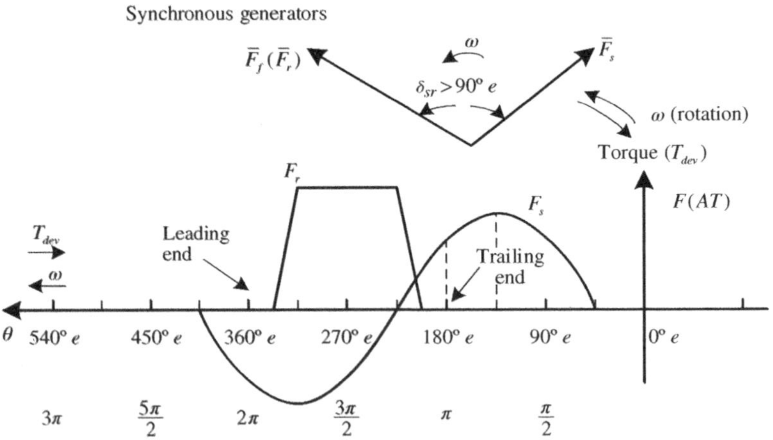

Figure 9.7 Stator and rotor MMF waveforms and corresponding space vectors, $\overline{F}_s$ and $\overline{F}_r$, relative space positions, and torque orientation versus orientation of rotor rotation for synchronous generators.

MMFs. Next, shown in this figure is the orientation of the developed electromagnetic/electromechanical torque, which is, naturally, opposite to rotation for this generator case. Furthermore, shown in this figure is the "leading-end" and the "trailing-end" of the north-pole on the rotor. Here, it is of a paramount importance to observe that the stator MMF, F_s, tends to magnetize the trailing-end of the north-pole on the rotor, while it tends to demagnetize the leading-end of the north-pole on the rotor, as will be verified a little later on in this section from the finite-element (FE) analysis of the magnetic field and resulting airgap flux density, $\overline{B}$, waveforms obtained from such FE modeling and computations, see Refs. [6–12].

In this section, Figure 9.8 can be developed from the earlier phasor-space-vector diagram, Figure 7.16, for an overexcited synchronous motor, to generate the corresponding relative positions of the stator MMF waveform, F_s, and rotor MMF waveform, F_r, and the consequent corresponding space vectors, $\overline{F}_s$ and $\overline{F}_r$. All are shown for counterclockwise rotation of the rotor and the rotating MMFs, and, here, the developed torque is, naturally, in the same counterclockwise direction as the rotor rotation. It should further be observed, based on the relative positions of the stator MMF, waveform, F_s, with respect to the rotor MMF, waveform, F_r, that F_s tends to magnetize the leading-end of the north-pole on the rotor, while tending to demagnetize the trailing-end of this north-pole on the rotor.

At this stage, we will introduce the actual cross-sectional configuration of a 2-pole, 3-phase, salient-pole synchronous generator, with 5 damper-bars per-pole, and 30-stator slots, see Figure 9.9, with a double-layer stator winding. Also, shown

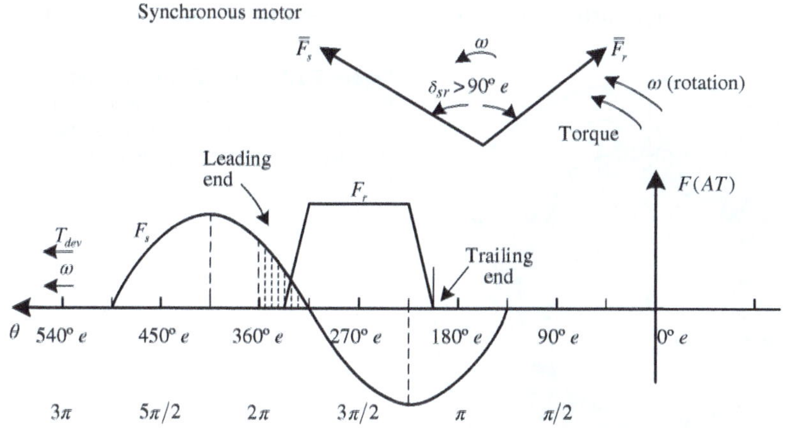

Figure 9.8 Stator and rotor MMF waveforms and corresponding space vectors, $\overline{F}_s$ and $\overline{F}_r$, relative space positions, and torque orientation versus orientation of rotor rotation for synchronous generators.

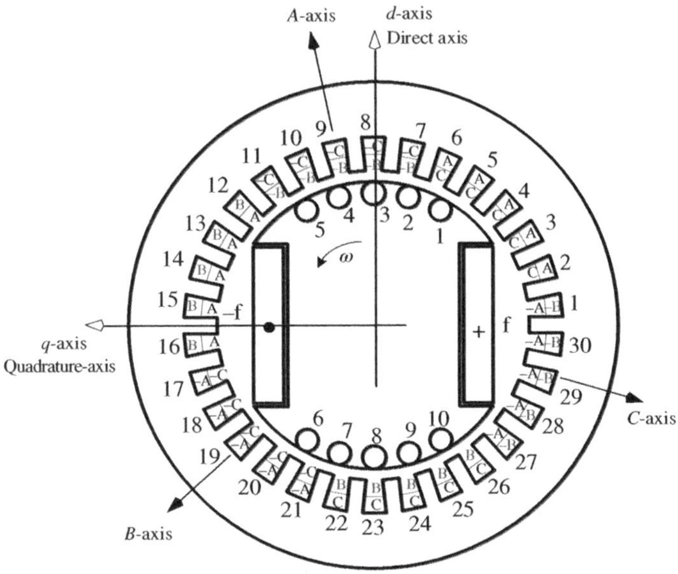

Figure 9.9 A 2-pole, 3-phase, 30-stator slots salient-pole synchronous machine, see Refs. [8–12].

in this figure are the magnetic axes of phases (A), (B), and (C), as well as the direct and quadrature axes, that is, the d and q axes on the salient-pole rotor. Note that the q-axis is always taken to lead the d-axis by ($\pi/2$) electrical radians, that is, $90°e$, in the counterclockwise direction. Note also the coil-sides of the field-excitation winding, (f) and ($-f$). Furthermore, Figure 9.10 shows the FE grid for modeling

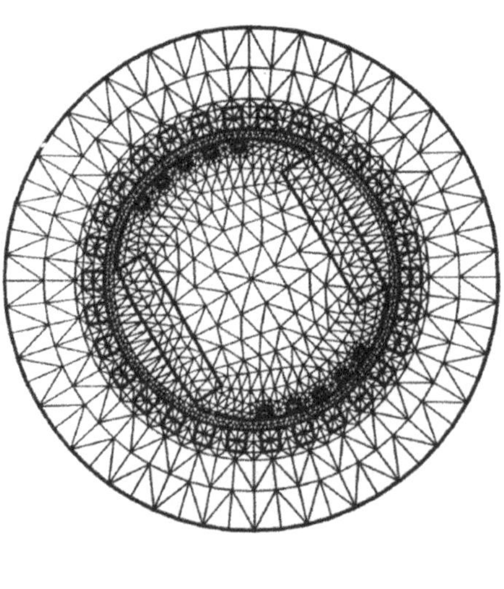

Figure 9.10 The FE grid for magnetic field computation of its distribution in a 2-pole salient-pole, 30-stator slots, synchronous generator, see Refs. [8–12].

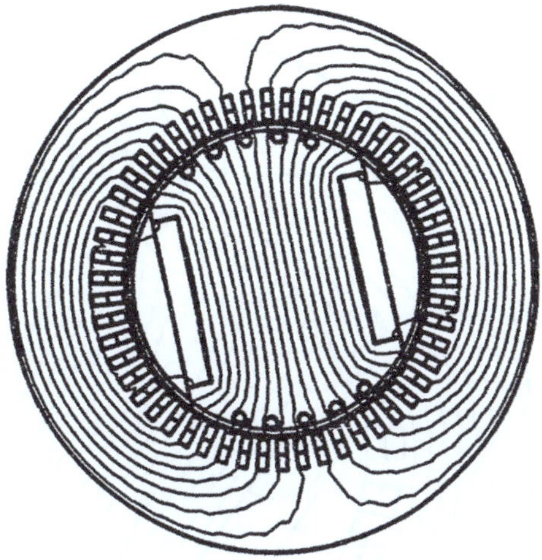

Figure 9.11 FE-computed flux plot in the 2-pole synchronous generator at no-load rated terminal voltage condition, see Refs. [8–12].

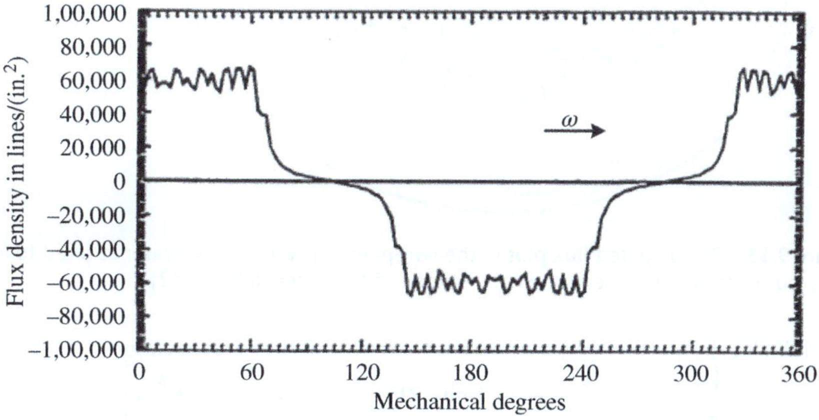

Figure 9.12 The FE-computed mid-airgap flux density waveform of the two-pole synchronous generator at no-load rated terminal voltage condition, see Refs. [8–12].

and computing the magnetic field distribution, at no-load and any other load condition within the rated kilo volt-ampere capability of this machine as a generator. The resulting FE-computed magnetic flux-plot under rated terminal voltage no-load condition is shown in Figure 9.11, with the corresponding no-load mid-airgap flux density waveform given in Figure 9.12, see Refs [8–12].

Meanwhile, the FE-computed magnetic flux-plot under rated terminal voltage, rated kilo volt-ampere unity power factor load condition is shown in Figure 9.13, with the corresponding mid-airgap flux density wave shown in Figure 9.14. The reader should notice the lack of symmetry in the waveform around the *d*-axis of

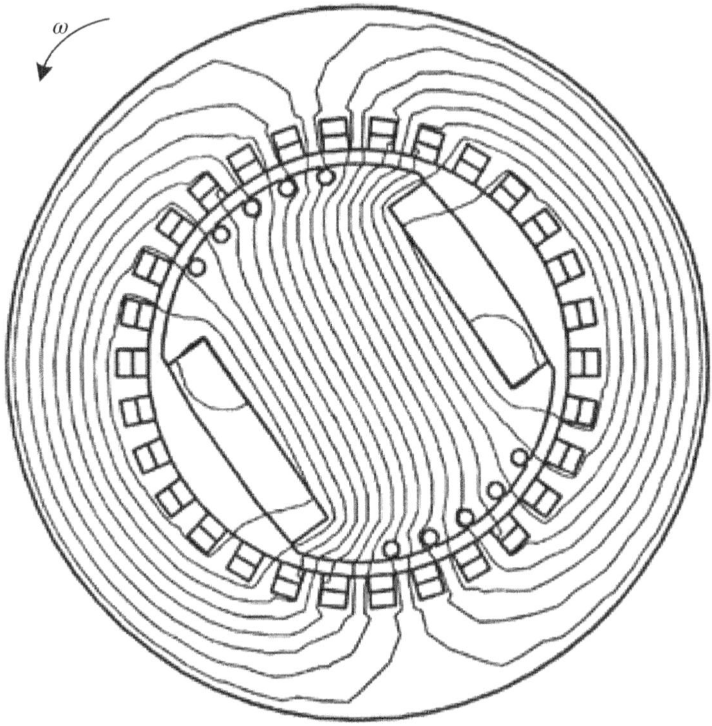

Figure 9.13 FE-computed flux plot in the two-pole synchronous generator at rated-load unity-power factor and rated terminal voltage condition, see Refs. [8–12].

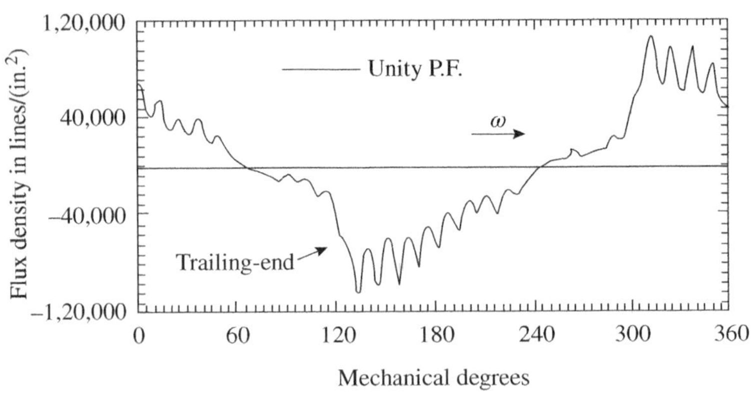

Figure 9.14 The FE-computed mid-airgap flux density waveform of the two-pole synchronous generator at rated-load unity-power factor and rated terminal voltage condition, see Refs. [8–12].

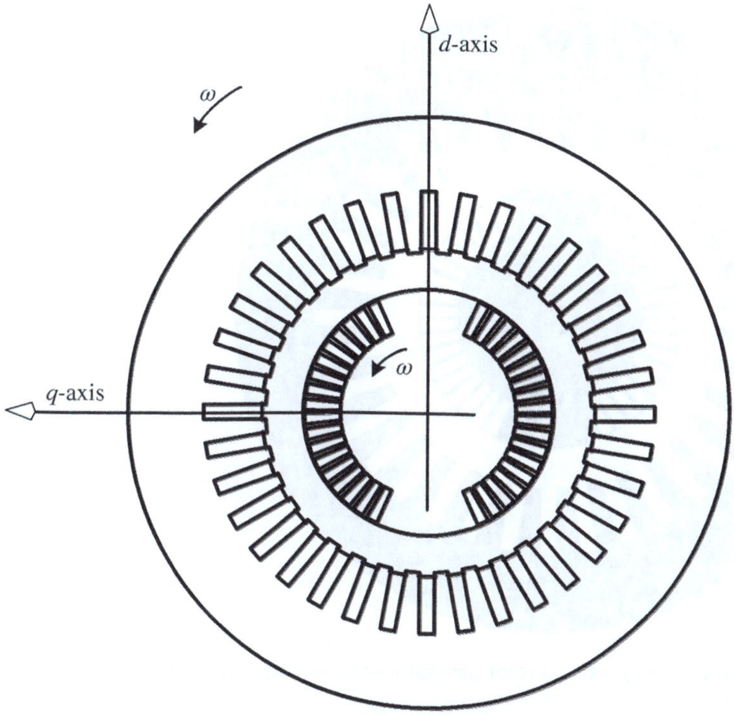

Figure 9.15 A 2-pole cylindrical rotor generator with 36 stator-slots, three-phase stator, see Refs. [6, 7].

the machine underload and should further notice the higher flux density values on the trailing-side of a pole span (pole-pitch), as discussed earlier in relation to the comments accompanying analysis of the meaning of the spatial positions of F_s and F_r MMFs in Figure 9.7.

Meanwhile, Figures 9.15 and 9.16, see Refs. [6, 7], show the cross-section of a 2-pole cylindrical rotor synchronous generator, with a 36 stator slots stator which houses a 3-phase, double-layer armature winding. The FE-grid used to compute the flux plots of this machine under no-load and load conditions is shown in Figure 9.17, with a detailed quadrant of this FE grid shown in Figure 9.18. The FE-computed flux plot under no-load rated terminal voltage condition is shown in Figure 9.19, while the corresponding mid-airgap radial flux density waveform at rated voltage and no-load condition is shown in Figure 9.20.

Meanwhile, the FE-computed flux plot under rated voltage and rated load condition is shown in Figure 9.21, and its corresponding mid-airgap radial flux density waveform over one plot-pitch is shown in Figure 9.22. Once again, recalling the earlier discussion associated with Figure 9.7, the reader should notice the higher

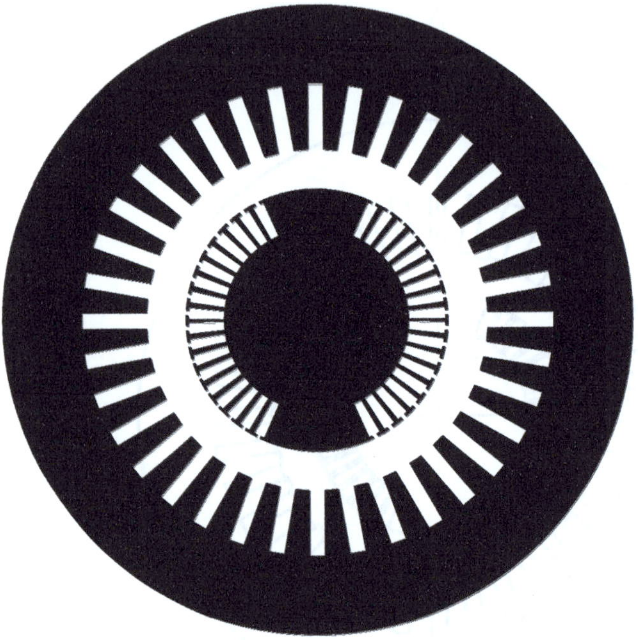

Figure 9.16 A 2-pole cylindrical rotor generator with 36 stator-slots, three-phase stator, see Refs. [6, 7].

flux density amplitudes appearing on the "trailing-side" of a pole pitch as shown in Figure 9.22. Meanwhile, the FE-computed phase self and mutual inductance profiles for L_{aa} and L_{ab}, as well as the field winding self-inductance, L_{ff}, under no-load are shown in Figures 9.23, 9.24, and 9.25, respectively. Furthermore, the same set of inductance profiles for, L_{aa}, L_{ab}, and L_{ff}, was FE-computed under load and is shown in Figures 9.26, 9.27, and 9.28, respectively. Now, with a better understanding of the magnetic field distribution under no-load and load conditions, we proceed to the two-reaction theory $dq0$ transformation to eliminate the dependence of the inductance matrix of the model of Eqs. (9.30) and (9.31) on the rotor position angle, σ, as given in Sections 9.4–9.7.

9.4 Park's *DQ0* Frame of Reference Transformation

The essence of the main idea behind Park's transformation is the resolution of the MMFs of the windings of the three phases, $\mathbf{F}_a$, $\mathbf{F}_b$, and $\mathbf{F}_c$, of phases (a), (b), and (c), along the direct "d-axis," and the quadrature "q-axis" of a salient-pole rotor of a synchronous machine, see Figure 9.29. In Figure 9.29, taking the d-axis as

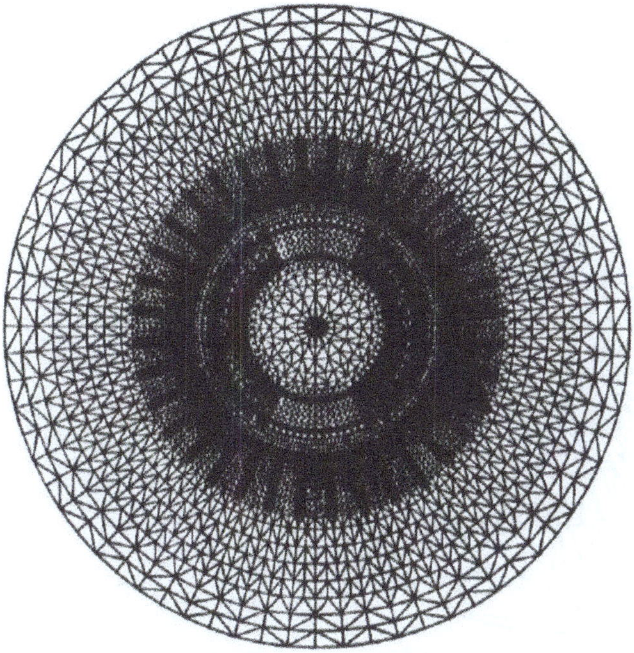

Figure 9.17 The FE-grid of a 2-pole cylindrical rotor generator with 36 stator-slots, see Refs. [6, 7].

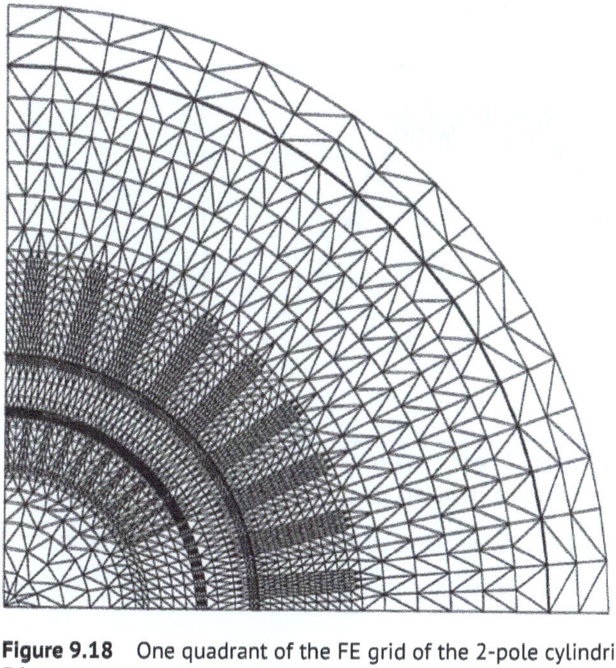

Figure 9.18 One quadrant of the FE grid of the 2-pole cylindrical rotor generator with 36 stator-slots, see Refs. [6, 7].

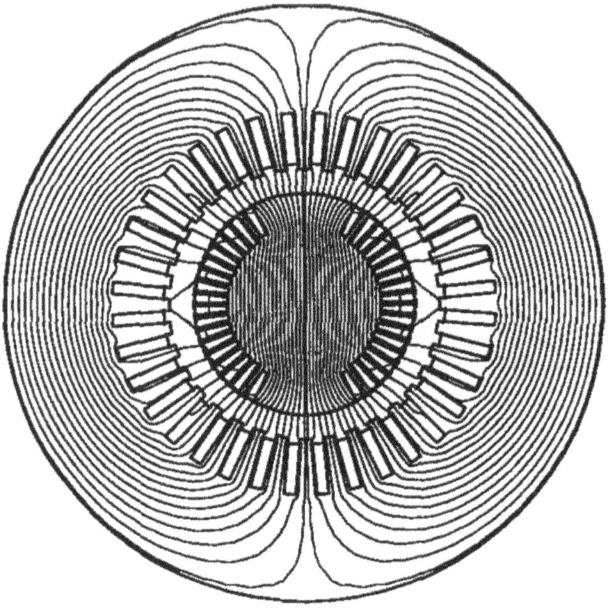

Figure 9.19 The rated voltage FE-computed flux plot at no-load, see Refs. [6, 7].

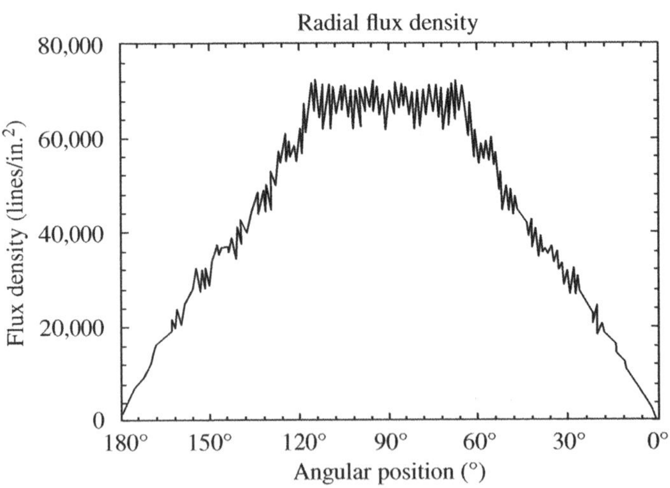

Figure 9.20 Mid-airgap radial flux density at rated voltage no-load, see Refs. [6, 7].

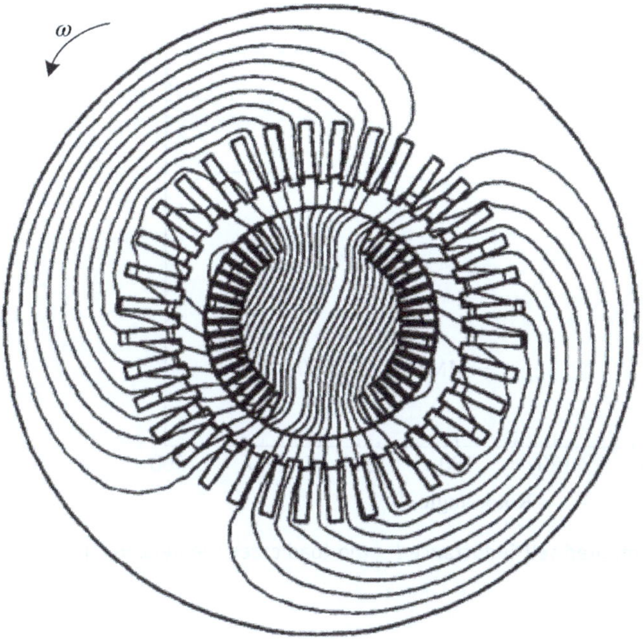

Figure 9.21 The rated load and rated voltage FE-computed flux plot, see Refs. [6, 7].

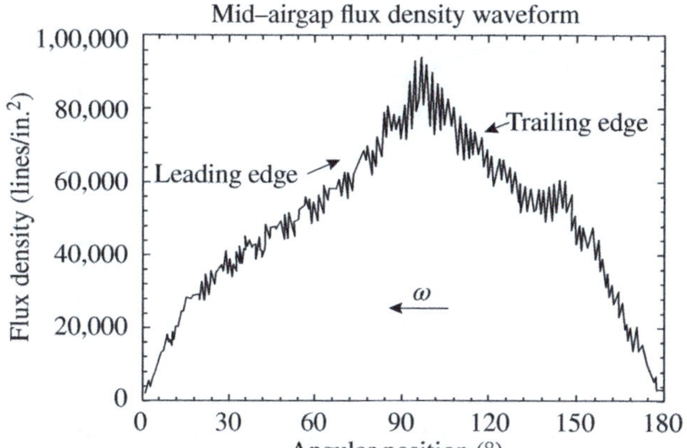

Figure 9.22 Mid-airgap radial flux density waveform at 0.85 power factor full-load, see Refs. [6, 7].

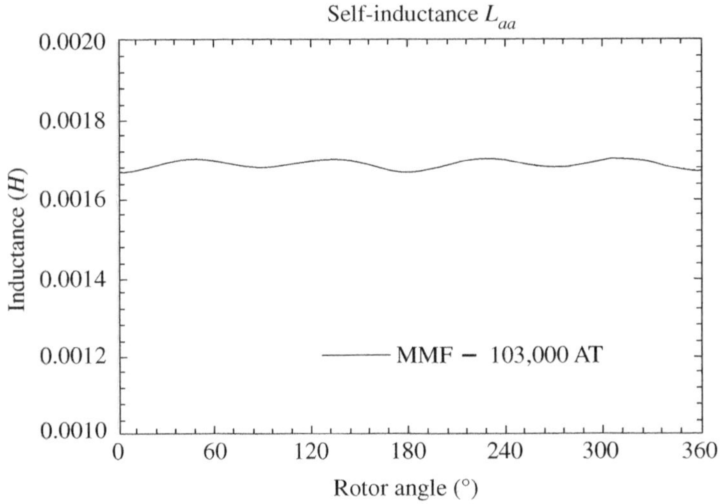

Figure 9.23 2D-FE-computed self-inductance L_{aa} (no-load case), see Refs. [6, 7].

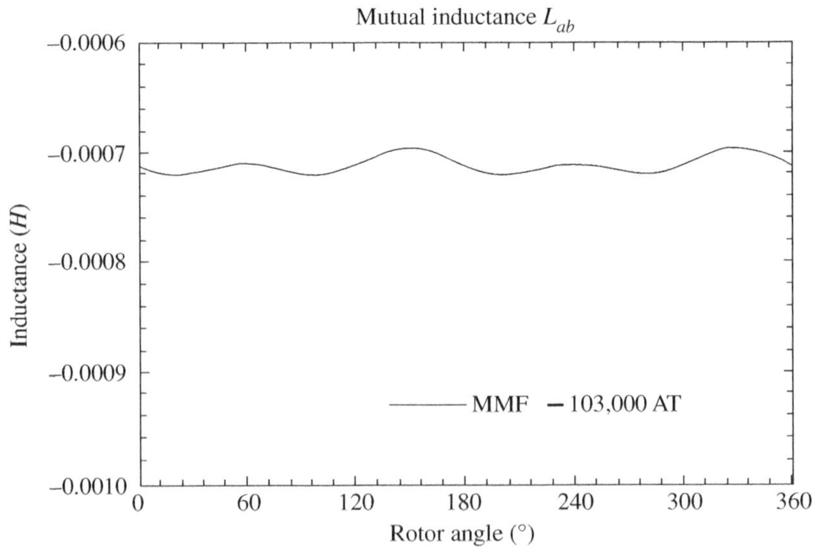

Figure 9.24 2D-FE-computed mutual inductance L_{ab} (no-load case), see Refs. [6, 7].

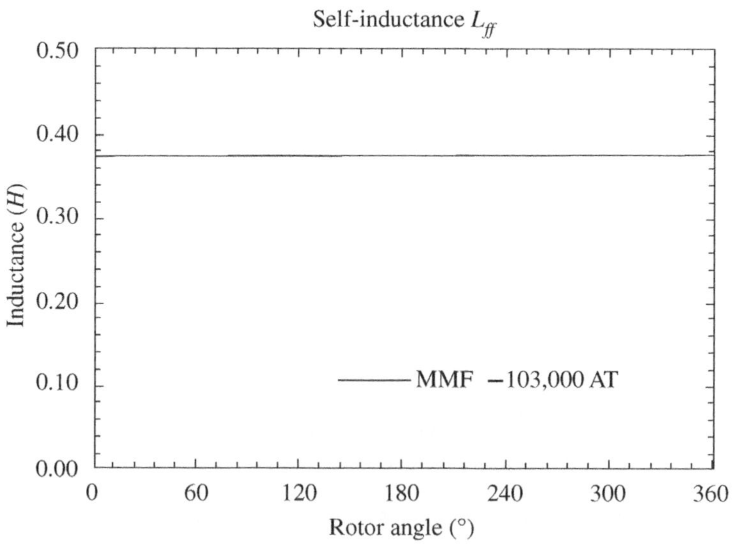

Figure 9.25 2D-FE-computed self-inductance L_{ff}, see Refs. [6, 7].

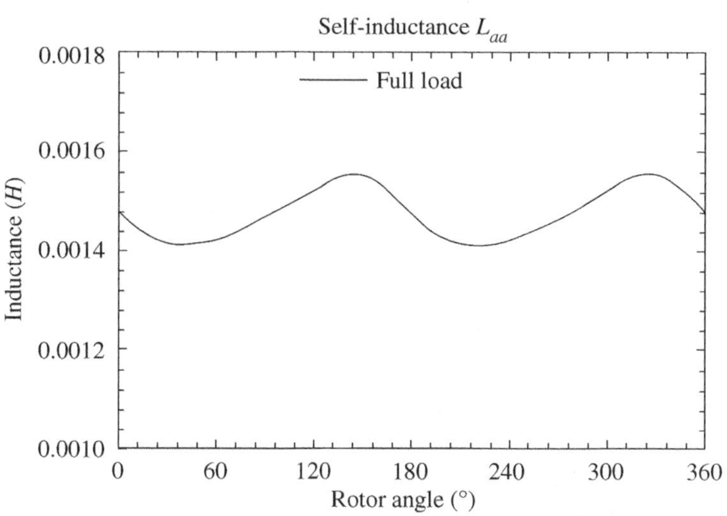

Figure 9.26 2D-FE-computed self-inductance L_{aa} (full-load case), see Refs. [6, 7].

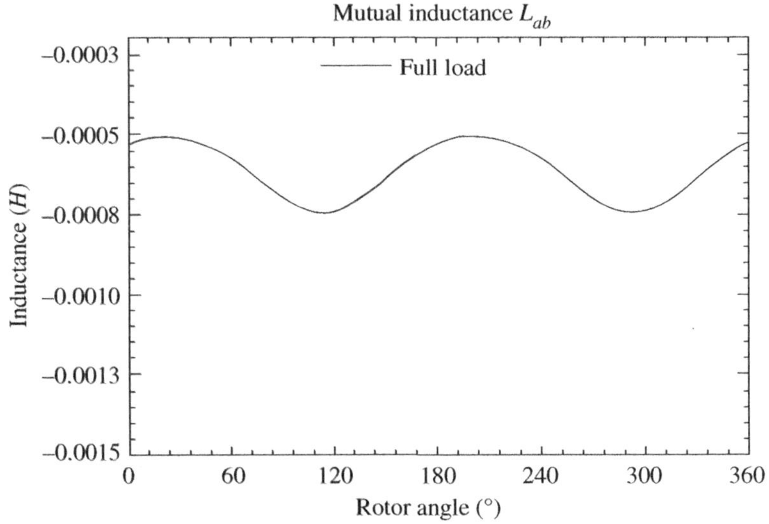

Figure 9.27 2D-FE-computed mutual inductance L_{ab} (full-load case), see Refs. [6, 7].

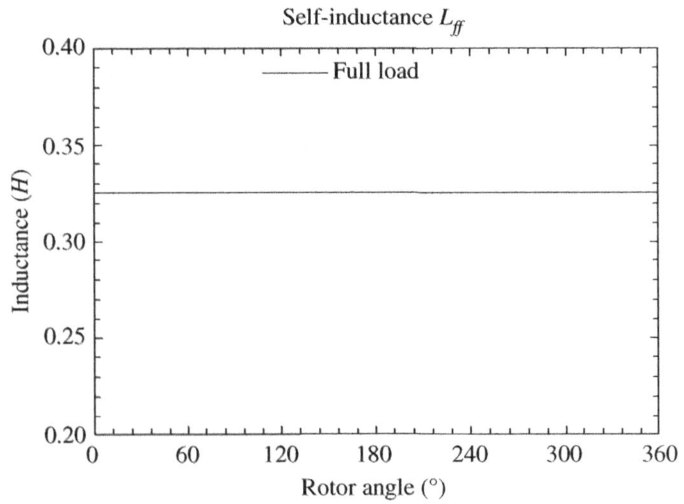

Figure 9.28 2D-FE-computed self-inductance L_{ff}, see Refs. [6, 7].

the reference and measuring all angular positions from this reference axis, in a counterclockwise direction, the position of the a, b, and c phase axes, σ_a, σ_b, and σ_c, can be written as follows:

$$\sigma_a = -\sigma, \sigma_b = (2\pi/3) - \sigma, \sigma_c = (4\pi/3) - \sigma \tag{9.50}$$

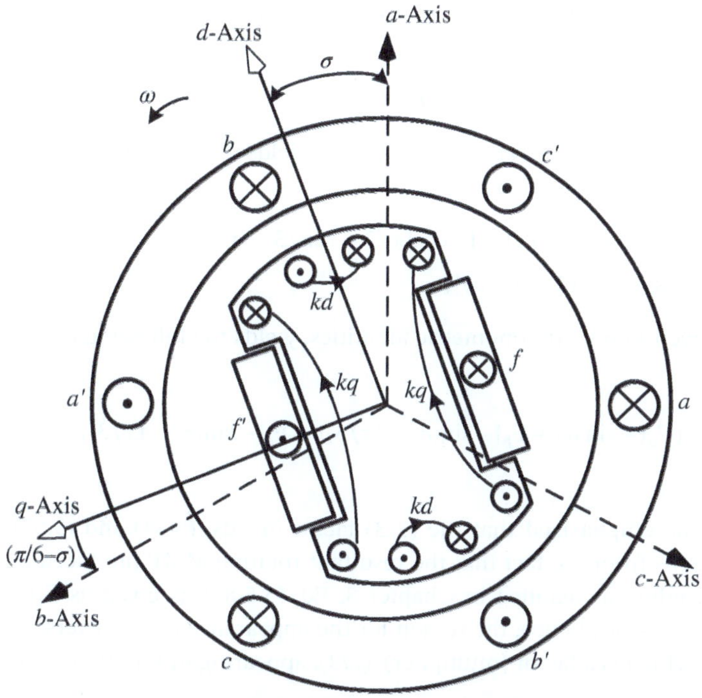

Figure 9.29 The development of Park's transformation from the geometric nature of salient-pole rotors of three-phase synchronous machines.

Accordingly, the resolution of the MMFs, $\mathbf{F}_a$, $\mathbf{F}_b$, and $\mathbf{F}_c$, along the d-axis to produce the MMF, $\mathbf{F}_d$, can be written as follows:

$$\mathbf{F}_d = (2/3)[\mathbf{F}_a \cos\sigma_a + \mathbf{F}_b \cos\sigma_b + \mathbf{F}_c \cos\sigma_c] \tag{9.51}$$

That is, substituting Eq. (9.50) into Eq. (9.51) yields the following:

$$\mathbf{F}_d = (2/3)\{\mathbf{F}_a \cos(-\sigma) + \mathbf{F}_b \cos[(2\pi/3) - \sigma] + \mathbf{F}_c \cos[(4\pi/3) - \sigma]\} \tag{9.52}$$

Hence, from well-known trigonometric identifies, Eq. (9.52) can be rewritten as follows:

$$\mathbf{F}_d = (2/3)[\mathbf{F}_a \cos(\sigma) + \mathbf{F}_b \cos(\sigma - 2\pi/3) + \mathbf{F}_c \cos(\sigma - 4\pi/3)] \tag{9.53}$$

Referring again to Figure 9.29, the positions of the a, b, and c axes, ψ_a, ψ_b, and ψ_c, with respect to the q-axis, can be written as follows:

$$\left.\begin{aligned}\psi_a &= [-(\sigma + \pi/2)]\\ \psi_b &= \psi_a + 2\pi/3 = [(2\pi/3 - \sigma) - \pi/2)]\\ \psi_c &= \psi_a + 4\pi/3 = [(4\pi/3 - \sigma) - \pi/2)]\end{aligned}\right\} \tag{9.54}$$

Accordingly, the resolution of $\mathbf{F}_a$, $\mathbf{F}_b$, and $\mathbf{F}_c$, along the q-axis to produce the MMF, $\mathbf{F}_q$, can be written as follows:

$$\mathbf{F}_q = (2/3)[\mathbf{F}_a \cos \psi_a + \mathbf{F}_b \cos \psi_b + \mathbf{F}_c \cos \psi_c] \tag{9.55}$$

That is, substituting Eq. (9.54) into Eq. (9.55) yields the following, using well-known trigonometric identities:

$$\begin{aligned}\mathbf{F}_q = (2/3)\{&\mathbf{F}_a \cos(\sigma + \pi/2) + \mathbf{F}_b \cos[\pi/2 - (2\pi/3 - \sigma)] \\ &+ \mathbf{F}_c \cos[\pi/2 - (4\pi/3 - \sigma)]\}\end{aligned} \tag{9.56}$$

which, through well-known trigonometric identities, yields the following for the MMF, $\mathbf{F}_q$:

$$\mathbf{F}_q = (2/3)\{\mathbf{F}_a(-\sin \sigma) + \mathbf{F}_b[-\sin(\sigma - 2\pi/3)] + \mathbf{F}_c[-\sin(\sigma - 4\pi/3)]\} \tag{9.57}$$

Here, it must be emphasized that the (2/3) factor in Eqs. (9.53) and (9.57), for $\mathbf{F}_d$ and $\mathbf{F}_q$ stems from the fact that the resultant rotating MMF produced by a three-phase winding as detailed in Chapter 5, Eq. (5.5-83), is given as $\mathbf{F}_1(\theta, t) = (3/2)F_{\text{max}} \cos(\theta - \omega t)$, hence, the reason for the appearance of the inverse of this (3/2) factor, which is a factor (multiplier), (2/3), appearing in Eqs. (9.53) and (9.57).

Here, in order to generate a "non-singular" linear transformation, the three MMFs, $\mathbf{F}_a$, $\mathbf{F}_b$, and $\mathbf{F}_c$, Park added to the replacement MMFS, for $\mathbf{F}_d$ and $\mathbf{F}_q$, a third MMF, $\mathbf{F}_0$, stemming from the concept of zero-sequence components of currents and voltages, borrowed from unsymmetrical fault analysis used in power systems, thus leading to the so-called zero-component MMF, $\mathbf{F}_0$, formulated as follows:

$$\mathbf{F}_0 = (1/3)[\mathbf{F}_a + \mathbf{F}_b + \mathbf{F}_c] \tag{9.58}$$

Accordingly, Eqs. (9.53), (9.57), and (9.58) can be summarized in matrix form as follows:

$$\begin{bmatrix} \mathbf{F}_d \\ \mathbf{F}_q \\ \mathbf{F}_0 \end{bmatrix} = \frac{2}{3} \begin{bmatrix} \cos(\sigma) & \cos(\sigma - 2\pi/3) & \cos(\sigma - 4\pi/3) \\ -\sin(\sigma) & -\sin(\sigma - 2\pi/3) & -\sin(\sigma - 4\pi/3) \\ \frac{1}{2} & \frac{1}{2} & \frac{1}{2} \end{bmatrix} \begin{bmatrix} \mathbf{F}_a \\ \mathbf{F}_b \\ \mathbf{F}_c \end{bmatrix} \tag{9.59}$$

Hence, arises the so-called *DQ*0 or *dq*0 Park's transformation given earlier in the (3×3) square matrix in Eq. (9.59), which plays a crucial and paramount part in the theory and design, as well as the control of electric machinery in adjustable speed drives (ASDs), and other-related systems containing AC machines as components in such systems.

In summary, Park's transformation matrix, $\underline{T}$, and its inverse, $\underline{T}^{-1}$, are given by the following (3×3) square matrices:

$$\underline{T} = (2/3)\begin{bmatrix} \cos(\sigma) & \cos(\sigma-(2\pi/3)) & \cos(\sigma-(4\pi/3)) \\ -\sin(\sigma) & -\sin(\sigma-(2\pi/3)) & -\sin(\sigma-(4\pi/3)) \\ (1/2) & (1/2) & (1/2) \end{bmatrix} \tag{9.60}$$

and

$$\underline{T}^{-1} = \begin{bmatrix} \cos(\sigma) & -\sin(\sigma) & 1 \\ \cos(\sigma-(2\pi/3)) & -\sin(\sigma-(2\pi/3)) & 1 \\ \cos(\sigma-(4\pi/3)) & -\sin(\sigma-(4\pi/3)) & 1 \end{bmatrix} \tag{9.61}$$

The reader is urged to verify that the product $[(\underline{T})\cdot(\underline{T}^{-1})]$ produces a square (3×3) identity matrix with the numbers, 1, 1, and 1, on the diagonal and zeros everywhere else in the resulting (3×3) matrix. We now proceed to applying Park's transformation in Section 9.5.

9.5 Applying Park's *DQ0* Transformation to the *ABC* State-Space Model Voltages, Currents, and Flux Linkages

In this section, we apply Park's *dq*0 transformation to the currents, voltages, and flux linkages originally formulated in the *abc* state-space model presented in Section 9.2 for the three-phase synchronous machine. Here, we define the armature (stator) current vector in the *dq*0 frame, $\underline{I}_{dq0}$, as follows:

$$\underline{I}_{dq0} = \underline{T}\cdot\underline{I}_{abc} \tag{9.62}$$

where the transformation matrix, $\underline{T}$, is as defined by Eq. (9.60). Therefore, one can write the following:

$$\begin{bmatrix} i_d \\ i_q \\ i_0 \end{bmatrix} = \frac{2}{3}\begin{bmatrix} \cos\sigma & \cos(\sigma-2\pi/3) & \cos(\sigma-4\pi/3) \\ -\sin\sigma & -\sin(\sigma-2\pi/3) & -\sin(\sigma-4\pi/3) \\ \frac{1}{2} & \frac{1}{2} & \frac{1}{2} \end{bmatrix}\begin{bmatrix} i_a \\ i_b \\ i_c \end{bmatrix} \tag{9.63}$$

Hence, it follows that

$$\underline{I}_{abc} = \underline{T}^{-1}\cdot\underline{I}_{dq0} \tag{9.64}$$

That is, in expanded matrix form, with $\underline{T}^{-1}$, given in Eq. (9.61), one can write the following:

$$\begin{bmatrix} i_a \\ i_b \\ i_c \end{bmatrix} = \begin{bmatrix} \cos\sigma & -\sin\sigma & 1 \\ \cos(\sigma-2\pi/3) & -\sin(\sigma-2\pi/3) & 1 \\ \cos(\sigma-4\pi/3) & -\sin(\sigma-4\pi/3) & 1 \end{bmatrix}\begin{bmatrix} i_d \\ i_q \\ i_0 \end{bmatrix} \tag{9.65}$$

Similarly, for the voltage vectors, $\underline{V}_{dq0}$ and $\underline{V}_{abc}$, one can write the following:

$$\underline{V}_{dq0} = \underline{T} \cdot \underline{V}_{abc} \tag{9.66}$$

and

$$\underline{V}_{abc} = \underline{T}^{-1} \cdot \underline{V}_{dq0} \tag{9.67}$$

The same type of definition applies to the flux linkage vector, hence, one can write the following:

$$\underline{\Lambda}_{dq0} = \underline{T} \cdot \underline{\Lambda}_{abc} \tag{9.68}$$

and

$$\underline{\Lambda}_{abc} = \underline{T}^{-1} \cdot \underline{\Lambda}_{dq0} \tag{9.69}$$

Now, consider the flux linkage, $\underline{\Lambda}_{abc}$, in Eqs. (9.45)–(9.49). It follows that $\underline{\Lambda}_{abc}$ in the synchronous machine model, from earlier developments in Section 9.2, is given as follows:

$$\underline{\Lambda}_{abc} = \underline{L}_{ss} \cdot \underline{I}_{abc} + \underline{L}_{sr} \cdot \underline{I}_{fk_dk_q} \tag{9.70}$$

Hence, from Eq. (9.69), one can write the following:

$$\underline{T}^{-1} \cdot \underline{\Lambda}_{dq0} = \underline{L}_{ss} \cdot \underline{T}^{-1} \cdot \underline{I}_{dq0} + \underline{L}_{sr} \cdot \underline{I}_{fk_dk_q} \tag{9.71}$$

Premultiplying Eq. (9.71) by $\underline{T}$, one can obtain the following:

$$\underline{\Lambda}_{dq0} = (\underline{T} \cdot \underline{L}_{ss} \cdot \underline{T}^{-1}) \cdot \underline{I}_{dq0} + (\underline{T} \cdot \underline{L}_{sr}) \cdot \underline{I}_{fk_dk_q} \tag{9.72}$$

It should further be noticed from Eq. (9.32) that

$$\underline{\Lambda}_{fk_dk_q} = \underline{L}_{sr}^{t} \cdot \underline{I}_{abc} + \underline{L}_{rr} \cdot \underline{I}_{fk_dk_q} \tag{9.73}$$

However, recalling from Eq. (9.64) that $\underline{I}_{abc} = \underline{T}^{-1} \cdot \underline{I}_{dq0}$, the flux linkage in Eq. (9.73) can be rewritten as follows:

$$\underline{\Lambda}_{fk_dk_q} = \left(\underline{L}_{sr}^{t} \cdot \underline{T}^{-1}\right) \cdot \underline{I}_{dq0} + (\underline{L}_{rr})\underline{I}_{fk_dk_q} \tag{9.74}$$

Next, we will detail the effects of the $DQ0$ transformation matrix, $\underline{T}$, and its inverse, $\underline{T}^{-1}$, on the new transformed inductance matrices $(\underline{T} \cdot \underline{L}_{ss} \cdot \underline{T}^{-1})$, $(\underline{T} \cdot \underline{L}_{sr})$, and $\left(\underline{L}_{sr}^{t} \cdot \underline{T}^{-1}\right)$.

Recall that

$$\underline{L}_{ss} = \begin{bmatrix} L_{aa} & L_{ab} & L_{ac} \\ L_{ab} & L_{bb} & L_{bc} \\ L_{ac} & L_{bc} & L_{cc} \end{bmatrix} \tag{9.75}$$

where

$$\left.\begin{aligned} L_{aa} &= L_{sa} + L_{sv}\cos 2\sigma \\ L_{bb} &= L_{sa} + L_{sv}\cos(2\sigma - 4\pi/3) \\ L_{cc} &= L_{sa} + L_{sv}\cos(2\sigma - 2\pi/3) \\ L_{ab} &= -L_{ma} + L_{mv}\cos(2\sigma - 2\pi/3) \\ L_{ac} &= -L_{ma} + L_{mv}\cos(2\sigma - 4\pi/3) \\ L_{bc} &= -L_{ma} + L_{mv}\cos(2\sigma) \end{aligned}\right\} \tag{9.76}$$

See derivations given in Section 9.2. Accordingly, from Eqs. (9.75) and (9.76), one can write the following:

$$\underline{L}_{ss} = \begin{bmatrix} (L_{sa} + L_{sv}\cos(2\sigma)) & (-L_{ma} + L_{mv}\cos(2\sigma - 2\pi/3)) & (-L_{ma} + L_{mv}\cos(2\sigma - 4\pi/3)) \\ (-L_{ma} + L_{mv}\cos(2\sigma - 2\pi/3)) & (L_{sa} + L_{sv}\cos(2\sigma - 4\pi/3)) & (-L_{ma} + L_{mv}\cos(2\sigma)) \\ (-L_{ma} + L_{mv}\cos(2\sigma - 4\pi/3)) & (-L_{ma} + L_{mv}\cos(2\sigma)) & (L_{sa} + L_{sv}\cos(2\sigma - 2\pi/3)) \end{bmatrix} \tag{9.77}$$

Therefore, recalling $\underline{L}_{ss}$ from Eq. (9.75), it follows that

$$\underline{L}_{ss} \cdot \underline{T}^{-1} = \begin{bmatrix} L_{aa} & L_{ab} & L_{ac} \\ L_{ab} & L_{bb} & L_{bc} \\ L_{ac} & L_{bc} & L_{cc} \end{bmatrix} \begin{bmatrix} \cos(\sigma) & -\sin(\sigma) & 1 \\ \cos(\sigma - (2\pi/3)) & -\sin(\sigma - (2\pi/3)) & 1 \\ \cos(\sigma - (4\pi/3)) & -\sin(\sigma - (4\pi/3)) & 1 \end{bmatrix} \tag{9.78}$$

Hence, it follows that

$$\underline{L}_{ss} \cdot \underline{T}^{-1} = \left[\begin{matrix} \{L_{aa}\cos(\sigma) + L_{ab}\cos(\sigma - (2\pi/3)) + L_{ac}\cos(\sigma - (4\pi/3))\} \\ \{L_{ab}\cos(\sigma) + L_{bb}\cos(\sigma - (2\pi/3)) + L_{bc}\cos(\sigma - (4\pi/3))\} \\ \{L_{ac}\cos(\sigma) + L_{bc}\cos(\sigma - (2\pi/3)) + L_{cc}\cos(\sigma - (4\pi/3))\} \end{matrix}\right.$$
$$\begin{matrix} \{-L_{aa}\sin(\sigma) - L_{ab}\sin(\sigma - (2\pi/3)) - L_{ac}\sin(\sigma - (4\pi/3))\} \\ \{-L_{ab}\sin(\sigma) - L_{bb}\sin(\sigma - (2\pi/3)) - L_{bc}\sin(\sigma - (4\pi/3))\} \\ \{-L_{ac}\sin(\sigma) - L_{bc}\sin(\sigma - (2\pi/3)) - L_{cc}\sin(\sigma - (4\pi/3))\} \end{matrix}$$
$$\left.\begin{matrix} \{L_{aa} + L_{ab} + L_{ac}\} \\ \{L_{ab} + L_{bb} + L_{bc}\} \\ \{L_{ac} + L_{bc} + L_{cc}\} \end{matrix}\right] \tag{9.79}$$

Accordingly, the transformed matrix, $(\underline{T} \cdot \underline{L}_{ss} \cdot \underline{T}^{-1})$, can be derived as follows:

$$\underline{T} \cdot \underline{L}_{ss} \cdot \underline{T}^{-1} = \frac{2}{3}\begin{bmatrix} \cos\sigma & \cos(\sigma - 2\pi/3) & \cos(\sigma - 4\pi/3) \\ -\sin\sigma & -\sin(\sigma - 2\pi/3) & -\sin(\sigma - 4\pi/3) \\ \frac{1}{2} & \frac{1}{2} & \frac{1}{2} \end{bmatrix}$$

$$\cdot \begin{bmatrix} \{L_{aa}\cos(\sigma) + L_{ab}\cos(\sigma - (2\pi/3)) + L_{ac}\cos(\sigma - (4\pi/3))\} \\ \{L_{ab}\cos(\sigma) + L_{bb}\cos(\sigma - (2\pi/3)) + L_{bc}\cos(\sigma - (4\pi/3))\} \\ \{L_{cc}\cos(\sigma) + L_{bc}\cos(\sigma - (2\pi/3)) + L_{cc}\cos(\sigma - (4\pi/3))\} \end{bmatrix} \tag{9.80}$$

$$\begin{matrix} \{-L_{aa}\sin(\sigma) + L_{ab}\sin(\sigma - (2\pi/3)) + L_{ac}\sin(\sigma - (4\pi/3))\} \\ \{-L_{ab}\sin(\sigma) + L_{bb}\sin(\sigma - (2\pi/3)) + L_{bc}\sin(\sigma - (4\pi/3))\} \\ \{-L_{cc}\sin(\sigma) + L_{bc}\sin(\sigma - (2\pi/3)) + L_{cc}\sin(\sigma - (4\pi/3))\} \end{matrix}$$

$$\left.\begin{matrix} \{L_{aa} + L_{ab} + L_{ac}\} \\ \{L_{ab} + L_{bb} + L_{bc}\} \\ \{L_{ac} + L_{bc} + L_{cc}\} \end{matrix}\right] = \begin{bmatrix} L_{dd} & L_{dq} & L_{d0} \\ L_{qd} & L_{qq} & L_{d0} \\ L_{0d} & L_{0d} & L_{00} \end{bmatrix}$$

It will be left to the student to substitute for the self-inductance terms, L_{aa}, L_{bb}, L_{cc}, and mutual inductance terms, $L_{ab} = L_{ba}$, $L_{bc} = L_{cb}$, and $L_{ca} = L_{ac}$, Eq. (9.76) into Eq. (9.80) to show that we will arrive at the following:

$$L_{dd} = L_{sa} + L_{sv}/2 + L_{ma} + L_{mv} \tag{9.81}$$

where all the terms on the right-hand side of Eq. (9.81) are defined in Section 9.2, and

$$L_{dq} = 0 \tag{9.82}$$

Similarly,

$$L_{qd} = 0 \tag{9.83}$$

in addition,

$$L_{qq} = L_{sa} + L_{ma} - L_{sv}/2 - L_{mv} \tag{9.84}$$

and

$$L_{00} = L_{sa} - 2L_{ma} \tag{9.85}$$

furthermore,

$$\left.\begin{aligned} L_{d0} &= (L_{sv} - L_{mv})\cos t(3\sigma) \\ L_{0d} &= [(L_{sv} - L_{mv})/2]\cos t(3\sigma) \end{aligned}\right\} \tag{9.86}$$

However, experimental testing and FE inductance computations show that the differences between the terms L_{sv} and L_{mv} are negligible to almost vanishing, in other words, $L_{sv} = L_{mv}$, thus,

$$L_{sv} = L_{mv} = L_v \tag{9.87}$$

and, hence,

$$L_{0d} = L_{d0} = 0 \tag{9.88}$$

and the student should also attempt to show that, $L_{q0} = (L_{mv} - L_{sv})\sin(3\sigma)$, and $L_{0q} = [(L_{mv} - L_{sv})/2]\sin(3\sigma)$, hence, if $L_{mv} = L_{sv}$, it follows that:

$$L_{0q} = L_{q0} = 0 \tag{9.89}$$

Accordingly, one can write the following:

$$(\underline{T} \cdot \underline{L}_{ss} \cdot \underline{T}^{-1}) = \begin{bmatrix} L_{dd} & L_{dq} & L_{d0} \\ L_{qd} & L_{qq} & L_{q0} \\ L_{0d} & L_{0q} & L_{00} \end{bmatrix} = \begin{bmatrix} (L_{sa} + L_{ma} + 3L_v/2) & 0 & 0 \\ 0 & (L_{sa} + L_{ma} - 3L_v/2) & 0 \\ 0 & 0 & (L_{sa} - 2L_{ma}) \end{bmatrix} \tag{9.90}$$

Now, consider the two matrices $[\underline{T} \cdot \underline{L}_{sr}]$ and $[\underline{L}^t_{sr} \cdot \underline{T}^{-1}]$. First, let us consider $[\underline{T} \cdot \underline{L}_{sr}]$:

$$\underline{T} \cdot \underline{L}_{sr} = \frac{2}{3}\begin{bmatrix} \cos\sigma & \cos(\sigma - 2\pi/3) & \cos(\sigma - 4\pi/3) \\ -\sin\sigma & -\sin(\sigma - 2\pi/3) & -\sin(\sigma - 4\pi/3) \\ \frac{1}{2} & \frac{1}{2} & \frac{1}{2} \end{bmatrix} \cdot \begin{bmatrix} L_{afm}\cos(\sigma) & L_{akdm}\cos(\sigma) & -L_{akqm}\sin(\sigma) \\ L_{afm}\cos(\sigma - 2\pi/3) & L_{akdm}\cos(\sigma - 2\pi/3) & -L_{akqm}\sin(\sigma - 2\pi/3) \\ L_{afm}\cos(\sigma - 4\pi/3) & L_{akdm}\cos(\sigma - 4\pi/3) & -L_{akqm}\sin(\sigma - 4\pi/3) \end{bmatrix} \tag{9.91}$$

Let

$$\sigma_1 = \sigma, \sigma_2 = (\sigma - 2\pi/3), \sigma_3 = (\sigma - 4\pi/3) \tag{9.92}$$

Thus, the matrix product in Eq. (9.91) can be written as follows:

$$\underline{T} \cdot \underline{L}_{sr} = \begin{bmatrix} \left(\frac{2}{3}L_{afm}\sum_{i=1}^{3}\cos^2\sigma_i\right) & \left(\frac{2}{3}L_{akdm}\sum_{i=1}^{3}\cos^2\sigma_i\right) & \left(-\frac{2}{3}L_{akqm}\sum_{i=1}^{3}\sin\sigma_i\cos\sigma_i\right) \\ \left(-\frac{2}{3}L_{afm}\sum_{i=1}^{3}\sin\sigma_i\cos\sigma_i\right) & \left(-\frac{2}{3}L_{akdm}\sum_{i=1}^{3}\sin\sigma_i\cos\sigma_i\right) & \left(\frac{2}{3}L_{akqm}\sum_{i=1}^{3}\sin^2\sigma_i\right) \\ \left(\frac{1}{3}L_{afm}\sum_{i=1}^{3}\cos\sigma_i\right) & \left(\frac{1}{3}L_{akdm}\sum_{i=1}^{3}\cos\sigma_i\right) & \left(-\frac{1}{3}L_{akqm}\sum_{i=1}^{3}\sin\sigma_i\right) \end{bmatrix} \tag{9.93}$$

But, note that

$$\sum_{i=1}^{3}\cos\sigma_i = \sum_{i=1}^{3}\sin\sigma_i = 0 \tag{9.94}$$

and

$$\sum_{i=1}^{3} \sin \sigma_i \cos \sigma_i = \sum_{i=1}^{3} \frac{1}{2} \sin 2\sigma_i = 0 \tag{9.95}$$

also,

$$\sum_{i=1}^{3} \sin^2 \sigma_i = \sum_{i=1}^{3} \left(\frac{1}{2} - \frac{1}{2} \cos 2\sigma_i \right) = \frac{3}{2} \tag{9.96}$$

Finally,

$$\sum_{i=1}^{3} \cos^2 \sigma_i = \sum_{i=1}^{3} \left(\frac{1}{2} + \frac{1}{2} \cos 2\sigma_i \right) = \frac{3}{2} \tag{9.97}$$

By substituting Eqs. (9.94)–(9.97) into Eq. (9.93), the following is deduced:

$$\underline{T} \cdot \underline{L}_{sr} = \begin{bmatrix} L_{afm} & L_{akdm} & 0 \\ 0 & 0 & L_{akqm} \\ 0 & 0 & 0 \end{bmatrix} \tag{9.98}$$

It is left to the student to prove that

$$\underline{L}_{sr}^{t} \cdot \underline{T}^{-1} = \begin{bmatrix} \frac{3}{2} L_{afm} & 0 & 0 \\ \frac{3}{2} L_{akdm} & 0 & 0 \\ 0 & \frac{3}{2} L_{akqm} & 0 \end{bmatrix} \tag{9.99}$$

Now the flux linkages can be examined in the $dq0$ frame of reference.

Let us summarize the formulation of the flux linkage vectors, $\underline{\Lambda}_{dq0}$ and $\underline{\Lambda}_{fkdkq}$, in the $dq0$ frame of reference as follows:

$$\begin{bmatrix} \underline{\Lambda}_{dq0} \\ \underline{\Lambda}_{fkdkq} \end{bmatrix} = \begin{bmatrix} (\underline{T} \cdot \underline{L}_{ss} \cdot \underline{T}^{-1}) & (\underline{T} \cdot \underline{L}_{sr}) \\ (\underline{L}_{sr}^{t} \cdot \underline{T}^{-1}) & (\underline{L}_{rr}) \end{bmatrix} \begin{bmatrix} \underline{I}_{dq0} \\ \underline{I}_{fkdkq} \end{bmatrix} \tag{9.100}$$

The expanded matrix form of Eq. (9.100) can be rewritten as follows:

$$\begin{bmatrix} \lambda_d \\ \lambda_q \\ \lambda_0 \\ \lambda_f \\ \lambda_{kd} \\ \lambda_{kq} \end{bmatrix} = \begin{bmatrix} L_{dd} & 0 & 0 & L_{afm} & L_{akdm} & 0 \\ 0 & L_{qq} & 0 & 0 & 0 & L_{akqm} \\ 0 & 0 & L_{00} & 0 & 0 & 0 \\ \frac{3}{2} L_{afm} & 0 & 0 & L_{ff} & L_{fkd} & 0 \\ \frac{3}{2} L_{akdm} & 0 & 0 & L_{fkd} & L_{kdkd} & 0 \\ 0 & \frac{3}{2} L_{akqm} & 0 & 0 & 0 & L_{kqkq} \end{bmatrix} \begin{bmatrix} i_d \\ i_q \\ i_0 \\ i_f \\ i_{kd} \\ i_{kq} \end{bmatrix} \tag{9.101}$$

where

$$L_{dd} = \left(L_{sa} + L_{ma} + \frac{3}{2} L_v \right) \tag{9.102}$$

$$L_{qq} = \left(L_{sa} + L_{ma} - \frac{3}{2}L_v\right) \tag{9.103}$$

and

$$L_{00} = (L_{sa} - 2L_{ma}) \tag{9.104}$$

Here,

$$L_v = L_{sv} = L_{mv} \tag{9.105}$$

9.6 State-Space Modeling Derivation in the *DQ0* Reference Frame of a Synchronous Machine Including Saliency Effects – The Inductance and Flux Linkage Matrices and Vectors

Let us consider the synchronous machine model, which is given in Section 9.2, Eq. (9.32), restated here as follows:

$$\begin{bmatrix} \underline{V}_{abc} \\ \underline{V}_{fkdkq} \end{bmatrix} = \begin{bmatrix} \underline{R}_{ss} & \underline{0} \\ \underline{0} & \underline{R}_{rr} \end{bmatrix} \begin{bmatrix} \underline{I}_{abc} \\ \underline{I}_{fkdkq} \end{bmatrix} + \frac{d}{dt}\left\{ \begin{bmatrix} \underline{L}_{ss} & \underline{L}_{sr} \\ \underline{L}_{sr}^t & \underline{L}_{rr} \end{bmatrix} \begin{bmatrix} \underline{I}_{abc} \\ \underline{I}_{fkdkq} \end{bmatrix} \right\} \tag{9.106}$$

which can be rewritten in flux linkage terms as follows:

$$\begin{bmatrix} \underline{V}_{abc} \\ \underline{V}_{fkdkq} \end{bmatrix} = \begin{bmatrix} \underline{R}_{ss} & \underline{0} \\ \underline{0} & \underline{R}_{rr} \end{bmatrix} \begin{bmatrix} \underline{I}_{abc} \\ \underline{I}_{fkdkq} \end{bmatrix} + \frac{d}{dt} \begin{bmatrix} \underline{\Lambda}_{abc} \\ \underline{\Lambda}_{fkdkq} \end{bmatrix} \tag{9.107}$$

Applying Park's transformation, discussed in Sections 9.4 and 9.5, to Eq. (9.107), one can write the following:

$$\begin{bmatrix} \underline{T} \cdot \underline{V}_{abc} \\ \underline{V}_{fkdkq} \end{bmatrix} = \begin{bmatrix} \underline{T} \cdot \underline{R}_{ss} \cdot \underline{T}^{-1} & \underline{0} \\ \underline{0} & \underline{R}_{rr} \end{bmatrix} \begin{bmatrix} \underline{T} \cdot \underline{I}_{abc} \\ \underline{I}_{fkdkq} \end{bmatrix} + \begin{bmatrix} \underline{T} & 0 \\ 0 & \underline{U} \end{bmatrix} \frac{d}{dt} \begin{bmatrix} \underline{\Lambda}_{abc} \\ \underline{\Lambda}_{fkdkq} \end{bmatrix} \tag{9.108}$$

where $\underline{U}$ is the identity matrix. However,

$$\underline{\Lambda}_{abc} = \underline{T}^{-1}\underline{\Lambda}_{dq0} \tag{9.109}$$

where according to Eq. (9.100), we have

$$\underline{\Lambda}_{dq0} = (\underline{T} \cdot \underline{L}_{ss} \cdot \underline{T}^{-1}) \cdot \underline{I}_{dq0} + (\underline{T} \cdot \underline{L}_{sr}) \cdot \underline{I}_{fkdkq} \tag{9.110}$$

Hence, the following is deduced from Eqs. (9.108) and (9.109):

$$\underline{V}_{dq0} = (\underline{T} \cdot \underline{R}_{ss} \cdot \underline{T}^{-1}) \cdot \underline{I}_{dq0} + \underline{T} \cdot \frac{d}{dt}(\underline{T}^{-1} \cdot \underline{\Lambda}_{dq0}) \tag{9.111}$$

and

$$\underline{V}_{fkdkq} = (\underline{R}_{rr}) \cdot \underline{I}_{fkdkq} + \frac{d}{dt}(\underline{\Lambda}_{fkdkq}) \tag{9.112}$$

One must recall that

$$\underline{\Lambda}_{fkdkq} = (\underline{L}_{sr}^{t}) \cdot \underline{I}_{abc} + (\underline{L}_{rr}) \cdot \underline{I}_{fkdkq} \tag{9.113}$$

After using Park's transformation, Eq. (9.113) can be rewritten as follows:

$$\underline{\Lambda}_{fkdkq} = (\underline{L}_{sr}^{t} \cdot \underline{T}^{-1}) \cdot \underline{I}_{dq0} + (\underline{L}_{rr}) \cdot \underline{I}_{fkdkq} \tag{9.114}$$

Consider the following matrix product:

$$\begin{aligned} \underline{T} \cdot \underline{R}_{ss} \cdot \underline{T}^{-1} &= \underline{T} \cdot r_s \begin{bmatrix} 1 & 0 & 0 \\ 0 & 1 & 0 \\ 0 & 0 & 1 \end{bmatrix} \underline{T}^{-1} = r_s \underline{T} \cdot \underline{U} \cdot \underline{T}^{-1} = r_s \underline{T} \cdot \underline{T}^{-1} \\ &= r_s \underline{U} = \begin{bmatrix} r_s & 0 & 0 \\ 0 & r_s & 0 \\ 0 & 0 & r_s \end{bmatrix} = \underline{R}_{ss} \end{aligned} \tag{9.115}$$

Consider the vector $\underline{T}\frac{d}{dt}(\underline{T}^{-1} \cdot \underline{\Lambda}_{dq0})$, which can be rewritten as follows:

$$\underline{T} \cdot \frac{d}{dt}(\underline{T}^{-1} \cdot \underline{\Lambda}_{dq0}) = \underline{T} \cdot \frac{d}{dt}(\underline{\Lambda}_{abc}) \tag{9.116}$$

Consider the fact that $\underline{\Lambda}_{dq0} = \underline{T} \cdot \underline{\Lambda}_{abc}$, it follows that

$$\frac{d}{dt}(\underline{\Lambda}_{dq0}) = \frac{d}{dt}(\underline{T} \cdot \underline{\Lambda}_{abc}) = \underline{T} \cdot \frac{d}{dt}(\underline{\Lambda}_{abc}) + \left(\frac{d}{dt}\underline{T}\right) \cdot \underline{\Lambda}_{abc} \tag{9.117}$$

from which, one can write the following:

$$\underline{T} \cdot \frac{d}{dt}(\underline{\Lambda}_{abc}) = \frac{d}{dt}(\underline{\Lambda}_{dq0}) - \left(\frac{d}{dt}\underline{T}\right) \cdot \underline{T}^{-1} \cdot \underline{\Lambda}_{dq0} \tag{9.118}$$

Hence, one can rewrite $\underline{\Lambda}_{abc}$ in Eq. (9.118) as $\underline{\Lambda}_{abc} = \underline{T}^{-1} \cdot \underline{\Lambda}_{dq0}$. Accordingly, it follows that

$$\underline{T} \cdot \frac{d}{dt}(\underline{T}^{-1} \cdot \underline{\Lambda}_{dq0}) = \frac{d}{dt}(\underline{\Lambda}_{dq0}) - \left(\frac{d}{dt}\underline{T}\right)(\underline{T}^{-1} \cdot \underline{\Lambda}_{dq0}) \tag{9.119}$$

The student should prove that

$$\left(\frac{d}{dt}\underline{T}\right) = \omega \underline{G} \cdot \underline{T} \tag{9.120}$$

where

$$\underline{G} = \begin{bmatrix} 0 & 1 & 0 \\ -1 & 0 & 0 \\ 0 & 0 & 0 \end{bmatrix} \tag{9.121}$$

Thus, Eq. (9.119) can be rewritten using Eq. (9.120) as follows:

$$\underline{T} \cdot \frac{d}{dt}(\underline{T}^{-1} \cdot \underline{\Lambda}_{dq0}) = \frac{d}{dt}(\underline{\Lambda}_{dq0}) - (\omega \cdot \underline{G} \cdot \underline{T} \cdot \underline{T}^{-1} \cdot \underline{\Lambda}_{dq0}) = \frac{d}{dt}(\underline{\Lambda}_{dq0}) - (\omega \cdot \underline{G} \cdot \underline{\Lambda}_{dq0}) \tag{9.122}$$

Thus, Eqs. (9.111) and (9.112) can be rewritten as follows:

$$\underline{V}_{dq0} = R_{ss} \cdot \underline{I}_{dq0} + \dot{\underline{\Lambda}}_{dq0} - \omega \cdot \underline{G} \cdot \underline{\Lambda}_{dq0} \tag{9.123}$$

and

$$\underline{V}_{fkdkq} = R_{rr} \cdot \underline{I}_{fkdkq} + \dot{\underline{\Lambda}}_{fkdkq} \tag{9.124}$$

In addition, we can state that

$$\dot{\underline{\Lambda}}_{dq0} = (\underline{T} \cdot \underline{L}_{ss} \cdot \underline{T}^{-1}) \cdot \dot{\underline{I}}_{dq0} + (\underline{T} \cdot \underline{L}_{sr}) \cdot \dot{\underline{I}}_{fkdkq} \tag{9.125}$$

Because $(\underline{T} \cdot \underline{L}_{ss} \cdot \underline{T}^{-1})$ and $(\underline{T} \cdot \underline{L}_{sr})$ are time independent. Also,

$$\dot{\underline{\Lambda}}_{fkdkq} = \left(\underline{L}_{sr}^{t} \cdot \underline{T}^{-1}\right) \cdot \dot{\underline{I}}_{dq0} + (\underline{L}_{rr}) \cdot \dot{\underline{I}}_{fkdkq} \tag{9.126}$$

Therefore, the complete model is as follows:

$$\begin{aligned}\underline{V}_{dq0} = R_{ss} \cdot \underline{I}_{dq0} + (\underline{T} \cdot \underline{L}_{ss} \cdot \underline{T}^{-1}) \cdot \dot{\underline{I}}_{dq0} + (\underline{T} \cdot \underline{L}_{sr}) \cdot \dot{\underline{I}}_{fkdkq} \\ - \omega \cdot \underline{G} \cdot (\underline{T} \cdot \underline{L}_{ss} \cdot \underline{T}^{-1}) \cdot \underline{I}_{dq0} - \omega \cdot \underline{G} \cdot (\underline{T} \cdot \underline{L}_{sr}) \cdot \underline{I}_{fkdkq}\end{aligned} \tag{9.127}$$

and

$$\underline{V}_{fkdkq} = (R_{rr}) \cdot \underline{I}_{fkdkq} + \left(\underline{L}_{sr}^{t} \cdot \underline{T}^{-1}\right) \cdot \dot{\underline{I}}_{dq0} + (\underline{L}_{rr}) \cdot \dot{\underline{I}}_{fkdkq} \tag{9.128}$$

Thus, in compact matrix form, the complete model becomes

$$\begin{aligned}\begin{bmatrix} \underline{V}_{dq0} \\ \underline{V}_{fkdkq} \end{bmatrix} = \begin{bmatrix} \underline{R}_{ss} & \underline{0} \\ \underline{0} & \underline{R}_{rr} \end{bmatrix} \begin{bmatrix} \underline{I}_{dq0} \\ \underline{I}_{fkdkq} \end{bmatrix} + \begin{bmatrix} (\underline{T} \cdot \underline{L}_{ss} \cdot \underline{T}^{-1}) & (\underline{T} \cdot \underline{L}_{sr}) \\ (\underline{L}_{sr}^{t} \cdot \underline{T}^{-1}) & (\underline{L}_{rr}) \end{bmatrix} \begin{bmatrix} \dot{\underline{I}}_{dq0} \\ \dot{\underline{I}}_{fkdkq} \end{bmatrix} \\ - \omega \begin{bmatrix} \underline{G} \cdot (\underline{T} \cdot \underline{L}_{ss} \cdot \underline{T}^{-1}) & \underline{G} \cdot (\underline{T} \cdot \underline{L}_{sr}) \\ \underline{0} & \underline{0} \end{bmatrix} \begin{bmatrix} \underline{I}_{dq0} \\ \underline{I}_{fkdkq} \end{bmatrix}\end{aligned} \tag{9.129}$$

Let us evaluate $\underline{G} \cdot (\underline{T} \cdot \underline{L}_{ss} \cdot \underline{T}^{-1})$ and $\underline{G} \cdot (\underline{T} \cdot \underline{L}_{sr})$, which yields the following:

$$\begin{aligned}\underline{G} \cdot (\underline{T} \cdot \underline{L}_{ss} \cdot \underline{T}^{-1}) &= \begin{bmatrix} 0 & 1 & 0 \\ -1 & 0 & 0 \\ 0 & 0 & 0 \end{bmatrix} \begin{bmatrix} L_{dd} & 0 & 0 \\ 0 & L_{qq} & 0 \\ 0 & 0 & L_{00} \end{bmatrix} = \begin{bmatrix} 0 & L_{qq} & 0 \\ -L_{dd} & 0 & 0 \\ 0 & 0 & 0 \end{bmatrix} \\ &= \begin{bmatrix} 0 & \left(L_{sa} + L_{ma} - \frac{3}{2}L_{v}\right) & 0 \\ -\left(L_{sa} + L_{ma} + \frac{3}{2}L_{v}\right) & 0 & 0 \\ 0 & 0 & 0 \end{bmatrix}\end{aligned} \tag{9.130}$$

and

$$\underline{G} \cdot (\underline{T} \cdot \underline{L}_{sr}) = \begin{bmatrix} 0 & 1 & 0 \\ -1 & 0 & 0 \\ 0 & 0 & 0 \end{bmatrix} \begin{bmatrix} L_{afm} & L_{akdm} & 0 \\ 0 & 0 & L_{akqm} \\ 0 & 0 & 0 \end{bmatrix} = \begin{bmatrix} 0 & 0 & L_{akqm} \\ -L_{afm} & -L_{akdm} & 0 \\ 0 & 0 & 0 \end{bmatrix} \tag{9.131}$$

Hence, the complete model in expanded form is as follows:

$$\begin{bmatrix} v_d \\ v_q \\ v_0 \\ \hdashline v_f \\ 0 \\ 0 \end{bmatrix} = \left[\begin{array}{ccc:ccc} r_s & 0 & 0 & 0 & 0 & 0 \\ 0 & r_s & 0 & 0 & 0 & 0 \\ 0 & 0 & r_s & 0 & 0 & 0 \\ \hdashline 0 & 0 & 0 & r_f & 0 & 0 \\ 0 & 0 & 0 & 0 & r_{kd} & 0 \\ 0 & 0 & 0 & 0 & 0 & r_{kq} \end{array}\right] \begin{bmatrix} i_d \\ i_q \\ i_0 \\ \hdashline i_f \\ i_{kd} \\ i_{kq} \end{bmatrix}$$
$$+ \left[\begin{array}{ccc:ccc} 0 & -\omega\left(L_{sa} + L_{ma} + \frac{3}{2}L_v\right) & 0 & 0 & 0 & -\omega L_{akqm} \\ \omega\left(L_{sa} + L_{ma} + \frac{3}{2}L_v\right) & 0 & 0 & \omega L_{afm} & \omega L_{akdm} & 0 \\ 0 & 0 & 0 & 0 & 0 & 0 \\ \hdashline 0 & 0 & 0 & 0 & 0 & 0 \\ 0 & 0 & 0 & 0 & 0 & 0 \\ 0 & 0 & 0 & 0 & 0 & 0 \end{array}\right] \begin{bmatrix} i_d \\ i_q \\ i_0 \\ \hdashline i_f \\ i_{kd} \\ i_{kq} \end{bmatrix}$$
$$+ \left[\begin{array}{ccc:ccc} (L_{sa} + L_{ma} + 3L_v/2) & 0 & 0 & L_{afm} & L_{akdm} & 0 \\ 0 & (L_{sa} + L_{ma} - 3L_v/2) & 0 & 0 & 0 & L_{akqm} \\ 0 & 0 & (L_{sa} - 2L_{ma}) & 0 & 0 & 0 \\ \hdashline \frac{3}{2}L_{afm} & 0 & 0 & L_{ff} & L_{fkd} & 0 \\ \frac{3}{2}L_{akdm} & 0 & 0 & L_{fkd} & L_{kdkd} & 0 \\ 0 & \frac{3}{2}L_{akqm} & 0 & 0 & 0 & L_{kqkq} \end{array}\right] \begin{bmatrix} \dot{i}_d \\ \dot{i}_q \\ \dot{i}_0 \\ \hdashline \dot{i}_f \\ \dot{i}_{kd} \\ \dot{i}_{kq} \end{bmatrix} \tag{9.132}$$

Here, in order to rewrite the complete model of Eq. (9.132), in compact matrix form, we introduce definitions of the following:

$$\underline{V} = \begin{bmatrix} v_d \\ v_q \\ v_0 \\ v_f \\ 0 \\ 0 \end{bmatrix} \tag{9.133}$$

$$\underline{I} = \begin{bmatrix} i_d \\ i_q \\ i_0 \\ i_f \\ 0 \\ 0 \end{bmatrix} \tag{9.134}$$

$$\dot{\underline{I}} = \frac{d}{dt}\begin{bmatrix} i_d \\ i_q \\ i_0 \\ i_f \\ 0 \\ 0 \end{bmatrix} = \begin{bmatrix} \dot{i}_d \\ \dot{i}_q \\ \dot{i}_0 \\ \dot{i}_f \\ 0 \\ 0 \end{bmatrix} \tag{9.135}$$

$$R = \left[\begin{array}{ccc|ccc} r_s & 0 & 0 & 0 & 0 & 0 \\ 0 & r_s & 0 & 0 & 0 & 0 \\ 0 & 0 & r_s & 0 & 0 & 0 \\ \hline 0 & 0 & 0 & r_f & 0 & 0 \\ 0 & 0 & 0 & 0 & r_{kd} & 0 \\ 0 & 0 & 0 & 0 & 0 & r_{kq} \end{array}\right] \tag{9.136}$$

$$\underline{L}_1 = \left[\begin{array}{ccc|ccc} 0 & -L_{qq} & 0 & 0 & 0 & -L_{akqm} \\ L_{dd} & 0 & 0 & L_{afm} & L_{akdm} & 0 \\ 0 & 0 & 0 & 0 & 0 & 0 \\ \hline 0 & 0 & 0 & 0 & 0 & 0 \\ 0 & 0 & 0 & 0 & 0 & 0 \\ 0 & 0 & 0 & 0 & 0 & 0 \end{array}\right] \tag{9.137}$$

and

$$\underline{L}_2 \left[\begin{array}{ccc|ccc} L_{dd} & 0 & 0 & L_{afm} & L_{akdm} & 0 \\ 0 & L_{qq} & 0 & 0 & 0 & L_{akqm} \\ 0 & 0 & L_{00} & 0 & 0 & 0 \\ \hline \frac{3}{2}L_{afm} & 0 & 0 & L_{ff} & L_{fkd} & 0 \\ \frac{3}{2}L_{akdm} & 0 & 0 & L_{fkd} & L_{kdkd} & 0 \\ 0 & \frac{3}{2}L_{akqm} & 0 & 0 & 0 & L_{kqkq} \end{array}\right] \tag{9.138}$$

where

$$L_{dd} = \left(L_{sa} + L_{ma} + \frac{3}{2}L_v\right) \tag{9.139}$$

$$L_{qq} = \left(L_{sa} + L_{ma} - \frac{3}{2}L_v\right) \tag{9.140}$$

and

$$L_{00} = (L_{sa} - 2L_{ma}) \tag{9.141}$$

Using the above definitions in Eqs. (9.133)–(9.141), Eq. (9.132) can be rewritten in compact matrix form as follows:

$$\underline{V} = \underline{R} \cdot \underline{I} + (\omega \underline{L}_1) \cdot \underline{I} + \underline{L}_2 \cdot \dot{\underline{I}} \quad (9.142)$$

where $\underline{R}$, $\underline{L}_1$, and $\underline{L}_2$ are time-independent matrices, if magnetic linearity is assumed. Accordingly, Eq. (9.142) can be written as follows:

$$\dot{\underline{I}} = -\underline{L}_2^{-1}(\underline{R} + \omega \underline{L}_1) \cdot \underline{I} + \underline{L}_2^{-1} \cdot \underline{V} \quad (9.143)$$

Here, Eq. (9.143) is of the well-known state-space form $\dot{\underline{X}} = \underline{A} \cdot \underline{X} + \underline{B} \cdot \underline{V}$, where $\underline{X}$ is the state-space variables vector, and $\underline{V}$ is the input vector, where the $\underline{A}$ and $\underline{B}$ matrices are time-independent system coefficient matrices.

9.7 Synchronous Machine *DQ*0 State-Space Model in the *DQ*0 Flux Linkage Frame of Reference

What preceded in Section 9.7, Eqs. (9.132)–(9.143) is the synchronous machine's $dq0$ state-space model in the current frame of reference. We now introduce the synchronous machine's $dq0$ state-space model in the flux linkage frame of reference, which is far more suitable when magnetic nonlinearities, due to ferrous material saturation, cannot be ignored and therefore must be taken into full account in the formulation and modeling process.

Let us merge Eqs. (9.123) and (9.124) into one system of matrix–vector equations, which will yield the following system:

$$\begin{bmatrix} \underline{V}_{dq0} \\ \underline{V}_{fkdkq} \end{bmatrix} = \begin{bmatrix} \underline{R}_{ss} & \underline{0} \\ \underline{0} & \underline{R}_{rr} \end{bmatrix} \begin{bmatrix} \underline{I}_{dq0} \\ \underline{I}_{fkdkq} \end{bmatrix} + \begin{bmatrix} \dot{\underline{\Lambda}}_{dq0} \\ \dot{\underline{\Lambda}}_{fkdkq} \end{bmatrix} - \omega \begin{bmatrix} \underline{G} \cdot \underline{\Lambda}_{dq0} \\ \underline{0} \end{bmatrix} \quad (9.144)$$

where $\underline{0}$ is a (3×3) zero matrix,

$$\underline{V}_{dq0} = [v_d \; v_q \; v_0]^t \quad (9.145)$$

$$\underline{V}_{fkdkq} = \left[v_f \; v_{kd} \; v_{kq}\right]^t \quad (9.146)$$

$$\underline{I}_{dq0} = \left[i_d \; i_q \; i_0\right]^t \quad (9.147)$$

$$\underline{I}_{fkdkq} = \left[i_f \; i_{kd} \; i_{kq}\right]^t \quad (9.148)$$

$$\underline{I}_{dq0fkdkq} = \left[i_d \; i_q \; i_0 \; i_f \; i_{kd} \; i_{kq}\right]^t \quad (9.149)$$

$$\underline{R}_{ss} = \begin{bmatrix} r_s & 0 & 0 \\ 0 & r_s & 0 \\ 0 & 0 & r_s \end{bmatrix} \quad (9.150)$$

where r_s is the stator winding resistance per phase

$$\underline{R}_{rr} = \begin{bmatrix} r_f & 0 & 0 \\ 0 & r_{kd} & 0 \\ 0 & 0 & r_{kq} \end{bmatrix} \tag{9.151}$$

where r_f is the field winding resistance, and r_{kd} and r_{kq} are the damper winding's direct and quadrature axes equivalent resistances, respectively, and

$$\underline{\Lambda}_{dq0} = \begin{bmatrix} \lambda_d & \lambda_q & \lambda_0 \end{bmatrix}^t \tag{9.152}$$

$$\underline{\Lambda}_{fkdkq} = \begin{bmatrix} \lambda_f & \lambda_{kd} & \lambda_{kq} \end{bmatrix}^t \tag{9.153}$$

$$\underline{\Lambda}_{dq0fkdkq} = \begin{bmatrix} \lambda_d & \lambda_q & \lambda_0 & \lambda_f & \lambda_{kd} & \lambda_{kq} \end{bmatrix}^t \tag{9.154}$$

Hence,

$$\underline{\Lambda}_{dq0fkdkq} = \begin{bmatrix} \underline{\Lambda}_{dq0} \\ \underline{\Lambda}_{fkdkq} \end{bmatrix} \begin{bmatrix} \lambda_d \\ \lambda_q \\ \lambda_0 \\ \lambda_f \\ \lambda_{kd} \\ \lambda_{kq} \end{bmatrix} = \left[\begin{array}{ccc|ccc} L_{dd} & 0 & 0 & L_{afm} & L_{akdm} & 0 \\ 0 & L_{qq} & 0 & 0 & 0 & L_{akqm} \\ 0 & 0 & L_{00} & 0 & 0 & 0 \\ \hline \frac{3}{2}L_{afm} & 0 & 0 & L_{ff} & L_{fkd} & 0 \\ \frac{3}{2}L_{akdm} & 0 & 0 & L_{fkd} & L_{kdkd} & 0 \\ 0 & \frac{3}{2}L_{akqm} & 0 & 0 & 0 & L_{kqkq} \end{array}\right] \begin{bmatrix} i_d \\ i_q \\ i_0 \\ i_f \\ i_{kd} \\ i_{kq} \end{bmatrix} \tag{9.155}$$

and

$$\underline{L}_{dq0fkdkq} = \underline{L}_2 \tag{9.156}$$

where L_{dd}, L_{qq}, and L_{00} are as given in Eqs. (9.139), (9.140), and (9.141), respectively, and $\underline{L}_2$ is as defined in Eq. (9.138).

Accordingly, using the definitions given in Eqs. (9.149), (9.154)–(9.156), one can express $\underline{I}_{dq0fkdkq}$ as follows:

$$\underline{I}_{dq0fkdkq} = \begin{bmatrix} \underline{I}_{dq0} \\ \underline{I}_{fkdkq} \end{bmatrix} = \underline{L}_{dq0fkdkq}^{-1} \begin{bmatrix} \underline{\Lambda}_{dq0} \\ \underline{\Lambda}_{fkdkq} \end{bmatrix} \tag{9.157}$$

or

$$\underline{I}_{dq0fkdkq} = \begin{bmatrix} \underline{I}_{dq0} \\ \underline{I}_{fkdkq} \end{bmatrix} = \underline{L}_2^{-1} \begin{bmatrix} \underline{\Lambda}_{dq0} \\ \underline{\Lambda}_{fkdkq} \end{bmatrix} \tag{9.158}$$

In order to derive the $dq0$ state-space model in the flux linkage frame of reference, we rearrange Eq. (9.144) as follows:

$$\begin{bmatrix} \dot{\underline{\Lambda}}_{dq0} \\ \dot{\underline{\Lambda}}_{fkdkq} \end{bmatrix} = \begin{bmatrix} \underline{V}_{dq0} \\ \underline{V}_{fkdkq} \end{bmatrix} - \begin{bmatrix} \underline{R}_{ss} & \underline{0} \\ \underline{0} & \underline{R}_{rr} \end{bmatrix} \begin{bmatrix} \underline{I}_{dq0} \\ \underline{I}_{fkdkq} \end{bmatrix} + \omega \begin{bmatrix} \underline{G} \cdot \underline{\Lambda}_{dq0} \\ \underline{0} \end{bmatrix} \tag{9.159}$$

or, in other way, one can rewrite Eq. (9.159) as follows:

$$\begin{bmatrix} \dot{\underline{\Lambda}}_{dq0} \\ \dot{\underline{\Lambda}}_{fkdkq} \end{bmatrix} = \begin{bmatrix} \underline{V}_{dq0} \\ \underline{V}_{fkdkq} \end{bmatrix} - \begin{bmatrix} \underline{R}_{ss} & \underline{0} \\ \underline{0} & \underline{R}_{rr} \end{bmatrix} \underline{L}^{-1}_{dq0fkdkq} \begin{bmatrix} \underline{\Lambda}_{dq0} \\ \underline{\Lambda}_{fkdkq} \end{bmatrix} + \omega \begin{bmatrix} \underline{G} & \underline{0} \\ \underline{0} & \underline{0} \end{bmatrix} \begin{bmatrix} \underline{\Lambda}_{dq0} \\ \underline{\Lambda}_{fkdkq} \end{bmatrix} \tag{9.160}$$

Accordingly, the state-space model formulation in the flux linkage form of Eq. (9.160) can be rewritten as follows:

$$\begin{bmatrix} \dot{\underline{\Lambda}}_{dq0} \\ \dot{\underline{\Lambda}}_{fkdkq} \end{bmatrix} = \left[\omega \begin{bmatrix} \underline{G} & \underline{0} \\ \underline{0} & \underline{0} \end{bmatrix} - \begin{bmatrix} \underline{R}_{ss} & \underline{0} \\ \underline{0} & \underline{R}_{rr} \end{bmatrix} \underline{L}^{-1}_{dq0fkdkq} \right] \begin{bmatrix} \underline{\Lambda}_{dq0} \\ \underline{\Lambda}_{fkdkq} \end{bmatrix} + \begin{bmatrix} \underline{V}_{dq0} \\ \underline{V}_{fkdkq} \end{bmatrix} \tag{9.161}$$

where $\underline{G}$ is as defined in Eq. (9.121), that is,

$$\underline{G} = \begin{bmatrix} 0 & 1 & 0 \\ -1 & 0 & 0 \\ 0 & 0 & 0 \end{bmatrix} \tag{9.162}$$

Accordingly, in expanded sixth-order state-space $dq0$ flux linkage form, Eq. (9.161) yields the following:

$$\begin{bmatrix} \dot{\lambda}_d \\ \dot{\lambda}_q \\ \dot{\lambda}_0 \\ \dot{\lambda}_f \\ \dot{\lambda}_{kd} \\ \dot{\lambda}_{kq} \end{bmatrix} = \left\{ \begin{bmatrix} 0 & \omega & 0 & 0 & 0 & 0 \\ -\omega & 0 & 0 & 0 & 0 & 0 \\ 0 & 0 & 0 & 0 & 0 & 0 \\ 0 & 0 & 0 & 0 & 0 & 0 \\ 0 & 0 & 0 & 0 & 0 & 0 \\ 0 & 0 & 0 & 0 & 0 & 0 \end{bmatrix} - \begin{bmatrix} r_s & 0 & 0 & 0 & 0 & 0 \\ 0 & r_s & 0 & 0 & 0 & 0 \\ 0 & 0 & r_s & 0 & 0 & 0 \\ 0 & 0 & 0 & r_f & 0 & 0 \\ 0 & 0 & 0 & 0 & r_{kd} & 0 \\ 0 & 0 & 0 & 0 & 0 & r_{kq} \end{bmatrix} \cdot \begin{bmatrix} L_{dd} & 0 & 0 & L_{afm} & L_{akdm} & 0 \\ 0 & L_{qq} & 0 & 0 & 0 & L_{akqm} \\ 0 & 0 & L_{00} & 0 & 0 & 0 \\ \frac{3}{2}L_{afm} & 0 & 0 & L_{ff} & L_{fkd} & 0 \\ \frac{3}{2}L_{akdm} & 0 & 0 & L_{fkd} & L_{kdkd} & 0 \\ 0 & \frac{3}{2}L_{akqm} & 0 & 0 & 0 & L_{kqkq} \end{bmatrix}^{-1} \right\} \begin{bmatrix} \lambda_d \\ \lambda_q \\ \lambda_0 \\ \lambda_f \\ \lambda_{kd} \\ \lambda_{kq} \end{bmatrix} + \begin{bmatrix} v_d \\ v_q \\ v_0 \\ v_f \\ 0 \\ 0 \end{bmatrix} \tag{9.163}$$

Again, Eq. (9.163) is the $dq0$ state-space formulation of the synchronous machine in the flux linkage frame of reference, which is the counter-part of Eqs. (9.132) and (9.143) in the current frame of reference.

9.8 Power and Torque Formulations for a Synchronous Machine in the *DQ0* Frame of Reference and *ABC* Frame of Reference

The instantaneous power, p_t, at the *abc* terminals of a synchronous machine can be expressed in the *abc* frame of reference as follows:

$$p_t = i_a v_a + i_b v_b + i_c v_c \tag{9.164}$$

or, in linear algebra vector form, one can express, P_t, as follows:

$$p_t = \begin{bmatrix} i_a & i_b & i_c \end{bmatrix} \begin{bmatrix} v_a \\ v_b \\ v_c \end{bmatrix} \tag{9.165}$$

That is,

$$p_t = \underline{I}^t_{abc} \cdot \underline{V}_{abc} \tag{9.166}$$

However, from the definitions given in Eqs. (9.64) and (9.67), one can rewrite Eq. (9.166) as follows:

$$p_t = (\underline{T}^{-1} \cdot \underline{I}_{dq0})^t (\underline{T}^{-1} \cdot \underline{V}_{dq0}) \tag{9.167}$$

That is,

$$p_t = \underline{I}^t_{dq0} \cdot (\underline{T}^{-1})^t \cdot (\underline{T}^{-1}) \cdot \underline{V}_{dq0} \tag{9.168}$$

Now, it will be left to the student to show that,

$$(\underline{T}^{-1})^t \cdot (\underline{T}^{-1}) = \begin{bmatrix} 3/2 & 0 & 0 \\ 0 & 3/2 & 0 \\ 0 & 0 & 3 \end{bmatrix} \tag{9.169}$$

Substituting for the matrix product, $(\underline{T}^{-1})^t \cdot (\underline{T}^{-1})$, into Eq. (9.168) yields the following for p_t:

$$p_t = \begin{bmatrix} i_d & i_q & i_0 \end{bmatrix} \begin{bmatrix} 3/2 & 0 & 0 \\ 0 & 3/2 & 0 \\ 0 & 0 & 3 \end{bmatrix} \begin{bmatrix} v_d \\ v_q \\ v_0 \end{bmatrix} \tag{9.170}$$

Hence, it follows that

$$p_t = (3/2) i_d v_d + (3/2) i_q v_q + 3 i_0 v_0 \tag{9.171}$$

That is,

$$p_t = (3/2)[i_d v_d + i_q v_q + 2i_0 v_0] \tag{9.172}$$

Now, regarding the developed electromagnetic/electromechanical torque, we need to seek the rotational voltages in a synchronous machine. These are the rotational voltage components, e_d and e_q, in the $dq0$ frame model given earlier in the second matrix term of Eq. (9.132), from which one can write the following:

$$e_d = -\omega(L_{sa} + L_{ma} - (3/2)L_v)i_q - \omega L_{akqm} i_{kq} \tag{9.173}$$

That is,

$$e_d = -\omega L_{qq} i_q - \omega L_{akqm} i_{kq} \tag{9.174}$$

However, recall that the q-axis flux linkage, λ_q, is given by

$$\lambda_q = L_{qq} i_q + L_{akqm} i_{kq} \tag{9.175}$$

It follows that e_d in Eq. (9.174) can be rewritten as follows:

$$e_d = -\omega(L_{qq} i_q + L_{akqm} i_{kq}) = -\omega\lambda_q \tag{9.176}$$

Also, based on the second matrix term of Eq. (9.132), one can write the following for the rotational voltage term, e_q:

$$e_q = \omega(L_{sa} + L_{ma} + (3/2)L_v)i_d + \omega L_{afm} i_f + \omega L_{akdm} i_{kd} \tag{9.177}$$

That is,

$$e_q = \omega L_{dd} i_d + \omega L_{afm} i_f + \omega L_{akdm} i_{kd} \tag{9.178}$$

However, recall that the d-axis flux linkage, λ_d, is given by

$$\lambda_d = L_{dd} i_d + L_{afm} i_f + L_{akdm} i_{kd} \tag{9.179}$$

It follows that e_q in Eq. (9.178) can be rewritten as follows:

$$e_q = \omega(L_{dd} i_d + L_{afm} i_f + L_{akdm} i_{kd}) = \omega\lambda_d \tag{9.180}$$

and, in the $dq0$ frame, the zero-sequence rotational voltage component, e_0, is zero. That is,

$$e_0 = 0 \tag{9.181}$$

Accordingly, the developed instantaneous electromagnetic/electromechanical power, P_{em}, is given by

$$P_{em} = (3/2)[i_d e_d + i_q e_q + 2i_0 e_0] \tag{9.182}$$

Substituting for e_d, e_q, and e_0 from Eqs. (9.176), (9.180), and (9.181), respectively, into Eq. (9.182), one can obtain the following for the developed instantaneous electromagnetic/electromechanical power, P_{em}:

$$P_{em} = (3/2)[i_d(-\omega\lambda_q) + i_q(\omega\lambda_d) + 2i_0(0)] \quad (9.183)$$

That is,

$$P_{em} = (3/2)\omega[\lambda_d i_q - \lambda_q i_d] \quad (9.184)$$

In order to obtain the developed torque, T_{em}, one needs to divide the P_{em} term in Eq. (9.184) by the mechanical speed, in mechanical radians per second, ω_m, where

$$\omega_m = \omega/(p/2) \quad (9.185)$$

where p is the number of poles in the machine. Hence, the torque, T_{em}, can be expressed using Eqs. (9.184) and (9.185) as follows:

$$T_{em} = P_{em}/\omega_m = (p/2)(3/2)[\lambda_d i_q - \lambda_q i_d] \quad (9.186)$$

where p is the number of poles in a synchronous machine.

Here, it is very important to point out that the developed torque Eq. (9.186) forms the basis for programming the controller of any power electronic drive energizing a poly-phase AC machine, in which the main objective of the switching of the power-electronic switches in such a drive is to maximize the q-axis component of the stator armature current component, i_q, while minimizing the d-axis component of the armature current, i_d, or even driving this component toward a value as close to zero as possible.

It is important to recall here that λ_q is given in Eq. (9.175), and λ_d is given in Eq. (9.179). Substituting for λ_q and λ_d from these equations into the developed torque, T_{em}, Eq. (9.186), yields the following:

$$T_{em} = P_{em}/\omega_m = (p/2)(3/2)[L_{dd} i_d i_q + L_{afm} i_f i_q + L_{akdm} i_{kd} i_q - L_{qq} i_q i_d - L_{akqm} i_{kq} i_d] \quad (9.187)$$

Rearranging, the developed torque, T_{em}, can be written as follows:

$$T_{em} = P_{em}/\omega_m = (p/2)(3/2)[L_{afm} i_f i_q + (L_{dd} - L_{qq}) i_q i_d + L_{akdm} i_{kd} i_q - L_{akqm} i_{kq} i_d] \quad (9.188)$$

Here, it should be pointed out that the $[(p/2)(3/2)L_{afm} i_f i_q]$ term represents the so-called synchronous (field) torque term, while the $[(p/2)(3/2)(L_{dd} - L_{qq}) i_q i_d]$ term represents the so-called synchronous reluctance torque term, which exists as long as $L_{dd} \neq L_{qq}$ due to rotor geometrically caused, or magnetic saturation-caused, saliency exists, and as a rotor approaches ideal cylindrical conditions, under which, $L_{dd} = L_{qq}$, this synchronous reluctance torque term will vanish. That is, it becomes equal to zero. The $[(p/2)(3/2)(L_{akdm}i_{kd}i_q - L_{akqm}i_{kq}i_d)]$ term represents the damper (or damping) winding term, which exists in both synchronous and induction machines.

At this stage, we will proceed to the derivation of the synchronous machine's phasor and space-vector diagrams in the $dq0$ frame of reference.

9.9 Derivation of the Synchronous Machine Phasor and Space-Vector Diagrams in the *DQ*0 Frame of Reference

Referring to the $dq0$ state-space model of synchronous machines given in Eq. (9.132), one can write the following for v_d and v_q:**Equation Section (Next)**

$$v_d = r_s i_d - \omega L_{qq} i_q - \omega L_{akqm} i_{kq} + L_{dd}(di_d/dt) + L_{afm}(di_f/dt) + L_{akdm}(di_{kd}/dt) \tag{9.189}$$

and

$$v_q = r_s i_q + \omega L_{dd} i_d + \omega L_{afm} i_f + \omega L_{akdm} i_{kd} + L_{qq}(di_q/dt) + L_{akqm}(di_{kq}/dt) \tag{9.190}$$

However, under steady-state conditions at synchronous speed, the damper winding currents, $i_{kd} = 0$ and $i_{kq} = 0$, and, hence, $(di_{kd}/dt) = 0$, $(di_{kq}/dt) = 0$, i_f = constant, and, therefore, $(di_f/dt) = 0$. Therefore, Eqs. (9.189) and (9.190) can be rewritten as follows:

$$v_d = r_s i_d - \omega L_{qq} i_q + L_{dd}(di_d/dt) \tag{9.191}$$

and

$$v_q = r_s i_q + \omega L_{dd} i_d + L_{qq}(di_q/dt) + \omega L_{afm} i_f \tag{9.192}$$

Also, under state-state conditions at synchronous speed, ω, in electrical radians per second, the d-axis position angle, σ, at any time, t, can be expressed as follows:

$$\sigma = \sigma_0 + \omega t \tag{9.193}$$

where σ_0 is the d-axis position angle at time, $t = 0$. Also, one must recall here that

$$\underline{I}_{dq0} = \underline{T} \cdot \underline{I}_{abc} \tag{9.194}$$

$$\underline{I}_{abc} = \underline{T}^{-1} \cdot \underline{I}_{dq0} \tag{9.195}$$

$$\underline{V}_{dq0} = \underline{T} \cdot \underline{V}_{abc} \tag{9.196}$$

$$\underline{V}_{abc} = \underline{T}^{-1} \cdot \underline{V}_{dq0} \tag{9.197}$$

and $\underline{T}$ and $\underline{T}^{-1}$ are as given in Section 9.4, Eqs. (9.60) and (9.61), respectively. Here, also, under steady-state, we define the terminal voltage vector, $\underline{V}_{abc}$, as follows:

$$\underline{V}_{abc} = \begin{bmatrix} \sqrt{2}V\cos(\omega t) \\ \sqrt{2}V\cos(\omega t - 2\pi/3) \\ \sqrt{2}V\cos(\omega t - 4\pi/3) \end{bmatrix} \tag{9.198}$$

Hence, applying the $dq0$ transformation, one can write the following:

$$\underline{V}_{dq0} = (2/3)\begin{bmatrix} \cos(\sigma) & \cos(\sigma - (2\pi/3)) & \cos(\sigma - (4\pi/3)) \\ -\sin(\sigma) & -\sin(\sigma - (2\pi/3)) & -\sin(\sigma - (4\pi/3)) \\ (1/2) & (1/2) & (1/2) \end{bmatrix}\begin{bmatrix} \sqrt{2}V\cos(\omega t) \\ \sqrt{2}V\cos(\omega t - 2\pi/3) \\ \sqrt{2}V\cos(\omega t - 4\pi/3) \end{bmatrix} \tag{9.199}$$

The student should find out by working the details of the algebraic and trigonometric operations in Eq. (9.199) that the following is the resulting $dq0$ terminal voltage of the machine:

$$\underline{V}_{dq0} = \begin{bmatrix} v_d \\ v_q \\ v_0 \end{bmatrix} = \begin{bmatrix} \sqrt{2}V\cos\sigma_0 \\ -\sqrt{2}V\sin\sigma_0 \\ 0 \end{bmatrix} \tag{9.200}$$

Here also, the abc currents vector, $\underline{I}_{abc}$, is as follows:

$$\underline{I}_{abc} = \begin{bmatrix} \sqrt{2}I\cos(\omega t + \psi) \\ \sqrt{2}I\cos(\omega t + \psi - 2\pi/3) \\ \sqrt{2}I\cos(\omega t + \psi - 4\pi/3) \end{bmatrix} \tag{9.201}$$

where ψ is the phase angle of the current of phase (a) with respect to the terminal voltage of this phase.

Accordingly, applying the $dq0$ transformation, one can write

$$\underline{I}_{dq0} = (2/3)\begin{bmatrix} \cos(\sigma) & \cos(\sigma - (2\pi/3)) & \cos(\sigma - (4\pi/3)) \\ -\sin(\sigma) & -\sin(\sigma - (2\pi/3)) & -\sin(\sigma - (4\pi/3)) \\ (1/2) & (1/2) & (1/2) \end{bmatrix} \times \begin{bmatrix} \sqrt{2}I\cos(\omega t + \psi) \\ \sqrt{2}I\cos(\omega t + \psi - 2\pi/3) \\ \sqrt{2}I\cos(\omega t + \psi - 4\pi/3) \end{bmatrix} \tag{9.202}$$

Again, the student should find out by working the details of the algebraic and trigonometric operations in Eq. (9.202) that the following is the resulting $dq0$ vector of armature currents of the machine:

$$\underline{I}_{dq0} = \begin{bmatrix} i_d \\ i_q \\ i_0 \end{bmatrix} = \begin{bmatrix} \sqrt{2}I\cos(\sigma_0 - \psi) \\ -\sqrt{2}I\sin(\sigma_0 - \psi) \\ 0 \end{bmatrix} \quad (9.203)$$

Hence, it follows that

$$\dot{i}_d = (di_d/dt) = d(\sqrt{2}I\cos(\sigma_0 - \psi))/dt = 0 \quad (9.204)$$

and

$$\dot{i}_q = (di_q/dt) = d(-\sqrt{2}I\sin(\sigma_0 - \psi))/dt = 0 \quad (9.205)$$

Substituting Eqs. (9.204) and (9.205) for (di_d/dt) and (di_q/dt), respectively, into Eqs. (9.191) and (9.192) yields the following, respectively, for v_d and v_q:

$$v_d = r_s i_d - \omega L_{qq} i_q \quad (9.206)$$

and

$$v_q = r_s i_q + \omega L_{dd} i_d + \omega L_{afm} i_f \quad (9.207)$$

Furthermore, substituting for i_d and i_q from Eq. (9.203) into Eqs. (9.206) and (9.207) yields the following for the two voltage components, v_d and v_q:

$$v_d = r_s[\sqrt{2}I\cos(\sigma_0 - \psi)] - \omega L_{qq}[-\sqrt{2}I\sin(\sigma_0 - \psi)]$$

That is,

$$v_d = \sqrt{2}Ir_s\cos(\sigma_0 - \psi) + \sqrt{2}I(\omega L_{qq})\sin(\sigma_0 - \psi) \quad (9.208)$$

and

$$v_q = r_s[-\sqrt{2}I\sin(\sigma_0 - \psi)] + \omega L_{dd}[\sqrt{2}I\cos(\sigma_0 - \psi)] + \omega L_{afm} I_f$$

That is,

$$v_q = -\sqrt{2}Ir_s\sin(\sigma_0 - \psi) + \sqrt{2}I(\omega L_{dd})\cos(\sigma_0 - \psi) + \omega L_{afm} I_f \quad (9.209)$$

where the field winding current, I_f, is given by the following:

$$I_f = (V_f/r_f) \quad (9.210)$$

where V_f is the terminal DC voltage applied by the exciter to the terminals of the field excitation winding, the resistance of which is r_f.

Now, we must recall that $\underline{V}_{abc} = \underline{T}^{-1} \cdot \underline{V}_{dq0}$, hence,

$$\begin{bmatrix} v_a \\ v_b \\ v_c \end{bmatrix} = \begin{bmatrix} \cos\sigma & -\sin\sigma & (1/2) \\ \cos(\sigma - 2\pi/3) & -\sin(\sigma - 2\pi/3) & (1/2) \\ \cos(\sigma - 4\pi/3) & -\sin(\sigma - 4\pi/3) & (1/2) \end{bmatrix} \begin{bmatrix} v_d \\ v_q \\ 0 \end{bmatrix} \quad (9.211)$$

Equation (9.211) yields the following for v_a:

$$v_a = v_d \cos\sigma - v_q \sin\sigma \tag{9.212}$$

Substituting for v_d and v_q from Eqs. (9.208) and (9.209), respectively, into Eq. (9.212), and subsequent algebraic as well as trigonometric manipulations, one can write the following:

$$\begin{aligned} v_a = \sqrt{2}V\cos(\omega t) &= r_s(\sqrt{2}I)\cos(\omega t+\psi) + (\omega L_{qq})[\sqrt{2}I\sin(\sigma_0-\psi)]\cos(\omega t+\sigma_0)+ \\ &(\omega L_{dd})[\sqrt{2}I\cos(\sigma_0-\psi)]\cos(\omega t+\sigma_0+\pi/2) + (\omega L_{afm})[I_f]\cos(\omega t+\sigma_0+\pi/2) \end{aligned} \tag{9.213}$$

Applying Euler's well-known formulation states that: $[\varepsilon^{j\theta} = \cos\theta + j\sin\theta]$ and, hence, $[\varepsilon^{j(\omega t+\beta)} = \varepsilon^{j\omega t}\cdot\varepsilon^{j\beta} = \cos(\omega t+\beta) + j\sin(\omega t+\beta)]$, it follows that any term of the form $[k\cos(\omega t+\beta)]$ can be rewritten as $[k\,\mathrm{Re}\{\varepsilon^{j(\omega t+\beta)}\}]$. Hence, applying this terminology to Eq. (9.213), this equation can be rewritten as follows:

$$\begin{aligned} v_a = \sqrt{2}VRe[\varepsilon^{j\omega t}] &= r_s(\sqrt{2}I)[\varepsilon^{j\omega t}\cdot\varepsilon^{j\psi}] + (\omega L_{qq})[\sqrt{2}I\sin(\sigma_0-\psi)]\left[\varepsilon^{j\omega t}\cdot\varepsilon^{j\sigma_0}\right] \\ &+ (\omega L_{dd})[\sqrt{2}I\cos(\sigma_0-\psi)]\left[\varepsilon^{j\omega t}\cdot\varepsilon^{j(\sigma_0+\pi/2)}\right] + (\omega L_{afm})[I_f]\left[\varepsilon^{j\omega t}\cdot\varepsilon^{j(\sigma_0+\pi/2)}\right] \end{aligned} \tag{9.214}$$

Meanwhile, dropping the common factor $(\sqrt{2})$ from both sides of Eq. (9.214) and recognizing that in complex algebraic notation, $\varepsilon^{j(\pi/2)} = j$, Eq. (9.214) can be rewritten as follows:

$$\begin{aligned} V(\varepsilon^{j\omega t}) &= r_s I\varepsilon^{j\psi}(\varepsilon^{j\omega t}) + (\omega L_{qq})[I\sin(\sigma_0-\psi)]\varepsilon^{j\sigma_0}(\varepsilon^{j\omega t}) \\ &+ j(\omega L_{dd})[I\cos(\sigma_0-\psi)]\varepsilon^{j\sigma_0}(\varepsilon^{j\omega t}) + j(\omega L_{afm})[I_f/\sqrt{2}]\varepsilon^{j\sigma_0}(\varepsilon^{j\omega t}) \end{aligned} \tag{9.215}$$

Dropping the common factor $(\varepsilon^{j\omega t})$ from both left- and right-hand sides of Eq. (9.215), one can write the following:

$$\begin{aligned} V &= r_s(I\varepsilon^{j\psi}) + (\omega L_{qq})[I\sin(\sigma_0-\psi)]\left(\varepsilon^{j\sigma_0}\right) \\ &+ j(\omega L_{dd})[I\cos(\sigma_0-\psi)]\left(\varepsilon^{j\sigma_0}\right) + j(\omega L_{afm}I_f/\sqrt{2})\left(\varepsilon^{j\sigma_0}\right) \end{aligned} \tag{9.216}$$

Recalling that in phasor notation, $\left(\varepsilon^{j\sigma_0}\right) = \angle\sigma_0$ and $j = \angle\pi/2$ and pure real numbers have an inherent phase angle equal to zero, that is, $\angle 0°$, Eq. (9.216) can be rewritten in root mean square (RMS) phasor notation as follows:

$$\begin{aligned} \overline{V}_{at} = V\angle 0° &= r_s I\angle\psi + (\omega L_{qq})[I\sin(\sigma_0-\psi)]\angle\sigma_0 \\ &+ (\omega L_{dd})[I\cos(\sigma_0-\psi)]\angle(\sigma_0+\pi/2) + (\omega L_{afm}I_f/\sqrt{2})\angle(\sigma_0+\pi/2) \end{aligned} \tag{9.217}$$

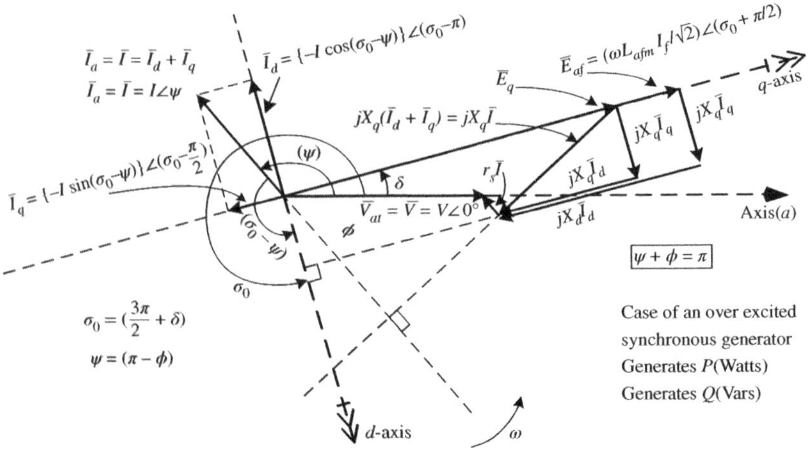

Figure 9.30 The D–Q phasor diagram of an overexcited synchronous generator.

Equation (9.217) can be rewritten in phasor notation as follows:

$$\begin{aligned}\overline{V}_{at} = V\angle 0° &= r_s I\angle\psi + j\omega L_{qq}\{[I\sin(\sigma_0-\psi)]\angle(\sigma_0-\pi/2)\}\\ &+ \omega L_{dd}[-I\cos(\sigma_0-\psi)]\angle(\sigma_0-\pi) + (\omega L_{afm} I_f/\sqrt{2})\angle(\sigma_0+\pi/2)\end{aligned} \tag{9.218}$$

Examining the phasor diagram corresponding to Eq. (9.218), shown in Figure 9.30, for an overexcited synchronous generator, one can write the armature current phasor, $\overline{I}_a$, as follows:

$$\overline{I}_a = I\angle\psi \tag{9.219}$$

and its q-axis phasor component, $\overline{I}_q$, is as follows:

$$\overline{I}_q = [I\sin(\sigma_0-\psi)]\angle(\sigma_0-\pi/2) \tag{9.220}$$

and its d-axis phasor component, $\overline{I}_d$, is as follows:

$$\overline{I}_d = [-I\cos(\sigma_0-\psi)]\angle(\sigma_0-\pi) \tag{9.221}$$

and the EMF phasor, $\overline{E}_{af}$, induced in the armature by the field excitation winding current, I_f, is as follows:

$$\overline{E}_{af} = [\omega L_{afm} I_f/\sqrt{2}]\angle(\sigma_0+\pi/2) \tag{9.222}$$

Furthermore, we define here $j\omega L_{qq} = jX_q$ as the q-axis synchronous reactance and $j\omega L_{dd} = jX_d$ as the d-axis synchronous reactance, thus, we can rewrite Eq. (9.218) in the following consumer (load) notation phasor form:

$$\overline{V}_{at} = V\angle 0° = r_s\overline{I} + jX_q\overline{I}_q + jX_d\overline{I}_d + \overline{E}_{af} \tag{9.223}$$

In Figure 9.30, it should be pointed out that the tip of the phasor, $\overline{E}_q$, always locates a point connecting it to the origin of the phasor diagram, where $\overline{E}_q$ is related to the reference terminal voltage phasor, $\overline{V}_{at} = \overline{V}$, by the following relationship:

$$\overline{V}_{at} = \overline{V} = V\angle 0^\circ = \overline{E}_q + r_s\overline{I} + jX_q(\overline{I}_d + \overline{I}_q) \tag{9.224}$$

However, here, $\overline{I} = \overline{I}_a = (\overline{I}_d + \overline{I}_q)$, and, hence,

$$\overline{V}_{at} = \overline{V} = V\angle 0^\circ = \overline{E}_q + r_s\overline{I} + jX_q\overline{I} \tag{9.225}$$

Therefore, the phasor relationship expressed in Eq. (9.225) is a key equation in helping locate the q-axis, and, hence, the so-called power angle, δ, subtended between the terminal voltage, $\overline{V}_{at} = \overline{V}$, and the EMF, $\overline{E}_{af}$, and, hence, the location of the d-axis, where the q-axis leads the d-axis, by $(\pi/2)$ electrical radians or $90^\circ e$ in the counterclockwise direction, and, hence, the angle, σ_0, can be written as follows:

$$\sigma_0 = (\delta + 3\pi/2) \tag{9.226}$$

and phase angle, ψ, of armature current phasor, $\overline{I}_a = \overline{I}$, is related to the power factor, PF, by

$$PF = \cos(\pi - \psi) \tag{9.227}$$

where $(\pi - \psi)$ is the angle, ϕ, see the phasor diagrams of Figures 9.30 and 9.31. In Figure 9.31, the spatial location of the magnetic axes of the windings of phases (A), (B), and (C), that is, the A-axis, B-axis, and C-axis, is clearly designated.

We now proceed to the derivation of the complex power, $\overline{S}$, associated with an overexcited synchronous generator, utilizing the geometry of the $dq0$ phasor diagrams of Figures 9.30 and 9.31. Here, one can write the following:

$$\overline{S} = 3\overline{V}_{at}\overline{I}_a^* = 3\overline{V}_{at}\overline{I}^* = 3\overline{V}_{at}\left(\overline{I}_d^* + \overline{I}_q^*\right) = P + jQ \tag{9.228}$$

since,

$$\overline{V}_{at} = V_{at}\angle 0^\circ = V\angle 0^\circ \tag{9.229}$$

the complex power, $\overline{S}$, can be rewritten from Eqs. (9.228) and (9.229) as follows:

$$\overline{S} = 3V_{at}\left(\overline{I}_d^* + \overline{I}_q^*\right) \tag{9.230}$$

However, from the $dq0$ phasor diagrams of Figures 9.30 and 9.31, it follows that

$$\overline{I}_d = I_d\angle(\delta + \pi/2) = I_d\cos(\delta + \pi/2) + jI_d\sin(\delta + \pi/2) \tag{9.231}$$

That is,

$$\overline{I}_d = -I_d\sin\delta + jI_d\cos\delta \tag{9.232}$$

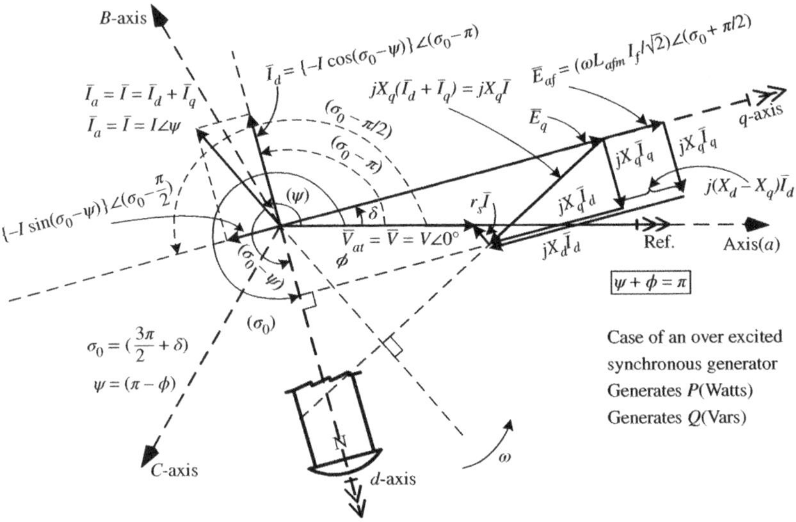

Figure 9.31 The D–Q phasor diagram of an overexcited synchronous generator including the spatial location of the magnetic axes of phases A, B, and C.

Hence,

$$\overline{I}_d^* = -I_d \sin\delta - jI_d \cos\delta \tag{9.233}$$

Also, here, we have,

$$\overline{I}_q = I_q\angle(\pi + \delta) = I_q \cos(\pi + \delta) + jI_q \sin(\pi + \delta) = -I_q \cos\delta - jI_q \sin\delta \tag{9.234}$$

Hence, it follows that

$$\overline{I}_q^* = -I_q \cos\delta + jI_q \sin\delta \tag{9.235}$$

Substituting for $\overline{I}_d^*$ and $\overline{I}_q^*$ from Eqs. (9.233) and (9.235) into Eq. (9.230), it follows that the complex power, $\overline{S}$, can be written as follows:

$$\overline{S} = -(3V_{at}I_d \sin\delta + 3V_{at}I_q \cos\delta) - j(3V_{at}I_d \cos\delta - 3V_{at}I_q \sin\delta) \tag{9.236}$$

However, from the geometries of the phasor diagrams of Figures 9.30 and 9.31, one can see that

$$X_d I_d = E_{af} - V_{at} \cos\delta \tag{9.237}$$

hence,

$$I_d = (E_{af} - V_{at} \cos\delta)/X_d \tag{9.238}$$

and that

$$X_q I_q = V_{at} \sin \delta \tag{9.239}$$

hence,

$$I_q = (V_{at} \sin \delta)/X_q \tag{9.240}$$

Substituting into Eq. (9.236) of the complex power, $\overline{S} = P + jQ$, for I_d and I_q from Eqs. (9.238) and (9.240), respectively, one can show that

$$P = -\left[(3V_{at}E_{af}/X_d)\sin \delta + \left(3V_{at}^2(X_d - X_q)/2X_dX_q\right)\sin 2\delta\right] \tag{9.241}$$

and

$$Q = -\left[(3V_{at}E_{af}/X_d)\cos \delta - \left(3V_{at}^2\right)\{(\cos^2\delta/X_d) + (\sin^2\delta/X_q)\}\right] \tag{9.242}$$

The P and Q negative signs mean generating watts and vars, respectively, as is always the case for most power plant synchronous generators operating in an overexcited mode. Next, we discuss using the *d-q* phasor diagram to obtain initial conditions for state-space transient/dynamic modeling of synchronous machines.

9.10 Initial Conditions in Transient Synchronous Machinery State-Space Time-Domain Simulations

Based on the *dq*0 phasor-space-vector diagrams of Figures 9.30 and 9.31 drawn at time, $t = 0$, one can write the following under overexcited synchronous generator load conditions:

In the *dq*0 frame of reference:

$$\sigma_0 = \left(\frac{3\pi}{2} + \delta\right) \text{ and } \overline{E}_{af} = E_{af}\angle\delta$$

$$\psi = (\pi - \phi)$$

$$\begin{bmatrix} i_d \\ i_q \\ i_0 \\ i_f \\ i_{kd} \\ i_{kq} \end{bmatrix}_{t\le 0} = \begin{bmatrix} \sqrt{2}I\cos(\sigma_0 - \psi) \\ -\sqrt{2}I\sin(\sigma_0 - \psi) \\ 0 \\ (\sqrt{2}E_{af}/\omega L_{afm}) \\ 0 \\ 0 \end{bmatrix} \tag{9.243}$$

$$\begin{bmatrix} v_d \\ v_q \\ v_0 \\ v_f \\ v_{kd} \\ v_{kq} \end{bmatrix}_{t<0} = \begin{bmatrix} \sqrt{2}V\cos\sigma_0 \\ -\sqrt{2}V\sin\sigma_0 \\ 0 \\ r_f(\sqrt{2}E_{af}/\omega L_{afm}) \\ 0 \\ 0 \end{bmatrix} \tag{9.244}$$

and

In the abc frame of reference:

$$\sigma_0 = \left(\frac{3\pi}{2} + \delta\right) \text{ and } \overline{E}_{af} = E_{af}\angle\delta$$

$$\psi = (\pi - \phi)$$

$$\begin{bmatrix} i_a \\ i_b \\ i_c \\ i_f \\ i_{kd} \\ i_{kq} \end{bmatrix}_{t=0} = \begin{bmatrix} \sqrt{2}I\cos(\psi) \\ \sqrt{2}I\cos(\psi - 2\pi/3) \\ \sqrt{2}I\cos(\psi - 4\pi/3) \\ (\sqrt{2}E_{af}/\omega L_{afm}) \\ 0 \\ 0 \end{bmatrix} \tag{9.245}$$

$$\begin{bmatrix} v_a \\ v_b \\ v_c \\ v_f \\ v_{kd} \\ v_{kq} \end{bmatrix}_{t<0} = \begin{bmatrix} \sqrt{2}V\cos(\omega t) \\ \sqrt{2}V\cos(\omega t - 2\pi/3) \\ \sqrt{2}V\cos(\omega t - 4\pi/3) \\ r_f(\sqrt{2}E_{af}/\omega L_{afm}) \\ 0 \\ 0 \end{bmatrix} \tag{9.246}$$

and for a generator operating under no-load, the initial conditions for the $dq0$ frame and abc frame are as follows:

$$\sigma_0 = \frac{3\pi}{2} \text{ and } \overline{E}_{af} = V\angle 0^\circ \ \delta = 0^\circ$$

$dq0$ frame:

$$\begin{bmatrix} i_d \\ i_q \\ i_0 \\ i_f \\ i_{kd} \\ i_{kq} \end{bmatrix}_{t\leq 0} = \begin{bmatrix} 0 \\ 0 \\ 0 \\ \sqrt{2}V/\omega L_{afm} \\ 0 \\ 0 \end{bmatrix} \tag{9.247}$$

$$v_f\Big|_{t<0} = r_f(\sqrt{2}V/\omega L_{afm}) \tag{9.248}$$

abc frame:

$$\begin{bmatrix} i_a \\ i_b \\ i_c \\ i_f \\ i_{kd} \\ i_{kq} \end{bmatrix}_{t\le 0} = \begin{bmatrix} 0 \\ 0 \\ 0 \\ \sqrt{2}V/\omega L_{afm} \\ 0 \\ 0 \end{bmatrix} \tag{9.249}$$

$$v_f\Big|_{t<0} = r_f(\sqrt{2}V/\omega L_{afm}) \tag{9.250}$$

Next, we will deal with modeling of PM-excited synchronous machines.

9.11 Synchronous Machine ABC Frame State-Space Models with Permanent Magnet Excitations

Consider the case of a PM machine in which the damping effect is negligible, and the permanent magnets are surface mounted on the rotor, Figure 9.32a,b, while the armature is mounted on the stator. For now, treating the permanent magnet excitation as a regular electromagnet (field winding) excitation, based on Figures 9.32a,b and 9.33, the following state-space model can be written, which governs the dynamics of the machine:

$$\begin{bmatrix} v_a \\ v_b \\ v_c \\ \hline v_f \end{bmatrix} = \left[\begin{array}{ccc|c} r_a & 0 & 0 & 0 \\ 0 & r_b & 0 & 0 \\ 0 & 0 & r_c & 0 \\ \hline 0 & 0 & 0 & r_f \end{array}\right] \begin{bmatrix} i_a \\ i_b \\ i_c \\ \hline i_f \end{bmatrix} + \frac{d}{dt}\left\{ \left[\begin{array}{ccc|c} L_{aa} & L_{ab} & L_{ac} & L_{af} \\ L_{ba} & L_{bb} & L_{bc} & L_{bf} \\ L_{ca} & L_{cb} & L_{cc} & L_{cf} \\ \hline L_{fa} & L_{fb} & L_{fc} & L_{ff} \end{array}\right] \begin{bmatrix} i_a \\ i_b \\ i_c \\ \hline i_f \end{bmatrix} \right\} \tag{9.251}$$

Equation (9.251) can be rewritten in expanded form as follows, considering only the armature circuit:

$$\begin{bmatrix} v_a \\ v_b \\ v_c \end{bmatrix} = \begin{bmatrix} r_a & 0 & 0 \\ 0 & r_b & 0 \\ 0 & 0 & r_c \end{bmatrix} \begin{bmatrix} i_a \\ i_b \\ i_c \end{bmatrix} + \frac{d}{dt}\left\{ \begin{bmatrix} L_{aa} & L_{ab} & L_{ac} \\ L_{ba} & L_{bb} & L_{bc} \\ L_{ca} & L_{cb} & L_{cc} \end{bmatrix} \begin{bmatrix} i_a \\ i_b \\ i_c \end{bmatrix} \right\} + \frac{d}{dt}\begin{bmatrix} L_{af}\cdot i_f \\ L_{bf}\cdot i_f \\ L_{cf}\cdot i_f \end{bmatrix} \tag{9.252}$$

As will be shown later, the field coil (winding), which in this case is a permanent magnet, can be replaced with an equivalent coil energized from a constant current source. Hence, there will be no dynamic (differential) equation corresponding to a field winding. This means that there are only three state variables for such a machine. These variables are the a, b, and c phase currents.

There are two types of permanent magnet machine designs. One such design has a radially oriented magnet structure, that is the flow of flux in the permanent

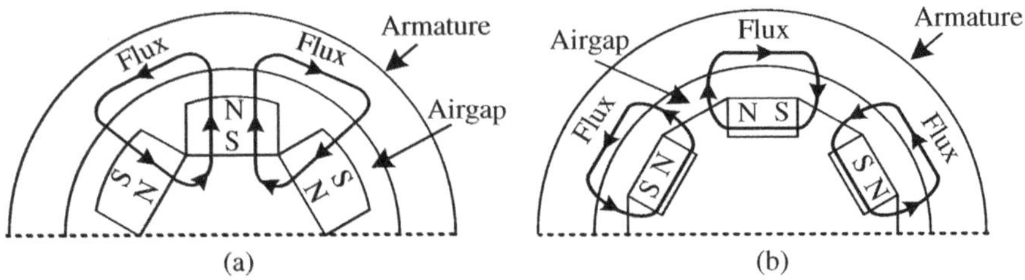

Figure 9.32 Radial and tangential (circumferential) surface-mounted permanent magnet machine designs. (a) Radial permanent magnet design. (b) Tangential permanent magnet design.

magnet is in the radial direction. The other design has a tangentially (or circumferentially) oriented magnet structure, that is the flow of flux in the permanent magnets is in the tangential (or circumferential) direction. Figure 9.32a,b depicts both types.

In the radially oriented permanent magnet machine design, the effect of saliency may be neglected because of the approximately equal large airgaps opposite the d-axis and q-axis, respectively. Notice that permanent magnets that are most commonly used are those made of rare earth (samarium–cobalt or neodymium–iron–born) compounds or strontium–ferrite compounds, both of which have a permeability nearly equal to that of air. Thus, the airgap along the direct and quadrature axes is effectively of the same length and nearly has the same value of magnetic permeance (reluctance). This causes the armature self and mutual inductances to be almost independent of the rotor position (rotation). That is, saliency effects are negligible, if any. Accordingly, for radially oriented surface-mounted permanent magnet machine designs, the state-model of Eq. (9.252) can be rewritten with the knowledge that

$$L_{aa} = L_{bb} = L_{cc} = L_{sa} \tag{9.253}$$

and

$$L_{ab} = L_{ba} = L_{bc} = L_{cb} = L_{ca} = L_{ac} = L_{ma} \tag{9.254}$$

where L_{sa} and L_{ma} are as defined in Section 9.2 on wound-field synchronous machines.

Furthermore, the common types of permanent magnet mentioned earlier have high electrical resistivities and are axially segmented, which reduces effects of any eddy current flows in the magnets under asynchronous conditions.

This means that damping effects in the magnets, and in the rotor in general, are negligible. Hence, the complete state model for such samarium–cobalt or neodymium–iron–boron or strontium–ferrite permanent magnet machines can be written as follows:

$$\begin{bmatrix} v_a \\ v_b \\ v_c \end{bmatrix} = \begin{bmatrix} r_s & 0 & 0 \\ 0 & r_s & 0 \\ 0 & 0 & r_s \end{bmatrix} \begin{bmatrix} i_a \\ i_b \\ i_c \end{bmatrix} + \begin{bmatrix} L_{sa} & L_{ma} & L_{ma} \\ L_{ma} & L_{sa} & L_{ma} \\ L_{ma} & L_{ma} & L_{sa} \end{bmatrix} \cdot \frac{d}{dt} \begin{bmatrix} i_a \\ i_b \\ i_c \end{bmatrix} + \frac{d}{dt} \begin{bmatrix} L_{af} \cdot i_f \\ L_{bf} \cdot i_f \\ L_{cf} \cdot i_f \end{bmatrix} \tag{9.255}$$

For the convenience of the reader, we repeat here, from Figures 8.33 and 8.34, the "to-scale" cross-sections of two 15-hp six-pole, samarium-cobalt based, and strontium–ferrite based, designs of two PM brushless DC motors, see Figures 9.33 and 9.34, respectively, albeit of larger axial length for the strontium-ferrite-based design because of its weaker PMs, see Refs. [1–3, 13, 14]. Also, given in Figures 9.35 and 9.36 are the corresponding FE grids for magnetic field computation in both machines. Meanwhile, the corresponding FE-computed flux plots at no-load are given for both machines in Figures 9.37 and 9.38, respectively. Also, the radial airgap flux density FE-computed waveforms at no-load, for samarium–cobalt

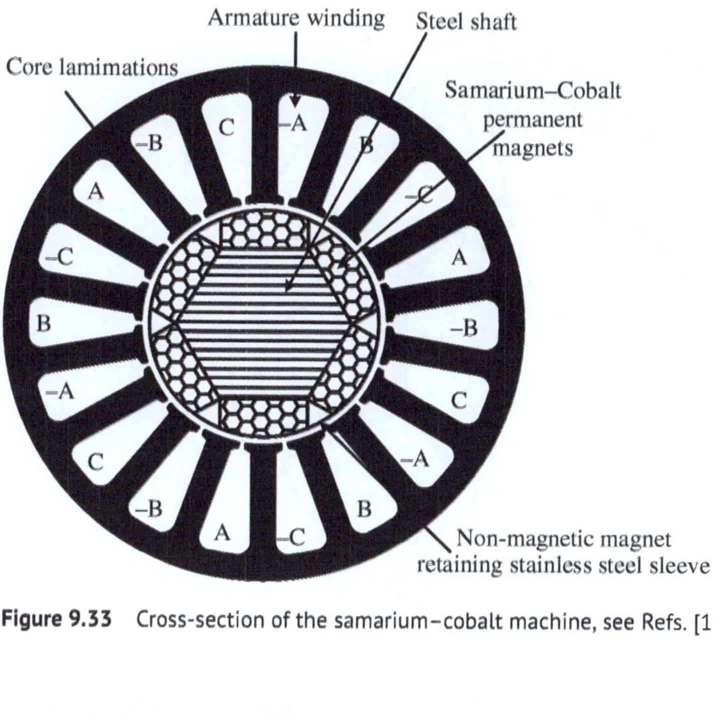

Figure 9.33 Cross-section of the samarium–cobalt machine, see Refs. [1–3, 13, 14].

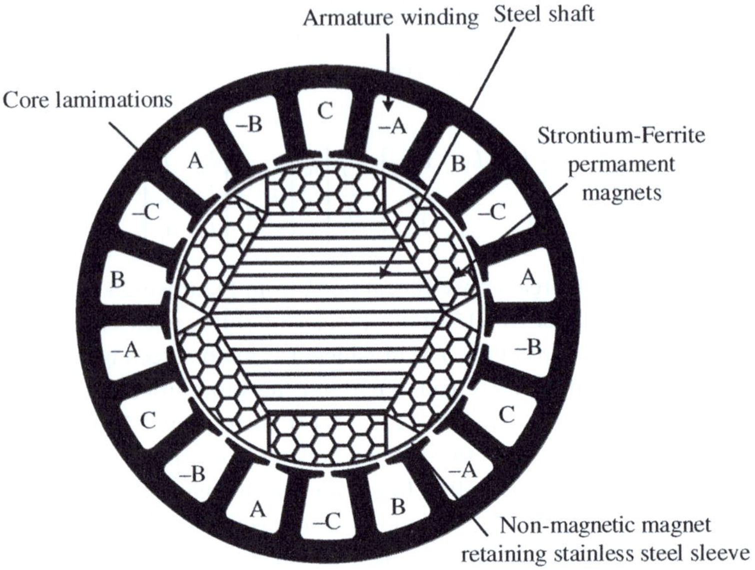

Figure 9.34 Cross-section of the strontium–ferrite machine, see Refs. [1–3, 13, 14].

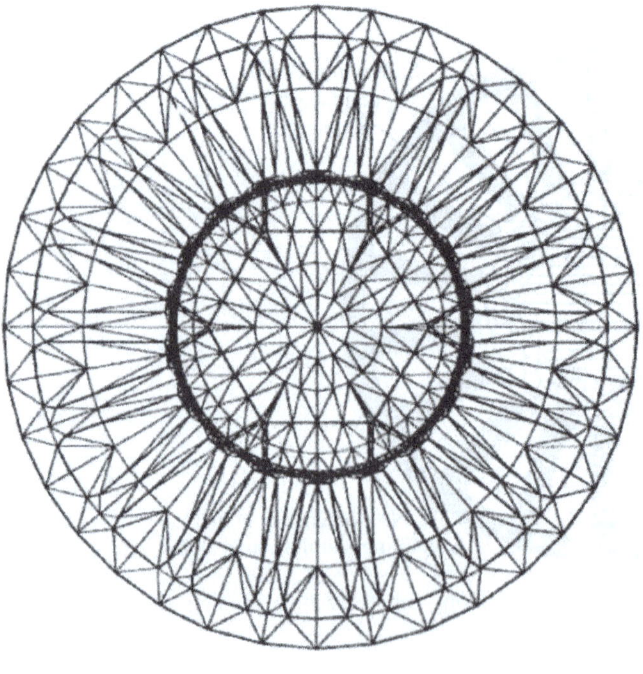

Figure 9.35 Finite-element grid of samarium–cobalt machine at a given rotor position, see Refs. [1–3, 13, 14].

Figure 9.36 Finite-element grid of strontium–ferrite machine at a given rotor position, see Refs. [1–3, 13, 14].

and strontium–ferrite-based PM machines, is given in Figures 9.39 and 9.40, respectively.

In order to understand the proper approach to represent permanent magnets in state-space modeling of electric machines, and FE modeling and computation of magnetic fields in electric machines, let us examine the effects of a PM in a magnetic circuit, such as in Figure 9.41. Applying Ampere's law, $\oint \overline{H} \cdot d\bar{l} = \sum$ enclosed Currents, with a core of an assumed infinite permeability and an airgap, one obtains

$$H_m l_m + H_g l_g = 0 \tag{9.256}$$

and if leakage as well as airgap fringing are assumed negligible, one can write the following:

$$\phi_m = \phi_g = B_m A = B_g A \tag{9.257}$$

where H_g and H_m are the magnetic field intensities in the airgap and permanent magnet, respectively, also, B_g and B_m are the magnetic flux densities in the airgap

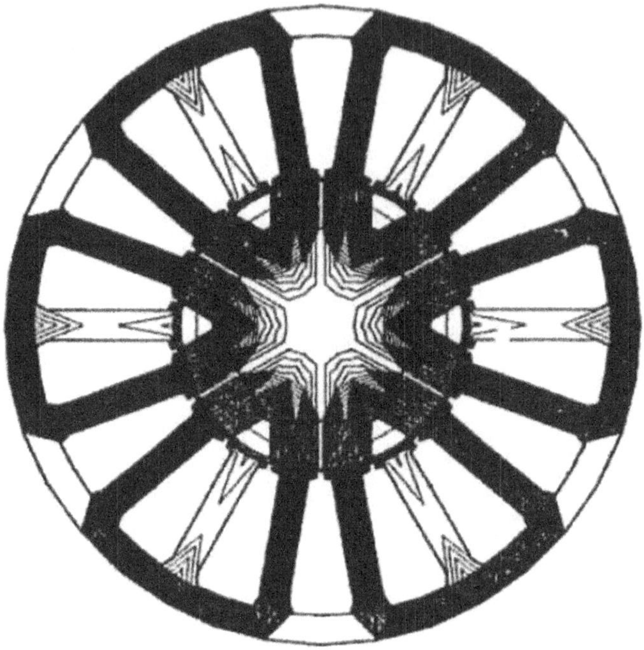

Figure 9.37 No-load equal VP contours (flux plot) of the samarium–cobalt machine for a given rotor position, see Refs. [1–3, 13, 14].

and permanent magnet, respectively, while ϕ_m and ϕ_g are the fluxes in the permanent magnet and airgap, respectively, and A is the cross-sectional area of the magnetic circuit's core.

Thus,

$$B_m = B_g \tag{9.258}$$

Also, from Eq. (9.256), one can write the following:

$$H_m = -\frac{l_g}{l_m} H_g \tag{9.259}$$

That is,

$$H_m = -\frac{B_g}{\mu_0}\frac{l_g}{l_m} = -\frac{B_m}{\mu_0}\frac{l_g}{l_m} \tag{9.260}$$

However, H_m and B_m are related by the demagnetization characteristic of the permanent magnet material, such as shown in Figure 9.42. Meanwhile, Eq. (9.260) states that

$$B_m = -\mu_0 \frac{l_m}{l_g} H_m \tag{9.261}$$

Expressing Eq. (9.261) graphically, one obtains from the schematic in Figure 9.42 the graphical construction shown in Figure 9.43.

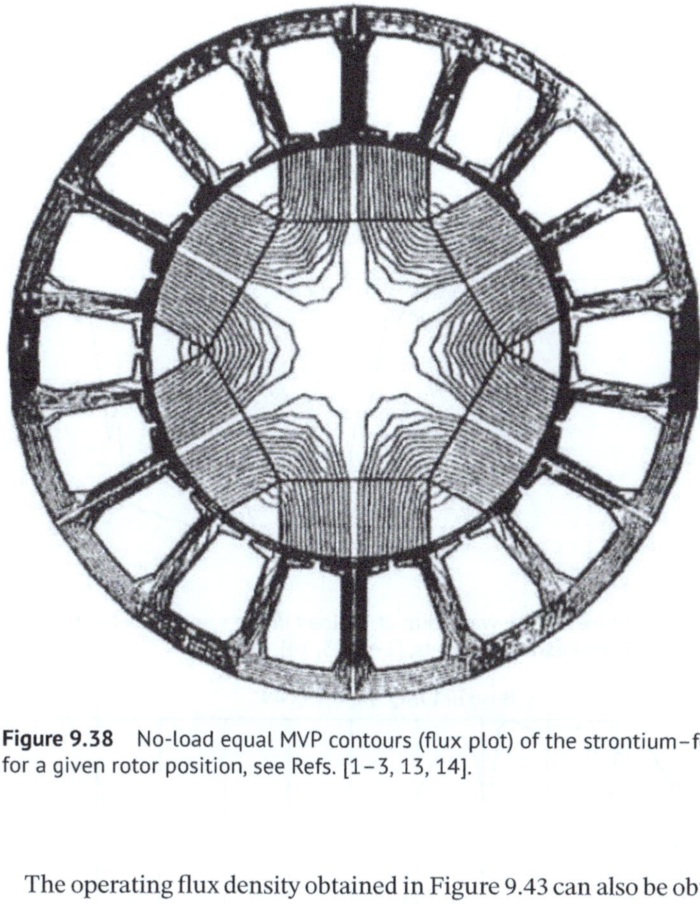

Figure 9.38 No-load equal MVP contours (flux plot) of the strontium–ferrite machine for a given rotor position, see Refs. [1–3, 13, 14].

The operating flux density obtained in Figure 9.43 can also be obtained in a magnetic circuit with an excitation MMF of $(l_m H_c)$ and an airgap height, l_g, as shown in Figure 9.44.

Hence, the permanent magnet can be replaced by an equivalent field winding whose field current is constant at all times. That is, in amperes:

$$i_f = H_c l_m \tag{9.262}$$

and whose number of turns is equal to one turn. Hence, the state equation (9.255) can be rewritten as follows:

$$\begin{bmatrix} v_a \\ v_b \\ v_c \end{bmatrix} = \begin{bmatrix} r_s & 0 & 0 \\ 0 & r_s & 0 \\ 0 & 0 & r_s \end{bmatrix} \cdot \begin{bmatrix} i_a \\ i_b \\ i_c \end{bmatrix} + \begin{bmatrix} L_{sa} & L_{ma} & L_{ma} \\ L_{ma} & L_{sa} & L_{ma} \\ L_{ma} & L_{ma} & L_{sa} \end{bmatrix} \cdot \begin{bmatrix} \dot{i}_a \\ \dot{i}_b \\ \dot{i}_c \end{bmatrix} + (H_c l_m) \frac{d}{dt} \begin{bmatrix} L_{afm}\cos(\sigma) \\ L_{afm}\cos\left(\sigma - \frac{2\pi}{3}\right) \\ L_{afm}\cos\left(\sigma - \frac{4\pi}{3}\right) \end{bmatrix} \tag{9.263}$$

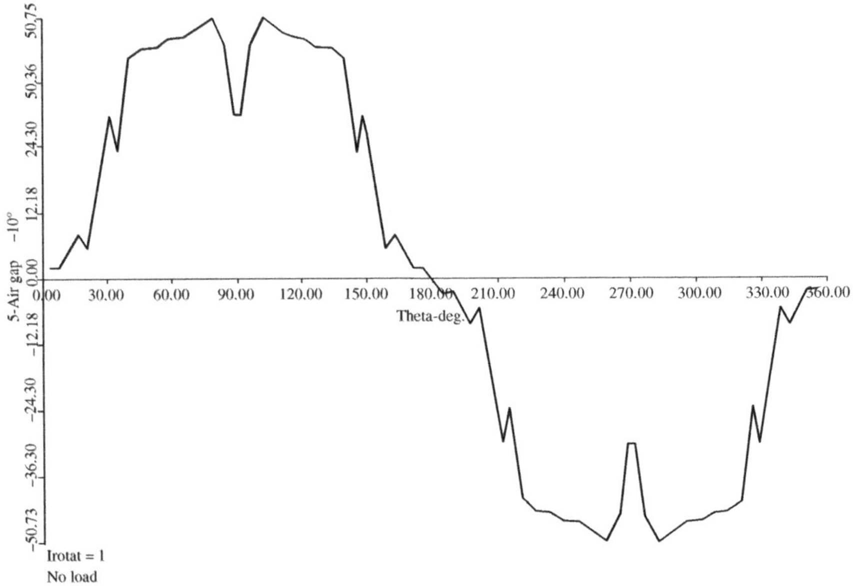

Figure 9.39 Mid-airgap flux density waveform at no-load in the samarium–cobalt machine-peak value 48,750 lines/in.^{2}, see Refs. [1–3, 13, 14].

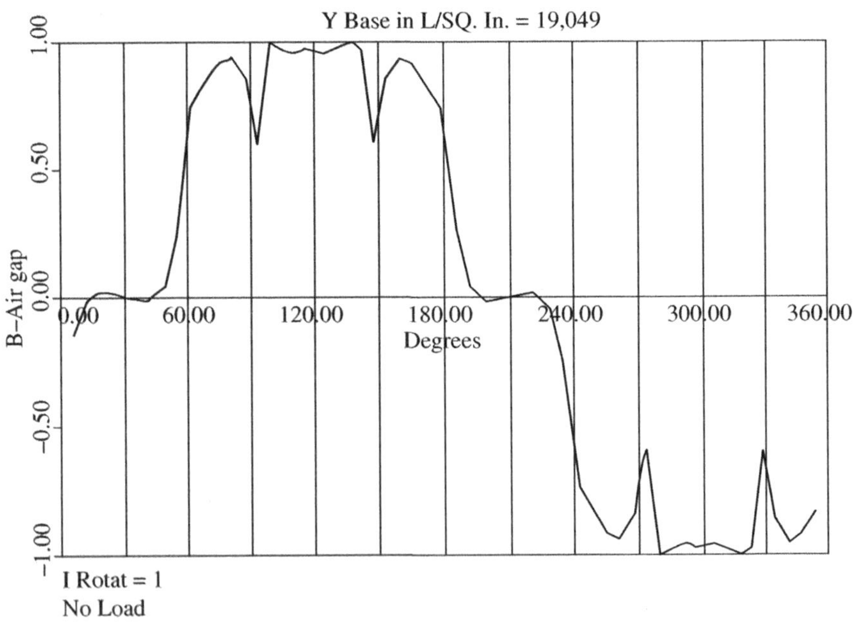

Figure 9.40 Mid-airgap flux density waveform at no-load in the strontium–ferrite machine-peak value 19,050 lines/in.^{2}, see Refs. [1–3, 13, 14].

Figure 9.41 Permanent magnet in a magnetic circuit with an airgap.

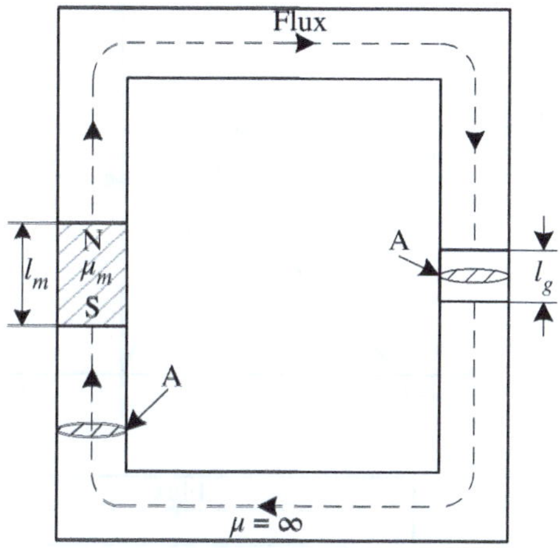

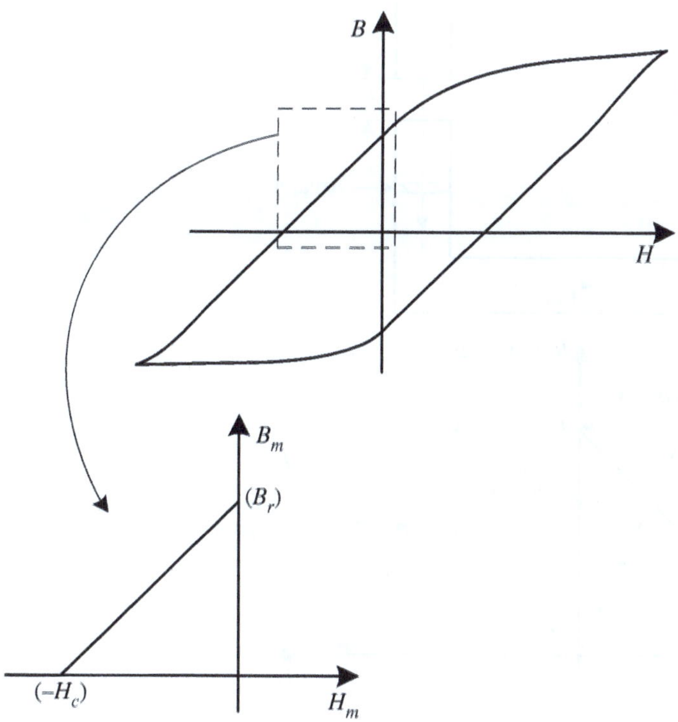

Figure 9.42 The demagnetization characteristic of a permanent magnet material (typical of samarium–cobalt, neodymium–iron–boron, or strontium–ferrite types).

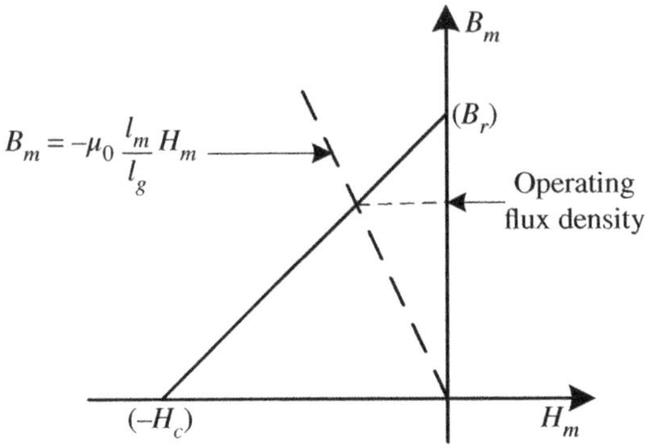

Figure 9.43 The operating point of a permanent magnet.

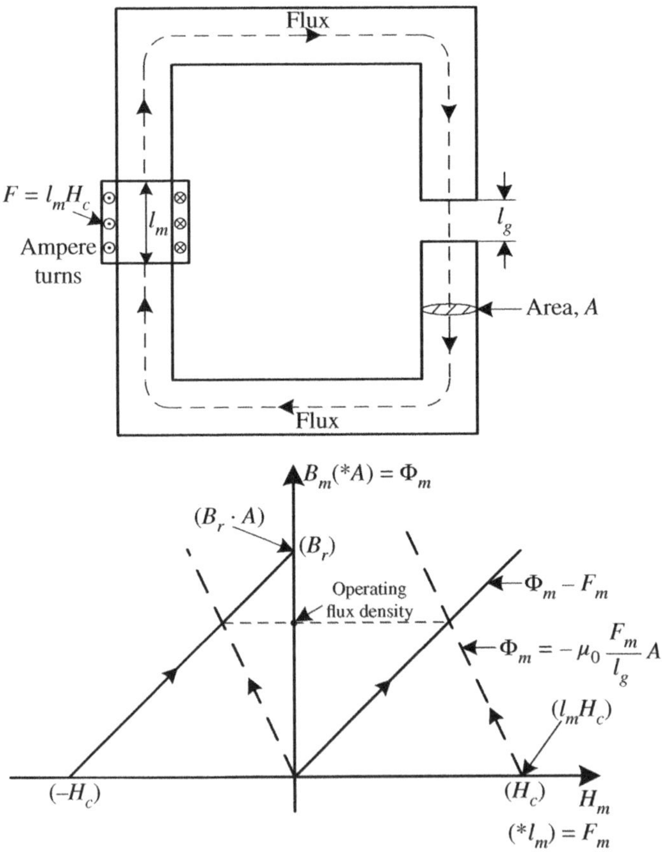

Figure 9.44 Equivalent magnetic circuit with electromagnet replacing the permanent magnet.

That is, with $(d\sigma/dt) = \dot{\sigma}$, we can write the following:

$$\begin{bmatrix} v_a \\ v_b \\ v_c \end{bmatrix} = \begin{bmatrix} r_s & 0 & 0 \\ 0 & r_s & 0 \\ 0 & 0 & r_s \end{bmatrix} \cdot \begin{bmatrix} i_a \\ i_b \\ i_c \end{bmatrix} + \begin{bmatrix} L_{sa} & L_{ma} & L_{ma} \\ L_{ma} & L_{sa} & L_{ma} \\ L_{ma} & L_{ma} & L_{sa} \end{bmatrix} \cdot \frac{d}{dt} \begin{bmatrix} i_a \\ i_b \\ i_c \end{bmatrix} + (-\dot{\sigma} H_c l_m) \begin{bmatrix} L_{afm} \sin(\sigma) \\ L_{afm} \sin\left(\sigma - \frac{2\pi}{3}\right) \\ L_{afm} \sin\left(\sigma - \frac{4\pi}{3}\right) \end{bmatrix} \tag{9.264}$$

Here,

$$\begin{bmatrix} e_a \\ e_b \\ e_c \end{bmatrix} = (-\dot{\sigma} H_c l_m) \begin{bmatrix} L_{afm} \sin(\sigma) \\ L_{afm} \sin\left(\sigma - \frac{2\pi}{3}\right) \\ L_{afm} \sin\left(\sigma - \frac{4\pi}{3}\right) \end{bmatrix} = E_{\max} \begin{bmatrix} L_{afm} \sin(\sigma) \\ L_{afm} \sin\left(\sigma - \frac{2\pi}{3}\right) \\ L_{afm} \sin\left(\sigma - \frac{4\pi}{3}\right) \end{bmatrix} \tag{9.265}$$

where $E_{\max} = (-\dot{\sigma} H_c l_m) L_{afm}$

That is, a radial permanent magnet electric machine dynamic state model can be written as follows:

$$\underline{V}_{abc} = \underbrace{\underline{R} \cdot \underline{I}_{abc}}_{\text{Ohmic Voltage}} + \underbrace{\underline{L} \cdot \underline{\dot{I}}_{abc}}_{\text{Transformer Voltage}} + \underbrace{\underline{E}_{abc}}_{\text{Rotational Voltage}} \tag{9.266}$$

or

$$\underbrace{\underline{\dot{I}}_{abc}}_{\underline{\dot{X}}} = \underbrace{(-\underline{L}^{-1} \cdot \underline{R})}_{\underline{A}} \cdot \underbrace{\underline{I}_{abc}}_{\underline{X}} + \underbrace{\underline{L}^{-1}}_{\underline{B}} \cdot \underbrace{(\underline{V}_{abc} - \underline{E}_{abc})}_{\underline{U}} \tag{9.267}$$

That is of the form, $\underline{\dot{X}} = \underline{A} \cdot \underline{X} + \underline{B} \cdot \underline{U}$, which is the general format for state-space models.

Again, the electromagnetic power is as follows:

$$P_{abc} = \underline{I}^t_{abc} \cdot \underline{E}_{abc} \tag{9.268}$$

and the developed torque (associated with the electromagnetic power) can be expressed as follows:

$$T_{em} = P_{abc}/\omega \tag{9.269}$$

where ω is the angular speed in mechanical radians per second.

If any rotor damping effect is included and the PM is replaced with an "equivalent" field winding with a constant current source excitation, the resulting equivalent coils schematic is shown in Figure 9.45, yielding the following

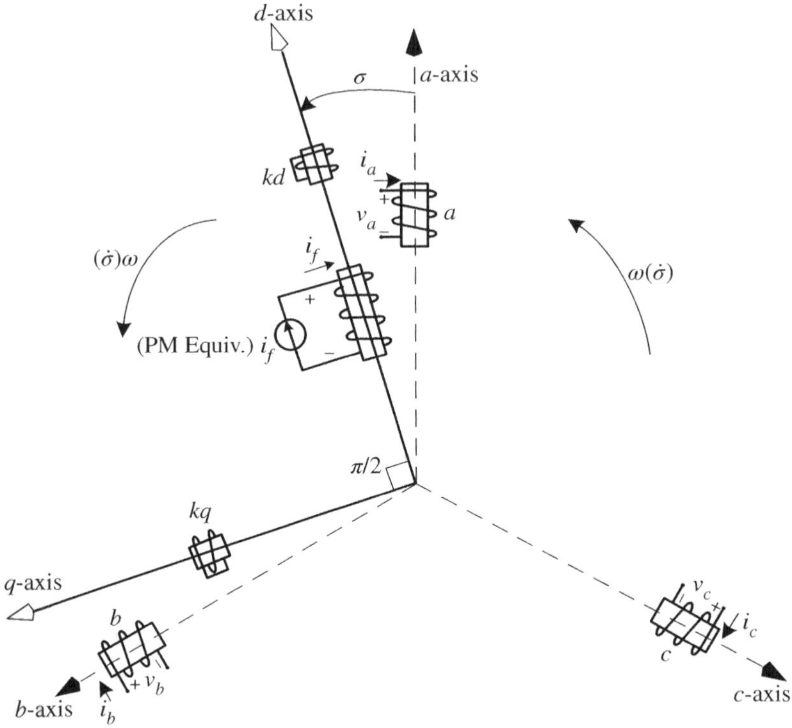

Figure 9.45 Schematic of armature and PM equivalent "field winding" (PM machine with damping).

state-space model:

$$\begin{bmatrix} v_a \\ v_b \\ v_c \\ 0 \\ 0 \end{bmatrix} = \begin{bmatrix} r_a & 0 & 0 & 0 & 0 \\ 0 & r_b & 0 & 0 & 0 \\ 0 & 0 & r_c & 0 & 0 \\ 0 & 0 & 0 & r_{kd} & 0 \\ 0 & 0 & 0 & 0 & r_{kq} \end{bmatrix} \begin{bmatrix} i_a \\ i_b \\ i_c \\ i_{kd} \\ i_{kq} \end{bmatrix} + \frac{d}{dt}\left\{ \begin{bmatrix} L_{aa} & L_{ab} & L_{ac} & L_{akd} & L_{akq} \\ L_{ba} & L_{bb} & L_{bc} & L_{bkd} & L_{bkq} \\ L_{ca} & L_{cb} & L_{cc} & L_{ckd} & L_{ckq} \\ L_{kda} & L_{kdb} & L_{kdc} & L_{kdkd} & 0 \\ L_{kqa} & L_{kqb} & L_{kqc} & 0 & L_{kqkq} \end{bmatrix} \begin{bmatrix} i_a \\ i_b \\ i_c \\ i_{kd} \\ i_{kq} \end{bmatrix} \right\}$$

$$+ \frac{d}{dt} \begin{bmatrix} L_{af} i_f \\ L_{bf} i_f \\ L_{cf} i_f \\ 0 \\ 0 \end{bmatrix} \tag{9.270}$$

In a practical situation with a three-phase armature and rotor-mounted permanent magnets, we have the following:

$$r_a = r_b = r_c = r_s$$
$$L_{ab} = L_{ba}, L_{bc} = L_{cb}, L_{ca} = L_{ac}$$
$$L_{akd} = L_{kda}, L_{bkd} = L_{kdb}, L_{ckd} = L_{kdc}$$
$$L_{akq} = L_{kqa}, L_{bkq} = L_{kqb}, L_{ckq} = L_{kqc}$$

and $\frac{d}{dt}(L_{af}i_f) = e_a$, $\frac{d}{dt}(L_{bf}i_f) = e_b$, $\frac{d}{dt}(L_{cf}i_f) = e_c$.

Hence, Eq. (9.271) can be rewritten as follows:

$$\begin{bmatrix} v_a \\ v_b \\ v_c \\ 0 \\ 0 \end{bmatrix} = \begin{bmatrix} r_s & 0 & 0 & 0 & 0 \\ 0 & r_s & 0 & 0 & 0 \\ 0 & 0 & r_s & 0 & 0 \\ 0 & 0 & 0 & r_{kd} & 0 \\ 0 & 0 & 0 & 0 & r_{kq} \end{bmatrix} \begin{bmatrix} i_a \\ i_b \\ i_c \\ i_{kd} \\ i_{kq} \end{bmatrix} + \begin{bmatrix} L_{aa} & L_{ab} & L_{ac} & L_{akd} & L_{akq} \\ L_{ab} & L_{bb} & L_{bc} & L_{bkd} & L_{bkq} \\ L_{ac} & L_{bc} & L_{cc} & L_{ckd} & L_{ckq} \\ L_{akd} & L_{bkd} & L_{ckd} & L_{kdkd} & 0 \\ L_{akq} & L_{bkq} & L_{ckq} & 0 & L_{kqkq} \end{bmatrix} \cdot \frac{d}{dt} \begin{bmatrix} i_a \\ i_b \\ i_c \\ i_{kd} \\ i_{kq} \end{bmatrix}$$
$$+ \underbrace{\left(\frac{d\sigma}{dt}\right)}_{\omega} \left\{ \frac{d}{d\sigma} \begin{bmatrix} L_{aa} & L_{ab} & L_{ac} & L_{akd} & L_{akq} \\ L_{ab} & L_{bb} & L_{bc} & L_{bkd} & L_{bkq} \\ L_{ac} & L_{bc} & L_{cc} & L_{ckd} & L_{ckq} \\ L_{akd} & L_{bkd} & L_{ckd} & L_{kdkd} & 0 \\ L_{akq} & L_{bkq} & L_{ckq} & 0 & L_{kqkq} \end{bmatrix} \right\} \begin{bmatrix} i_a \\ i_b \\ i_c \\ i_{kd} \\ i_{kq} \end{bmatrix} + \begin{bmatrix} e_a \\ e_b \\ e_c \\ 0 \\ 0 \end{bmatrix} \tag{9.271}$$

In abbreviated matrix notation, Eq. (9.271) can be rewritten as follows:

$$\underline{V} = \underline{R} \cdot \underline{I} + \underline{L} \cdot \dot{\underline{I}} + \omega \overbrace{\left(\frac{d\underline{L}}{d\sigma}\right)}^{\dot{\underline{L}}} \cdot \underline{I} + \underline{E} \tag{9.272}$$

or

$$\underbrace{\dot{\underline{I}}}_{\dot{\underline{X}}} = \underbrace{-\underline{L}^{-1} \cdot (\underline{R} + \dot{\underline{L}})}_{\underline{A}} \cdot \underbrace{\underline{I}}_{\underline{X}} + \underbrace{\underline{L}^{-1}}_{\underline{B}} \cdot \underbrace{(\underline{V} - \underline{E})}_{\underline{U}} \tag{9.273}$$

The inductance terms in Eqs. (9.270) and (9.271) can be expressed as follows in terms of d-axis position angle, σ, with respect to the magnetic axis of phase (a):

$$L_{aa} = L_{sa} + L_{sv}\cos(2\sigma)$$
$$L_{bb} = L_{sa} + L_{sv}\cos(2(\sigma - 2\pi/3)) = L_{sa} + L_{sv}\cos(2\sigma - 4\pi/3)$$
$$L_{cc} = L_{sa} + L_{sv}\cos(2(\sigma - 4\pi/3)) = L_{sa} + L_{sv}\cos(2\sigma - 2\pi/3)$$
$$L_{ab} = -L_{ma} + L_{mv}\cos(2\sigma - 2\pi/3) = L_{ba}$$
$$L_{bc} = -L_{ma} + L_{mv}\cos(2\sigma) = L_{cb}$$
$$L_{ca} = -L_{ma} + L_{mv}\cos(2\sigma - 4\pi/3) = L_{ac}$$

$$L_{akd} = L_{akdm}\cos(\sigma) = L_{kda}$$
$$L_{akq} = -L_{akqm}\sin(\sigma) = L_{kqa}$$
$$L_{bkd} = L_{akdm}\cos(\sigma - 2\pi/3) = L_{kdb}$$
$$L_{bkq} = -L_{akqm}\sin(\sigma - 2\pi/3) = L_{kqb}$$
$$L_{ckd} = L_{akdm}\cos(\sigma - 4\pi/3) = L_{kdc}$$
$$L_{ckq} = -L_{akqm}\sin(\sigma - 4\pi/3) = L_{kqc}$$

$$L_{kdkd} = \text{constant}_1 = L_{kdkd}$$
$$L_{kdkq} = 0 = L_{kqkd}$$
$$L_{kqkq} = \text{constant}_2 = L_{kqkq}$$

Equation (9.270) can be written in vector–matrix abbreviated form as follows:

$$\underline{V} = \underline{R}\cdot\underline{I} + \dot{\underline{\Lambda}} + \underline{E} \tag{9.274}$$

where

$$\underline{\Lambda} = \underline{L}\cdot\underline{I}$$

Hence,

$$\underline{I} = \underline{L}^{-1}\cdot\underline{\Lambda}$$

Therefore, Eq. (9.274) can be rewritten as follows:

$$\underline{V} = (\underline{R}\cdot\underline{L}^{-1})\cdot\underline{\Lambda} + \dot{\underline{\Lambda}} + \underline{E} \tag{9.275}$$

Hence, one can rewrite Eq. (9.275) as follows:

$$\dot{\underline{\Lambda}} = [-(\underline{R}\cdot\underline{L}^{-1})]\cdot\underline{\Lambda} + \underline{U}\cdot(\underline{V} - \underline{E}) \tag{9.276}$$

where $\underline{U}$ is a diagonal identity matrix.

Equation (9.276) is the state-space form of the governing differential equation in the flux-linkage frame of reference and is of the form $\dot{\underline{X}} = \underline{A}\cdot\underline{X} + \underline{B}\cdot\underline{V}$, so common in discrete time system formulation.

9.12 Synchronous Machine *DQ*0 Frame State-Space Models with Permanent Magnet Excitations

In this section, we start by rewriting Eq. (9.270) in short-hand matrix form as follows:

$$\begin{bmatrix}\underline{V}_{abc}\\ \underline{0}\end{bmatrix} = \begin{bmatrix}\underline{R}_{ss} & \underline{0}\\ \underline{0} & \underline{R}_{rr}\end{bmatrix}\begin{bmatrix}\underline{I}_{abc}\\ \underline{I}_{kdkq}\end{bmatrix} + \frac{d}{dt}\left\{\begin{bmatrix}\underline{L}_{ss} & \underline{L}_{sr}\\ \underline{L}_{rs} & \underline{L}_{rr}\end{bmatrix}\begin{bmatrix}\underline{I}_{abc}\\ \underline{I}_{kdkq}\end{bmatrix}\right\} + \frac{d}{dt}\left\{i_f\begin{bmatrix}\underline{L}_{sf}\\ \underline{0}\end{bmatrix}\right\} \tag{9.277}$$

That is,

$$\begin{bmatrix} \underline{V}_{abc} \\ \underline{0} \end{bmatrix} = \begin{bmatrix} \underline{R}_{ss} & \underline{0} \\ \underline{0} & \underline{R}_{rr} \end{bmatrix} \begin{bmatrix} \underline{I}_{abc} \\ \underline{I}_{kdkq} \end{bmatrix} + \frac{d}{dt}\left\{ \begin{bmatrix} \underline{L}_{ss} & \underline{L}_{sr} \\ \underline{L}_{rs} & \underline{L}_{rr} \end{bmatrix} \begin{bmatrix} \underline{I}_{abc} \\ \underline{I}_{kdkq} \end{bmatrix} \right\} + \begin{bmatrix} \underline{E}_{abc} \\ \underline{0} \end{bmatrix} \tag{9.278}$$

where

$$\underline{R}_{ss} = \begin{bmatrix} r_s & 0 & 0 \\ 0 & r_s & 0 \\ 0 & 0 & r_s \end{bmatrix} \tag{9.279}$$

$$\underline{R}_{rr} = \begin{bmatrix} r_{kd} & 0 \\ 0 & r_{kq} \end{bmatrix} \tag{9.280}$$

$$\underline{E}_{abc} = \begin{bmatrix} e_a & e_b & e_c \end{bmatrix}^t \tag{9.281}$$

$$\underline{L}_{ss} = \begin{bmatrix} L_{aa} & L_{ab} & L_{ac} \\ L_{ba} & L_{bb} & L_{bc} \\ L_{ca} & L_{cb} & L_{cc} \end{bmatrix} \tag{9.282}$$

$$\underline{L}_{sr} = \begin{bmatrix} L_{akd} & L_{akq} \\ L_{bkd} & L_{bkq} \\ L_{ckd} & L_{ckq} \end{bmatrix} \tag{9.283}$$

$$\underline{L}_{rs} = (\underline{L}_{sr})^t \tag{9.284}$$

$$\underline{L}_{rr} = \begin{bmatrix} L_{kdkd} & 0 \\ 0 & L_{kqkq} \end{bmatrix} \tag{9.285}$$

where all the inductance terms in Eqs. (9.282)–(9.285) are as defined in Section 9.11, and

$$\left.\begin{aligned} \underline{V}_{dq0} &= \underline{T} \cdot \underline{V}_{abc} \\ \underline{V}_{abc} &= \underline{T}^{-1} \cdot \underline{V}_{dq0} \\ \underline{I}_{dq0} &= \underline{T} \cdot \underline{I}_{abc} \\ \underline{I}_{abc} &= \underline{T}^{-1} \cdot \underline{I}_{dq0} \\ \underline{\Lambda}_{dq0} &= \underline{T} \cdot \underline{\Lambda}_{abc} \\ \underline{\Lambda}_{abc} &= \underline{T}^{-1} \cdot \underline{\Lambda}_{dq0} \end{aligned}\right\} \tag{9.286}$$

Also,

$$\begin{bmatrix} \underline{\Lambda}_{abc} \\ \underline{\Lambda}_{kdkq} \end{bmatrix} = \begin{bmatrix} \underline{L}_{ss} & \underline{L}_{sr} \\ \underline{L}_{rs} & \underline{L}_{rr} \end{bmatrix} \begin{bmatrix} \underline{I}_{abc} \\ \underline{I}_{kdkq} \end{bmatrix} \tag{9.287}$$

Thus, applying the transformations in Eq. (9.286)–Eq. (9.278) and the definitions of flux linkages in Eq. (9.287), and the fact that $\underline{E}_{dq0} = \underline{T} \cdot \underline{E}_{abc}$, one can write the following:

$$\begin{bmatrix} \underline{V}_{dq0} \\ \underline{0} \end{bmatrix} = \begin{bmatrix} \underline{T} \cdot \underline{R}_{ss} \cdot \underline{T}^{-1} & \underline{0} \\ \underline{0} & \underline{R}_{rr} \end{bmatrix} \begin{bmatrix} \underline{I}_{dq0} \\ \underline{I}_{kdkq} \end{bmatrix} + \begin{bmatrix} \underline{T} & \underline{0} \\ \underline{0} & \underline{U} \end{bmatrix} \frac{d}{dt} \begin{bmatrix} \underline{T}^{-1} \cdot \underline{\Lambda}_{dq0} \\ \underline{\Lambda}_{kdkq} \end{bmatrix} + \begin{bmatrix} \underline{E}_{dq0} \\ 0 \end{bmatrix} \tag{9.288}$$

From past $dq0$ transformation developments in Sections 9.6 and 9.7, we know that

$$\underline{T} \cdot \frac{d}{dt}(\underline{T}^{-1} \cdot \underline{\Lambda}_{dq0}) = \underline{\dot{\Lambda}}_{dq0} - (\underline{\dot{T}}) \cdot \underline{T}^{-1} \cdot \underline{\Lambda}_{dq0} \tag{9.289}$$

and

$$\underline{\dot{T}} = \omega \underline{G} \cdot \underline{T} \tag{9.290}$$

where

$$\underline{G} = \begin{bmatrix} 0 & 1 & 0 \\ -1 & 0 & 0 \\ 0 & 0 & 0 \end{bmatrix} \tag{9.291}$$

and

$$\underline{T} = \frac{2}{3} \begin{bmatrix} \cos(\sigma) & \cos(\sigma - 2\pi/3) & \cos(\sigma - 4\pi/3) \\ -\sin(\sigma) & -\sin(\sigma - 2\pi/3) & -\sin(\sigma - 4\pi/3) \\ 1/2 & 1/2 & 1/2 \end{bmatrix} \tag{9.292}$$

and

$$\underline{T}^{-1} = \begin{bmatrix} \cos\sigma & -\sin\sigma & 1/2 \\ \cos(\sigma - 2\pi/3) & -\sin(\sigma - 2\pi/3) & 1/2 \\ \cos(\sigma - 4\pi/3) & -\sin(\sigma - 4\pi/3) & 1/2 \end{bmatrix} \tag{9.293}$$

Expanding Eq. (9.288) yields

$$\left.\begin{aligned} \underline{V}_{dq0} &= (\underline{T} \cdot \underline{R}_{ss} \cdot \underline{T}^{-1})\underline{I}_{dq0} + \underline{T} \cdot \tfrac{d}{dt}(\underline{T}^{-1} \cdot \underline{\Lambda}_{dq0}) + \underline{E}_{dq0} \\ \underline{0} &= \underline{R}_{rr} \cdot \underline{I}_{kdkq} + \tfrac{d}{dt}(\underline{\Lambda}_{kdkq}) \end{aligned}\right\} \tag{9.294}$$

However, from Eqs. (9.289) and (9.290), one can write the following:

$$\underline{T} \cdot \frac{d}{dt}(\underline{T}^{-1} \cdot \underline{\Lambda}_{dq0}) = \underline{\dot{\Lambda}}_{dq0} - \omega(\underline{G} \cdot \underline{T}) \cdot \underline{T}^{-1} \cdot \underline{\Lambda}_{dq0}$$

Thus,

$$\underline{T} \cdot \frac{d}{dt}(\underline{T}^{-1} \cdot \underline{\Lambda}_{dq0}) = \underline{\dot{\Lambda}}_{dq0} - \omega \underline{G} \cdot \underline{\Lambda}_{dq0} \tag{9.295}$$

Substituting Eq. (9.295) into Eq. (9.294), one obtains the following:

$$\left.\begin{aligned} \underline{V}_{dq0} &= (\underline{T} \cdot \underline{R}_{ss} \cdot \underline{T}^{-1}) \cdot \underline{I}_{dq0} + \underline{\dot{\Lambda}}_{dq0} - \omega \underline{G} \cdot \underline{\Lambda}_{dq0} + \underline{E}_{dq0} \\ \underline{0} &= \underline{R}_{rr} \cdot \underline{I}_{kdkq} + \underline{\dot{\Lambda}}_{kdkq} \end{aligned}\right\} \tag{9.296}$$

However, from previous developments in Sections 9.6 and 9.7 on wound-field synchronous machines, we have the following:

$$\underline{\Lambda}_{dq0} = (\underline{T} \cdot \underline{L}_{ss} \cdot \underline{T}^{-1})\underline{I}_{dq0} + (\underline{T} \cdot \underline{L}_{sr})\underline{I}_{kdkq} \tag{9.297}$$

and

$$\underline{\Lambda}_{kdkq} = \left(\underline{L}_{sr}^{t} \cdot \underline{T}^{-1}\right) \underline{I}_{dq0} + (\underline{L}_{rr})\underline{I}_{kdkq} \tag{9.298}$$

In Eqs. (9.288) and (9.298), one should recall that

$$\underline{T} \cdot \underline{L}_{ss} \cdot \underline{T}^{-1} = \begin{bmatrix} \overbrace{\left(L_{sa} + L_{ma} + \frac{3}{2}L_{v}\right)}^{L_{dd} \text{ or } \left(L_{sa}+L_{ma}+\frac{1}{2}L_{sv}+L_{mv}\right)} & 0 & 0 \\ 0 & \overbrace{\left(L_{sa} + L_{ma} - \frac{3}{2}L_{v}\right)}^{L_{qq} \text{ or } \left(L_{sa}+L_{ma}-\frac{1}{2}L_{sv}-L_{mv}\right)} & 0 \\ 0 & 0 & \overbrace{(L_{sa} - 2L_{ma})}^{L_{00}} \end{bmatrix} \tag{9.299}$$

and

$$\underline{T} \cdot \underline{L}_{sr} = \begin{bmatrix} \underline{L}_{akdm} & 0 \\ 0 & \underline{L}_{akqm} \\ 0 & 0 \end{bmatrix} \tag{9.300}$$

and

$$\left(\underline{L}_{sr}^{t} \cdot \underline{T}^{-1}\right) = \begin{bmatrix} \frac{3}{2}\underline{L}_{akdm} & 0 & 0 \\ 0 & \frac{3}{2}\underline{L}_{akdm} & 0 \end{bmatrix} \tag{9.301}$$

Since all inductance matrices in Eqs. (9.297) and (9.298) are time-independent, it follows that

$$\dot{\underline{\Lambda}}_{dq0} = (\underline{T} \cdot \underline{L}_{ss} \cdot \underline{T}^{-1})\dot{\underline{I}}_{dq0} + (\underline{T} \cdot \underline{L}_{sr})\dot{\underline{I}}_{kdkq} \tag{9.302}$$

and

$$\dot{\underline{\Lambda}}_{kdkq} = \left(\underline{L}_{sr}^{t} \cdot \underline{T}^{-1}\right) \dot{\underline{I}}_{dq0} + (\underline{L}_{rr})\dot{\underline{I}}_{kdkq} \tag{9.303}$$

Substituting Eqs. (9.297), (9.298), (9.302), and (9.303) into Eq. (9.296) yields the following:

$$\begin{aligned} \underline{V}_{dq0} = {} & (\underline{T} \cdot \underline{R}_{ss} \cdot \underline{T}^{-1})\underline{I}_{dq0} + (\underline{T} \cdot \underline{L}_{ss} \cdot \underline{T}^{-1})\dot{\underline{I}}_{dq0} + (\underline{T} \cdot \underline{L}_{sr})\dot{\underline{I}}_{kdkq} \\ & - \omega\underline{G}(\underline{T} \cdot \underline{L}_{ss} \cdot \underline{T}^{-1})\underline{I}_{dq0} - \omega\underline{G}(\underline{T} \cdot \underline{L}_{sr})\underline{I}_{kdkq} + \underline{E}_{dq0} \end{aligned} \tag{9.304}$$

and

$$\underline{0} = \underline{R}_{rr} \cdot \underline{I}_{kdkq} + (\underline{L}^t_{sr} \cdot \underline{T}^{-1})\,\dot{\underline{I}}_{dq0} + (\underline{L}_{rr})\dot{\underline{I}}_{kdkq} \tag{9.305}$$

with the knowledge that

$$(\underline{T} \cdot \underline{R}_{ss} \cdot \underline{T}^{-1}) = \underline{R}_{ss} \tag{9.306}$$

Equations (9.304) and (9.305) can be rearranged and reassembled as follows:

$$\begin{bmatrix} \underline{V}_{dq0} \\ \underline{0} \end{bmatrix} = \begin{bmatrix} \underline{R}_{ss} & \underline{0} \\ \underline{0} & \underline{R}_{rr} \end{bmatrix} \begin{bmatrix} \underline{I}_{dq0} \\ \underline{I}_{kdkq} \end{bmatrix} + \begin{bmatrix} -\omega\underline{G}(\underline{T} \cdot \underline{L}_{ss} \cdot \underline{T}^{-1}) & -\omega\underline{G}(\underline{T} \cdot \underline{L}_{sr}) \\ \underline{0} & \underline{0} \end{bmatrix} \begin{bmatrix} \underline{I}_{dq0} \\ \underline{I}_{kdkq} \end{bmatrix}$$
$$+ \begin{bmatrix} (\underline{T} \cdot \underline{L}_{ss} \cdot \underline{T}^{-1}) & (\underline{T} \cdot \underline{L}_{sr}) \\ (\underline{L}^t_{sr} \cdot \underline{T}^{-1}) & (\underline{L}_{rr}) \end{bmatrix} \begin{bmatrix} \dot{\underline{I}}_{dq0} \\ \dot{\underline{I}}_{kdkq} \end{bmatrix} + \begin{bmatrix} \underline{E}_{dq0} \\ 0 \end{bmatrix} \tag{9.307}$$

Now

$$-\omega\underline{G}(\underline{T} \cdot \underline{L}_{ss} \cdot \underline{T}^{-1}) = -\omega \begin{bmatrix} 0 & 1 & 0 \\ -1 & 0 & 0 \\ 0 & 0 & 0 \end{bmatrix} \begin{bmatrix} L_{dd} & 0 & 0 \\ 0 & L_{qq} & 0 \\ 0 & 0 & L_{00} \end{bmatrix}$$
$$= -\omega \begin{bmatrix} 0 & L_{qq} & 0 \\ -L_{dd} & 0 & 0 \\ 0 & 0 & 0 \end{bmatrix} = \begin{bmatrix} 0 & -\omega L_{qq} & 0 \\ \omega L_{dd} & 0 & 0 \\ 0 & 0 & 0 \end{bmatrix}$$

Hence,

$$-\omega\underline{G}(\underline{T} \cdot \underline{L}_{ss} \cdot \underline{T}^{-1}) = \begin{bmatrix} 0 & -\omega L_{qq} & 0 \\ \omega L_{dd} & 0 & 0 \\ 0 & 0 & 0 \end{bmatrix} \tag{9.308}$$

and

$$-\omega\underline{G}(\underline{T} \cdot \underline{L}_{sr}) = -\omega \begin{bmatrix} 0 & 1 & 0 \\ -1 & 0 & 0 \\ 0 & 0 & 0 \end{bmatrix} \begin{bmatrix} \underline{L}_{akdm} & 0 \\ 0 & \underline{L}_{akqm} \\ 0 & 0 \end{bmatrix} = -\omega \begin{bmatrix} 0 & \underline{L}_{akqm} \\ -\underline{L}_{akdm} & 0 \\ 0 & 0 \end{bmatrix}$$
$$= \begin{bmatrix} 0 & -\omega\underline{L}_{akqm} \\ \omega\underline{L}_{akdm} & 0 \\ 0 & 0 \end{bmatrix} - \omega\underline{G}(\underline{T} \cdot \underline{L}_{sr}) = \begin{bmatrix} 0 & -\omega\underline{L}_{akqm} \\ \omega\underline{L}_{akdm} & 0 \\ 0 & 0 \end{bmatrix} \tag{9.309}$$

Substituting Eqs. (9.299), (9.300), (9.301), (9.308), and (9.309) into Eq. (9.307) and with the knowledge that

$$\underline{R}_{ss} = \begin{bmatrix} r_s & 0 & 0 \\ 0 & r_s & 0 \\ 0 & 0 & r_s \end{bmatrix} \text{ and } \underline{R}_{rr} = \begin{bmatrix} r_{kd} & 0 \\ 0 & r_{kq} \end{bmatrix} \tag{9.310}$$

one can rewrite Eq. (9.307) into expanded form as follows:

$$\underbrace{\begin{bmatrix} v_d \\ v_q \\ v_0 \\ 0 \\ 0 \end{bmatrix}}_{\underline{V}} = \underbrace{\begin{bmatrix} r_s & 0 & 0 & 0 & 0 \\ 0 & r_s & 0 & 0 & 0 \\ 0 & 0 & r_s & 0 & 0 \\ 0 & 0 & 0 & r_{kd} & 0 \\ 0 & 0 & 0 & 0 & r_{kq} \end{bmatrix}}_{\underline{R}} \underbrace{\begin{bmatrix} i_d \\ i_q \\ i \\ i \\ i \end{bmatrix}}_{\underline{I}} + \underbrace{\begin{bmatrix} 0 & -\omega L_{qq} & 0 & -\omega L_{akqm} & 0 \\ \omega L_{dd} & 0 & \omega L_{akdm} & 0 & 0 \\ 0 & 0 & 0 & 0 & 0 \\ 0 & 0 & 0 & 0 & 0 \\ 0 & 0 & 0 & 0 & 0 \end{bmatrix}}_{\omega \underline{L}_2} \underbrace{\begin{bmatrix} i_d \\ i_q \\ i_0 \\ i_{kd} \\ i_{kq} \end{bmatrix}}_{\underline{I}}$$
$$+ \underbrace{\begin{bmatrix} L_{dd} & 0 & 0 & L_{akdm} & 0 \\ 0 & L_{qq} & 0 & 0 & L_{akqm} \\ 0 & 0 & L_{00} & 0 & 0 \\ \frac{3}{2}L_{akdm} & 0 & 0 & L_{kdkd} & 0 \\ 0 & \frac{3}{2}L_{akqm} & 0 & 0 & L_{kqkq} \end{bmatrix}}_{\underline{L}_1} \underbrace{\begin{bmatrix} \dot{i}_d \\ \dot{i}_q \\ \dot{i}_0 \\ \dot{i}_{kd} \\ \dot{i}_{kq} \end{bmatrix}}_{\dot{\underline{I}}} + \underbrace{\begin{bmatrix} e_d \\ e_q \\ e_0 \\ 0 \\ 0 \end{bmatrix}}_{\underline{E}} \tag{9.311}$$

Hence, Eq. (9.311) can be rewritten in short-hand matrix form as follows:

$$\underline{V} = (\underline{R})\underline{I} + (\omega \underline{L}_2)\underline{I} + (\underline{L}_1)\dot{\underline{I}} + \underline{E} \tag{9.312}$$

Now, Eq. (9.312) can be rewritten as follows:

$$\dot{\underline{I}} = \underbrace{\left(-\underline{L}_1^{-1}\right)(\underline{R} + \omega \underline{L}_2)}_{\underline{A}}\underline{I} + \underbrace{\left(\underline{L}_1^{-1}\right)}_{\underline{B}}(\underline{V} - \underline{E}) \tag{9.313}$$

That is,

$$\dot{\underline{I}} = \underline{A} \cdot \underline{I} + \underline{B} \cdot (\underline{V} - \underline{E}) \tag{9.314}$$

where

$$\underline{A} = \left\{ -\begin{bmatrix} L_{dd} & 0 & 0 & L_{akdm} & 0 \\ 0 & L_{qq} & 0 & 0 & L_{akqm} \\ 0 & 0 & L_{00} & 0 & 0 \\ \frac{3}{2}L_{akqm} & 0 & 0 & L_{kdkd} & 0 \\ 0 & \frac{3}{2}L_{akqm} & 0 & 0 & L_{kqkq} \end{bmatrix}^{-1} \begin{bmatrix} r_s & -\omega L_{qq} & 0 & 0 & -\omega L_{akqm} \\ \omega L_{dd} & r_s & 0 & \omega L_{akdm} & 0 \\ 0 & 0 & r_s & 0 & 0 \\ 0 & 0 & 0 & r_{kd} & 0 \\ 0 & 0 & 0 & 0 & r_{kq} \end{bmatrix} \right\} \tag{9.315}$$

and

$$\underline{B} = \left\{ \left[\begin{array}{ccc|cc} L_{dd} & 0 & 0 & L_{akdm} & 0 \\ 0 & L_{qq} & 0 & 0 & L_{akqm} \\ 0 & 0 & L_{00} & 0 & 0 \\ \hline \frac{3}{2}L_{akdm} & 0 & 0 & L_{kdkd} & 0 \\ 0 & \frac{3}{2}L_{akqm} & 0 & 0 & L_{kqkq} \end{array} \right]^{-1} \right\} \tag{9.316}$$

Equation (9.311) can be rewritten in compact matrix form as follows:

$$\underline{V} = \underline{R} \cdot \underline{I} + \omega \underline{L}_2 \cdot \underline{I} + \frac{d}{dt}(\underline{L}_1 \cdot \underline{I}) + \underline{E} \tag{9.317}$$

where

$$\underline{\Lambda} = \underline{L}_1 \cdot \underline{I}$$

and

$$\underline{I} = \underline{L}_1^{-1} \cdot \underline{\Lambda}$$

and

$$\dot{\underline{\Lambda}} = \frac{d}{dt}(\underline{L}_1 \cdot \underline{I}) = \underline{L}_1 \cdot \dot{\underline{I}}$$

Thus, one can rewrite Eq. (9.317) as follows:

$$\dot{\underline{\Lambda}} = -\underline{R} \cdot \underline{L}_1^{-1} \cdot \underline{\Lambda} - \omega \underline{L}_2 \cdot \underline{L}_1^{-1} \cdot \underline{\Lambda} + \underline{V} - \underline{E} \tag{9.318}$$

or

$$\dot{\underline{\Lambda}} = (-\underline{R} - \omega \underline{L}_2) \cdot \underline{L}_1^{-1} \cdot \underline{\Lambda} + (\underline{V} - \underline{E}) \tag{9.319}$$

Hence,

$$\dot{\underline{\Lambda}} = \left\{ -(\underline{R} + \omega \underline{L}_2) \cdot \underline{L}_1^{-1} \right\} \cdot \underline{\Lambda} + (\underline{V} - \underline{E}) \tag{9.320}$$

Equation (9.320) is of the form $\dot{\underline{X}} = \underline{A} \cdot \underline{X} + \underline{B} \cdot \underline{U}$
where

$$\dot{\underline{X}} = \dot{\underline{\Lambda}}, \underline{X} = \underline{\Lambda} \tag{9.321}$$

and

$$\underline{A} = \left\{ -(\underline{R} + \omega \underline{L}_2) \cdot \underline{L}_1^{-1} \right\} \tag{9.322}$$

and

$$\underline{B} = \text{Identity Matrix}$$

where

$$\underline{U} = (\underline{V} - \underline{E}) \tag{9.323}$$

That is, the state-space model formulation in the flux-linkage frame.

9.13 Vector Control of Wound-Field Synchronous Machines

Although wound-field synchronous machines (WFSMs) have been intensively used for power generation, these machines have also received much attention as attractive motor candidates, due to the lack of need for costly rare earth magnets, broad speed regulation range, and flexibility in reactive power/power factor controls. In this section, we will focus on the vector control of such WFSMs functioning as motors.

Different from the vector control of PMSMs introduced in Section 7.7, the control of WFSMs lies in the regulation of two stator currents in rotor-field reference frame (I_d, I_q) and the rotor field excitation current (I_f). The airgap torque of a WFSM can be expressed as follows:

$$T_{ag} = 3PI_q(L_mI_f + (L_d - L_q)I_d) \tag{9.324}$$

where P refers to the number of poles, L_d and L_q refer to the d-axis inductance and q-axis inductance in a d–q rotor-field reference frame, respectively. Unlike the case for interior PMSMs, L_d is larger than L_q for WFSMs, while L_m is the mutual inductance between the stator phase and field windings. As can be seen, from Eq. (9.324), the airgap torque of a WFSM can be synergistically controlled by regulating I_d, I_q, and I_f.

A general vector control block diagram for a WFSM is shown in Figure 9.46. In this doubly closed loop control strategy, the outer loop is the rotor speed control loop, and the inner loop is the current control loop. The outer speed controller generates the torque reference (T_{ref}), which is used, in addition to the feedback

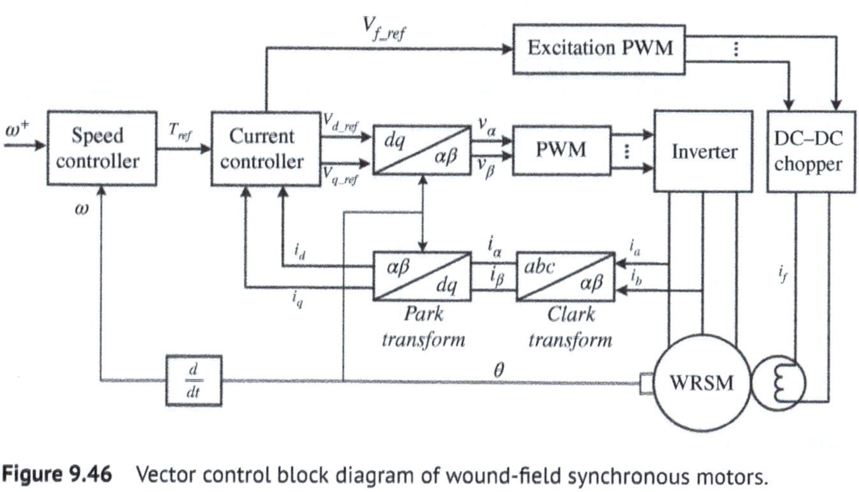

Figure 9.46 Vector control block diagram of wound-field synchronous motors.

stator dq currents (I_d, I_q) as the inputs for the inner current controller. The current controller provides the voltage references for the PWM generators of the inverter and the DC–DC chopper, so that the desired currents can be supplied to the armature and field windings of the WFSM.

Such WFSMs generally have broad speed operation ranges during the constant power region, which is very suitable for electric vehicles and other applications where the constant power speed range (CPSR) is a critical performance factor. At high speed, the back-EMF induced from the high rotating speed of the rotor will counteract the voltages supplied by the ASDs. To extend the speed range in the constant power region, the field current (I_f) needs to be reduced to weaken the flux. The operation of such a WFSM is illustrated in Figure 9.47. In this figure, the red circle is the current limit circle, which refers to the maximum current amplitude (i_{s_max}) of the stator current determined by the current ratings of the ASD power electronics and wire size of the WFSM stator windings. The current limit circle can be expressed as follows:

$$i_d^2 + i_q^2 = i_s^2 \le i_{s\_max}^2 \tag{9.325}$$

Likewise, the WFSM stator voltage is limited by the voltage output capability of the associated ASD and dielectric insulation ratings of the stator windings, which is illustrated as the blue ellipses in Figure 9.47. The overlapping area of the current

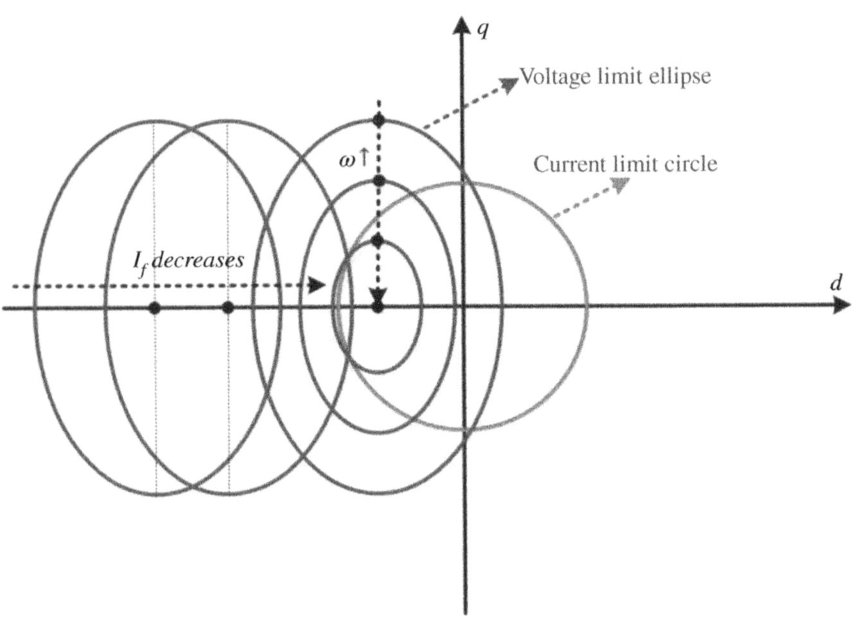

Figure 9.47 Current limit circle and voltage limit ellipses for WFSMs.

limit circle and the voltage limit ellipse defines the possible operating region of a WFSM. The mathematical expression of the voltage limit ellipse is derived as follows:

$$v_d^2 + v_q^2 = v_s^2 \leq v_{s\_max}^2 \tag{9.326}$$

where v_d and v_q refer to the voltages in the d-axis and q-axis, respectively. v_s is the stator voltage, and v_{s_max} is the maximum voltage provided by the associated ASD. During a steady state, if the stator winding resistive voltage drop can be neglected, such a voltage limitation in Eq. (9.326) can be expressed as follows:

$$\psi_d^2 + \psi_q^2 = \left(\frac{v_s}{\omega}\right)^2 \leq \left(\frac{v_{s\_max}}{\omega}\right)^2 \tag{9.327}$$

Equation (9.327) can be rewritten in terms of "apparent" inductances as follows:

$$(L_d I_d + L_m I_f)^2 + (L_q I_q)^2 = \left(\frac{v_s}{\omega}\right)^2 \leq \left(\frac{v_{s\_max}}{\omega}\right)^2 \tag{9.328}$$

Equation (9.328) can be rewritten as an analytical expression for an ellipse as the following:

$$\frac{(I_d + I_{ch})^2}{a^2} + \frac{I_q^2}{b^2} \leq 1 \tag{9.329}$$

where $I_{ch} = \frac{L_m}{L_d} I_f$ is defined as the characteristic current, determining the center $(-I_{ch}, 0)$ of the voltage ellipses. Obviously, as the field current decreases, the center shifts horizontally toward the origin. In Eq. (9.329), a and b refer to the ellipse's major semi-axis and minor semi-axis, respectively, which are defined as follows:

$$a = \frac{v_{s\_max}}{\omega L_d} \quad \text{and} \quad b = \frac{v_{s\_max}}{\omega L_q} \tag{9.330}$$

Equation (9.330) indicates that, as the rotor speed increases, the major semi-axis and minor semi-axis decrease, and the voltage limit ellipse keeps shrinking towards its center, as illustrated in Figure 9.47.

Additionally, one unique benefit associated with WFSMs is that the field excitation current (I_f) is fully controllable, which can be reduced to zero to avoid any overvoltage during an uncontrolled generator fault scenario, in which the related ASD is somehow tripped (e.g., by EMI) while the motor is spinning at high speed inducing significant back-EMFs. It should be noted that such a fault scenario is very challenging to address for PMSMs where the rotor flux of the permanent magnets is uncontrollable, and very high voltages could be induced and reflected on the inverter DC bus and may well damage the associated ASD components due to the considerable back-EMFs.

Another advantage for WFSMs is the flexible power factor control based on the regulation of the field current (I_f), and it is necessary to understand the operating points of achieving unity power factor in the dq plane. Unity power factor means zero reactive power. Assuming that the winding resistive voltage drop can be neglected, we can have the following:

$$Q_{var} = \frac{3}{2}(V_q I_d - V_d I_q) = 0 \tag{9.331}$$

that is,

$$\psi_d I_d + \psi_q I_q = 0 \tag{9.332}$$

By rewriting Eq. (9.332) in terms of the "apparent" inductances, one can obtain the following relationship:

$$L_d {I_d}^2 + L_q {I_q}^2 + L_m I_f I_d = 0 \tag{9.333}$$

Considering that the values of L_d, L_q, L_m, and I_f are non-negative, I_d has to be negative to satisfy Eq. (9.333). For a given value of I_f, Eq. (9.333) can be reformulated as an equation for an ellipse as follows:

$$\frac{(I_d + a_1)^2}{{a_1}^2} + \frac{(I_q)^2}{{b_1}^2} = 1 \tag{9.334}$$

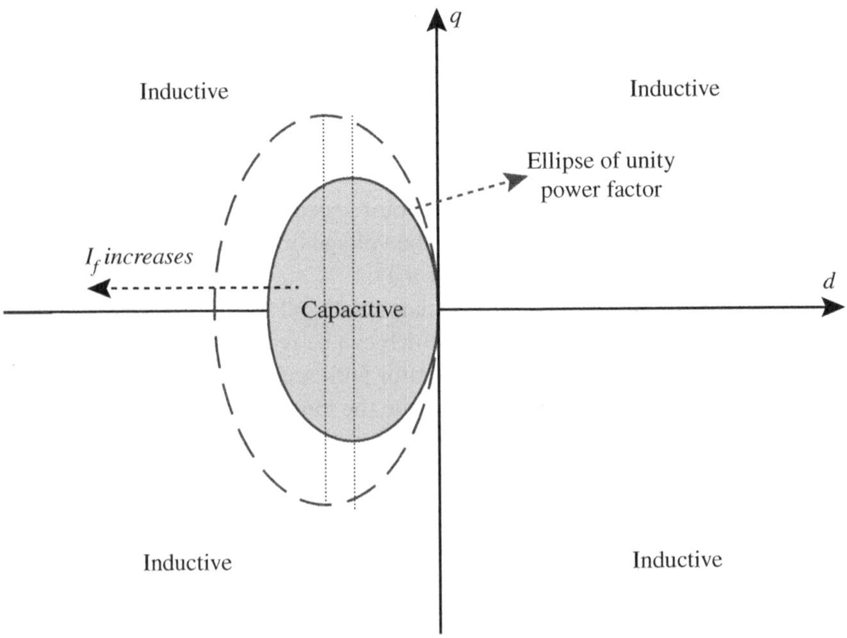

Figure 9.48 Unity power factor ellipses for WFSMs in *dq* plane.

where a_1 and b_1 are the ellipse's major semi-axis and minor semi-axis, respectively, and the ellipse's center is $(-a_1, 0)$. Here, a_1 and b_1 are defined as follows:

$$a_1 = \frac{L_m I_f}{2L_d} \qquad b_1 = \frac{L_m I_f}{2\sqrt{L_d L_q}} \tag{9.335}$$

As can be seen in Eqs. (9.334) and (9.335), as the field current I_f increases, the unity power factor ellipse expands, and vice versa. Figure 9.48 shows the changes of the unity power factor ellipses with the increase of the field current. Additionally, within the unity power factor ellipse, the WFSM becomes capacitive (i.e., leading power factor); likewise, outside of the ellipse, the WFSM becomes inductive (lagging power factor). To achieve unity power factor, the WFSM needs to be controlled along the loci of the unity power factor ellipse through controlling the ASDs.

Homework Problems

9.1 In this project, consider a 659-MVA, 60-Hz, 20-kV, 0.9-P.F., double-Y-connected, 4-pole, 3-phase turbogenerator (cylindrical-rotor synchronous machine), the various winding parameters (inductances and resistances) are given in SI units, using EECE 6210 Class Notes symbolism, as follows:

Stator and Rotor Resistances

$r_s = 0.741\,\text{m}\Omega,\ r_f = 86.00\,\text{m}\Omega$
$r_{kd1} = 0.1580\,\text{m}\Omega,\ r_{kd2} = 13.00\,\text{m}\Omega$
$r_{kq1} = 0.1227\,\text{m}\Omega,\ r_{kq2} = 17.40\,\text{m}\Omega$

Stator and Rotor Inductances

$L_{sa} = 1.95\,\text{mH},\ L_{sv} = 0.05\,\text{mH}$
$L_{ma} = 0.80\,\text{mH},\ L_{mv} = 0.05\,\text{mH}$
$L_{afm} = 26.10\,\text{mH},\ L_{ff} = 444\,\text{mH}$
$L_{fkd1} = 5.05\,\text{mH},\ L_{fkd2} = 12.38\,\text{mH}$
$L_{kd1f} = 6.29\,\text{mH},\ L_{kd2f} = 12.38\,\text{mH}$
$L_{akdm1} = 0.447\,\text{mH},\ L_{akdm2} = 0.790\,\text{mH}$
$L_{akqm1} = 0.670\,\text{mH},\ L_{akqm2} = 0.378\,\text{mH}$
$L_{kd1kd1} = 0.1254\,\text{mH},\ L_{kd1kd2} = 0.190\,\text{mH}$
$L_{ld2ld1} = 0.148\,\text{mH},\ L_{kd2kd2} = 0.4158\,\text{mH}$
$L_{kq1kq1} = 0.3762\,\text{mH},\ L_{kq1kq2} = 0.159\,\text{mH}$
$L_{kq2kq1} = 0.160\,\text{mH},\ L_{kq2kq2} = 0.1375\,\text{mH}$

Perform the following tasks:

(a) Verify that

$$L_{dd} = 2.825\,\text{mH},\ L_{qq} = 2.675\,\text{mH and } L_{oo} = 0.35\,\text{mH}$$

(b) At no-load, calculate the necessary field excitation current, i_f, in amperes to produce a rated terminal line-to-line voltage of 20 kV. Hence, obtain the necessary field excitation voltage, v_f, in volts, which would sustain this level of excitation current under steady-state condition.

(c) Under the earlier condition, find the initial value of the angular position of the *d*-axis of the machine, σ_0, taking the phase (*a*) voltage phasor as your reference.

(d) Find the initial values of the following state variables:
$i_a, i_b, i_c, i_d, i_q, i_o, i_f, i_{kd1}, i_{kd2}, i_{kq1}$, and i_{kq2}, as well as their corresponding flux linkages.

(e) Now, at time, $t = 0$, a three-phase to neutral short-circuit event occurs across the *a*, *b*, and *c* phase terminals of the machine, while v_f is held constant at its steady-state no-load value. Simulate this transient event using a "*dq0*" frame-of-reference, flux-linkage-based, state-space model, and obtain time-domain plots (profiles) of the following variables:
The flux linkages, $\lambda_d, \lambda_q, \lambda_o, \lambda_a, \lambda_b, \lambda_c, \lambda_f, \lambda_{kd1}, \lambda_{kd2}, \lambda_{kq1}$, and λ_{kq2}.
The currents, $i_d, i_q, i_o, i_a, i_b, i_c, i_f, i_{kd1}, i_{kd2}, i_{kq1}$, and i_{kq2}.
The developed electromagnetic torque. (Hint: 100 time increments/steps or samples are suggested here per one ac cycle of (1/60) seconds.)

(f) Repeat the work requested in (e), using an "*a-b-c*" frame of reference, flux-linkage-based, state-space model, and obtain time-domain plots (profiles) of the following variables:
The flux linkages, $\lambda_d, \lambda_q, \lambda_o, \lambda_a, \lambda_b, \lambda_c, \lambda_f, \lambda_{kd1}, \lambda_{kd2}, \lambda_{kq1}$, and λ_{kq2}.
The currents, $i_d, i_q, i_o, i_a, i_b, i_c, i_f, i_{kd1}, i_{kd2}, i_{kq1}$, and i_{kq2}.
The developed electromagnetic torque. (Hint: 100 time increments/steps or samples are suggested here per one ac cycle of (1/60) seconds and increase the sampling numbers to 200, 300, 400, 500, 600, … , etc. steps, until the obtained profiles become "almost identical" to those corresponding ones you obtained in (e).)

9.2 In this project, consider a 659-MVA, 60-Hz, 20-kV, 0.9-P.F., double-*Y*-connected, 4-pole, 3-phase turbogenerator (cylindrical-rotor synchronous machine), the various winding parameters (inductances and resistances) are given in SI units, using EECE 6210 Class Notes symbolism, as follows:

Stator and Rotor Resistances

$r_s = 0.741\ \text{m}\Omega$, $r_f = 86.00\ \text{m}\Omega$
$r_{kd1} = 0.1580\ \text{m}\Omega$, $r_{kd2} = 13.00\ \text{m}\Omega$
$r_{kq1} = 0.1227\ \text{m}\Omega$, $r_{kq2} = 17.40\ \text{m}\Omega$

Stator and Rotor Inductances

$L_{sa} = 1.95\,\text{mH}$, $L_{sv} = 0.05\,\text{mH}$
$L_{ma} = 0.80\,\text{mH}$, $L_{mv} = 0.05\,\text{mH}$
$L_{afm} = 26.10\,\text{mH}$, $L_{ff} = 444\,\text{mH}$
$L_{fkd1} = 5.05\,\text{mH}$, $L_{fkd2} = 12.38\,\text{mH}$
$L_{kd1f} = 6.29\,\text{mH}$, $L_{kd2f} = 12.38\,\text{mH}$
$L_{akdm1} = 0.447\,\text{mH}$, $L_{akdm2} = 0.790\,\text{mH}$
$L_{akqm1} = 0.670\,\text{mH}$, $L_{akqm2} = 0.378\,\text{mH}$
$L_{kd1kd1} = 0.1254\,\text{mH}$, $L_{kd1kd2} = 0.190\,\text{mH}$
$L_{kd2kd1} = 0.148\,\text{mH}$, $L_{kd2kd2} = 0.4158\,\text{mH}$
$L_{kq1kq1} = 0.3762\,\text{mH}$, $L_{kq1kq2} = 0.159\,\text{mH}$
$L_{kq2kq1} = 0.160\,\text{mH}$, $L_{kq2kq2} = 0.1375\,\text{mH}$

Perform the following tasks:

(a) The machine is operating at a full rated-load condition as an overexcited synchronous generator, that is delivering 659-MVA at 20-kV and 0.9-P.F. Calculate the necessary field excitation current, i_f, in amperes to produce a rated terminal line-to-line voltage of 20 kV. Hence, obtain the necessary field excitation voltage, v_f in volts, which would sustain this level of excitation current under steady-state condition.

(b) Under the earlier condition, find the initial value of the angular position of the *d*-axis of the machine, σ_o, taking the phase (*a*) voltage phasor as your reference.

(c) Find the initial values of the following state variables: i_a, i_b, i_c, i_d, i_q, i_o, i_f, i_{kd1}, i_{kd2}, i_{kq1}, and i_{kq2}, as well as their corresponding flux linkages.

(d) Now, from time, $t = 0$ to $t = (1/6)$ seconds, the machine operates at rated load condition, while v_f is held constant at its steady-state rated load value. Simulate the operation of the machine for the period $0 \le t \le (1/6)$ seconds, followed by a three-phase-to-neutral short-circuit for $(1/6) < t < (1/2)$ seconds. Use the "*dq*0" frame of reference, flux-linkage-based, state-space model, and obtain time-domain plots (profiles) of the following variables:

The flux linkages, λ_d, λ_q, λ_o, λ_a, λ_b, λ_c, λ_f, λ_{kd1}, λ_{kd2}, λ_{kq1}, and λ_{kq2}.

The currents, i_d, i_q, i_o, i_a, i_b, i_c, i_f, i_{kd1}, i_{kd2}, i_{kq1}, and i_{kq2}.

The developed electromagnetic torque. (Hint: 100 time increments/steps or samples are suggested here per one ac cycle of (1/60) seconds).

(e) Repeat the work requested in (d), using an "*a*-*b*-*c*" frame of reference, flux-linkage-based, state-space model, and obtain time-domain plots (profiles) of the variables you were requested to plot in (d).

(Hint: You may start using 100 time increments/steps or samples per one cycle of (1/60) seconds, and increase the sampling numbers to 200, 300, 400, 500, 600, … , etc., steps until the obtained profiles become "almost identical" to those corresponding ones you obtained in (d).)

(f) Compare the results and the CPU times of the simulations performed in (d) and (e) and discuss as well as analyze (comment on) your results.

9.3 Consider the six switches, Q_1 through Q_6, in the three-phase ac inverter, given in Chapter 8 on permanent magnet brushless dc (PMBLDC) motors. This type of inverter can be operated using a switching sequence whose conduction period per ac cycle per switch can be of (180°*e*) or (120°*e*) duration, as shown in Figure P3.1 of this project statement. The corresponding ac output voltage waveforms for the 180°*e* and 120°*e* conduction duration are shown in Figures P3.2 and P3.3, respectively. For PMBLDC motors, only the (120°*e*) conduction duration per switch per ac cycle is usually applied. In Figures P3.2 and P3.3, V is the dc input voltage to the inverter. Meanwhile, v_{an}, v_{bn}, and v_{cn}, are the line-to-neutral (phase) voltages on the ac output side of the three-phase inverter, and v_{ab} is a line-to-line voltage. Again, here the three-phase ac inverter switches, Q_1 through Q_6, are assumed to be operating using the (120°*e*) conduction duration per switch per ac cycle, under an *a*-*b*-*c* positive sequence regime.

Now consider the 15-hp, 8600-rev/min, 120-V, 6-pole, samarium–cobalt, radially oriented PMBLDC motor of Refs. [1, 2] given to you. This 15-hp PMBLDC motor is energized from a six-switch inverter, whose adjustable commutation (switch firing) angle, β, see Figure P3.4, can vary from 0°*e* to 30°*e*.

When the two-quadrant chopper is operated under a continuous "on" state, the chopper output voltage, which is the dc input voltage into the inverter, is equal to the dc-bus (source) voltage, v_s, see Figure P3.4. In this figure, σ is the instantaneous value of the position angle of the *d*-axis of the rotor with respect to the magnetic axis of the phase (*a*) winding. The relationship between the fundamental component of the induced back-EMF in phase (*a*), e_{a1}, due the PM flux, and the *d*-axis position angle, σ, is depicted in Figure P3.4. Here, $\sigma(t) = (\sigma_0 + \omega t)$, where ω is the rotor speed in electrical rad/s, which will be taken as a constant in the simulations, and σ_0 is the initial value of the rotor position angle, σ. The angular relationships between the inverter output voltages, v_{an}, v_{bn}, v_{cn}, and the rotor position angle, σ, as well as the inverter switch commutation (firing) angle, β, and the fundamental component of the phase (*a*) induced back-EMF, e_{a1}, are all depicted in Figure P3.4 and elaborated analytically in Table P3.1. Notice

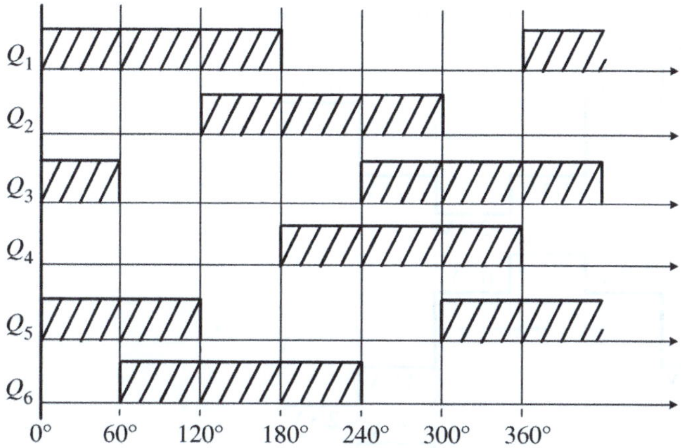

Inverter switching sequence for 180°e conduction.

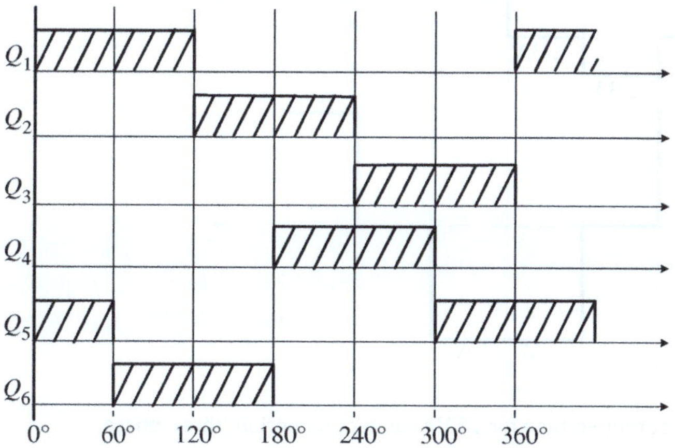

Inverter switching sequence for 120°e conduction.

Figure P3.1 Inverter switching sequence for 180°e and 120°e conduction.

that $\beta = 30°e$ for normal inverter switch firing (commutation) and $\beta = 0°e$ for a 30°e advanced inverter switch firing (commutation).

The induced EMFs in phases (a), (b), and (c) due to the PM flux were obtained in Fourier series form as a function of the angle, σ, and are given as follows:

$$e_a(\sigma) = \omega_m \sum_{n=1}^{13} A_n \cos[n(\sigma) - \Psi_n] \text{ V} \tag{9.336}$$

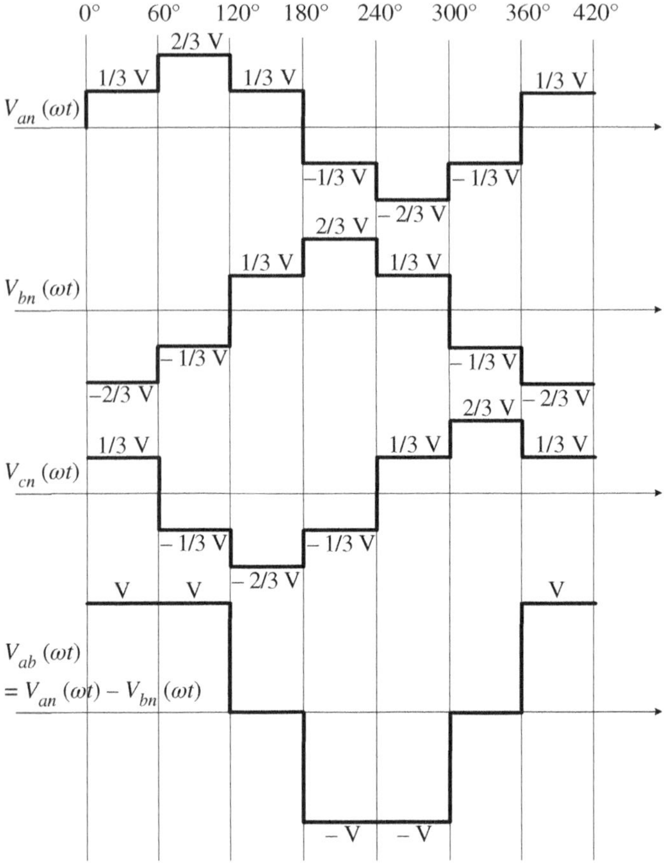

Figure P3.2 Inverter output voltage for 180°e conduction period (three-phase, six-switch inverter).

$$e_b(\sigma) = \omega_m \sum_{n=1}^{13} A_n \cos[n(\sigma - 2\pi/3) - \Psi_n]\,\text{V} \tag{9.337}$$

$$e_c(\sigma) = \omega_m \sum_{n=1}^{13} A_n \cos[n(\sigma - 4\pi/3) - \Psi_n]\,\text{V} \tag{9.338}$$

where ω_m is the rotor speed in mechanical radians per second (mech. rad/s), and the parameters, A_n and ψ_n, are given in Table P3.2. The three-phase armature winding's and PM sleeve's equivalent d and q damping resistances and self as well as mutual winding inductances are given for this 15-hp PMBLDC motor in Table P3.3.

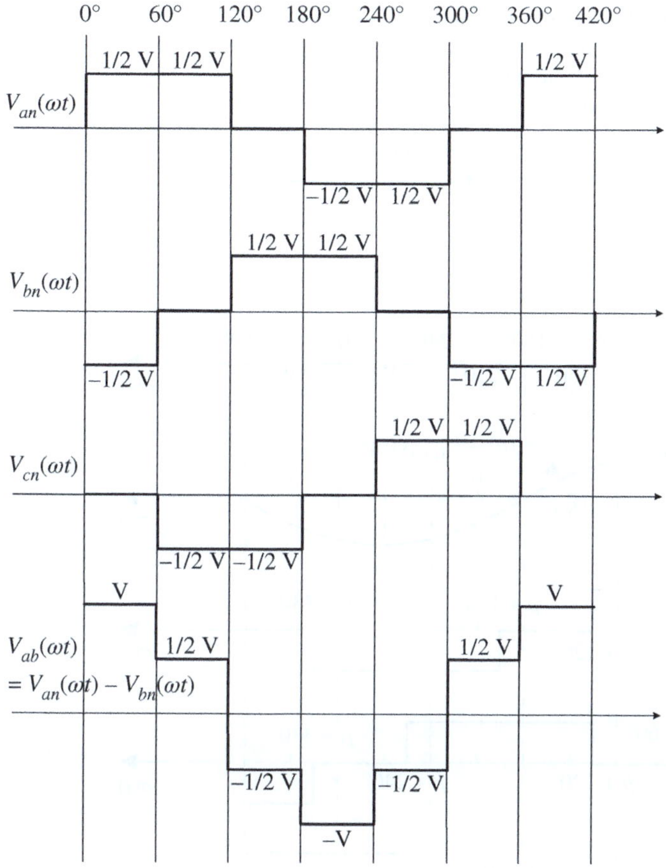

Figure P3.3 Inverter output voltage for 120°*e* conduction period (three-phase six-switch inverter).

Here, we have

$$L_{dd} = \underbrace{L_{sa} + L_{ma} + \frac{3}{2}L_v}_{(41.4644\times10^{-6}\text{H})} \text{ or } \underbrace{L_{sa} + L_{ma} + \frac{1}{2}L_{sv} + L_{mv}}_{(41.7398\times10^{-6}\text{H})}$$

$$L_{qq} = \underbrace{L_{sa} + L_{ma} - \frac{3}{2}L_v}_{(40.7617\times10^{-6}\text{H})} \text{ or } \underbrace{L_{sa} + L_{ma} - \frac{1}{2}L_{sv} - L_{mv}}_{(40.4862\times10^{-6}\text{H})}$$

$$L_{00} = \underbrace{L_{sa} - 2L_{ma}}_{(31.939\times10^{-6}\text{H})}$$

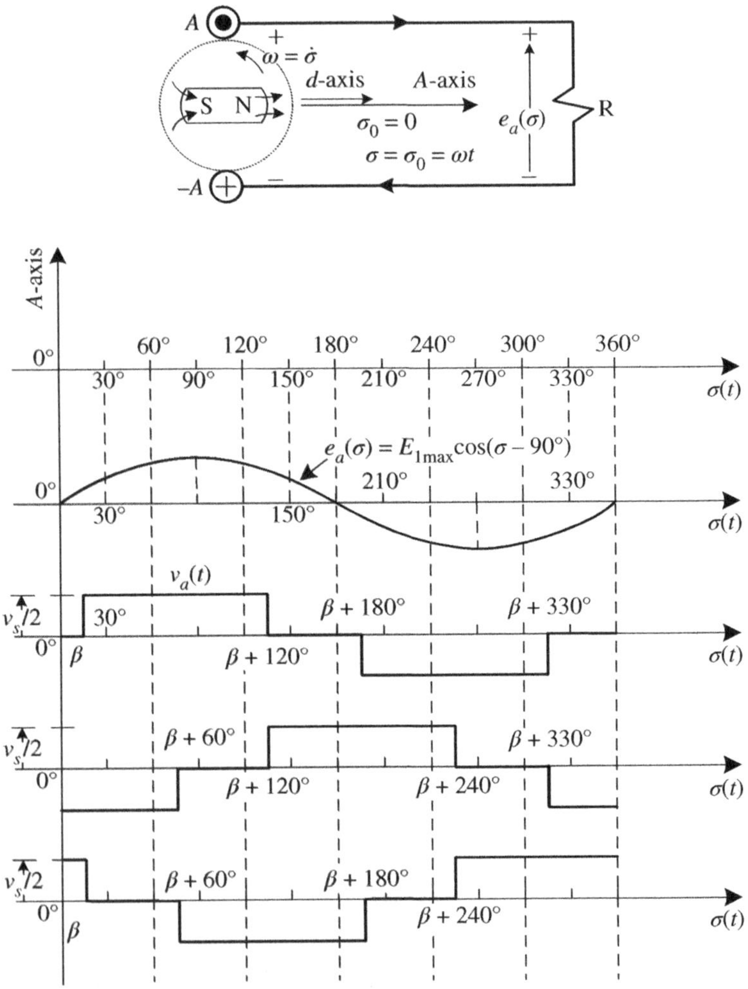

Figure P3.4 EMF in phase (A) and inverter output voltage, v_a, v_b, and v_c, for a 120°e three-phase, six-switch inverter.

where

$$L_{sa} = 38.055 \times 10^{-6}\text{H}$$

$$L_v = \frac{1}{2}(\underbrace{0.1518}_{L_{sv}} + \underbrace{0.5509}_{L_{mv}}) \times 10^{-6} = \frac{1}{2}(0.7027) \times 10^{-6}\text{H}$$

$$L_v = 0.35135 \times 10^{-6}\text{H}$$

$$L_{ma} = 3.058 \times 10^{-6}\text{H}$$

Table P3.1 Inverter output voltage functions and conduction periods.

The voltage outputs of the six-switch inverter:		
$v_a(t)$:	$v_a = 0$	$0 \le \sigma < \beta$
	$v_a = v_s/2$	$\beta \le \sigma \le \beta + 120^\circ$
	$v_a = 0$	$\beta + 120^\circ < \sigma < \beta + 180^\circ$
	$v_a = -v_s/2$	$\beta + 180^\circ \le \sigma \le \beta + 300^\circ$
	$v_a = 0$	$\beta + 300^\circ \le \sigma < 360^\circ$
$v_b(t)$:	$v_b = -v_s/2$	$0 \le \sigma \le \beta + 60^\circ$
	$v_b = 0$	$\beta + 60^\circ < \sigma < \beta + 120^\circ$
	$v_b = v_s/2$	$\beta + 120^\circ \le \sigma \le \beta + 240^\circ$
	$v_b = 0$	$\beta + 240^\circ < \sigma < \beta + 300^\circ$
	$v_b = -v_s/2$	$\beta + 300^\circ \le \sigma < 360^\circ$
$v_c(t)$:	$v_c = v_s/2$	$0 \le \sigma \le \beta$
	$v_c = 0$	$\beta < \sigma < \beta + 60^\circ$
	$v_c = -v_s/2$	$\beta + 60^\circ \le \sigma \le \beta + 180^\circ$
	$v_c = 0$	$\beta + 180^\circ < \sigma < \beta + 240^\circ$
	$v_c = v_s/2$	$\beta + 240^\circ \le \sigma < 360^\circ$

Table P3.2 EMF parameters.

n	A_n (Volts/mech. rad/s)	Ψ_n (radians)
1	68.7880×10^{-3}	$\pi/2$
3	1.3281×10^{-3}	$3\pi/2$
5	1.8833×10^{-3}	$-\pi/2$
7	0.0594×10^{-3}	$\pi/10$
9	0.1751×10^{-3}	$3\pi/2$
11	0.4055×10^{-3}	$-\pi/2$
13	0.3360×10^{-3}	$-\pi/2$

$$L_{akdm} = 2.9760 \times 10^{-6}\text{H}$$

$$L_{akqm} = 3.4550 \times 10^{-6}\text{H}$$

$$L_{kdkd} = 1.17 \times 10^{-6}\text{H}$$

$$L_{kqkq} = 1.35 \times 10^{-6}\text{H}$$

Table P3.3 15 hp brushless DC motor inductances and resistances.

Parameter	Expression
$r_a = r_b = r_c = r_s$	$2.35 \times 10^{-3}\,\Omega$
$r_{sd} = r_{sq}$	$61.8 \times 10^{-3}\,\Omega$
$r_{kd} = r_{kq}$	$2.26 \times 10^{-3}\,\Omega$
L_{aa}	$[38.055 + 0.1518\cos(2\sigma) + 0.0286\cos(2\sigma)] \times 10^{-6}$ H
L_{ab}	$[-3.058 - 0.5509\sin(2\sigma - \pi/3) - 0.08345\sin(4\sigma - \pi/3)] \times 10^{-6}$ H
L_{sdsd}	$[1.17] \times 10^{-6}$ H
L_{asd}	$[2.970\cos(\sigma) + 0.0069\cos(3\sigma) + 0.0154\cos(5\sigma)] \times 10^{-6}$ H
L_{sqsq}	$[1.35] \times 10^{-6}$ H
L_{asq}	$[3.4550\cos(\sigma + \pi/2) + 0.0179\cos(3\sigma + \pi/2) + 0.0134\cos(5\sigma + \pi/2)] \times 10^{-6}$ H
L_{kdkd}	$[1.15] \times 10^{-6}$ H
L_{akd}	$[2.9010\cos(\sigma) + 0.0067\cos(3\sigma) + 0.0150\cos(5\sigma)] \times 10^{-6}$ H
L_{kqkq}	$[1.20] \times 10^{-6}$ H
L_{akq}	$[3.1600\cos(\sigma + \pi/2) + 0.0163\cos(3\sigma + \pi/2) + 0.0123\cos(5\sigma + \pi/2)] \times 10^{-6}$ H
L_{sdkd}	$[1.12] \times 10^{-6}$ H
L_{sqkq}	$[1.25] \times 10^{-6}$ H

$$r_s = 2.35 \times 10^{-3}\,\Omega$$

$$r_{sd} = r_{sq} = 61.8 \times 10^{-3}\,\Omega$$

The student is requested to develop the state-space model of this motor in the *a-b-c* frame, current frame of reference, and in the *dq*0 frame, current frame of reference, for the purposes of carrying out the simulations requested in the following.

(1a) For a PMBLDC motor speed of 8350 = rev/min, at an inverter dc input supply voltage, $v_s = 116$ V, and a switch commutation (firing) angle, $\beta = 30°e$, obtain the waveforms of the "instantaneous steady-state" *a*, *b*, and *c* phase currents, as well as the developed torque profile, assuming that the initial rotor position angle, $\sigma(0) = \sigma_0 = 0°e$, and the initial values of the currents, $i_a(0) = i_b(0) = i_c(0) = 0$, using the following:

(1b) The *a-b-c* frame-current frame of reference state-space model, thus including all the effects of the space harmonics in the PM-induced EMFs, e_a, e_b, and e_c, and in all the winding inductances including the equivalent sleeve dampings, *sd* and *sq*.

(1c) The $dq0$ frame-current frame of reference state-space model, thus excluding all the effects of space harmonics that are caused by space harmonic terms higher than the second order in the PM-induced EMFs and in all the winding inductances.
In the simulations of (1b) and (1c),use at least 60 integration steps per ac cycle, and note that the simulation may require more than 10 ac cycles to reach the "instantaneous steady-state" condition. Instantaneous steady-state is considered to have been reached when any decaying dc transient current has nearly vanished, and the positive peak value of an armature phase current in two consecutive cycles remains nearly at the same value.

(1d) From the torque profiles obtained in (1b) and (1c), determine the magnitudes of the average developed torque in Nm and the average developed powers in kW and hp. Compare your results with those in Table (1) of Ref. [1] and comment on this comparison with whatever physical insights you may have gained from this experience. Also, compare your phase current profiles over the last complete ac cycle of simulation in (1b) and (1c) with those in the test oscillogram of Figure (5) and the simulation waveform of Figure (6) in Ref. [1], and comment on this comparison with your physical insights of the reasons for any discrepancies between your simulations and those of Ref. [1].

(2a) Repeat the simulations in (1b) for a value of the switching (firing) angle, $\beta = 0°e$, that is, a $30°e$ of advanced inverter transistor switching (firing), the so-called flux weakening operation. Compare the results of the maximum phase current magnitude, as well as the average developed torque, developed power, and percentage torque ripple magnitudes in (1) and (2), and comment with physical insight on the effects of advancing the inverter transistor switching (firing) angle on these PMBLDC motor performance characteristics.

(2b) Repeat the simulations in (1b) and (2a), when $v_s = 120\,\text{V}$. Hence, tabulate the results of the average developed torque, developed power, percentage torque ripple, and peak phase current magnitude, for dc bus voltage $v_s = 116\,\text{V}$ ($\beta = 30°e$, $0°e$) and dc bus voltage $v_s = 120\,\text{V}$ ($\beta = 30°e$, $0°e$), and comment on the effects of advancing the inverter switching and increasing the dc supply voltage, v_s, on the developed torque, developed power, and percentage torque ripple.

9.4 Consider the 15-hp, 8600-rev/min, 120-V, 6-pole, samarium–cobalt, radially oriented permanent magnet brushless dc (PMBLDC) motor of Project #3.

This PMBLDC motor was the subject of two studies detailed in Refs. [15, 16] in Project #3. This PMBLDC motor system including its associated electronic power conditioner (dc bus–chopper–inverter), see the circuit schematic of Figure P4.1, is to be modeled in its entirety, see the circuit model of the motor-power conditioner system of Figure P4.2. Use a software package of your choosing (P-Spice or MATLAB-SIMULINK, etc.). Develop the requested model for simulating the performance of the PMBLDC motor

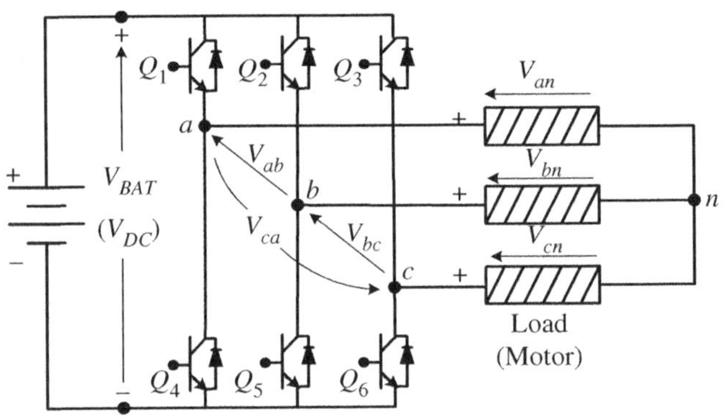

Figure P4.1 Voltage–source inverters schematic of switches.

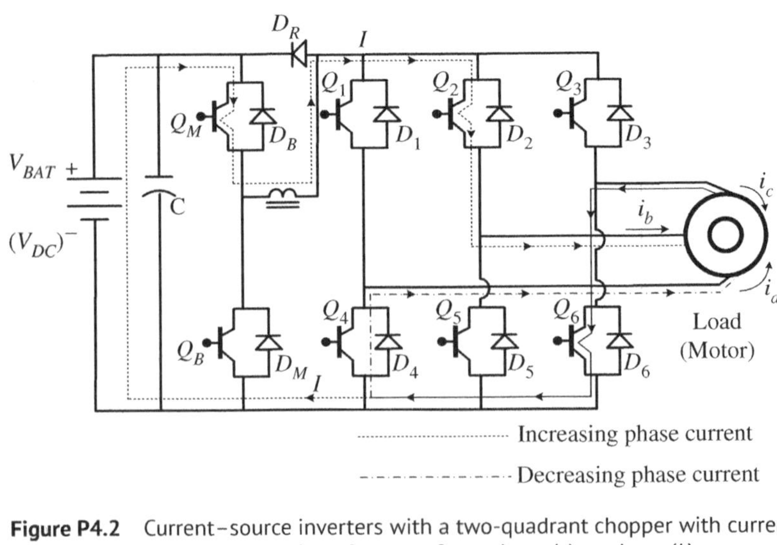

Figure P4.2 Current–source inverters with a two-quadrant chopper with current patterns during the commutation of current from phase (a) to phase (b).

system using the *abc* frame of reference. Use this model in the following simulations and studies:

(a) Repeat the work and analysis requested in item (a) of Project #3 in the *abc* frame including effects of all the space harmonics. That is, excluding the *dq*0 frame portion of the modeling, simulations, and associated comparisons.
(b) Repeat the work and analysis requested in item (b) of Project #3, using modeling and formulations based on the *abc* frame including effects of all the space harmonics.
(c) Repeat the work and analysis requested in item (c) of Project #3, using modeling and formulations based on the *abc* frame including effects of all the space harmonics.

Compare your results with the corresponding results in Ref. [15] of Project #3 and comment with proper insightful discussions, see the simulated current waveforms and oscillograms of the phase currents in Ref. [15].

References

1 Nehl, T.W., Fouad, F.A., and Demerdash, N.A. (1982). Determination of saturated values of rotating machinery incremental and apparent inductances by an energy perturbation method. *IEEE Transactions on Power Apparatus and Systems* PAS-101: 4441–4451.

2 Demerdash, N.A., Fouad, F.A., and Nehl, T.W. (1982). Determination of winding inductances in ferrite type permanent magnet electric machinery by finite elements. *IEEE Transactions on Magnetics* MAG-18: 1052–1054.

3 Demerdash, N.A., Hijazi, T.M., and Arkadan, A.A. (1988). Computation of winding inductances of permanent magnet brushless DC motors with damper windings by energy perturbation. *IEEE Transactions on Energy Conversion* EC-3 (3): 705–713.

4 Arkadan, A.A., Demerdash, N.A., Vaidya, J.G., and Shah, M.R. (1988). Impact of load on winding inductances of permanent magnet generators with multiple damping circuits using energy perturbation. *IEEE Transactions on Energy Conversion* EC-3 (4): 880–889.

5 Arakadan, A.A. and Demerdash, N.A. (1988). Modeling of transients in permanent magnet generators with multiple damping circuits using the natural ABC frame of reference. *IEEE Transactions on Energy Conversion* EC-3 (3): 715–722.

6 Chaudrhry, S.R., Ahmed-Zaid, S., and Demerdash, N.A. (1995). Coupled finite-element/state-space modeling of turbogenerators in the ABC frame of reference – the no-load case. *IEEE Transactions on Energy Conversion* 10 (1): 56–62.

7 Chaudrhry, S.R., Ahmed-Zaid, S., and Demerdash, N.A. (1995). Coupled finite/state-space modeling of turbogenerators in the ABC frame of reference – the short-circuit and load cases including saturated parameters. *IEEE Transactions on Energy Conversion* 10 (1): 63–70.

8 Deng, F. and Demerdash, N.A. (1996). A coupled finite element – state space approach for synchronous generators – Part I: model development. *IEEE Transactions on Aerospace and Electronic Systems* 32 (2): 775–784.

9 Deng, F., Demerdash, N.A., Vaidya, J.G., and Shah, J.J. (1996). A coupled finite element – state space approach for synchronous generators – Part II: applications. *IEEE Transactions on Aerospace and Electronic Systems* 32 (2): 785–794.

10 Deng, F. and Demerdash, N.A. (1997). CFE-SS approach for synchronous generators under unbalances – Part I: the unbalanced loads. *IEEE Transactions on Aerospace and Electronic Systems* 33 (1): 142–152.

11 Demerdash, N.A. and Deng, F. (1997). CFS-SS approach for synchronous generators under unbalances – Part II: the balanced and unbalanced rectifier loads. *IEEE Transactions on Aerospace and Electronic Systems* 33 (1): 152–162.

12 Deng, F. and Demerdash, N.A. (1998). Comprehensive salient-pole synchronous machine parametric design analysis using time-step finite element – state space modeling techniques. *IEEE Transactions on Energy Conversion* 13 (3): 221–229.

13 Demerdash, N.A. and Nehl, T.W. (1999). Electric machinery parameters and torques by current and energy perturbations from field computations – Part I: theory and formulation. *IEEE Transactions on Energy Conversion* 14 (4): 1507–1513.

14 Demerdash, N.A. and Nehl, T.W. (1999). Electric machinery parameters and torques by current and energy perturbations from field computations – Part II: applications and results. *IEEE Transactions on Energy Conversion* 14 (4): 1514–1522.

15 Hijazi, T.M. and Demerdash, N.A. (1988). Computer-aided modeling and experimental verification of the performance of power conditioner operated permanent magnet brushless DC motors including rotor damping effects. *IEEE Transactions on Energy Conversion* 3 (3): 714–721.

16 Hijazi, T.M. and Demerdash, N.A. (1988). Impact on the addition of a rotor-mounted damper cage on the performance of samarium – cobalt permanent magnet brushless DC motor systems. *IEEE Transactions on Energy Conversion* 3 (4): 890–898.

10

State-Space Modeling of Induction Machines

The constructural features of poly-phase induction machines, their AC steady-state T-equivalent circuit models, and other steady-state performance modeling and computations are previously covered in detail in Chapter 6. This includes their starting performance, as well as the various methods of torque–speed control, including scalar constant volts per hertz and vector control of induction motors in adjustable speed drives (ASDs). Meanwhile, in this chapter, we present the dynamic time-domain (instantaneous) state-space modeling and formulation of induction motors, including the mechanical dynamics of induction motor drives and associated loads. Coverage in this chapter will include derivation of the parameter of the state model in terms of the basic geometric dimensions such as the airgap height, mid-airgap diameter, specifics of the windings layouts, and slot geometries. We will start with the discrete representation of the windings of a wound-rotor induction machine in the so-called natural frame of reference of the A, B, and C phase windings on the stator, as well as the a, b, and c phase windings on the rotor.

10.1 Discrete Representation of the Windings of a Wound-Rotor Induction Machine in the Natural Frame of Reference

A schematic representation of the distributed nature of the three-phase windings in a stator and wound-rotor of an induction machine is shown in Figure 10.1. In this figure, a schematic layout of the three stator phase belts, (A, $-A$), (B, $-B$), and (C, $-C$), as well as the magnetic axes of these phases, A, B, and C, are illustrated. Also, shown in Figure 10.1 is a schematic layout of the three rotor phase belts, (a, $-a$), (b, $-b$), and (c, $-c$), as well as the magnetic axes of these phases, a, b, and c. Illustrated in this figure also, is the instantaneous angle, σ, always in electrical

Electric Machinery and Drives: An Electromagnetics Perspective, First Edition.
Nabeel A. O. Demerdash, JiangBiao He, and Hao Chen.

Companion website: www.wiley.com/go/demerdash/electricmachineryanddrives

measure (electrical radians), which is a function of an initial position and rotor speed, $d\sigma/dt = \dot{\sigma}$, that varies according to the mechanical dynamics of the rotating masses, including the rotor and the associated mechanical load, as will be detailed later on in this chapter.

A concentrated coil representation of distributed three-phase windings of the stator and wound-rotor is shown in Figure 10.2. The magnetic axes of phases, A, B, and C, on the stator, as well as the magnetic axes of phases, a, b, and c, on the rotor are included in this figure. Next, we shall use Figures 10.1 and 10.2 to develop the dynamic time-domain state-space model of an induction machine, with a wound-rotor winding, in the natural ABC–abc frame of reference.

10.2 State-Space Modeling Derivations of an Induction Machine in the Natural *ABC*–*abc* Frame of Reference for the Poly-phase Stator and Rotor Windings

Consider the three-phase wound-rotor induction machine shown schematically in Figure 10.1, whose stator and rotor phases, A, B, C, a, b, and c, are represented,

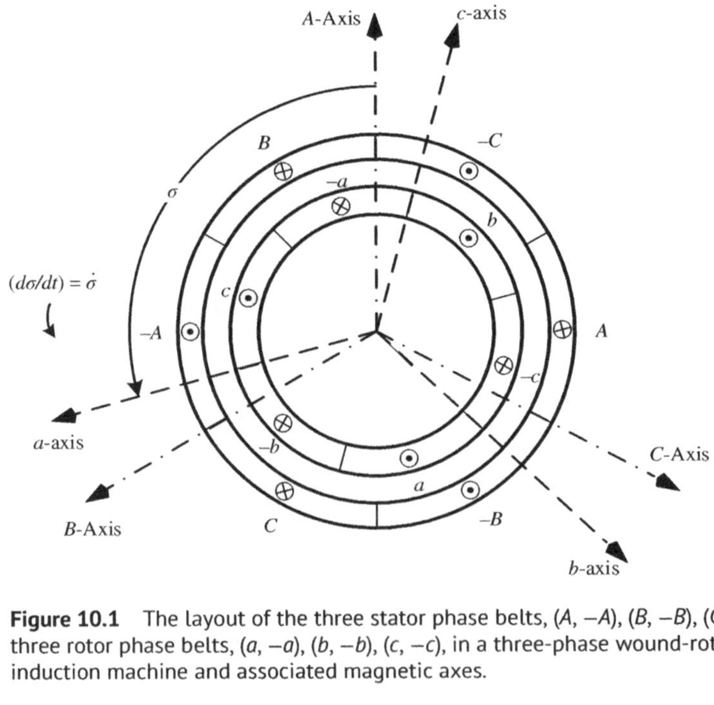

Figure 10.1 The layout of the three stator phase belts, $(A, -A)$, $(B, -B)$, $(C, -C)$, and the three rotor phase belts, $(a, -a)$, $(b, -b)$, $(c, -c)$, in a three-phase wound-rotor two-pole induction machine and associated magnetic axes.

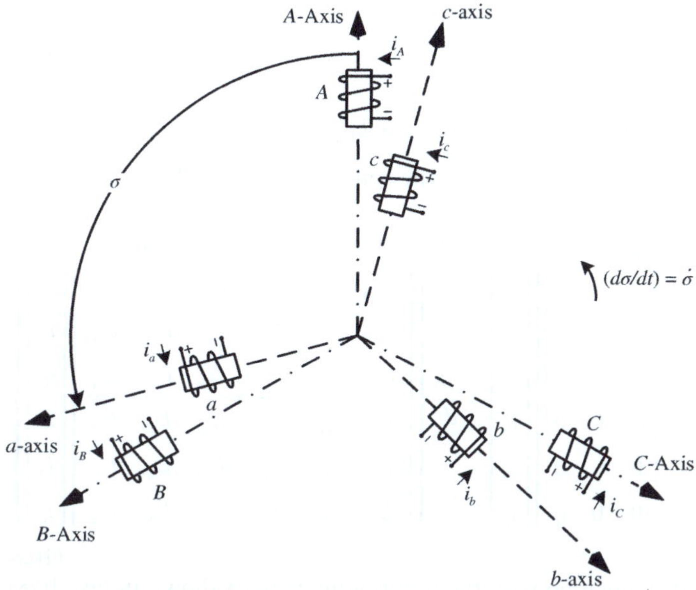

Figure 10.2 Three-phase equivalent coil schematic of the *A*, *B*, and *C* stator phases and the *a*, *b*, and *c* wound-rotor phases and their associated magnetic axes of an induction machine.

each by a coil, in Figure 10.2. The terminal voltage, v_A, of phase A, can be expressed using Faraday's law as follows:

$$v_A = r_A i_A + d\lambda_A/dt \tag{10.1}$$

where i_A is the current in phase A, and λ_A is the flux-linkage of phase A, which can be expressed in terms of the self and mutual inductances between phase A and the other phases, B, C, a, b, and c, as well as the currents in these phases, as follows:

$$\lambda_A = (L_{AA} i_A + L_{AB} i_B + L_{AC} i_C + L_{Aa} i_a + L_{Ab} i_b + L_{Ac} i_c) \tag{10.2}$$

where i_A, i_B, i_C, i_a, i_b, and i_c are the currents in the phases, L_{AA} is the self-inductance of phase A, and L_{AB}, L_{AC}, L_{Aa}, L_{Ab}, and L_{Ac} are the mutual inductances between phase A and the other phases B, C, a, b, and c, respectively. Hence, Eq. (10.1) can be rewritten as follows:

$$v_A = r_A i_A + \frac{d}{dt}(L_{AA} i_A + L_{AB} i_B + L_{AC} i_C + L_{Aa} i_a + L_{Ab} i_b + L_{Ac} i_c) \tag{10.3}$$

Accordingly, for phases B and C, similar expressions are written as follows:

$$v_B = r_B i_B + \frac{d}{dt}(\lambda_B) \tag{10.4}$$

$$v_B = r_B i_B + \frac{d}{dt}(L_{AB} i_A + L_{BB} i_B + L_{BC} i_C + L_{Ba} i_a + L_{Bb} i_b + L_{Bc} i_c) \tag{10.5}$$

Also,

$$v_C = r_C i_C + \frac{d}{dt}(\lambda_C) \tag{10.6}$$

$$v_C = r_C i_C + \frac{d}{dt}(L_{AC}i_A + L_{BC}i_B + L_{CC}i_C + L_{Ca}i_a + L_{Cb}i_b + L_{Cc}i_c) \tag{10.7}$$

Similar expressions can be written for the coils representing phases, a, b and c, on the rotor.

Hence, the resulting model can be written in matrix form as follows:

$$\begin{bmatrix} v_A \\ v_B \\ v_C \\ \hdashline v_a \\ v_b \\ v_c \end{bmatrix} = \left[\begin{array}{ccc:ccc} r_A & 0 & 0 & & & \\ 0 & r_B & 0 & & \underline{0} & \\ 0 & 0 & r_C & & & \\ \hdashline & & & r_a & 0 & 0 \\ & \underline{0} & & 0 & r_b & 0 \\ & & & 0 & 0 & r_c \end{array}\right] \begin{bmatrix} i_A \\ i_B \\ i_C \\ \hdashline i_a \\ i_b \\ i_c \end{bmatrix} + \frac{d}{dt}\left\{\left[\begin{array}{ccc:ccc} L_{AA} & L_{BA} & L_{AC} & L_{Aa} & L_{Ab} & L_{Ac} \\ L_{AB} & L_{BB} & L_{BC} & L_{Ba} & L_{Bb} & L_{Bc} \\ L_{AC} & L_{BC} & L_{CC} & L_{Ca} & L_{Cb} & L_{Cc} \\ \hdashline L_{Aa} & L_{Ba} & L_{Ca} & L_{aa} & L_{ab} & L_{ac} \\ L_{Ab} & L_{Bb} & L_{Cb} & L_{ab} & L_{bb} & L_{bc} \\ L_{Ac} & L_{Bc} & L_{Cc} & L_{ac} & L_{bc} & L_{cc} \end{array}\right] \begin{bmatrix} i_A \\ i_B \\ i_C \\ \hdashline i_a \\ i_b \\ i_c \end{bmatrix}\right\} \tag{10.8}$$

In compact matrix notation with the partitioning pattern shown in Eq. (10.8), one can write the following:

$$\begin{bmatrix} \underline{v}_{ABC} \\ \underline{v}_{abc} \end{bmatrix} = \left[\begin{array}{c:c} \underline{R}_{ss} & \underline{0} \\ \hdashline \underline{0} & \underline{R}_{ss} \end{array}\right] \begin{bmatrix} \underline{I}_{ABC} \\ \underline{I}_{abc} \end{bmatrix} + \frac{d}{dt}\left\{\left[\begin{array}{c:c} \underline{L}_{ss} & \underline{L}_{sr} \\ \hdashline \underline{L}^t_{sr} & \underline{L}_{rr} \end{array}\right] \begin{bmatrix} \underline{I}_{ABC} \\ \underline{I}_{abc} \end{bmatrix}\right\} \tag{10.9}$$

Now, one must examine the inductance coefficients which constitute the inductance matrices, $\underline{\mathbf{L}}_{ss}$, $\underline{\mathbf{L}}_{sr}$, $\underline{\mathbf{L}}^t_{sr}$, and $\underline{\mathbf{L}}_{rr}$. Consider the coefficients of inductances forming $\underline{\mathbf{L}}_{ss}$, from the geometries in Figure 10.1, the inductance coefficients, L_{AA}, $L_{BB,}$ and L_{CC}, are equal since the flux paths for the three windings, A, B, and C, are identical.

Also, these self-inductances are independent of the rotor position, σ. Hence, L_{AA}, $L_{BB,}$ and L_{CC} can the expressed as follows:

$$L_{AA} = L_{BB} = L_{CC} = L_{ss} \tag{10.10}$$

Similarly, it can be seen from Figure 10.1 that the mutual inductances, $L_{AB} = L_{BA}$, $L_{AC} = L_{CA}$, and $L_{BC} = L_{CB}$, are equal in magnitude and that they are independent of the rotor position, σ. Hence, these inductances can be expressed as follows:

$$L_{AB} = L_{BA} = L_{BC} = L_{CB} = L_{AC} = L_{CA} = L_{sm} \tag{10.11}$$

Hence, the matrix, $\underline{\mathbf{L}}_{ss}$, can be expressed as follows:

$$\underline{\mathbf{L}}_{ss} = \begin{bmatrix} L_{ss} & L_{sm} & L_{sm} \\ L_{sm} & L_{ss} & L_{sm} \\ L_{sm} & L_{sm} & L_{ss} \end{bmatrix} \tag{10.12}$$

A similar examination of the coefficients forming $\underline{\mathbf{L}}_{rr}$ and Figure 10.1 show that L_{aa}, L_{bb}, and L_{cc} are independent of the rotor position and, hence,

$$L_{aa} = L_{bb} = L_{cc} = L_{rr} \tag{10.13}$$

Similarly,

$$L_{ab} = L_{bc} = L_{ca} = L_{rm} \tag{10.14}$$

Hence, $\underline{\mathbf{L}}_{rr}$ can be expressed as follows:

$$\underline{\mathbf{L}}_{rr} = \begin{bmatrix} L_{rr} & L_{rm} & L_{rm} \\ L_{rm} & L_{rr} & L_{rm} \\ L_{rm} & L_{rm} & L_{rr} \end{bmatrix} \tag{10.15}$$

The coefficients of inductances forming the matrix, $\underline{\mathbf{L}}_{sr}$, contrary to the previous cases, are dependent on the angular positions of the rotor windings with respect to the stator windings. From the geometries of Figures 10.1 and 10.2, it can be easily deduced that all these coefficients vary cosinusoidally with the rotor angular position, σ, with some phase differences.

It can be easily seen from Figures 10.1 and 10.2 that one can write the following:

$$L_{Aa} = L_{srm}\cos(\sigma) = L_{Bb} = L_{Cc} \tag{10.16}$$

$$L_{Ab} = L_{srm}\cos(\sigma + 2\pi/3) = L_{Bc} = L_{Ca} \tag{10.17}$$

$$L_{Ac} = L_{srm}\cos(\sigma + 4\pi/3) = L_{Ba} = L_{Cb} \tag{10.18}$$

Hence, the matrix, $\underline{\mathbf{L}}_{sr}$, can be expressed as follows:

$$\underline{\mathbf{L}}_{sr} = L_{srm}\begin{bmatrix} \cos(\sigma) & \cos(\sigma + 2\pi/3) & \cos(\sigma + 4\pi/3) \\ \cos(\sigma + 4\pi/3) & \cos(\sigma) & \cos(\sigma + 2\pi/3) \\ \cos(\sigma + 2\pi/3) & \cos(\sigma + 4\pi/3) & \cos(\sigma) \end{bmatrix} \tag{10.19}$$

for a steady-state operating condition of constant rotor speed, ω, in electrical measure.

Hence, $\underline{\mathbf{L}}^t_{sr}$, can be written as follows:

$$\underline{\mathbf{L}}^t_{sr} = L_{srm}\begin{bmatrix} \cos(\sigma) & \cos(\sigma + 4\pi/3) & \cos(\sigma + 2\pi/3) \\ \cos(\sigma + 2\pi/3) & \cos(\sigma) & \cos(\sigma + 4\pi/3) \\ \cos(\sigma + 4\pi/3) & \cos(\sigma + 2\pi/3) & \cos(\sigma) \end{bmatrix} \tag{10.20}$$

again, for steady-state operating condition of constant rotor speed, ω. Also,

$$r_A = r_B = r_C = r_s \text{ and } r_a = r_b = r_c = r_r \tag{10.21}$$

Thus, substituting for all the inductance matrices, $\underline{\mathbf{L}}_{ss}$, $\underline{\mathbf{L}}_{rr}$, $\underline{\mathbf{L}}_{sr}$, and $\underline{\mathbf{L}}_{sr}^{t}$ into Eq. (10.9) gives the state-space model for the wound-rotor induction machine as follows:

$$
\begin{bmatrix} v_A \\ v_B \\ v_C \\ \hdashline v_a \\ v_b \\ v_c \end{bmatrix}
=
\left[\begin{array}{ccc:ccc}
r_s & 0 & 0 & & & \\
0 & r_s & 0 & & \underline{0} & \\
0 & 0 & r_s & & & \\
\hdashline
 & & & r_r & 0 & 0 \\
 & \underline{0} & & 0 & r_r & 0 \\
 & & & 0 & 0 & r_r
\end{array}\right]
\begin{bmatrix} i_A \\ i_B \\ i_C \\ \hdashline i_a \\ i_b \\ i_c \end{bmatrix}
+ \frac{d}{dt}\left\{
\left[\begin{array}{ccc:ccc}
L_{ss} & L_{sm} & L_{sm} & L_{srm}\cos(\sigma) & L_{srm}\cos(\sigma + 2\pi/3) & L_{srm}\cos(\sigma + 4\pi/3) \\
L_{sm} & L_{ss} & L_{sm} & L_{srm}\cos(\sigma + 4\pi/3) & L_{srm}\cos(\sigma) & L_{srm}\cos(\sigma + 2\pi/3) \\
L_{sm} & L_{sm} & L_{ss} & L_{srm}\cos(\sigma + 2\pi/3) & L_{srm}\cos(\sigma + 4\pi/3) & L_{srm}\cos(\sigma) \\
\hdashline
L_{srm}\cos(\sigma) & L_{srm}\cos(\sigma + 4\pi/3) & L_{srm}\cos(\sigma + 2\pi/3) & L_{rr} & L_{rm} & L_{rm} \\
L_{srm}\cos(\sigma + 2\pi/3) & L_{srm}\cos(\sigma) & L_{srm}\cos(\sigma + 4\pi/3) & L_{rm} & L_{rr} & L_{rm} \\
L_{srm}\cos(\sigma + 4\pi/3) & L_{srm}\cos(\sigma + 2\pi/3) & L_{srm}\cos(\sigma) & L_{rm} & L_{rm} & L_{rr}
\end{array}\right]
\begin{bmatrix} i_A \\ i_B \\ i_C \\ \hdashline i_a \\ i_b \\ i_c \end{bmatrix}
\right\}
\tag{10.22}
$$

Taking into account the cylindrical geometry of the magnetic circuit of the induction motor and rotating MMF considerations for balanced three-phase AC windings, the inductance parameters are discussed in Section 10.3 immediately after completion of the state-space model formulation.

For induction machines with squirrel-cage type of rotors, such squirrel-cages can be represented by two short-circuited equivalent coils, k_d and k_q, along the direct-axis (d-axis) and quadrature-axis (q-axis), see Figure 10.3. This results in the following formulation in Eq. (10.23) replacing Eq. (10.22) for the wound-rotor machine:

$$\begin{bmatrix} v_A \\ v_B \\ v_C \\ \hdashline v_a \\ v_b \\ v_c \end{bmatrix} = \left[\begin{array}{ccc:cc} r_s & 0 & 0 & 0 & 0 \\ 0 & r_s & 0 & 0 & 0 \\ 0 & 0 & r_s & 0 & 0 \\ \hdashline 0 & 0 & 0 & r_{kd} & 0 \\ 0 & 0 & 0 & 0 & r_{kq} \end{array}\right] \begin{bmatrix} i_A \\ i_B \\ i_C \\ \hdashline i_a \\ i_b \\ i_c \end{bmatrix}$$

$$+\frac{d}{dt}\left\{\left[\begin{array}{ccc:cc} L_{ss} & L_{sm} & L_{sm} & L_{Akm}\cos(\sigma) & -L_{Akm}\sin(\sigma) \\ L_{sm} & L_{ss} & L_{sm} & L_{Akm}\cos(\sigma-2\pi/3) & -L_{Akm}\sin(\sigma-2\pi/3) \\ L_{sm} & L_{sm} & L_{ss} & L_{Akm}\cos(\sigma-4\pi/3) & -L_{Akm}\sin(\sigma-4\pi/3) \\ \hdashline L_{Akm}\cos(\sigma) & L_{Akm}\cos(\sigma-2\pi/3) & L_{Akm}\cos(\sigma-4\pi/3) & L_{kdkd} & L_{kdkq} \\ -L_{Akm}\sin(\sigma) & -L_{Akm}\sin(\sigma-2\pi/3) & -L_{Akm}\sin(\sigma-4\pi/3) & L_{kqkd} & L_{kqkq} \end{array}\right] \begin{bmatrix} i_A \\ i_B \\ i_C \\ i_a \\ i_b \\ i_c \end{bmatrix}\right\}$$

(10.23)

Here, $L_{kdkd} = L_{kqkq} = L_{kk}$, and $L_{kdkq} = L_{kqkd} = 0$. Now, a brief discussion on the relationship between the inductance parameters, L_{ss}, L_{sm}, L_{srm}, L_{rm}, and L_{rr}, and those of the conventional T-equivalent circuit, L_{1l}, L_{ϕ}, and L_{2l}, is highly desirable as given in Section 10.3.

Furthermore, since $\underline{\mathbf{L}}_{sr}$ and $\underline{\mathbf{L}}_{sr}^{t}$ are time dependent, it follows that a state-space model in the form ($\dot{\underline{X}} = \underline{A} \cdot \underline{X} + \underline{B} \cdot \underline{U}$) is not readily available from Eqs. (10.22) or (10.23). Therefore, a linear transformation to change the frames of references which enable one to obtain the model formulation in the form aforementioned ($\dot{\underline{X}} = \underline{A} \cdot \underline{X} + \underline{B} \cdot \underline{U}$) is necessary. This leads to subsequent sections in this chapter in which we study possible types of transformations and their effects on the machine model. Meanwhile, once more a brief glance is in order at the corresponding model to Eq. (10.23) for a squirrel-cage three-phase induction motor, with a k_d and k_q squirrel-cage circuit representation similar to that of the damper windings of a synchronous machine, discussed and detailed in Chapter 9, Eq. (9.30).

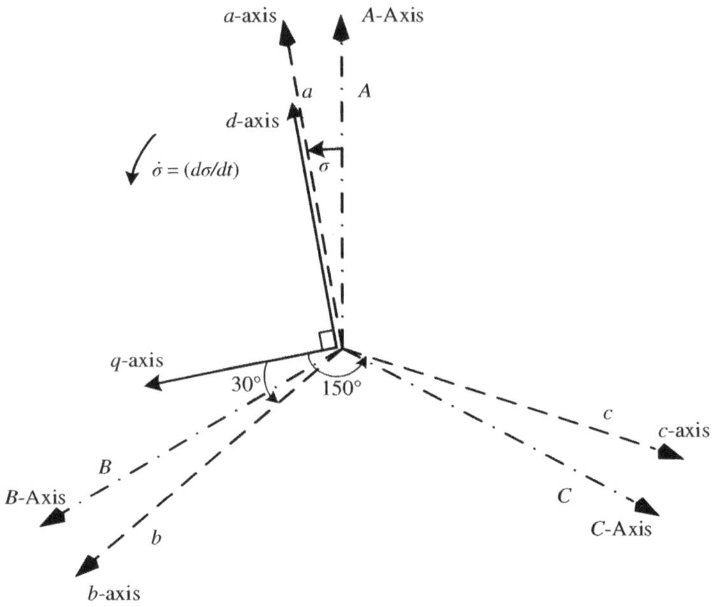

Figure 10.3 The choice of the locations of the *d*-axis along the axis of phase a of a wound-rotor machine.

10.3 On the Relationship Between the Terms of an Induction Machine's Inductance Matrix and the Inductances and Reactances of Its Conventional *T*-Equivalent Circuit

In this section, one must recall from earlier developments in Chapter 5, specifically Section 5.5, Eq. (5.102), that the resultant rotating MMF produced by a balanced three-phase AC armature can be expressed as:

$$\mathbf{F}_1(\theta, t) = (3/2)F_{\max}\cos(\theta - \omega t) \tag{10.24}$$

where

$$F_{\max} = (4\sqrt{2}/\pi)(k_W N_{ph} I/mp) \tag{10.25}$$

Here, $k_W = k_{W1}$ is the fundamental winding factor of the AC armature, N_{ph} is the number of turns per phase, I is the RMS current per phase, m is the number of parallel paths per phase, and p is the number of poles in the AC armature winding. Hence, one can write the following for $\mathbf{F}_1(\theta, t)$:

$$\mathbf{F}_1(\theta, t) = (6\sqrt{2}/\pi)(k_W N_{ph} I/mp)\cos(\theta - \omega t) \tag{10.26}$$

From Eq. (10.26), previous developments in Chapter 6, Section 6.4, Eqs. (6.54)–(6.67), the magnetizing inductance, L_ϕ, is given by the following expression:

$$L_\phi = (12/\pi)(N_{ph}k_w/mp)^2(\mu_0 Dl/h_g) \tag{10.27}$$

where all terms are as defined earlier and in Chapters 5 and 6.

Now, from Eq. (10.8), the flux linkage vector, $\underline{\Lambda}$, of the A, B, C, a, b, and c windings can be written as follows:

$$\underline{\Lambda} = [\lambda_A \ \lambda_B \ \lambda_C \ \lambda_a \ \lambda_b \ \lambda_c]^t \tag{10.28}$$

That is,

$$\underline{\Lambda} = \underline{\mathbf{L}} \cdot \underline{I} \tag{10.29}$$

where $\underline{I}$ is the current vector given by:

$$\underline{I} = \left[i_A \ i_B \ i_C \ i_a \ i_b \ i_c\right]^t \tag{10.30}$$

and $\underline{\mathbf{L}}$ is the apparent winding's inductance matrix, which can be rewritten here from Eq. (10.8), as follows:

$$\underline{\mathbf{L}} = \begin{bmatrix} L_{AA} & L_{AB} & L_{AC} & L_{Aa} & L_{Ab} & L_{Ac} \\ L_{BA} & L_{BB} & L_{BC} & L_{Ba} & L_{Bb} & L_{Bc} \\ L_{CA} & L_{CB} & L_{CC} & L_{Ca} & L_{Cb} & L_{Cc} \\ L_{aA} & L_{aB} & L_{aC} & L_{aa} & L_{ab} & L_{ac} \\ L_{bA} & L_{bB} & L_{bC} & L_{ba} & L_{bb} & L_{bc} \\ L_{cA} & L_{cB} & L_{cC} & L_{ca} & L_{cb} & L_{cc} \end{bmatrix} \tag{10.31}$$

From Eq. (10.29), one can express the current vector, $\underline{I}$, in terms of the flux linkage vector, $\underline{\Lambda}$, as follows:

$$\underline{I} = (\underline{\mathbf{L}}^{-1})\underline{\Lambda} \tag{10.32}$$

Therefore, Eq. (10.8) can be rewritten as follows:

$$\underline{V} = \underline{R} \cdot \underline{I} + \frac{d}{dt}(\underline{\Lambda}) \tag{10.33}$$

or,

$$\underline{V} = \underline{R} \cdot \underline{I} + \dot{\underline{\Lambda}} \tag{10.34}$$

where

$$\underline{V} = \left[v_A \ v_B \ v_C \ v_a \ v_b \ v_c\right]^t \tag{10.35}$$

and $\underline{R}$ is the (6×6) diagonal-type resistance matrix in Eq. (10.8).

Accordingly, from Eqs. (10.32) and (10.34), one can write the following:

$$\underline{V} = \underline{R} \cdot (\underline{\mathbf{L}}^{-1}) \cdot \underline{\Lambda} + \underline{\dot{\Lambda}} \tag{10.36}$$

or,

$$\frac{d}{dt}(\underline{\Lambda}) = \underline{\dot{\Lambda}} = (-\underline{R} \cdot \underline{\mathbf{L}}^{-1}) \cdot \underline{\Lambda} + \underline{V} \tag{10.37}$$

Equation (10.37) is the so-called state-space formulation for induction machines in the flux-linkage frame of reference.

The various inductance terms of the inductance matrix, $\underline{L}$, in Eq. (10.31) will be defined next. Given that N_s is the number of stator winding turns per phase, k_{ws} is the winding factor for the stator winding, N_r is the number of rotor winding turns per phase, k_{wr} is the winding factor for the rotor winding, L_{sl} is the leakage inductance of the stator winding per phase, L_{rl} is the leakage inductance of the rotor winding per phase, we can write the following for the various inductance terms in Eq. (10.31):

First,

$$L_{srm} = (2/3)(N_r k_{wr}/N_s k_{ws})L_\phi \tag{10.38}$$

where (2/3) factor is the inverse of (3/2) factor in Eq. (10.24), to correspond to the MMF amplitude of one phase, $F_{\max}$, not the resultant, $(3/2)F_{\max}$, and

$$L_{sm} = (2/3)L_\phi[\cos(120°\text{e})] = (-1/3)L_\phi \tag{10.39}$$

and,

$$L_{rm} = (2/3)(N_r k_{wr}/N_s k_{ws})^2[\cos(120°\text{e})]L_\phi$$

and the cosine argument of (120°e) is the electrical angle between the magnetic axes of any two adjacent phase in a three-phase armature winding, therefore,

$$L_{rm} = (-1/3)L_\phi(N_r k_{wr}/N_s k_{ws})^2 \tag{10.40}$$

This leads to the following for the inductance coefficients in the inductance matrix, $\underline{L}$, in Eq. (10.31):

$$L_{AA} = L_{BB} = L_{CC} = L_{ss} = L_{sl} + \left(\frac{N_s k_{ws}}{N_r k_{wr}}\right) L_{srm} \tag{10.41}$$

$$L_{AB} = L_{BA} = L_{BC} = L_{CB} = L_{CA} = L_{AC} = L_{sm} = -\frac{1}{2}\left(\frac{N_s k_{ws}}{N_r k_{wr}}\right) L_{srm} \tag{10.42}$$

$$L_{aa} = L_{bb} = L_{cc} = L_{rr} = L_{sl} + \left(\frac{N_s k_{ws}}{N_r k_{wr}}\right) L_{srm} \tag{10.43}$$

$$L_{ab} = L_{ba} = L_{bc} = L_{cb} = L_{ca} = L_{ac} = L_{rm} = -\frac{1}{2}\left(\frac{N_s k_{ws}}{N_r k_{wr}}\right) L_{srm} \tag{10.44}$$

$$L_{Aa} = L_{aA} = L_{Bb} = L_{bB} = L_{Cc} = L_{cC} = L_{srm}\cos(\sigma) \tag{10.45}$$

$$L_{Ab} = L_{bA} = L_{Bc} = L_{cB} = L_{Ca} = L_{aC} = L_{srm}\cos(\sigma - 4\pi/3) \tag{10.46}$$

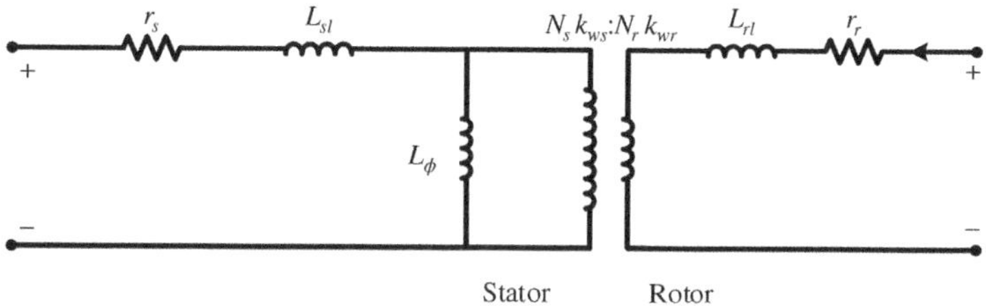

Figure 10.4 Per phase *T*-equivalent circuit of a three-phase induction machine.

and

$$L_{Ac} = L_{cA} = L_{Ba} = L_{aB} = L_{Cb} = L_{bC} = L_{srm}\cos(\sigma - 2\pi/3) \tag{10.47}$$

It is very much worthwhile here to recall the origins of the "*T*-Equivalent Circuit" of the induction machine, and more specifically in Chapter 6, Figures 6.12 and 6.14. Using the symbolism of the state-space model in Eq. (10.22), the "*T*-Equivalent Circuit" of an induction machine is shown in Figure 10.4. From earlier developments in Eqs. (10.38)–(10.47), one can write the following:

$$L_{ss} = L_{sl} + (2/3)L_\phi \tag{10.48}$$

$$L_{sm} = (-1/3)L_\phi \tag{10.49}$$

$$L_{rr} = L_{rl} + (2/3)(N_r k_{wr}/N_s k_{ws})^2 L_\phi \tag{10.50}$$

$$L_{rm} = (-1/3)(N_r k_{wr}/N_s k_{ws})^2 L_\phi \tag{10.51}$$

and

$$L_{srm} = (2/3)(N_r k_{wr}/N_s k_{ws})L_\phi \tag{10.52}$$

Finally, the various terms of the inductance matrix, $\underline{\mathbf{L}}$, in Eq. (10.31) can be rewritten, in terms of the *T*-equivalent circuit parameters in Figure 10.4, as follows:

$$L_{AA} = L_{BB} = L_{CC} = L_{ss} = L_{sl} + (2/3)L_\phi \tag{10.53}$$

$$L_{AB} = L_{BA} = L_{BC} = L_{CB} = L_{CA} = L_{AC} = (-1/3)L_\phi \tag{10.54}$$

$$L_{aa} = L_{bb} = L_{cc} = L_{rr} = L_{rl} + (2/3)(N_r k_{wr}/N_s k_{ws})L_\phi \tag{10.55}$$

$$L_{ab} = L_{ba} = L_{bc} = L_{cb} = L_{ca} = L_{ac} = L_{rm} = (-1/3)(N_r k_{wr}/N_s k_{ws})^2 L_\phi \tag{10.56}$$

$$\begin{aligned} L_{Aa} &= L_{aA} = L_{Bb} = L_{bB} = L_{Cc} = L_{cC} = L_{srm}\cos(\sigma) \\ &= (2/3)(N_r k_{wr}/N_s k_{ws})L_\phi \cos(\sigma) \end{aligned} \tag{10.57}$$

$$L_{Ab} = L_{bA} = L_{Bc} = L_{cB} = L_{Ca} = L_{aC} = L_{srm}\cos(\sigma - 4\pi/3)$$
$$= (2/3)(N_r k_{wr}/N_s k_{ws})L_\phi \cos(\sigma - 4\pi/3) \tag{10.58}$$

$$L_{Ac} = L_{cA} = L_{Ba} = L_{aB} = L_{Cb} = L_{bC} = L_{srm}\cos(\sigma - 2\pi/3)$$
$$= (2/3)(N_r k_{wr}/N_s k_{ws})L_\phi \cos(\sigma - 2\pi/3) \tag{10.59}$$

10.4 State-Space Model of an Induction Machine Using the Winding Flux Linkages as State Variables

Here, Eq. (10.9) can be rewritten as follows:

$$\begin{bmatrix} \underline{V}_{ABC} \\ \underline{V}_{abc} \end{bmatrix} = \begin{bmatrix} \underline{R}_{ss} & 0 \\ 0 & \underline{R}_{rr} \end{bmatrix} \begin{bmatrix} \underline{I}_{ABC} \\ \underline{I}_{abc} \end{bmatrix} + \frac{d}{dt}\begin{bmatrix} \underline{\Lambda}_{ABC} \\ \underline{\Lambda}_{abc} \end{bmatrix} \tag{10.60}$$

where

$$\underline{V}_{ABC} = \begin{bmatrix} v_A & v_B & v_C \end{bmatrix}^t \tag{10.61}$$

$$\underline{V}_{abc} = \begin{bmatrix} v_a & v_b & v_c \end{bmatrix}^t \tag{10.62}$$

$$\underline{I}_{ABC} = \begin{bmatrix} i_A & i_B & i_C \end{bmatrix}^t \tag{10.63}$$

$$\underline{I}_{abc} = \begin{bmatrix} i_a & i_b & i_c \end{bmatrix}^t \tag{10.64}$$

$$\underline{\Lambda}_{ABC} = \begin{bmatrix} \lambda_A & \lambda_B & \lambda_C \end{bmatrix}^t \tag{10.65}$$

$$\underline{\Lambda}_{abc} = \begin{bmatrix} \lambda_a & \lambda_b & \lambda_c \end{bmatrix}^t \tag{10.66}$$

$$\underline{R}_{ss} = \begin{bmatrix} r_s & 0 & 0 \\ 0 & r_s & 0 \\ 0 & 0 & r_s \end{bmatrix} \tag{10.67}$$

$$\underline{R}_{rr} = \begin{bmatrix} r_r & 0 & 0 \\ 0 & r_r & 0 \\ 0 & 0 & r_r \end{bmatrix} \tag{10.68}$$

and $\underline{0}$ is a (3×3) matrix of zeros. Also, using the definitions in Eqs. (10.12), (10.15), (10.19), and (10.20), one can write the following:

$$\begin{bmatrix} \underline{\Lambda}_{ABC} \\ \underline{\Lambda}_{abc} \end{bmatrix} = \begin{bmatrix} \underline{\mathbf{L}}_{ss} & \underline{\mathbf{L}}_{sr} \\ \underline{\mathbf{L}}_{sr}^t & \underline{\mathbf{L}}_{rr} \end{bmatrix} \begin{bmatrix} \underline{I}_{ABC} \\ \underline{I}_{abc} \end{bmatrix} \tag{10.69}$$

Accordingly, one can rewrite Eq. (10.69) in the following form:

$$\begin{bmatrix} \underline{I}_{ABC} \\ \underline{I}_{abc} \end{bmatrix} = \begin{bmatrix} \underline{\mathbf{L}}_{ss} & \underline{\mathbf{L}}_{sr} \\ \underline{\mathbf{L}}_{sr}^{t} & \underline{\mathbf{L}}_{rr} \end{bmatrix}^{-1} \cdot \begin{bmatrix} \underline{\Lambda}_{ABC} \\ \underline{\Lambda}_{abc} \end{bmatrix} \tag{10.70}$$

Therefore, substituting for the current vectors, $\underline{I}_{ABC}$ and $\underline{I}_{abc}$, from Eq. (10.70) into Eq. (10.60), one obtains the following state-space model:

$$\begin{bmatrix} \underline{V}_{ABC} \\ \underline{V}_{abc} \end{bmatrix} = \begin{bmatrix} \underline{R}_{ss} & 0 \\ 0 & \underline{R}_{rr} \end{bmatrix} \begin{bmatrix} \underline{\mathbf{L}}_{ss} & \underline{\mathbf{L}}_{sr} \\ \underline{\mathbf{L}}_{sr}^{t} & \underline{\mathbf{L}}_{rr} \end{bmatrix}^{-1} \cdot \begin{bmatrix} \underline{\Lambda}_{ABC} \\ \underline{\Lambda}_{abc} \end{bmatrix} + \frac{d}{dt} \begin{bmatrix} \underline{\Lambda}_{ABC} \\ \underline{\Lambda}_{abc} \end{bmatrix} \tag{10.71}$$

Rearranging the terms of Eq. (10.71), one obtains the following:

$$\frac{d}{dt} \begin{bmatrix} \underline{\Lambda}_{ABC} \\ \underline{\Lambda}_{abc} \end{bmatrix} = - \begin{bmatrix} \underline{R}_{ss} & 0 \\ 0 & \underline{R}_{rr} \end{bmatrix} \begin{bmatrix} \underline{\mathbf{L}}_{ss} & \underline{\mathbf{L}}_{sr} \\ \underline{\mathbf{L}}_{sr}^{t} & \underline{\mathbf{L}}_{rr} \end{bmatrix}^{-1} \cdot \begin{bmatrix} \underline{\Lambda}_{ABC} \\ \underline{\Lambda}_{abc} \end{bmatrix} + \begin{bmatrix} \underline{V}_{ABC} \\ \underline{V}_{abc} \end{bmatrix} \tag{10.72}$$

In short-hand matrix notation, Eq. (10.72) can be rewritten as follows:

$$\frac{d}{dt}[\underline{\Lambda}] = \dot{\underline{\Lambda}} = (-\underline{R} \cdot \underline{L}^{-1})\underline{\Lambda} + \underline{V} \tag{10.73}$$

which is the induction machine's state-space model formulation in the flux linkage frame of reference, where the various terms of Eq. (10.73) can be defined as follows:

$$\dot{\underline{\Lambda}} = \frac{d}{dt} \begin{bmatrix} \underline{\Lambda}_{ABC} \\ \underline{\Lambda}_{abc} \end{bmatrix} \tag{10.74}$$

$$\underline{R} = \begin{bmatrix} \underline{R}_{ss} & 0 \\ 0 & \underline{R}_{rr} \end{bmatrix} \tag{10.75}$$

$$\underline{L}^{-1} = \begin{bmatrix} \underline{\mathbf{L}}_{ss} & \underline{\mathbf{L}}_{sr} \\ \underline{\mathbf{L}}_{sr}^{t} & \underline{\mathbf{L}}_{rr} \end{bmatrix}^{-1} \tag{10.76}$$

and

$$\underline{V} = \begin{bmatrix} \underline{V}_{ABC} \\ \underline{V}_{abc} \end{bmatrix} \tag{10.77}$$

Eqs. (10.73)–(10.77) completely define the induction machine's state-space dynamic model in the flux linkage frame of reference. Now, it is appropriate to introduce the *DQ*–*dq*0 transformations to simplify the *ABC*–*abc*-based state-space models into a form more amenable to numerical simulations and solutions in the next section of this chapter.

10.5 Transformation of the State-Space Model of an Induction Machine from the *ABC*–*abc* Frame to the *DQ*0–*dq*0 Frame of Reference

We now recall the induction machine state-space model written in the *ABC*–*abc* frame of reference, Eq. (10.60) in Section 10.4. This model formulation can be rewritten as follows:

$$\underline{V}_{ABC} = \underline{R}_{ss} \cdot \underline{I}_{ABC} + \frac{d}{dt}(\underline{\Lambda}_{ABC}) \tag{10.78}$$

$$\underline{V}_{abc} = \underline{R}_{rr} \cdot \underline{I}_{abc} + \frac{d}{dt}(\underline{\Lambda}_{abc}) \tag{10.79}$$

where

$$\underline{\Lambda}_{ABC} = \underline{\mathbf{L}}_{ss} \cdot \underline{I}_{ABC} + \underline{\mathbf{L}}_{sr} \cdot \underline{I}_{abc} \tag{10.80}$$

and

$$\underline{\Lambda}_{abc} = \underline{\mathbf{L}}_{sr}^{t} \cdot \underline{I}_{ABC} + \underline{\mathbf{L}}_{rr} \cdot \underline{I}_{abc} \tag{10.81}$$

or

$$\begin{bmatrix} \underline{\Lambda}_{ABC} \\ \underline{\Lambda}_{abc} \end{bmatrix} = \begin{bmatrix} \underline{\mathbf{L}}_{ss} & \underline{\mathbf{L}}_{sr} \\ \underline{\mathbf{L}}_{sr}^{t} & \underline{\mathbf{L}}_{rr} \end{bmatrix} \begin{bmatrix} \underline{I}_{ABC} \\ \underline{I}_{abc} \end{bmatrix} \tag{10.82}$$

We choose a nonsingular transformation for the stator, $\underline{T}_s$, such that

$$\underline{I}_{DQ0} = \underline{T}_s \cdot \underline{I}_{ABC} \tag{10.83}$$

$$\underline{V}_{DQ0} = \underline{T}_s \cdot \underline{V}_{ABC} \tag{10.84}$$

and

$$\underline{\Lambda}_{DQ0} = \underline{T}_s \cdot \underline{\Lambda}_{ABC} \tag{10.85}$$

Also, we choose a nonsingular transformation for the rotor, $\underline{T}_r$, such that

$$\underline{I}_{dq0} = \underline{T}_r \cdot \underline{I}_{abc} \tag{10.86}$$

$$\underline{V}_{dq0} = \underline{T}_r \cdot \underline{V}_{abc} \tag{10.87}$$

and

$$\underline{\Lambda}_{dqo} = \underline{T}_r \cdot \underline{\Lambda}_{abc} \tag{10.88}$$

Hence, inverse relationships also exist because $\underline{T}_s$ and $\underline{T}_r$ are nonsingular transformations, and thus

$$\left.\begin{aligned} &\underline{I}_{ABC} = \underline{T}_s^{-1} \cdot \underline{I}_{DQ0}, \ \underline{V}_{ABC} = \underline{T}_s^{-1} \cdot \underline{V}_{DQ0} \\ &\underline{\Lambda}_{ABC} = \underline{T}_s^{-1} \cdot \underline{\Lambda}_{DQ0} \end{aligned}\right\} \tag{10.89}$$

Also,

$$\left.\begin{array}{l}\underline{I}_{abc} = \underline{T}_r^{-1} \cdot \underline{I}_{dq0},\ \underline{V}_{abc} = \underline{T}_r^{-1} \cdot \underline{V}_{dq0} \\ \underline{\Lambda}_{abc} = \underline{T}_r^{-1} \cdot \underline{\Lambda}_{dq0}\end{array}\right\} \tag{10.90}$$

Using the transformations in Eqs. (10.85) and (10.88), Eq. (5.123) can be rewritten as follows:

$$\begin{bmatrix}\underline{\Lambda}_{DQ0}\\ \underline{\Lambda}_{dq0}\end{bmatrix} = \begin{bmatrix}\underline{T}_s \cdot \mathbf{\underline{L}}_{ss} & \underline{T}_s \cdot \mathbf{\underline{L}}_{sr}\\ \underline{T}_r \cdot \mathbf{\underline{L}}_{sr}^t & \underline{T}_r \cdot \mathbf{\underline{L}}_{rr}\end{bmatrix}\begin{bmatrix}\underline{I}_{ABC}\\ \underline{I}_{abc}\end{bmatrix} \tag{10.91}$$

Substituting for $\underline{I}_{ABC}$ and $\underline{I}_{abc}$ from the inverse relationships (10.89) and (10.90), the flux linkage vectors in Eq. (10.91) can be rewritten as follows:

$$\begin{bmatrix}\underline{\Lambda}_{DQ0}\\ \underline{\Lambda}_{dq0}\end{bmatrix} = \begin{bmatrix}\left(\underline{T}_s \cdot \mathbf{\underline{L}}_{ss} \cdot \underline{T}_s^{-1}\right) & \left(\underline{T}_s \cdot \mathbf{\underline{L}}_{sr} \cdot \underline{T}_r^{-1}\right)\\ \left(\underline{T}_r \cdot \mathbf{\underline{L}}_{sr}^t \cdot \underline{T}_s^{-1}\right) & \left(\underline{T}_r \cdot \mathbf{\underline{L}}_{rr} \cdot \underline{T}_r^{-1}\right)\end{bmatrix}\begin{bmatrix}\underline{I}_{DQ0}\\ \underline{I}_{dqo}\end{bmatrix} \tag{10.92}$$

Applying the transformation, $\underline{T}_s$, to Eq. (10.78) gives

$$\underline{T}_s \cdot \underline{V}_{ABC} = \underline{T}_s \cdot R_{ss} \cdot \underline{I}_{ABC} + \underline{T}_s \cdot \frac{d}{dt}\underline{\Lambda}_{abc} \tag{10.93}$$

However, from Eqs. (10.83)–(10.85), Equation (10.93) can be rewritten as follows:

$$\underline{V}_{DQ0} = \underline{T}_s \cdot R_{ss} \cdot \underline{T}_s^{-1} \cdot \underline{I}_{DQ0} + \underline{T}_s \cdot \left(\frac{d}{dt}\underline{\Lambda}_{ABC}\right) \tag{10.94}$$

Meanwhile,

$$\frac{d}{dt}\underline{\Lambda}_{DQ0} = \frac{d}{dt}(\underline{T}_s \cdot \underline{\Lambda}_{ABC}) = \left(\frac{d}{dt}\underline{T}_s\right) \cdot \underline{\Lambda}_{ABC} + \underline{T}_s\left(\frac{d}{dt}\underline{\Lambda}_{ABC}\right) \tag{10.95}$$

$$\underline{T}_s \cdot \frac{d}{dt}(\underline{\Lambda}_{ABC}) = \frac{d}{dt}\underline{\Lambda}_{DQ0} - \left(\frac{d}{dt}\underline{T}_s\right) \cdot \underline{\Lambda}_{ABC} \tag{10.96}$$

$$\underline{T}_s \cdot \frac{d}{dt}(\underline{\Lambda}_{ABC}) = \frac{d}{dt}\underline{\Lambda}_{DQ0} - \left(\frac{d}{dt}\underline{T}_s\right) \cdot \underline{T}_s^{-1} \cdot \underline{\Lambda}_{DQ0} \tag{10.97}$$

Hence, substituting from Eq. (10.97) into Eq. (10.94) yields the following:

$$\underline{V}_{DQ0} = \left(\underline{T}_s \cdot R_{ss} \cdot \underline{T}_s^{-1}\right)\underline{I}_{DQ0} + \frac{d}{dt}\underline{\Lambda}_{DQ0} - \left(\frac{d}{dt}\underline{T}_s\right) \cdot \underline{T}_s^{-1} \cdot \underline{\Lambda}_{DQ0} \tag{10.98}$$

Similarly, using the rotor transformation, $\underline{T}_r$, and the transformation Eqs. (10.83)–(10.90), Eq. (10.79) is transformed as follows:

$$\underline{T}_r \cdot \underline{V}_{abc} = \underline{T}_r \cdot \underline{R}_{rr} \cdot \underline{I}_{abc} + \underline{T}_r \cdot \frac{d}{dt}\underline{\Lambda}_{abc} \tag{10.99}$$

Hence,

$$\underline{V}_{dq0} = \underline{T}_r \cdot \underline{R}_{rr} \cdot \underline{T}_r^{-1} \cdot \underline{I}_{dq0} + \underline{T}_r \cdot \frac{d}{dt}\underline{\Lambda}_{abc} \tag{10.100}$$

However,

$$\frac{d}{dt}\underline{\Lambda}_{dq0} = \frac{d}{dt}(\underline{T}_r \cdot \underline{\Lambda}_{abc}) = \left(\frac{d\underline{T}_r}{dt}\right) \cdot \underline{\Lambda}_{abc} + \underline{T}_r \cdot \left(\frac{d}{dt}\underline{\Lambda}_{abc}\right) \tag{10.101}$$

Hence,

$$\underline{T}_r \cdot \left(\frac{d}{dt}\underline{\Lambda}_{abc}\right) = \frac{d}{dt}\underline{\Lambda}_{dq0} - \left(\frac{d\underline{T}_r}{dt}\right) \cdot \underbrace{\underline{\Lambda}_{abc}}_{=\underline{T}_r^{-1} \cdot \underline{\Lambda}_{dq0}} \tag{10.102}$$

Therefore,

$$\underline{V}_{dq0} = (\underline{T}_r \cdot \underline{R}_{rr} \cdot \underline{T}_r^{-1}) \cdot \underline{I}_{dq0} + \frac{d}{dt}\underline{\Lambda}_{dq0} - \left(\frac{d\underline{T}_r}{dt}\right) \cdot \underline{T}_r^{-1} \cdot \underline{\Lambda}_{dq0} \tag{10.103}$$

Accordingly, the transformed induction machine model becomes

$$\begin{aligned}
\begin{bmatrix} \underline{V}_{DQ0} \\ \underline{V}_{dq0} \end{bmatrix} &= \begin{bmatrix} (\underline{T}_s \cdot R_{SS} \cdot \underline{T}_s^{-1}) & \underline{0} \\ \underline{0} & (\underline{T}_r \cdot \underline{R}_{rr} \cdot \underline{T}_r^{-1}) \end{bmatrix} \begin{bmatrix} \underline{I}_{DQ0} \\ \underline{I}_{dq0} \end{bmatrix} + \frac{d}{dt}\begin{bmatrix} \underline{\Lambda}_{DQ0} \\ \underline{\Lambda}_{dq0} \end{bmatrix} \\
&- \begin{bmatrix} \left(\frac{d}{dt}\underline{T}_s\right) \cdot \underline{T}_s^{-1} & \underline{0} \\ \underline{0} & \left(\frac{d}{dt}\underline{T}_r\right) \cdot \underline{T}_r^{-1} \end{bmatrix} \begin{bmatrix} \underline{\Lambda}_{DQ0} \\ \underline{\Lambda}_{dq0} \end{bmatrix}
\end{aligned} \tag{10.104}$$

Again, Eq. (10.104) represents the transformed model of the induction machine. One further step can be carried by substituting for the flux linkage in terms of the inductance matrices and current vectors as in Eq. (10.92). This gives the model of the induction machine as

$$\begin{aligned}
\begin{bmatrix} \underline{V}_{DQ0} \\ \underline{V}_{dq0} \end{bmatrix} &= \begin{bmatrix} (\underline{T}_s \cdot R_{SS} \cdot \underline{T}_s^{-1}) & \underline{0} \\ \underline{0} & (\underline{T}_r \cdot \underline{R}_{rr} \cdot \underline{T}_r^{-1}) \end{bmatrix} \begin{bmatrix} \underline{I}_{DQ0} \\ \underline{I}_{dq0} \end{bmatrix} \\
&+ \frac{d}{dt}\left\{ \begin{bmatrix} (\underline{T}_s \cdot \underline{\mathbf{L}}_{SS} \cdot \underline{T}_s^{-1}) & (\underline{T}_s \cdot \underline{\mathbf{L}}_{sr} \cdot \underline{T}_r^{-1}) \\ (\underline{T}_r \cdot \underline{\mathbf{L}}_{sr}^t \cdot \underline{T}_s^{-1}) & (\underline{T}_r \cdot \underline{\mathbf{L}}_{rr} \cdot \underline{T}_r^{-1}) \end{bmatrix} \begin{bmatrix} \underline{I}_{DQ0} \\ \underline{I}_{dq0} \end{bmatrix} \right\} \\
&- \left\{ \begin{bmatrix} (\dot{\underline{T}}_s \cdot \underline{T}_s^{-1}) & \underline{0} \\ \underline{0} & (\dot{\underline{T}}_r \cdot \underline{T}_r^{-1}) \end{bmatrix} \begin{bmatrix} (\underline{T}_s \cdot \underline{\mathbf{L}}_{SS} \cdot \underline{T}_s^{-1}) & (\underline{T}_s \cdot \underline{\mathbf{L}}_{sr} \cdot \underline{T}_r^{-1}) \\ (\underline{T}_r \cdot \underline{\mathbf{L}}_{sr}^t \cdot \underline{T}_s^{-1}) & (\underline{T}_r \cdot \underline{\mathbf{L}}_{rr} \cdot \underline{T}_r^{-1}) \end{bmatrix} \begin{bmatrix} \underline{I}_{DQ0} \\ \underline{I}_{dq0} \end{bmatrix} \right\}
\end{aligned} \tag{10.105}$$

Hence, the induction machine model becomes

$$\begin{aligned}
\begin{bmatrix} \underline{V}_{DQ0} \\ \underline{V}_{dq0} \end{bmatrix} &= \begin{bmatrix} (\underline{T}_s \cdot R_{SS} \cdot \underline{T}_s^{-1}) & \underline{0} \\ \underline{0} & (\underline{T}_r \cdot \underline{R}_{rr} \cdot \underline{T}_r^{-1}) \end{bmatrix} \begin{bmatrix} \underline{I}_{DQ0} \\ \underline{I}_{dq0} \end{bmatrix} \\
&+ \frac{d}{dt}\left\{ \begin{bmatrix} (\underline{T}_s \cdot \underline{\mathbf{L}}_{SS} \cdot \underline{T}_s^{-1}) & (\underline{T}_s \cdot \underline{\mathbf{L}}_{sr} \cdot \underline{T}_r^{-1}) \\ (\underline{T}_r \cdot \underline{\mathbf{L}}_{sr}^t \cdot \underline{T}_s^{-1}) & (\underline{T}_r \cdot \underline{\mathbf{L}}_{rr} \cdot \underline{T}_r^{-1}) \end{bmatrix} \begin{bmatrix} \underline{I}_{DQ0} \\ \underline{I}_{dq0} \end{bmatrix} \right\} \\
&- \begin{bmatrix} (\dot{\underline{T}}_s \cdot \underline{\mathbf{L}}_{SS} \cdot \underline{T}_s^{-1}) & (\dot{\underline{T}}_s \cdot \underline{\mathbf{L}}_{sr} \cdot \underline{T}_r^{-1}) \\ (\dot{\underline{T}}_r \cdot \underline{\mathbf{L}}_{sr}^t \cdot \underline{T}_s^{-1}) & (\dot{\underline{T}}_r \cdot \underline{\mathbf{L}}_{rr} \cdot \underline{T}_r^{-1}) \end{bmatrix} \begin{bmatrix} \underline{I}_{DQ0} \\ \underline{I}_{dq0} \end{bmatrix}
\end{aligned} \tag{10.106}$$

10.6 Three Types of DQ0–dq0 Frame of Reference Transformation and the Resulting State-Space Models

There are three types of $DQ0$–$dq0$ transformations that will be treated here. The first is Case 10.1, in which the d–q axes frame of references is rotating counterclockwise at the same speed of the rotor, and the d-axis (or D-axis) is aligned with the magnetic axis of the phase (a) winding of a 3-phase wound-rotor, while the q-axis (or Q-axis) leads the d-axis by $(\pi/2)$ electrical radians, that is, $90°e$, in a counterclockwise direction. The second is Case 10.2, in which the d–q (or D–Q) axes frame of reference is stationary, that is, fixed with respect to the stator, with the d-axis (or D-axis) aligned with the magnetic axis of phase (A) on the stator, while the q-axis (or Q-axis) leads the d-axis by $(\pi/2)$ electrical radians, that is, $90°e$, in a counterclockwise direction. The third is Case 10.3, in which the d–q (or D–Q) axes frame of reference is rotating at the synchronous speed in a counterclockwise direction, with the d-axis (or D-axis) leading the A-axis in a counterclockwise direction by an angle, σ_s ($\sigma_{synchronous}$), that is a function of time, and leading the a-axis also in a counterclockwise direction by an angle, σ, that is also a function of time. Hence, in this Case 10.3, the d-axis (or D-axis) leads the a-axis in a counterclockwise direction by an angle, $(\sigma_s - \sigma)$, that is also a function of time. Meanwhile, the q-axis (or Q-axis) leads the d-axis (or D-axis) by $(\pi/2)$ electrical radians, that is, $90°e$, in a counterclockwise direction. Now, we detail the effects of these three cases of d–q (or D–Q) reference frame transformations on the resulting state-space models and formulations of induction machines.

Case 10.1 ***The* d–q *(or* D–Q*) frame rotating at the rotor speed, $\dot{\sigma} = (d\sigma/dt)$, see Figure 10.5.*** Recall the transformed state-space model of an induction machine, Eq. (10.106), repeated once more for the convenience of the reader:

$$\begin{bmatrix} \underline{V}_{DQ0} \\ \underline{V}_{dq0} \end{bmatrix} = \begin{bmatrix} (\underline{T}_s \cdot R_{ss} \cdot \underline{T}_s^{-1}) & \underline{0} \\ \underline{0} & (\underline{T}_r \cdot \underline{R}_{rr} \cdot \underline{T}_r^{-1}) \end{bmatrix} \begin{bmatrix} \underline{I}_{DQ0} \\ \underline{I}_{dq0} \end{bmatrix}$$
$$+ \frac{d}{dt}\left\{ \begin{bmatrix} (\underline{T}_s \cdot \underline{\mathbf{L}}_{ss} \cdot \underline{T}_s^{-1}) & (\underline{T}_s \cdot \underline{\mathbf{L}}_{sr} \cdot \underline{T}_r^{-1}) \\ (\underline{T}_r \cdot \underline{\mathbf{L}}_{sr}^t \cdot \underline{T}_s^{-1}) & (\underline{T}_r \cdot \underline{\mathbf{L}}_{rr} \cdot \underline{T}_r^{-1}) \end{bmatrix} \begin{bmatrix} \underline{I}_{DQ0} \\ \underline{I}_{dq0} \end{bmatrix} \right\} \quad (10.107)$$
$$- \begin{bmatrix} (\dot{\underline{T}}_s \cdot \underline{\mathbf{L}}_{ss} \cdot \underline{T}_s^{-1}) & (\dot{\underline{T}}_s \cdot \underline{\mathbf{L}}_{sr} \cdot \underline{T}_r^{-1}) \\ (\dot{\underline{T}}_r \cdot \underline{\mathbf{L}}_{sr}^t \cdot \underline{T}_s^{-1}) & (\dot{\underline{T}}_r \cdot \underline{\mathbf{L}}_{rr} \cdot \underline{T}_r^{-1}) \end{bmatrix} \begin{bmatrix} \underline{I}_{DQ0} \\ \underline{I}_{dq0} \end{bmatrix}$$

In this case, using the geometry in Figure 10.5, recalling earlier developments in Chapter 9 with regard to synchronous machines, we develop the transformation

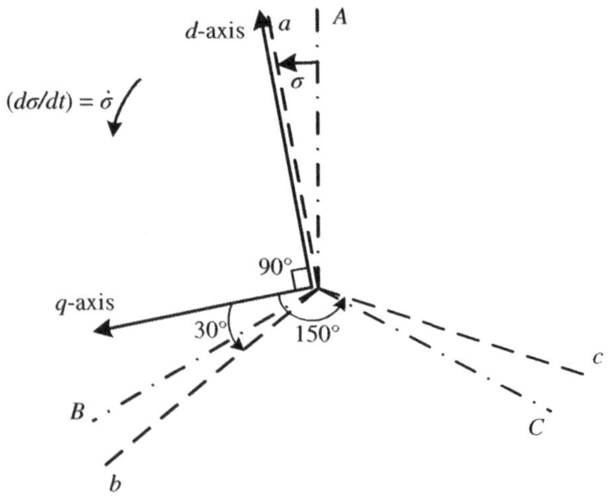

Figure 10.5 The d–q (or D–Q) frame of reference rotating at the rotor speed, $(d\sigma/dt) = \dot{\sigma}$.

matrices, $\underline{T}_s$ and $\underline{T}_r$, as follows:

$$\underline{T}_s = \frac{2}{3}\begin{bmatrix} \cos\sigma & \cos(\sigma - 2\pi/3) & \cos(\sigma - 4\pi/3) \\ -\sin\sigma & -\sin(\sigma - 2\pi/3) & -\sin(\sigma - 4\pi/3) \\ 1/2 & 1/2 & 1/2 \end{bmatrix} \tag{10.108}$$

and

$$\underline{T}_r = \frac{2}{3}\begin{bmatrix} \cos 0° & \cos 120° & \cos 240° \\ \cos -90° & \cos 30° & \cos 150° \\ 1/2 & 1/2 & 1/2 \end{bmatrix} = \frac{2}{3}\begin{bmatrix} 1 & -1/2 & -1/2 \\ 0 & \sqrt{3}/2 & -\sqrt{3}/2 \\ 1/2 & 1/2 & 1/2 \end{bmatrix} \tag{10.109}$$

Furthermore, the reader should verify that here we have

$$\underline{T}_s^{-1} = \begin{bmatrix} \cos\sigma & -\sin\sigma & 1 \\ \cos(\sigma - 2\pi/3) & -\sin(\sigma - 2\pi/3) & 1 \\ \cos(\sigma - 4\pi/3) & -\sin(\sigma - 4\pi/3) & 1 \end{bmatrix} \tag{10.110}$$

and

$$\underline{T}_r^{-1} = \begin{bmatrix} 1 & 0 & 1 \\ -1/2 & \sqrt{3}/2 & 1 \\ -1/2 & -\sqrt{3}/2 & 1 \end{bmatrix} \tag{10.111}$$

by showing that

$$I_s \cdot I_s^{-1} = \underline{U} = \begin{bmatrix} 1 & 0 & 0 \\ 0 & 1 & 0 \\ 0 & 0 & 1 \end{bmatrix} = (3 \times 3) \text{ identity matrix} \tag{10.112}$$

and that

$$\underline{T}_r \cdot \underline{T}_r^{-1} = \underline{U} = \begin{bmatrix} 1 & 0 & 0 \\ 0 & 1 & 0 \\ 0 & 0 & 1 \end{bmatrix} \tag{10.113}$$

Based on Eqs. (10.108)–(10.113), it follows that one can write

$$\underline{T}_s \cdot \underline{R}_{ss} \cdot \underline{T}_s^{-1} = \underline{T}_s(r_s\underline{U})\underline{T}_s^{-1} = r_s \cdot \underline{U} \tag{10.114}$$

Hence,

$$\underline{T}_s \cdot \underline{R}_{ss} \cdot \underline{T}_s^{-1} = \begin{bmatrix} r_s & 0 & 0 \\ 0 & r_s & 0 \\ 0 & 0 & r_s \end{bmatrix} \tag{10.115}$$

Similarly,

$$\underline{T}_r \cdot \underline{R}_{rr} \cdot \underline{T}_r^{-1} = \begin{bmatrix} r_r & 0 & 0 \\ 0 & r_r & 0 \\ 0 & 0 & r_r \end{bmatrix} \tag{10.116}$$

Now, we consider the term $(\underline{T}_s \cdot \underline{\mathbf{L}}_{ss} \cdot \underline{T}_s^{-1})$, thus,

$$\begin{aligned} \underline{T}_s \cdot \underline{\mathbf{L}}_{ss} \cdot \underline{T}_s^{-1} = \frac{2}{3} &\begin{bmatrix} \cos\sigma & \cos(\sigma - 2\pi/3) & \cos(\sigma - 4\pi/3) \\ -\sin\sigma & -\sin(\sigma - 2\pi/3) & -\sin(\sigma - 4\pi/3) \\ 1/2 & 1/2 & 1/2 \end{bmatrix} \\ &\cdot \begin{bmatrix} L_{ss} & L_{sm} & L_{sm} \\ L_{sm} & L_{ss} & L_{sm} \\ L_{sm} & L_{sm} & L_{ss} \end{bmatrix} \cdot \begin{bmatrix} \cos\sigma & -\sin\sigma & 1 \\ \cos(\sigma - 2\pi/3) & -\sin(\sigma - 2\pi/3) & 1 \\ \cos(\sigma - 4\pi/3) & -\sin(\sigma - 4\pi/3) & 1 \end{bmatrix} \end{aligned} \tag{10.117}$$

which, after trigonometric and algebraic manipulations, leads to

$$\begin{aligned} \underline{T}_s \cdot \underline{\mathbf{L}}_{ss} \cdot \underline{T}_s^{-1} = \frac{2}{3} &\begin{bmatrix} (L_{ss} - L_{sm})\cos\sigma & (L_{ss} - L_{sm})\cos(\sigma - 2\pi/3) & (L_{ss} - L_{sm})\cos(\sigma - 4\pi/3) \\ -(L_{ss} - L_{sm})\sin\sigma & -(L_{ss} - L_{sm})\sin(\sigma - 2\pi/3) & -(L_{ss} - L_{sm})\sin(\sigma - 4\pi/3) \\ \left(\frac{1}{2}L_{ss} + L_{sm}\right) & \left(\frac{1}{2}L_{ss} + L_{sm}\right) & \left(\frac{1}{2}L_{ss} + L_{sm}\right) \end{bmatrix} \\ &\cdot \begin{bmatrix} \cos\sigma & -\sin\sigma & 1 \\ \cos(\sigma - 2\pi/3) & -\sin(\sigma - 2\pi/3) & 1 \\ \cos(\sigma - 4\pi/3) & -\sin(\sigma - 4\pi/3) & 1 \end{bmatrix} = \begin{bmatrix} (L_{ss} - L_{sm}) & 0 & 0 \\ 0 & (L_{ss} - L_{sm}) & 0 \\ 0 & 0 & (L_{ss} + 2L_{sm}) \end{bmatrix} \end{aligned} \tag{10.118}$$

Now, consider the term $(\underline{T}_r \cdot \underline{\mathbf{L}}_{rr} \cdot \underline{T}_r^{-1})$, as follows:

$$\underline{T}_r \cdot \underline{\mathbf{L}}_{rr} \cdot \underline{T}_r^{-1} = \frac{2}{3} \begin{bmatrix} 1 & -1/2 & -1/2 \\ 0 & \sqrt{3}/2 & -\sqrt{3}/2 \\ 1/2 & 1/2 & 1/2 \end{bmatrix} \cdot \begin{bmatrix} L_{rr} & L_{rm} & L_{rm} \\ L_{rm} & L_{rr} & L_{rm} \\ L_{rm} & L_{rm} & L_{rr} \end{bmatrix} \cdot \begin{bmatrix} 1 & 0 & 1 \\ -1/2 & \sqrt{3}/2 & 1 \\ -1/2 & -\sqrt{3}/2 & 1 \end{bmatrix} \tag{10.119}$$

Hence,

$$\underline{T}_r \cdot \underline{\mathbf{L}}_{rr} \cdot \underline{T}_r^{-1} = \frac{2}{3}\begin{bmatrix} (L_{rr}-L_{rm}) & 1/2(L_{rm}-L_{rr}) & 1/2(L_{rm}-L_{rr}) \\ 0 & \sqrt{3}/2(L_{rr}-L_{rm}) & -\sqrt{3}/2(L_{rm}-L_{rr}) \\ 1/2(L_{rr}+L_{rm}) & 1/2(L_{rr}+L_{rm}) & 1/2(L_{rr}+L_{rm}) \end{bmatrix} \cdot \begin{bmatrix} 1 & 0 & 1 \\ -1/2 & \sqrt{3}/2 & 1 \\ -1/2 & -\sqrt{3}/2 & 1 \end{bmatrix} \tag{10.120}$$

which yields the following, after algebraic manipulations,

$$\underline{T}_r \cdot \underline{\mathbf{L}}_{rr} \cdot \underline{T}_r^{-1} = \begin{bmatrix} (L_{rr}-L_{rm}) & 0 & 0 \\ 0 & (L_{rr}-L_{rm}) & 0 \\ 0 & 0 & (L_{rr}+2L_{rm}) \end{bmatrix} \tag{10.121}$$

We now consider the term $(\underline{T}_s \cdot \underline{\mathbf{L}}_{sr} \cdot \underline{T}_r^{-1})$, hence, the following can be written:

$$\underline{T}_s \cdot \underline{\mathbf{L}}_{sr} \cdot \underline{T}_r^{-1} = \begin{bmatrix} \cos\sigma & \cos(\sigma-2\pi/3) & \cos(\sigma-4\pi/3) \\ -\sin\sigma & -\sin(\sigma-2\pi/3) & -\sin(\sigma-4\pi/3) \\ 1/2 & 1/2 & 1/2 \end{bmatrix} \cdot \begin{bmatrix} L_{srm}\cos\sigma & L_{srm}\cos(\sigma+2\pi/3) & L_{srm}\cos(\sigma+4\pi/3) \\ L_{srm}\cos(\sigma+4\pi/3) & L_{srm}\cos\sigma & L_{srm}\cos(\sigma+2\pi/3) \\ L_{srm}\cos(\sigma+2\pi/3) & L_{srm}\cos(\sigma+4\pi/3) & L_{srm}\cos\sigma \end{bmatrix} \cdot \begin{bmatrix} 1 & 0 & 1 \\ -1/2 & \sqrt{3}/2 & 1 \\ -1/2 & -\sqrt{3}/2 & 1 \end{bmatrix} \tag{10.122}$$

which, after trigonometric and algebraic manipulations, yields the following:

$$\underline{T}_s \cdot \underline{\mathbf{L}}_{sr} \cdot \underline{T}_r^{-1} = \frac{2}{3}L_{srm}\begin{bmatrix} 9/4 & 0 & 0 \\ 0 & 9/4 & 0 \\ 0 & 0 & 0 \end{bmatrix} = \begin{bmatrix} \frac{3}{2}L_{srm} & 0 & 0 \\ 0 & \frac{3}{2}L_{srm} & 0 \\ 0 & 0 & 0 \end{bmatrix} \tag{10.123}$$

Now, consider the term $(\underline{T}_r \cdot \underline{\mathbf{L}}_{sr}^t \cdot \underline{T}_s^{-1})$, where one can write the following:

$$\underline{T}_r \cdot \underline{\mathbf{L}}_{sr}^t \cdot \underline{T}_s^{-1} = \frac{2}{3}\begin{bmatrix} 1 & -1/2 & -1/2 \\ 0 & \sqrt{3}/2 & -\sqrt{3}/2 \\ 1/2 & 1/2 & 1/2 \end{bmatrix} \cdot L_{srm}\begin{bmatrix} \cos\sigma & \cos(\sigma+4\pi/3) & \cos(\sigma+2\pi/3) \\ \cos(\sigma+2\pi/3) & \cos\sigma & \cos(\sigma+4\pi/3) \\ \cos(\sigma+4\pi/3) & \cos(\sigma+2\pi/3) & \cos\sigma \end{bmatrix} \cdot \begin{bmatrix} \cos\sigma & -\sin\sigma & 1 \\ \cos(\sigma-2\pi/3) & -\sin(\sigma-2\pi/3) & 1 \\ \cos(\sigma-4\pi/3) & -\sin(\sigma-4\pi/3) & 1 \end{bmatrix} \tag{10.124}$$

which yields, through trigonometry and algebraic manipulations, the following:

$$\left(\underline{T}_r \cdot \underline{\mathbf{L}}_{sr}^t \cdot \underline{T}_s^{-1}\right) = \frac{2}{3}L_{srm}\begin{bmatrix} 9/4 & 0 & 0 \\ 0 & 9/4 & 0 \\ 0 & 0 & 0 \end{bmatrix} = \begin{bmatrix} \frac{3}{2}L_{srm} & 0 & 0 \\ 0 & \frac{3}{2}L_{srm} & 0 \\ 0 & 0 & 0 \end{bmatrix} \tag{10.125}$$

Now, consider $\dot{\underline{T}}_s = \frac{d}{dt}(\underline{T}_s)$, we obtain the following:

$$\dot{\underline{T}}_s = \frac{2}{3}(\dot{\sigma})\begin{bmatrix} -\sin(\sigma) & -\sin(\sigma - 2\pi/3) & -\sin(\sigma - 4\pi/3) \\ -\cos(\sigma) & -\cos(\sigma - 2\pi/3) & -\cos(\sigma - 4\pi/3) \\ 0 & 0 & 0 \end{bmatrix} \tag{10.126}$$

and

$$\dot{\underline{T}}_r = \frac{d}{dt}(\underline{T}_r) = \begin{bmatrix} 0 & 0 & 0 \\ 0 & 0 & 0 \\ 0 & 0 & 0 \end{bmatrix} \tag{10.127}$$

Also, consider here the term $(\dot{\underline{T}}_s \cdot \underline{\mathbf{L}}_{ss} \cdot \underline{T}_s^{-1})$, one can write the following:

$$(\dot{\underline{T}}_s \cdot \underline{\mathbf{L}}_{ss} \cdot \underline{T}_s^{-1}) = \frac{2}{3}(\dot{\sigma})\begin{bmatrix} -\sin(\sigma) & -\sin(\sigma - 2\pi/3) & -\sin(\sigma - 4\pi/3) \\ -\cos(\sigma) & -\cos(\sigma - 2\pi/3) & -\cos(\sigma - 4\pi/3) \\ 0 & 0 & 0 \end{bmatrix}$$
$$\cdot \begin{bmatrix} L_{ss} & L_{sm} & L_{sm} \\ L_{sm} & L_{ss} & L_{sm} \\ L_{sm} & L_{sm} & L_{ss} \end{bmatrix} \cdot \begin{bmatrix} \cos\sigma & -\sin\sigma & 1 \\ \cos(\sigma - 2\pi/3) & -\sin(\sigma - 2\pi/3) & 1 \\ \cos(\sigma - 4\pi/3) & -\sin(\sigma - 4\pi/3) & 1 \end{bmatrix} \tag{10.128}$$

After utilizing the appropriate trigonometric identities and algebraic manipulations, we obtain the following:

$$(\dot{\underline{T}}_s \cdot \underline{\mathbf{L}}_{ss} \cdot \underline{T}_s^{-1}) = -\frac{2}{3}\dot{\sigma}(L_{ss} - L_{sm})\begin{bmatrix} 0 & -3/2 & 0 \\ 3/2 & 0 & 0 \\ 0 & 0 & 0 \end{bmatrix} \tag{10.129}$$

which yields the following:

$$(\dot{\underline{T}}_s \cdot \underline{\mathbf{L}}_{ss} \cdot \underline{T}_s^{-1}) = \begin{bmatrix} 0 & \dot{\sigma}(L_{ss} - L_{sm}) & 0 \\ -\dot{\sigma}(L_{ss} - L_{sm}) & 0 & 0 \\ 0 & 0 & 0 \end{bmatrix} \tag{10.130}$$

Now, consider the term $(\dot{\underline{T}}_s \cdot \underline{\mathbf{L}}_{sr} \cdot \underline{T}_r^{-1})$, one can write the following:

$$(\dot{\underline{T}}_s \cdot \underline{\mathbf{L}}_{sr} \cdot \underline{T}_r^{-1}) = -\frac{2}{3}L_{srm}\begin{bmatrix} \sin(\sigma) & \sin(\sigma - 2\pi/3) & \sin(\sigma - 4\pi/3) \\ \cos(\sigma) & \cos(\sigma - 2\pi/3) & \cos(\sigma - 4\pi/3) \\ 0 & 0 & 0 \end{bmatrix}$$
$$\cdot \begin{bmatrix} \cos(\sigma) & \cos(\sigma + 2\pi/3) & \cos(\sigma + 4\pi/3) \\ \cos(\sigma + 4\pi/3) & \cos(\sigma + 2\pi/3) & \cos(\sigma + 2\pi/3) \\ \cos(\sigma + 2\pi/3) & \cos(\sigma + 4\pi/3) & \cos(\sigma) \end{bmatrix}$$
$$\cdot \begin{bmatrix} 1 & 0 & 1 \\ -1/2 & \sqrt{3}/2 & 1 \\ -1/2 & -\sqrt{3}/2 & 1 \end{bmatrix} \tag{10.131}$$

which yields, after trigonometric and algebraic manipulations, the following:

$$\left(\dot{\underline{T}}_s \cdot \underline{\mathbf{L}}_{sr} \cdot \underline{T}_r^{-1}\right) = -\frac{2}{3}\dot{\sigma}L_{srm}\begin{bmatrix} 0 & (-3\sqrt{3}/4) & (3\sqrt{3}/4) \\ (3/2) & (-3/4) & (-3/4) \\ 0 & 0 & 0 \end{bmatrix}\begin{bmatrix} 1 & 0 & 1 \\ (-1/2) & (\sqrt{3}/2) & 1 \\ (1/2) & (-\sqrt{3}/2) & 1 \end{bmatrix} \tag{10.132}$$

which gives the following:

$$\left(\dot{\underline{T}}_s \cdot \underline{\mathbf{L}}_{sr} \cdot \underline{T}_r^{-1}\right) = \begin{bmatrix} 0 & (2/3)\dot{\sigma}L_{srm} & 0 \\ -(2/3)\dot{\sigma}L_{srm} & 0 & 0 \\ 0 & 0 & 0 \end{bmatrix} \tag{10.133}$$

Also, since

$$\dot{\underline{T}}_r = \begin{bmatrix} 0 & 0 & 0 \\ 0 & 0 & 0 \\ 0 & 0 & 0 \end{bmatrix} \tag{10.134}$$

It follows that we can write the following for $\left(\dot{\underline{T}}_r \cdot \underline{\mathbf{L}}_{sr}^t \cdot \underline{T}_s^{-1}\right)$:

$$\left(\dot{\underline{T}}_r \cdot \underline{\mathbf{L}}_{sr}^t \cdot \underline{T}_s^{-1}\right) = \left(\dot{\underline{T}}_r \cdot \underline{\mathbf{L}}_{rr} \cdot \underline{T}_r^{-1}\right) = \begin{bmatrix} 0 & 0 & 0 \\ 0 & 0 & 0 \\ 0 & 0 & 0 \end{bmatrix} \tag{10.135}$$

Substituting from Eqs. (10.115), (10.116), (10.118), (10.121), (10.123), (10.125), (10.130), (10.133), and (10.135) into Eq. (10.107) and expanding yields the following:

$$\begin{bmatrix} v_D \\ v_Q \\ v_0 \\ \hdashline v_d \\ v_q \\ v_0 \end{bmatrix} = \left[\begin{array}{ccc:ccc} r_s & 0 & 0 & 0 & 0 & 0 \\ 0 & r_s & 0 & 0 & 0 & 0 \\ 0 & 0 & r_s & 0 & 0 & 0 \\ \hdashline 0 & 0 & 0 & r_r & 0 & 0 \\ 0 & 0 & 0 & 0 & r_r & 0 \\ 0 & 0 & 0 & 0 & 0 & r_r \end{array}\right]\begin{bmatrix} i_D \\ i_Q \\ i_0 \\ \hdashline i_d \\ i_q \\ i_0 \end{bmatrix}$$
$$+\frac{d}{dt}\left\{\left[\begin{array}{ccc:ccc} (L_{ss}-L_{sm}) & 0 & 0 & \frac{3}{2}L_{srm} & 0 & 0 \\ 0 & (L_{ss}-L_{sm}) & 0 & 0 & \frac{3}{2}L_{srm} & 0 \\ 0 & 0 & (L_{ss}+2L_{sm}) & 0 & 0 & 0 \\ \hdashline \frac{3}{2}L_{srm} & 0 & 0 & (L_{rr}-L_{rm}) & 0 & 0 \\ 0 & \frac{3}{2}L_{srm} & 0 & 0 & (L_{rr}-L_{rm}) & 0 \\ 0 & 0 & 0 & 0 & 0 & (L_{rr}+2L_{rm}) \end{array}\right]\begin{bmatrix} i_D \\ i_Q \\ i_0 \\ \hdashline i_d \\ i_q \\ i_0 \end{bmatrix}\right\}$$
$$+\dot{\sigma}\left[\begin{array}{ccc:ccc} 0 & -(L_{ss}-L_{sm}) & 0 & 0 & -(3/2)L_{srm} & 0 \\ (L_{ss}-L_{sm}) & 0 & 0 & (3/2)L_{srm} & 0 & 0 \\ 0 & 0 & 0 & 0 & 0 & 0 \\ \hdashline 0 & 0 & 0 & 0 & 0 & 0 \\ 0 & 0 & 0 & 0 & 0 & 0 \\ 0 & 0 & 0 & 0 & 0 & 0 \end{array}\right]\begin{bmatrix} i_D \\ i_Q \\ i_0 \\ \hdashline i_d \\ i_q \\ i_0 \end{bmatrix} \tag{10.136}$$

In abbreviated/short-hand matrix form, Eq. (10.136) can be written as follows:

$$\underline{V} = \underline{R} \cdot \underline{I} + \frac{d}{dt}(\underline{\mathbf{L}}_1 \cdot \underline{I}) + \dot{\sigma}\underline{\mathbf{L}}_2 \cdot \underline{I} \tag{10.137}$$

and since the matrix $\underline{\mathbf{L}}_1$ has time-independent inductance terms, Eq. (10.137) can be rewritten as follows:

$$\underline{V} = \underline{R} \cdot \underline{I} + \underline{\mathbf{L}}_1 \cdot \dot{\underline{I}} + \dot{\sigma}\underline{\mathbf{L}}_2 \cdot \underline{I} \tag{10.138}$$

or

$$\dot{\underline{I}} = -\underline{\mathbf{L}}_1^{-1}(\underline{R} + \dot{\sigma}\underline{\mathbf{L}}_2)\underline{I} + \underline{\mathbf{L}}_1^{-1} \cdot \underline{V} \tag{10.139}$$

which is of the state-space standard form $\dot{\underline{X}} = \underline{A} \cdot \underline{X} + \underline{B} \cdot \underline{U}$ and is the induction machine's DQ0–dq0 frame in the current-frame of reference.

From Eqs. (10.136) and (10.137), one can write the following:

$$\underline{\Lambda} = \underbrace{\begin{bmatrix} \lambda_D \\ \lambda_Q \\ \lambda_0 \\ \lambda_d \\ \lambda_q \\ \lambda_0 \end{bmatrix}}_{\underline{\Lambda}} = \underbrace{\left[\begin{array}{ccc|ccc} (L_{ss} - L_{sm}) & 0 & 0 & \frac{3}{2}L_{srm} & 0 & 0 \\ 0 & (L_{ss} - L_{sm}) & 0 & 0 & \frac{3}{2}L_{srm} & 0 \\ 0 & 0 & (L_{ss} + 2L_{sm}) & 0 & 0 & 0 \\ \hline \frac{3}{2}L_{srm} & 0 & 0 & (L_{rr} - L_{rm}) & 0 & 0 \\ 0 & \frac{3}{2}L_{srm} & 0 & 0 & (L_{rr} - L_{rm}) & 0 \\ 0 & 0 & 0 & 0 & 0 & (L_{rr} - L_{rm}) \end{array}\right]}_{\underline{L}_1} \underbrace{\begin{bmatrix} i_D \\ i_Q \\ i_0 \\ i_d \\ i_q \\ i_0 \end{bmatrix}}_{\underline{I}} \tag{10.140}$$

Hence,

$$\underline{I} = [\underline{\mathbf{L}}_1]^{-1} \cdot \underline{\Lambda} \tag{10.41}$$

Also, Eq. (10.137) can be rewritten as follows:

$$\underline{V} = \underline{R} \cdot \underline{I} + \dot{\underline{\Lambda}} + \dot{\sigma}\underline{\mathbf{L}}_2 \cdot \underline{I} \tag{10.42}$$

Substituting from (10.141) into (10.142), one obtains:

$$\underline{V} = \underline{R} \cdot [\underline{\mathbf{L}}_1]^{-1} \cdot \underline{\Lambda} + \dot{\underline{\Lambda}} + \dot{\sigma}\underline{\mathbf{L}}_2 \cdot [\underline{\mathbf{L}}_1]^{-1} \cdot \underline{\Lambda} \tag{10.43}$$

or

$$\dot{\underline{\Lambda}} = -\left\{(\underline{R} + \dot{\sigma}\underline{\mathbf{L}}_2) \cdot [\underline{\mathbf{L}}_1]^{-1}\right\} \cdot \underline{\Lambda} + \underline{V} \tag{10.44}$$

Eq. (10.144) is the induction motor state-space model in the *DQ*0–*dq*0 flux linkage frame of reference in which $\underline{R}$ is defined in Eq. (10.136) and $\underline{\mathbf{L}}_1$ is defined in Eqs. (10.136) and (10.140), and $\underline{\mathbf{L}}_2$ is also defined in Eq. (10.136) from which one can write the following:

$$\underline{\mathbf{L}}_2 = \left[\begin{array}{ccc|ccc} 0 & -(L_{ss}-L_{sm}) & 0 & 0 & -\frac{3}{2}L_{srm} & 0 \\ (L_{ss}-L_{sm}) & 0 & 0 & \frac{3}{2}L_{srm} & 0 & 0 \\ 0 & 0 & 0 & 0 & 0 & 0 \\ \hline 0 & 0 & 0 & 0 & 0 & 0 \\ 0 & 0 & 0 & 0 & 0 & 0 \\ 0 & 0 & 0 & 0 & 0 & 0 \end{array}\right] \tag{10.145}$$

Again, in summary, the induction motor's flux-linkage frame-of-reference DQ0–dq0 state-space model can be rewritten as follows:

$$\dot{\underline{\Lambda}} = -\left\{(\underline{R} + \dot{\sigma}\underline{\mathbf{L}}_2)\cdot\underline{\mathbf{L}}_1^{-1}\right\}\cdot\underline{\Lambda} + \underline{V} \tag{10.146}$$

and

$$\underline{I} = \underline{\mathbf{L}}_1^{-1}\cdot\underline{V} \tag{10.147}$$

Case 10.2 ***The stationary DQ0 stator reference frame, see Figure 10.6*** The model expressed in stationary stator reference frame can also be derived from Equation (10.106), except that the transformation matrices, $\underline{T}_s$ and $\underline{T}_r$, are different

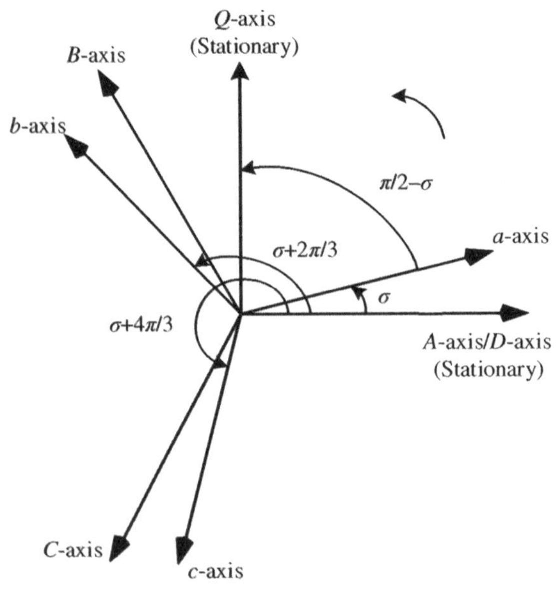

Figure 10.6 The stationary *DQ0* stator reference frame with a rotor speed $(d\sigma/dt) = \dot{\sigma}$ in e.rad/s.

and are developed next, Eq. (10.106) is restated here for convenience as follows:

$$\begin{bmatrix} \underline{V}_{DQ0} \\ \underline{V}_{dq0} \end{bmatrix} = \begin{bmatrix} (\underline{T}_s \cdot R_{SS} \cdot \underline{T}_s^{-1}) & \underline{0} \\ \underline{0} & (\underline{T}_r \cdot \underline{R}_{rr} \cdot \underline{T}_r^{-1}) \end{bmatrix} \begin{bmatrix} \underline{I}_{DQ0} \\ \underline{I}_{dq0} \end{bmatrix}$$
$$+ \frac{d}{dt}\left\{ \begin{bmatrix} (\underline{T}_s \cdot \underline{\mathbf{L}}_{SS} \cdot \underline{T}_s^{-1}) & (\underline{T}_s \cdot \underline{\mathbf{L}}_{sr} \cdot \underline{T}_r^{-1}) \\ (\underline{T}_r \cdot \underline{\mathbf{L}}_{sr}^t \cdot \underline{T}_s^{-1}) & (\underline{T}_r \cdot \underline{\mathbf{L}}_{rr} \cdot \underline{T}_r^{-1}) \end{bmatrix} \begin{bmatrix} \underline{I}_{DQ0} \\ \underline{I}_{dq0} \end{bmatrix} \right\} \quad (10.105)$$
$$- \begin{bmatrix} (\dot{\underline{T}}_s \cdot \underline{\mathbf{L}}_{SS} \cdot \underline{T}_s^{-1}) & (\dot{\underline{T}}_s \cdot \underline{\mathbf{L}}_{sr} \cdot \underline{T}_r^{-1}) \\ (\dot{\underline{T}}_r \cdot \underline{\mathbf{L}}_{sr}^t \cdot \underline{T}_s^{-1}) & (\dot{\underline{T}}_r \cdot \underline{\mathbf{L}}_{rr} \cdot \underline{T}_r^{-1}) \end{bmatrix} \begin{bmatrix} \underline{I}_{DQ0} \\ \underline{I}_{dq0} \end{bmatrix}$$

From Figure 10.6, we derive the transformation matrices, $\underline{T}_s$ and $\underline{T}_r$, as follows:

$$\underline{T}_s = \frac{2}{3}\begin{bmatrix} \cos(0) & \cos(2\pi/3) & \cos(4\pi/3) \\ \cos(-\pi/2) & \cos(2\pi/3-\pi/2) & \cos(4\pi/3-\pi/2) \\ 1/2 & 1/2 & 1/2 \end{bmatrix} = \frac{2}{3}\begin{bmatrix} 1 & -1/2 & -1/2 \\ 0 & \sqrt{3}/2 & -\sqrt{3}/2 \\ 1/2 & 1/2 & 1/2 \end{bmatrix} \quad (10.148)$$

and

$$\underline{T}_r = \frac{2}{3}\begin{bmatrix} \cos(-\sigma) & \cos(-\sigma-2\pi/3) & \cos(-\sigma-4\pi/3) \\ \cos(\pi/2-\sigma) & \cos(\pi/2-(2\pi/3+\sigma)) & \cos(\pi/2-(4\pi/3+\sigma)) \\ 1/2 & 1/2 & 1/2 \end{bmatrix} \quad (10.149)$$

That is,

$$\underline{T}_r = \frac{2}{3}\begin{bmatrix} \cos(\sigma) & \cos(\sigma+2\pi/3) & \cos(\sigma+4\pi/3) \\ \sin(\sigma) & \sin(\sigma+2\pi/3) & \sin(\sigma+4\pi/3) \\ 1/2 & 1/2 & 1/2 \end{bmatrix} \quad (10.150)$$

Hence,

$$\underline{T}_s^{-1} = \begin{bmatrix} 1 & 0 & 1 \\ -1/2 & \sqrt{3}/2 & 1 \\ -1/2 & -\sqrt{3}/2 & 1 \end{bmatrix} \quad (10.151)$$

and

$$(\underline{T}_s^{-1})^t = \begin{bmatrix} 1 & -1/2 & -1/2 \\ 0 & \sqrt{3}/2 & -\sqrt{3}/2 \\ 1 & 1 & 1 \end{bmatrix} \quad (10.152)$$

It follows that

$$(\underline{T}_s^{-1})^t \cdot \underline{T}_s^{-1} = \begin{bmatrix} 1 & -1/2 & -1/2 \\ 0 & \sqrt{3}/2 & -\sqrt{3}/2 \\ 1 & 1 & 1 \end{bmatrix} \cdot \begin{bmatrix} 1 & 0 & 1 \\ -1/2 & \sqrt{3}/2 & 1 \\ -1/2 & -\sqrt{3}/2 & 1 \end{bmatrix} = \begin{bmatrix} 3/2 & 0 & 0 \\ 0 & 3/2 & 0 \\ 0 & 0 & 3 \end{bmatrix} = K_S \cdot \underline{G}_s \quad (10.153)$$

where

$$K_s = 3/2 \tag{10.154}$$

and

$$\underline{G}_s = \begin{bmatrix} 1 & 0 & 0 \\ 0 & 1 & 0 \\ 0 & 0 & 2 \end{bmatrix} \tag{10.155}$$

Also,

$$\underline{T}_r^{-1} = \begin{bmatrix} \cos(\sigma) & \sin(\sigma) & 1 \\ \cos(\sigma + 2\pi/3) & \sin(\sigma + 2\pi/3) & 1 \\ \cos(\sigma + 4\pi/3) & \sin(\sigma + 4\pi/3) & 1 \end{bmatrix} \tag{10.156}$$

and

$$\left(\underline{T}_r^{-1}\right)^t = \begin{bmatrix} \cos(\sigma) & \cos(\sigma + 2\pi/3) & \cos(\sigma + 4\pi/3) \\ \sin(\sigma) & \sin(\sigma + 2\pi/3) & \sin(\sigma + 4\pi/3) \\ 1 & 1 & 1 \end{bmatrix} \tag{10.157}$$

Hence,

$$\begin{aligned} \left(\underline{T}_r^{-1}\right)^t \cdot \underline{T}_r^{-1} = &\begin{bmatrix} \cos(\sigma) & \cos(\sigma + 2\pi/3) & \cos(\sigma + 4\pi/3) \\ \sin(\sigma) & \sin(\sigma + 2\pi/3) & \sin(\sigma + 4\pi/3) \\ 1 & 1 & 1 \end{bmatrix} \\ &\cdot \begin{bmatrix} \cos(\sigma) & \sin(\sigma) & 1 \\ \cos(\sigma + 2\pi/3) & \sin(\sigma + 2\pi/3) & 1 \\ \cos(\sigma + 4\pi/3) & \sin(\sigma + 4\pi/3) & 1 \end{bmatrix} \end{aligned} \tag{10.158}$$

which yields the following:

$$\left(\underline{T}_r^{-1}\right)^t \cdot \underline{T}_r^{-1} = \begin{bmatrix} 3/2 & 0 & 0 \\ 0 & 3/2 & 0 \\ 0 & 0 & 3 \end{bmatrix} = K_r \cdot \underline{G}_r \tag{10.159}$$

where

$$K_r = 3/2 \tag{10.160}$$

and

$$\underline{G}_r = \begin{bmatrix} 1 & 0 & 0 \\ 0 & 1 & 0 \\ 0 & 0 & 2 \end{bmatrix} \tag{10.161}$$

Also, similar to Case 10.1, it can be shown that

$$\underline{T}_s \cdot \underline{R}_{ss} \cdot \underline{T}_s^{-1} = \begin{bmatrix} r_s & 0 & 0 \\ 0 & r_s & 0 \\ 0 & 0 & r_s \end{bmatrix} \tag{10.162}$$

and

$$\underline{T}_r \cdot \underline{R}_{rr} \cdot \underline{T}_r^{-1} = \begin{bmatrix} r_r & 0 & 0 \\ 0 & r_r & 0 \\ 0 & 0 & r_r \end{bmatrix} \tag{10.163}$$

Following steps similar to those performed in Case 10.1 earlier in this section, it can be shown that

$$\underline{T}_s \cdot \underline{\mathbf{L}}_{ss} \cdot \underline{T}_s^{-1} = \begin{bmatrix} L_{ss} - L_{sm} & 0 & 0 \\ 0 & L_{ss} - L_{sm} & 0 \\ 0 & 0 & L_{ss} + 2L_{sm} \end{bmatrix} \tag{10.164}$$

and

$$\underline{T}_r \cdot \underline{\mathbf{L}}_{rr} \cdot \underline{T}_r^{-1} = \begin{bmatrix} L_{rr} - L_{rm} & 0 & 0 \\ 0 & L_{rr} - L_{rm} & 0 \\ 0 & 0 & L_{rr} + 2L_{rm} \end{bmatrix} \tag{10.165}$$

Also,

$$\underline{T}_s \cdot \underline{\mathbf{L}}_{sr} \cdot \underline{T}_r^{-1} = \begin{bmatrix} \frac{3}{2}L_{srm} & 0 & 0 \\ 0 & \frac{3}{2}L_{srm} & 0 \\ 0 & 0 & 0 \end{bmatrix} \tag{10.166}$$

and

$$\underline{T}_r \cdot \underline{\mathbf{L}}_{sr}^t \cdot \underline{T}_s^{-1} = \begin{bmatrix} \frac{3}{2}L_{srm} & 0 & 0 \\ 0 & \frac{3}{2}L_{srm} & 0 \\ 0 & 0 & 0 \end{bmatrix} \tag{10.167}$$

Recalling that

$$\underline{T}_r = \frac{2}{3} \begin{bmatrix} \cos(\sigma) & \cos(\sigma + 2\pi/3) & \cos(\sigma + 4\pi/3) \\ \sin(\sigma) & \sin(\sigma + 2\pi/3) & \sin(\sigma + 4\pi/3) \\ 1/2 & 1/2 & 1/2 \end{bmatrix} \tag{10.168}$$

it follows that

$$\dot{\underline{T}}_r = \frac{2}{3}\dot{\sigma} \begin{bmatrix} -\sin(\sigma) & -\sin(\sigma + 2\pi/3) & -\sin(\sigma + 4\pi/3) \\ \cos(\sigma) & \cos(\sigma + 2\pi/3) & \cos(\sigma + 4\pi/3) \\ 0 & 0 & 0 \end{bmatrix} \tag{10.169}$$

where

$$\dot{\sigma} = (d\sigma/dt) \tag{10.170}$$

Also,

$$\dot{\underline{T}}_s = [0] = \underline{0} \text{ matrix} \tag{10.171}$$

Therefore, the following matrices can be obtained:

$$\underline{\dot{T}}_s \cdot \underline{\mathbf{L}}_{ss} \cdot \underline{T}_s^{-1} = [0] = \underline{0} \text{ matrix} \tag{10.172}$$

and

$$\underline{\dot{T}}_s \cdot \underline{\mathbf{L}}_{sr} \cdot \underline{T}_r^{-1} = [0] = \underline{0} \text{ matrix} \tag{10.173}$$

Also, it can be shown that

$$\underline{\dot{T}}_r \cdot \underline{\mathbf{L}}_{sr}^t \cdot \underline{T}_s^{-1} = \dot{\sigma}\begin{bmatrix} 0 & -(3/2)L_{srm} & 0 \\ (3/2)L_{srm} & 0 & 0 \\ 0 & 0 & 0 \end{bmatrix} \tag{10.174}$$

and

$$\underline{\dot{T}}_r \cdot \underline{\mathbf{L}}_{rr} \cdot \underline{T}_r^{-1} = \dot{\sigma}\begin{bmatrix} 0 & -(L_{rr}-L_{rm}) & 0 \\ (L_{rr}-L_{rm}) & 0 & 0 \\ 0 & 0 & 0 \end{bmatrix} \tag{10.175}$$

where, again, $\dot{\sigma} = (d\sigma/dt)$.

Substituting from Eqs. (10.162)–(10.175) into Eq. (10.106) yields the following, with the knowledge that all the transformed inductance matrices are independent of the rotor position angle, σ, and, hence, are independent of time, t:

$$\begin{aligned}
\begin{bmatrix} v_D \\ v_Q \\ v_0 \\ \hdashline v_d \\ v_q \\ v_0 \end{bmatrix} &= \left[\begin{array}{ccc:ccc} r_s & 0 & 0 & 0 & 0 & 0 \\ 0 & r_s & 0 & 0 & 0 & 0 \\ 0 & 0 & r_s & 0 & 0 & 0 \\ \hdashline 0 & 0 & 0 & r_r & 0 & 0 \\ 0 & 0 & 0 & 0 & r_r & 0 \\ 0 & 0 & 0 & 0 & 0 & r_r \end{array}\right] \begin{bmatrix} i_D \\ i_Q \\ i_0 \\ \hdashline i_d \\ i_q \\ i_0 \end{bmatrix} \\
&+ \frac{d}{dt}\left\{\left[\begin{array}{ccc:ccc} (L_{ss}-L_{sm}) & 0 & 0 & \frac{3}{2}L_{srm} & 0 & 0 \\ 0 & (L_{ss}-L_{sm}) & 0 & 0 & \frac{3}{2}L_{srm} & 0 \\ 0 & 0 & (L_{ss}+2L_{sm}) & 0 & 0 & 0 \\ \hdashline \frac{3}{2}L_{srm} & 0 & 0 & (L_{rr}-L_{rm}) & 0 & 0 \\ 0 & \frac{3}{2}L_{srm} & 0 & 0 & (L_{rr}-L_{rm}) & 0 \\ 0 & 0 & 0 & 0 & 0 & (L_{rr}+2L_{rm}) \end{array}\right] \begin{bmatrix} i_D \\ i_Q \\ i_0 \\ \hdashline i_d \\ i_q \\ i_0 \end{bmatrix}\right\} \\
&+ \dot{\sigma}\left[\begin{array}{ccc:ccc} 0 & -(L_{ss}-L_{sm}) & 0 & 0 & -(3/2)L_{srm} & 0 \\ (L_{ss}-L_{sm}) & 0 & 0 & (3/2)L_{srm} & 0 & 0 \\ 0 & 0 & 0 & 0 & 0 & 0 \\ \hdashline 0 & 0 & 0 & 0 & 0 & 0 \\ 0 & 0 & 0 & 0 & 0 & 0 \\ 0 & 0 & 0 & 0 & 0 & 0 \end{array}\right] \begin{bmatrix} i_D \\ i_Q \\ i_0 \\ \hdashline i_d \\ i_q \\ i_0 \end{bmatrix}
\end{aligned} \tag{10.176}$$

Here, we define the following:

$$\underline{V} = \begin{bmatrix} v_D \\ v_Q \\ v_0 \\ v_d \\ v_q \\ v_0 \end{bmatrix} \tag{10.177}$$

$$\underline{I} = \begin{bmatrix} i_D \\ i_Q \\ i_0 \\ i_d \\ i_q \\ i_0 \end{bmatrix} \tag{10.178}$$

$$\underline{\dot{I}} = \begin{bmatrix} \dot{i}_D \\ \dot{i}_Q \\ \dot{i}_0 \\ \hline \dot{i}_d \\ \dot{i}_q \\ \dot{i}_0 \end{bmatrix} \tag{10.179}$$

Also, we define the following:

$$\underline{\mathbf{L}}_1 = \left[\begin{array}{ccc|ccc} L_{ss} - L_{sm} & 0 & 0 & \frac{3}{2}L_{srm} & 0 & 0 \\ 0 & L_{ss} - L_{sm} & 0 & 0 & \frac{3}{2}L_{srm} & 0 \\ 0 & 0 & L_{ss} + 2L_{sm} & 0 & 0 & 0 \\ \hline \frac{3}{2}L_{srm} & 0 & 0 & L_{rr} - L_{rm} & 0 & 0 \\ 0 & \frac{3}{2}L_{srm} & 0 & 0 & L_{rr} - L_{rm} & 0 \\ 0 & 0 & 0 & 0 & 0 & L_{rr} + 2L_{rm} \end{array}\right] \tag{10.180}$$

$$\underline{\mathbf{L}}_2 = \left[\begin{array}{ccc|ccc} 0 & 0 & 0 & 0 & 0 & 0 \\ 0 & 0 & 0 & 0 & 0 & 0 \\ 0 & 0 & 0 & 0 & 0 & 0 \\ \hline 0 & \frac{3}{2}L_{srm} & 0 & 0 & (L_{rr} - L_{rm}) & 0 \\ -\frac{3}{2}L_{srm} & 0 & 0 & -(L_{rr} - L_{rm}) & 0 & 0 \\ 0 & 0 & 0 & 0 & 0 & 0 \end{array}\right] \tag{10.181}$$

and

$$\underline{R} = \left[\begin{array}{ccc|ccc} r_s & 0 & 0 & 0 & 0 & 0 \\ 0 & r_s & 0 & 0 & 0 & 0 \\ 0 & 0 & r_s & 0 & 0 & 0 \\ \hline 0 & 0 & 0 & r_r & 0 & 0 \\ 0 & 0 & 0 & 0 & r_r & 0 \\ 0 & 0 & 0 & 0 & 0 & r_r \end{array}\right] \tag{10.182}$$

Accordingly, the flux linkage rector, $\underline{\Lambda}$, can be expressed as follows:

$$\underline{\Lambda} = \underline{\mathbf{L}}_1 \cdot \underline{I} \tag{10.183}$$

and since $\underline{\mathbf{L}}_1$ is time independent, it follows that

$$\underline{\mathbf{L}}_1 \cdot \underline{\dot{I}} = d(\underline{\mathbf{L}}_1 \cdot \underline{I})/dt = \underline{\dot{\Lambda}} \tag{10.184}$$

Hence, Eq. (10.176) can be rewritten in abbreviated matrix/vector form as follows:

$$\underline{V} = \underline{R} \cdot \underline{I} + \underline{\mathbf{L}}_1 \cdot \underline{\dot{I}} + \dot{\sigma}\underline{\mathbf{L}}_2 \cdot \underline{I} \tag{10.185}$$

Equation (10.185) can be rewritten as follows:

$$\underline{\dot{I}}\left\{-\underline{\mathbf{L}}_1^{-1}(\underline{R} + \dot{\sigma}\underline{\mathbf{L}}_2)\right\} \cdot \underline{I} + \left\{\underline{\mathbf{L}}_1^{-1}\right\} \cdot \underline{V} \tag{10.186}$$

which is of the state-space form $\underline{\dot{X}} = \underline{A} \cdot \underline{X} + \underline{B} \cdot \underline{U}$. Thus, Eq. (10.186) constitutes the *DQ*0–*dq*0 state-space model of induction machines in the machine currents frame of reference.

Furthermore, utilizing the definitions given in Eqs. (10.179)–(10.184), Eq. (10.176) can also be rewritten in abbreviated matrix/vector form as follows:

$$\underline{V} = \underline{R} \cdot \underline{I} + \underline{\dot{\Lambda}} + \dot{\sigma}\underline{\mathbf{L}}_2\underline{I} \tag{10.187}$$

However,

$$\underline{I} = \underline{\mathbf{L}}_1^{-1} \cdot \underline{\Lambda} \tag{10.188}$$

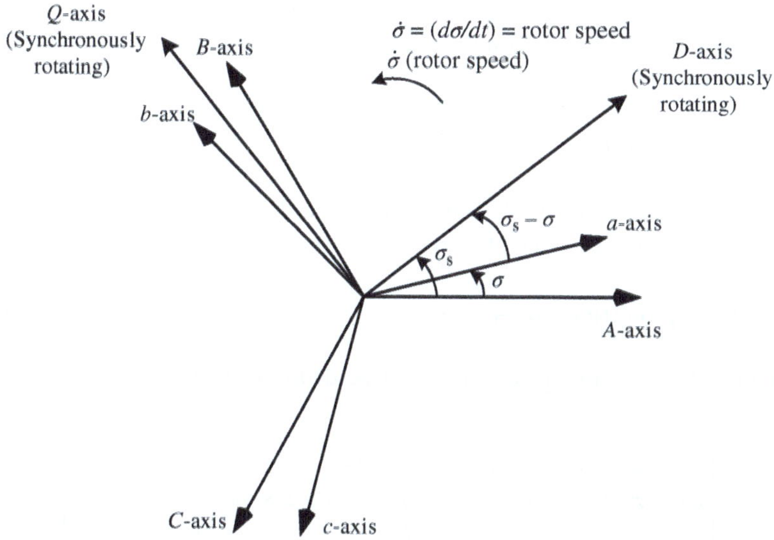

Figure 10.7 The synchronously rotating *DQ*0 frame with a rotor speed $(d\sigma/dt) = \dot{\sigma}$ in e.rad/s.

Hence, Eq. (10.187) can be rewritten, substituting for $\underline{I}$ from Eq. (10.188), as follows:

$$\underline{V} = \underline{R} \cdot \underline{\mathbf{L}}_1^{-1} \cdot \underline{\Lambda} + \dot{\underline{\Lambda}} + \dot{\sigma}\underline{\mathbf{L}}_2 \cdot \left(\underline{\mathbf{L}}_1^{-1}\right) \underline{\Lambda} \tag{10.189}$$

Rearranging the terms in Eq. (10.189), one obtains the following:

$$\dot{\underline{\Lambda}} = \left\{-(\underline{R} + \dot{\sigma} \cdot \underline{\mathbf{L}}_2)\left(\underline{\mathbf{L}}_1^{-1}\right)\right\} \cdot \underline{\Lambda} + \underline{V} \tag{10.190}$$

which is of the state-space form $\dot{\underline{X}} = \underline{A} \cdot \underline{X} + \underline{B} \cdot \underline{U}$. Therefore, Eq. (10.190) constitutes the *DQ*0–*dq*0 state-space model of induction machines in the machine flux linkage frame of reference.

Case 10.3 ***The synchronously rotating* DQ0 *reference frame, see Figure 10.7***
Again, the model expressed in the synchronously rotating reference frame can be derived from Eq. (10.106), restated here again for convenience as follows:

$$\begin{aligned}
\begin{bmatrix} \underline{V}_{DQ0} \\ \underline{V}_{dq0} \end{bmatrix} &= \begin{bmatrix} \left(\underline{T}_s \cdot R_{ss} \cdot \underline{T}_s^{-1}\right) & 0 \\ 0 & \left(\underline{T}_r \cdot \underline{R}_{rr} \cdot \underline{T}_r^{-1}\right) \end{bmatrix} \begin{bmatrix} \underline{I}_{DQ0} \\ \underline{I}_{dq0} \end{bmatrix} \\
&+ \frac{d}{dt}\left\{ \begin{bmatrix} \left(\underline{T}_s \cdot \underline{\mathbf{L}}_{ss} \cdot \underline{T}_s^{-1}\right) & \left(\underline{T}_s \cdot \underline{\mathbf{L}}_{sr} \cdot \underline{T}_r^{-1}\right) \\ \left(\underline{T}_r \cdot \underline{\mathbf{L}}_{sr}^t \cdot \underline{T}_s^{-1}\right) & \left(\underline{T}_r \cdot \underline{\mathbf{L}}_{rr} \cdot \underline{T}_r^{-1}\right) \end{bmatrix} \begin{bmatrix} \underline{I}_{DQ0} \\ \underline{I}_{dq0} \end{bmatrix} \right\} \\
&- \begin{bmatrix} \left(\dot{\underline{T}}_s \cdot \underline{\mathbf{L}}_{ss} \cdot \underline{T}_s^{-1}\right) & \left(\dot{\underline{T}}_s \cdot \underline{\mathbf{L}}_{sr} \cdot \underline{T}_r^{-1}\right) \\ \left(\dot{\underline{T}}_r \cdot \underline{\mathbf{L}}_{sr}^t \cdot \underline{T}_s^{-1}\right) & \left(\dot{\underline{T}}_r \cdot \underline{\mathbf{L}}_{rr} \cdot \underline{T}_r^{-1}\right) \end{bmatrix} \begin{bmatrix} \underline{I}_{DQ0} \\ \underline{I}_{dq0} \end{bmatrix}
\end{aligned} \tag{10.105}$$

Here, we define $\dot{\sigma}$ as the rotor speed, and ω_s as the synchronous speed.

Also, here, σ is the angle between the a-axis on the rotor and the A-axis on the stator which is equal to

$$\sigma = \sigma_0 + \int_0^t \dot{\sigma}(t)dt \tag{10.191}$$

Here, σ_s is the angle between the stator A-axis and the synchronously rotating D-axis, and σ_{s0} is the initial value of σ_s at time, $t = 0$.

$$\sigma_s = \sigma_{s0} + \int_0^t \omega_s(t)dt = \sigma_{s0} + \omega_s t \tag{10.192}$$

From Figure 10.7, $\underline{T}_s$ and $\underline{T}_r$ matrixes can be deduced, which are written as follows:

$$\underline{T}_s = \frac{2}{3}\begin{bmatrix} \cos(-\sigma_s) & \cos(-\sigma_s + 2\pi/3) & \cos(-\sigma_s + 4\pi/3) \\ \cos(-\pi/2 - \sigma_s) & \cos(2\pi/3 - \pi/2 - \sigma_s) & \cos(4\pi/3 - \pi/2 - \sigma_s) \\ 1/2 & 1/2 & 1/2 \end{bmatrix} \tag{10.193}$$

That is,

$$\underline{T}_s = \frac{2}{3}\begin{bmatrix} \cos(-\sigma_s) & \cos(\sigma_s - 2\pi/3) & \cos(\sigma_s - 4\pi/3) \\ \sin(-\sigma_s) & \sin(2\pi/3 - \sigma_s) & \sin(4\pi/3 - \sigma_s) \\ 1/2 & 1/2 & 1/2 \end{bmatrix} \tag{10.194}$$

and, hence,

$$\underline{T}_s = \frac{2}{3}\begin{bmatrix} \cos(\sigma_s) & \cos(\sigma_s - 2\pi/3) & \cos(\sigma_s - 4\pi/3) \\ -\sin(-\sigma_s) & -\sin(2\pi/3 - \sigma_s) & -\sin(4\pi/3 - \sigma_s) \\ 1/2 & 1/2 & 1/2 \end{bmatrix} \tag{10.195}$$

Furthermore,

$$\underline{T}_r = \frac{2}{3}\begin{bmatrix} \cos(-(\sigma_s - \sigma)) & \cos(2\pi/3 - (\sigma_s - \sigma)) & \cos(4\pi/3 - (\sigma_s - \sigma)) \\ \cos(-\pi/2 - (\sigma_s - \sigma)) & \cos(2\pi/3 - \pi/2 - (\sigma_s - \sigma)) & \cos(4\pi/3 - \pi/2 - (\sigma_s - \sigma)) \\ 1/2 & 1/2 & 1/2 \end{bmatrix} \tag{10.196}$$

which yields

$$\underline{T}_r = \frac{2}{3}\begin{bmatrix} \cos(\sigma_s - \sigma) & \cos((\sigma_s - \sigma) - 2\pi/3) & \cos((\sigma_s - \sigma) - 4\pi/3) \\ -\sin(\sigma_s - \sigma) & -\sin((\sigma_s - \sigma) - 2\pi/3) & -\sin((\sigma_s - \sigma) - 4\pi/3) \\ 1/2 & 1/2 & 1/2 \end{bmatrix} \tag{10.197}$$

Here,

$$\underline{T}_s^{-1} = \frac{2}{3}\begin{bmatrix} \cos(\sigma_s) & -\sin(\sigma_s) & 1 \\ \cos(\sigma_s - 2\pi/3) & -\sin(\sigma_s - 2\pi/3) & 1 \\ \cos(\sigma_s - 4\pi/3) & -\sin(\sigma_s - 4\pi/3) & 1 \end{bmatrix} \tag{10.198}$$

This yields

$$\left(\underline{T}_s^{-1}\right)^t \cdot \underline{T}_s^{-1} = \begin{bmatrix} \cos(\sigma_s) & \cos(\sigma_s - 2\pi/3) & \cos(\sigma_s - 4\pi/3) \\ -\sin(\sigma_s) & -\sin(\sigma_s - 2\pi/3) & -\sin(\sigma_s - 4\pi/3) \\ 1 & 1 & 1 \end{bmatrix} \cdot \begin{bmatrix} \cos(\sigma_s) & -\sin(\sigma_s) & 1 \\ \cos(\sigma_s - 2\pi/3) & -\sin(\sigma_s - 2\pi/3) & 1 \\ \cos(\sigma_s - 4\pi/3) & -\sin(\sigma_s - 4\pi/3) & 1 \end{bmatrix} \tag{10.199}$$

that is,

$$\left(\underline{T}_s^{-1}\right)^t \cdot \underline{T}_s^{-1} = \begin{bmatrix} 3/2 & 0 & 0 \\ 0 & 3/2 & 0 \\ 0 & 0 & 3 \end{bmatrix} = K_s \cdot \underline{G}_s \tag{10.200}$$

where

$$K_s = 3/2 \tag{10.201}$$

and

$$\underline{G}_s = \begin{bmatrix} 1 & 0 & 0 \\ 0 & 1 & 0 \\ 0 & 0 & 2 \end{bmatrix} \tag{10.202}$$

Also, in this case, we have the following:

$$\underline{T}_r^{-1} = \begin{bmatrix} \cos(\sigma_s - \sigma) & -\sin(\sigma_s - \sigma) & 1 \\ \cos((\sigma_s - \sigma) - 2\pi/3) & -\sin((\sigma_s - \sigma) - 2\pi/3) & 1 \\ \cos((\sigma_s - \sigma) - 4\pi/3) & -\sin((\sigma_s - \sigma) - 4\pi/3) & 1 \end{bmatrix} \tag{10.203}$$

and

$$\left(\underline{T}_r^{-1}\right)^t = \begin{bmatrix} \cos(\sigma_s - \sigma) & \cos((\sigma_s - \sigma) - 2\pi/3) & \cos((\sigma_s - \sigma) - 4\pi/3) \\ -\sin(\sigma_s - \sigma) & -\sin((\sigma_s - \sigma) - 2\pi/3) & -\sin((\sigma_s - \sigma) - 4\pi/3) \\ 1 & 1 & 1 \end{bmatrix} \tag{10.204}$$

Thus,

$$\left(\underline{T}_r^{-1}\right)^t \cdot \underline{T}_r^{-1} = \begin{bmatrix} \cos(\sigma_s - \sigma) & \cos((\sigma_s - \sigma) - 2\pi/3) & \cos((\sigma_s - \sigma) - 4\pi/3) \\ -\sin(\sigma_s - \sigma) & -\sin((\sigma_s - \sigma) - 2\pi/3) & -\sin((\sigma_s - \sigma) - 4\pi/3) \\ 1 & 1 & 1 \end{bmatrix} \cdot \begin{bmatrix} \cos(\sigma_s - \sigma) & -\sin(\sigma_s - \sigma) & 1 \\ \cos((\sigma_s - \sigma) - 2\pi/3) & -\sin((\sigma_s - \sigma) - 2\pi/3) & 1 \\ \cos((\sigma_s - \sigma) - 4\pi/3) & -\sin((\sigma_s - \sigma) - 4\pi/3) & 1 \end{bmatrix} \tag{10.205}$$

and, hence,

$$\left(\underline{T}_r^{-1}\right)^t \cdot \underline{T}_r^{-1} = \begin{bmatrix} 3/2 & 0 & 0 \\ 0 & 3/2 & 0 \\ 0 & 0 & 3 \end{bmatrix} = K_r \cdot \underline{G}_r \tag{10.206}$$

where

$$K_r = 3/2 \tag{10.207}$$

and

$$\underline{G}_r = \begin{bmatrix} 1 & 0 & 0 \\ 0 & 1 & 0 \\ 0 & 0 & 2 \end{bmatrix} \tag{10.208}$$

Furthermore,

$$\dot{\underline{T}}_s = \frac{2}{3}\omega_s \begin{bmatrix} -\sin(\sigma_s) & -\sin(\sigma_s - 2\pi/3) & -\sin(\sigma_s - 4\pi/3) \\ -\cos(\sigma_s) & -\cos(\sigma_s - 2\pi/3) & -\cos(\sigma_s - 4\pi/3) \\ 0 & 0 & 0 \end{bmatrix} \tag{10.209}$$

and

$$\dot{\underline{T}}_r = \frac{2}{3}(\omega_s - \dot{\sigma}) \begin{bmatrix} -\sin(\sigma_s - \sigma) & -\sin((\sigma_s - \sigma) - 2\pi/3) & -\sin((\sigma_s - \sigma) - 4\pi/3) \\ -\cos(\sigma_s - \sigma) & -\cos((\sigma_s - \sigma) - 2\pi/3) & -\cos((\sigma_s - \sigma) - 4\pi/3) \\ 0 & 0 & 0 \end{bmatrix} \tag{10.210}$$

Similar to Case 10.1, it can be shown that

$$\underline{T}_s \cdot \underline{R}_{ss} \cdot \underline{T}_s^{-1} = \begin{bmatrix} r_s & 0 & 0 \\ 0 & r_s & 0 \\ 0 & 0 & r_s \end{bmatrix} \tag{10.211}$$

and

$$\underline{T}_r \cdot \underline{R}_{rr} \cdot \underline{T}_r^{-1} = \begin{bmatrix} r_r & 0 & 0 \\ 0 & r_r & 0 \\ 0 & 0 & r_r \end{bmatrix} \tag{10.212}$$

Following steps similar to those performed in Case 10.1 earlier in this section, it can be shown that

$$\underline{T}_s \cdot \underline{\mathbf{L}}_{ss} \cdot \underline{T}_s^{-1} = \begin{bmatrix} L_{ss} - L_{sm} & 0 & 0 \\ 0 & L_{ss} - L_{sm} & 0 \\ 0 & 0 & L_{ss} + 2L_{sm} \end{bmatrix} \tag{10.213}$$

$$\underline{T}_r \cdot \underline{\mathbf{L}}_{rr} \cdot \underline{T}_r^{-1} = \begin{bmatrix} L_{rr} - L_{rm} & 0 & 0 \\ 0 & L_{rr} - L_{rm} & 0 \\ 0 & 0 & L_{rr} + 2L_{rm} \end{bmatrix} \tag{10.214}$$

$$\underline{T}_s \cdot \underline{\mathbf{L}}_{sr} \cdot \underline{T}_r^{-1} = \begin{bmatrix} \frac{3}{2}L_{srm} & 0 & 0 \\ 0 & \frac{3}{2}L_{srm} & 0 \\ 0 & 0 & 0 \end{bmatrix} \tag{10.215}$$

$$\underline{T}_r \cdot \underline{\mathbf{L}}_{sr}^t \cdot \underline{T}_s^{-1} = \begin{bmatrix} \frac{3}{2}L_{srm} & 0 & 0 \\ 0 & \frac{3}{2}L_{srm} & 0 \\ 0 & 0 & 0 \end{bmatrix} \tag{10.216}$$

$$\dot{\underline{T}}_s \cdot \underline{\mathbf{L}}_{ss} \cdot \underline{T}_s^{-1} = \begin{bmatrix} 0 & \omega_s(L_{ss} - L_{sm}) & 0 \\ -\omega_s(L_{ss} - L_{sm}) & 0 & 0 \\ 0 & 0 & 0 \end{bmatrix} \tag{10.217}$$

$$\dot{\underline{T}}_r \cdot \underline{\mathbf{L}}_{rr} \cdot \underline{T}_r^{-1} = \begin{bmatrix} 0 & (\omega_s - \dot{\sigma})(L_{ss} - L_{sm}) & 0 \\ -(\omega_s - \dot{\sigma})(L_{ss} - L_{sm}) & 0 & 0 \\ 0 & 0 & 0 \end{bmatrix} \tag{10.218}$$

$$\dot{\underline{T}}_s \cdot \underline{\mathbf{L}}_{sr} \cdot \underline{T}_r^{-1} = \begin{bmatrix} 0 & \frac{3}{2}\omega_s L_{srm} & 0 \\ -\frac{3}{2}\omega_s L_{srm} & 0 & 0 \\ 0 & 0 & 0 \end{bmatrix} \tag{10.219}$$

and

$$\dot{\underline{T}}_r \cdot \underline{\mathbf{L}}_{sr}^t \cdot \underline{T}_s^{-1} = \begin{bmatrix} 0 & \frac{3}{2}(\omega_s - \dot{\sigma})L_{srm} & 0 \\ -\frac{3}{2}(\omega_s - \dot{\sigma})L_{srm} & 0 & 0 \\ 0 & 0 & 0 \end{bmatrix} \tag{10.220}$$

where $\dot{\sigma} = (d\sigma/dt)$.

Substituting from Eqs. (10.211)–(10.220) in Eq. (10.106) yields the following, with the knowledge that all the transformed inductance matrices are independent of the rotor position angle, σ, and, hence, are independent of time, t:

$$\begin{bmatrix} v_D \\ v_Q \\ v_0 \\ \hline v_d \\ v_q \\ v_0 \end{bmatrix} = \left[\begin{array}{ccc|ccc} r_s & 0 & 0 & 0 & 0 & 0 \\ 0 & r_s & 0 & 0 & 0 & 0 \\ 0 & 0 & r_s & 0 & 0 & 0 \\ \hline 0 & 0 & 0 & r_r & 0 & 0 \\ 0 & 0 & 0 & 0 & r_r & 0 \\ 0 & 0 & 0 & 0 & 0 & r_r \end{array}\right] \begin{bmatrix} i_D \\ i_Q \\ i_0 \\ \hline i_d \\ i_q \\ i_0 \end{bmatrix}$$

$$+\left\{\left[\begin{array}{ccc|ccc} (L_{ss}-L_{sm}) & 0 & 0 & \frac{3}{2}L_{srm} & 0 & 0 \\ 0 & (L_{ss}-L_{sm}) & 0 & 0 & \frac{3}{2}L_{srm} & 0 \\ 0 & 0 & (L_{ss}+2L_{sm}) & 0 & 0 & 0 \\ \hline \frac{3}{2}L_{srm} & 0 & 0 & (L_{rr}-L_{rm}) & 0 & 0 \\ 0 & \frac{3}{2}L_{srm} & 0 & 0 & (L_{rr}-L_{rm}) & 0 \\ 0 & 0 & 0 & 0 & 0 & (L_{rr}+2L_{rm}) \end{array}\right]\left[\begin{array}{c} \dot{i}_D \\ \dot{i}_Q \\ \dot{i}_0 \\ \hline \dot{i}_d \\ \dot{i}_q \\ \dot{i}_0 \end{array}\right]\right\}$$
$$+\left\{\left[\begin{array}{ccc|ccc} 0 & -\omega_s(L_{ss}-L_{sm}) & 0 & 0 & -(3/2)L_{srm} & 0 \\ \omega_s(L_{ss}-L_{sm}) & 0 & 0 & (3/2)L_{srm} & 0 & 0 \\ 0 & 0 & 0 & 0 & 0 & 0 \\ \hline 0 & -\frac{3}{2}(\omega_s-\dot{\sigma})L_{srm} & 0 & 0 & -(\omega_s-\dot{\sigma})(L_{rr}-L_{rm}) & 0 \\ \frac{3}{2}(\omega_s-\dot{\sigma})L_{srm} & 0 & 0 & (\omega_s-\dot{\sigma})(L_{rr}-L_{rm}) & 0 & 0 \\ 0 & 0 & 0 & 0 & 0 & 0 \end{array}\right]\left[\begin{array}{c} i_D \\ i_Q \\ i_0 \\ \hline i_d \\ i_q \\ i_0 \end{array}\right]\right\} \tag{10.221}$$

Again, here we define the following:

$$\underline{V} = \left[\begin{array}{c} v_D \\ v_Q \\ v_0 \\ \hline v_d \\ v_q \\ v_0 \end{array}\right] \tag{10.222}$$

$$\underline{I} = \begin{bmatrix} i_D \\ i_Q \\ i_0 \\ i_d \\ i_q \\ i_0 \end{bmatrix} \tag{10.223}$$

$$\dot{\underline{I}} = \begin{bmatrix} \dot{i}_D \\ \dot{i}_Q \\ \dot{i}_0 \\ \dot{i}_d \\ \dot{i}_q \\ \dot{i}_0 \end{bmatrix} \tag{10.224}$$

and

$$\underline{R} = \left[\begin{array}{ccc|ccc} r_s & 0 & 0 & 0 & 0 & 0 \\ 0 & r_s & 0 & 0 & 0 & 0 \\ 0 & 0 & r_s & 0 & 0 & 0 \\ \hline 0 & 0 & 0 & r_r & 0 & 0 \\ 0 & 0 & 0 & 0 & r_r & 0 \\ 0 & 0 & 0 & 0 & 0 & r_r \end{array}\right] \tag{10.225}$$

Also,

$$\underline{\mathbf{L}}_1 = \left[\begin{array}{ccc|ccc} L_{ss} - L_{sm} & 0 & 0 & \frac{3}{2}L_{srm} & 0 & 0 \\ 0 & L_{ss} - L_{sm} & 0 & 0 & \frac{3}{2}L_{srm} & 0 \\ 0 & 0 & L_{ss} + 2L_{sm} & 0 & 0 & 0 \\ \hline \frac{3}{2}L_{srm} & 0 & 0 & L_{rr} - L_{rm} & 0 & 0 \\ 0 & \frac{3}{2}L_{srm} & 0 & 0 & L_{rr} - L_{rm} & 0 \\ 0 & 0 & 0 & 0 & 0 & L_{rr} + 2L_{rm} \end{array}\right] \tag{10.226}$$

$$\underline{\hat{\mathbf{L}}}_2 = \left[\begin{array}{ccc|ccc} 0 & -\omega_s(L_{ss} - L_{sm}) & 0 & 0 & -\frac{3}{2}\omega_s L_{srm} & 0 \\ \omega_s(L_{ss} - L_{sm}) & 0 & 0 & \frac{3}{2}\omega_s L_{srm} & 0 & 0 \\ 0 & 0 & 0 & 0 & 0 & 0 \\ \hline 0 & -\frac{3}{2}(\omega_s - \dot{\sigma})L_{srm} & 0 & 0 & -(\omega_s - \dot{\sigma})(L_{rr} - L_{rm}) & 0 \\ \frac{3}{2}(\omega_s - \dot{\sigma})L_{srm} & 0 & 0 & (\omega_s - \dot{\sigma})(L_{rr} - L_{rm}) & 0 & 0 \\ 0 & 0 & 0 & 0 & 0 & 0 \end{array}\right] \tag{10.227}$$

Accordingly, the flux linkage vector, $\underline{\Lambda}$, can be expressed as follows:

$$\underline{\Lambda} = \underline{\mathbf{L}}_1 \cdot \underline{I} \text{ and } \underline{I} = \underline{\mathbf{L}}_1^{-1} \cdot \underline{\Lambda} \tag{10.228}$$

and since $\underline{\mathbf{L}}_1$ is time independent, it follows that

$$\underline{\mathbf{L}}_1 \cdot \dot{\underline{I}} = d(\underline{\mathbf{L}}_1 \cdot \underline{I})/dt = d\underline{\Lambda}/dt = \dot{\underline{\Lambda}} \tag{10.229}$$

Hence, Eq. (10.221) can be rewritten in abbreviated matrix form as follows:

$$\underline{V} = \underline{R} \cdot \underline{I} + \underline{\mathbf{L}}_1 \cdot \dot{\underline{I}} + \underline{\hat{\mathbf{L}}}_2 \cdot \underline{I} \tag{10.230}$$

Accordingly, rearranging Eq. (10.230), we can write the following:

$$\underline{\mathbf{L}}_1 \cdot \dot{\underline{I}} = -(\underline{R} + \underline{\hat{\mathbf{L}}}_2)\underline{I} + \underline{V} \tag{10.231}$$

Hence,

$$\underline{\dot{I}} = \left\{ -\underline{\mathbf{L}}_1^{-1} \cdot (\underline{R} + \underline{\hat{\mathbf{L}}}_2) \right\} \cdot \underline{I} + \{\underline{\mathbf{L}}_1^{-1}\} \cdot \underline{V} \tag{10.232}$$

Equation (10.232) is the transformed state-space model of the induction machine in the current frame of reference. However, Eq. (10.230) can be rewritten in terms of the flux linkage vector, $\underline{\Lambda}$, using Eq. (10.228), as follows:

$$\underline{V} = (\underline{R} + \underline{\mathbf{L}}_1^{-1}) \cdot \underline{\Lambda} + \underline{\dot{\Lambda}} + \underline{\hat{\mathbf{L}}}_2 \cdot (\underline{\mathbf{L}}_1^{-1} \cdot \underline{\Lambda}) \tag{10.233}$$

Hence, we can write

$$\underline{\dot{\Lambda}} = \left\{ -(\underline{R} + \underline{\hat{\mathbf{L}}}_2) \cdot (\underline{\mathbf{L}}_1^{-1}) \right\} \cdot \underline{\Lambda} + \underline{V} \tag{10.234}$$

That is,

$$\underline{\dot{\Lambda}} = \left\{ -(\underline{R} + \underline{\hat{\mathbf{L}}}_2) \cdot \underline{\mathbf{L}}_1^{-1} \right\} \cdot \underline{\Lambda} + \underline{V} \tag{10.235}$$

Equations (10.234) and (10.235) are the induction machine's state-space formulations in the transformed flux linkage frame of reference for the synchronously rotating $DQ0$–$dq0$ frame. We now proceed to machine torque formulations for the three $DQ0$–$dq0$ frames of reference, Case 10.4–10.6.

10.7 On Torque and Power Computations in an Induction Machine in the *DQ0*–*dq0* and *ABC*–*abc* Reference Frames

The input power, P_{in}, into a wound-rotor induction machine can generally be expressed as follows:

$$P_{in} = \underline{I}_{ABC}^t \cdot \underline{V}_{ABC} + \underline{I}_{abc}^t \cdot \underline{V}_{abc} \tag{10.236}$$

where P_{in} is the "input power" from both the stator ABC terminals and the rotor *abc* terminals.

However, the portion of P_{in} involved in the development and production of an electromagnetic/electromechanical torque, T_{dev}, is the Faraday portion of the voltage in both the stator and rotor windings, which is the electromagnetic (or airgap) power, P_{em}. Here, P_{em} can be expressed as follows:

$$P_{em} = \underline{I}_{ABC}^t \cdot \left\{ \frac{d}{dt}(\underline{\Lambda}_{ABC}) \right\} + \underline{I}_{abc}^t \cdot \left\{ \frac{d}{dt}(\underline{\Lambda}_{abc}) \right\} \tag{10.237}$$

Applying the stator $DQ0$ and rotor $dq0$ transformations, $\underline{T}_s$ and $\underline{T}_r$ introduced earlier in Eqs. (10.83)–(10.90), results in the following:

$$P_{em} = \left(\underline{T}_s^{-1} \cdot \underline{I}_{DQ0} \right)^t \cdot \left\{ \frac{d}{dt}(\underline{\Lambda}_{ABC}) \right\} + \left(\underline{T}_r^{-1} \cdot \underline{I}_{dq0} \right)^t \cdot \left\{ \frac{d}{dt}(\underline{\Lambda}_{abc}) \right\} \tag{10.238}$$

That is,

$$P_{em} = \underline{I}_{DQ0}^{t} \cdot \left(\underline{T}_{s}^{-1}\right)^{t} \cdot \left\{ \frac{d}{dt}(\underline{\Lambda}_{ABC}) \right\} + \underline{I}_{dq0}^{t} \left(\underline{T}_{r}^{-1}\right)^{t} \cdot \left\{ \frac{d}{dt}(\underline{\Lambda}_{abc}) \right\} \quad (10.239)$$

However, from earlier developments in Eqs. (10.97) and (10.102), the following can be written for the terms $(d(\underline{\Lambda}_{ABC})/dt)$ and $(d(\underline{\Lambda}_{abc})/dt)$:

$$\frac{d}{dt}(\underline{\Lambda}_{ABC}) = \left\{ \underline{T}_{s}^{-1} \cdot \frac{d}{dt}(\underline{\Lambda}_{DQ0}) - \underline{T}_{s}^{-1} \cdot \dot{\underline{T}}_{s} \cdot \underline{T}_{s}^{-1} \cdot \underline{\Lambda}_{DQ0} \right\} \quad (10.240)$$

and

$$\frac{d}{dt}(\underline{\Lambda}_{abc}) = \left\{ \underline{T}_{r}^{-1} \cdot \frac{d}{dt}(\underline{\Lambda}_{dq0}) - \underline{T}_{r}^{-1} \cdot \dot{\underline{T}}_{r} \cdot \underline{T}_{r}^{-1} \cdot \underline{\Lambda}_{dq0} \right\} \quad (10.241)$$

Substituting from Eqs. (10.240) and (10.241) into Eqs. (10.238) and (10.239) yields the following for P_{em}:

$$\begin{aligned} P_{em} &= \underline{I}_{DQ0}^{t} \cdot \left[\left(\underline{T}_{s}^{-1}\right)^{t} \cdot \underline{T}_{s}^{-1} \cdot \tfrac{d}{dt}(\underline{\Lambda}_{DQ0}) - \left(\underline{T}_{s}^{-1}\right)^{t} \cdot \underline{T}_{s}^{-1} \cdot \dot{\underline{T}}_{s} \cdot \underline{T}_{s}^{-1} \cdot \underline{\Lambda}_{DQ0} \right] \\ &\quad + \underline{I}_{dq0}^{t} \cdot \left[\left(\underline{T}_{r}^{-1}\right)^{t} \cdot \underline{T}_{r}^{-1} \cdot \tfrac{d}{dt}(\underline{\Lambda}_{dq0}) - \left(\underline{T}_{r}^{-1}\right)^{t} \cdot \underline{T}_{r}^{-1} \cdot \dot{\underline{T}}_{r} \cdot \underline{T}_{r}^{-1} \cdot \underline{\Lambda}_{dq0} \right] \end{aligned} \quad (10.242)$$

However, from earlier developments in Case 10.1, for the *DQ*0–*dq*0 frame rotating at the rotor speed, $\dot{\sigma} = (d\sigma/dt)$, see Figure 10.5, we have the following involving the rotational voltages and the resulting electromagnetic power, P_{em}:

$$\begin{aligned} P_{em} = &-\underline{I}_{DQ0}^{t} \cdot \left\{ \left(\underline{T}_{s}^{-1}\right)^{t} \cdot \underline{T}_{s}^{-1} \right\} \cdot \dot{\underline{T}}_{s} \cdot \underline{T}_{s}^{-1} \cdot \underline{\Lambda}_{DQ0} \\ &-\underline{I}_{dq0}^{t} \cdot \left\{ \left(\underline{T}_{r}^{-1}\right)^{t} \cdot \underline{T}_{r}^{-1} \right\} \cdot \dot{\underline{T}}_{r} \cdot \underline{T}_{r}^{-1} \cdot \underline{\Lambda}_{dq0} \end{aligned} \quad (10.243)$$

Here, we will be treating three cases of *DQ*0–*dq*0 transformations, namely Case 10.1 is that with a reference frame rotating at the rotor speed, $\dot{\sigma} = (d\sigma/dt)$, see Figure 10.5; Case 10.2 is that with a reference frame that is stationary and fixed on the stator; and Case 10.3 is that with a reference frame rotating at the synchronous speed, ω_s, Accordingly, we will deal here with Cases 10.4–10.6, respectively.

Case 10.4 ***A reference frame rotating at the rotor speed,*** $\dot{\sigma} = (d\sigma/dt)$ ***in e.rad/s.*** Here, from Eq. (10.140), one can write the following:

$$\underline{\Lambda}_{DQ0} = \begin{bmatrix} \lambda_D \\ \lambda_Q \\ \lambda_0 \end{bmatrix} = \begin{bmatrix} (L_{ss} - L_{sm}) & 0 & 0 \\ 0 & (L_{ss} - L_{sm}) & 0 \\ 0 & 0 & (L_{ss} + 2L_{sm}) \end{bmatrix} \begin{bmatrix} i_D \\ i_Q \\ i_0 \end{bmatrix} + \begin{bmatrix} \frac{3}{2}L_{srm} & 0 & 0 \\ 0 & \frac{3}{2}L_{srm} & 0 \\ 0 & 0 & 0 \end{bmatrix} \begin{bmatrix} i_d \\ i_q \\ i_0 \end{bmatrix} \quad (10.244)$$

and

$$\underline{\Lambda}_{dqo} = \begin{bmatrix} \lambda_d \\ \lambda_q \\ \lambda_0 \end{bmatrix} = \begin{bmatrix} \frac{3}{2}L_{srm} & 0 & 0 \\ 0 & \frac{3}{2}L_{srm} & 0 \\ 0 & 0 & 0 \end{bmatrix} \begin{bmatrix} i_D \\ i_Q \\ i_0 \end{bmatrix} + \begin{bmatrix} (L_{rr} - L_{rm}) & 0 & 0 \\ 0 & (L_{rr} - L_{rm}) & 0 \\ 0 & 0 & (L_{rr} + 2L_{rm}) \end{bmatrix} \begin{bmatrix} i_d \\ i_q \\ i_0 \end{bmatrix} \tag{10.245}$$

Thus, in Case 10.4, we have

$$\underline{T}_s^{-1} = \begin{bmatrix} \cos\sigma & -\sin\sigma & 1 \\ \cos(\sigma - 2\pi/3) & -\sin(\sigma - 2\pi/3) & 1 \\ \cos(\sigma - 4\pi/3) & -\sin(\sigma - 4\pi/3) & 1 \end{bmatrix} \tag{10.246}$$

Hence,

$$\left(\underline{T}_s^{-1}\right)^t = \begin{bmatrix} \cos\sigma & \cos(\sigma - 2\pi/3) & \cos(\sigma - 4\pi/3) \\ -\sin\sigma & -\sin(\sigma - 2\pi/3) & -\sin(\sigma - 4\pi/3) \\ 1 & 1 & 1 \end{bmatrix} \tag{10.247}$$

Hence,

$$\left(\underline{T}_s^{-1}\right)^t \left(\underline{T}_s^{-1}\right) = \begin{bmatrix} 3/2 & 0 & 0 \\ 0 & 3/2 & 0 \\ 0 & 0 & 3 \end{bmatrix} = \frac{3}{2}\begin{bmatrix} 1 & 0 & 0 \\ 0 & 1 & 0 \\ 0 & 0 & 2 \end{bmatrix} = K_s \underline{G}_s \tag{10.248}$$

where

$$K_s = 3/2, \quad \underline{G}_s = \begin{bmatrix} 1 & 0 & 0 \\ 0 & 1 & 0 \\ 0 & 0 & 2 \end{bmatrix} \tag{10.249}$$

Also,

$$\underline{T}_r^{-1} = \begin{bmatrix} 1 & 0 & 1 \\ -1/2 & \sqrt{3}/2 & 1 \\ -1/2 & -\sqrt{3}/2 & 1 \end{bmatrix} \tag{10.250}$$

Hence,

$$\left(\underline{T}_r^{-1}\right)^t = \begin{bmatrix} 1 & -\frac{1}{2} & -\frac{1}{2} \\ 0 & \frac{\sqrt{3}}{2} & -\frac{\sqrt{3}}{2} \\ 1 & 1 & 1 \end{bmatrix} \tag{10.251}$$

Hence,

$$\left(\underline{T}_r^{-1}\right)^t \cdot \left(\underline{T}_r^{-1}\right) = \begin{bmatrix} 3/2 & 0 & 0 \\ 0 & 3/2 & 0 \\ 0 & 0 & 3 \end{bmatrix} = \frac{3}{2}\begin{bmatrix} 1 & 0 & 0 \\ 0 & 1 & 0 \\ 0 & 0 & 2 \end{bmatrix} = K_r \underline{G}_r \tag{10.252}$$

where

$$K_r = 3/2,\quad \underline{G}_r = \begin{bmatrix} 1 & 0 & 0 \\ 0 & 1 & 0 \\ 0 & 0 & 2 \end{bmatrix} \tag{10.253}$$

Here,

$$\underline{\dot{T}}_s = -\frac{2}{3}\dot{\sigma}\begin{bmatrix} \sin\sigma & \sin(\sigma - 2\pi/3) & \sin(\sigma - 4\pi/3) \\ \cos\sigma & \cos(\sigma - 2\pi/3) & \cos(\sigma - 4\pi/3) \\ 0 & 0 & 0 \end{bmatrix} \tag{10.254}$$

Therefore,

$$\underline{\dot{T}}_s \cdot \underline{T}_s^{-1} = -\frac{2}{3}\dot{\sigma}\begin{bmatrix} 0 & -3/2 & 0 \\ 3/2 & 0 & 0 \\ 0 & 0 & 0 \end{bmatrix} = \begin{bmatrix} 0 & \dot{\sigma} & 0 \\ \dot{\sigma} & 0 & 0 \\ 0 & 0 & 0 \end{bmatrix} \tag{10.255}$$

Since $\underline{T}_r$ is a matrix, the elements of which are constants, it follows that $\underline{\dot{T}}_r = (d\underline{T}_r/dt)$ is a (3×3) zero matrix, that is, $\underline{\dot{T}}_r = \underline{0}$, and, therefore,

$$\underline{\dot{T}}_r \cdot \underline{T}_r^{-1} = \begin{bmatrix} 0 & 0 & 0 \\ 0 & 0 & 0 \\ 0 & 0 & 0 \end{bmatrix} \tag{10.256}$$

Hence, the term, P_{em}, in Eq. (10.243) is reduced to the following, using Eqs. (10.248) and (10.249):

$$P_{em} = -\underline{I}_{DQ0}^t \cdot \{K_s\underline{G}_s\} \cdot \{\underline{\dot{T}}_s \cdot \underline{T}_s^{-1}\} \cdot \underline{\Lambda}_{DQ0} \tag{10.257}$$

Accordingly, the developed electromagnetic torque, T_{dev}, is given by the following:

$$T_{dev} = (P_{em}/\dot{\sigma_m}) = \left[-\underline{I}_{DQ0}^t \cdot \{K_s\underline{G}_s\} \cdot \{\underline{\dot{T}}_s \cdot \underline{T}_s^{-1}\} \cdot \underline{\Lambda}_{DQ0}\right] \Big/ \dot{\sigma_m} \tag{10.258}$$

where $\dot{\sigma_m} = \dot{\sigma}/(p/2)$, and p is the number of poles in the machine. Substituting from Eqs. (10.249) and (10.255) into Eq. (10.258) and expanding gives the following:

$$T_{dev} = -\frac{3/2}{\dot{\sigma_m}}\begin{bmatrix} i_D & i_Q & i_0 \end{bmatrix}\begin{bmatrix} 1 & 0 & 0 \\ 0 & 1 & 0 \\ 0 & 0 & 2 \end{bmatrix}\begin{bmatrix} 0 & \dot{\sigma} & 0 \\ \dot{\sigma} & 0 & 0 \\ 0 & 0 & 0 \end{bmatrix}\begin{bmatrix} \lambda_D \\ \lambda_Q \\ \lambda_0 \end{bmatrix} \tag{10.259}$$

That is,

$$T_{dev} = -\frac{3/2}{\dot{\sigma_m}}\begin{bmatrix} i_D & i_Q & i_O \end{bmatrix}\begin{bmatrix} 1 & 0 & 0 \\ 0 & 1 & 0 \\ 0 & 0 & 2 \end{bmatrix}\begin{bmatrix} \dot{\sigma}\lambda_Q \\ -\dot{\sigma}\lambda_D \\ 0 \end{bmatrix} \tag{10.260}$$

Equation (10.260) yields the following:

$$T_{dev} = -\frac{3/2}{\dot{\sigma_m}} \begin{bmatrix} i_D & i_Q & i_0 \end{bmatrix} \begin{bmatrix} \dot{\sigma}\lambda_Q \\ -\dot{\sigma}\lambda_D \\ 0 \end{bmatrix} \tag{10.261}$$

However, $\dot{\sigma}_m = \dot{\sigma}/(p/2)$, hence, in terms of flux linkages and currents one can write

$$T_{dev} = (3/2)(p/2)(i_Q\lambda_D - i_D\lambda_Q) \tag{10.262}$$

In terms of current, one can write the following:

$$\lambda_D = (L_{ss} - L_{sm})i_D + \left(\frac{3}{2}L_{srm}\right) i_d \tag{10.263}$$

$$\lambda_Q = (L_{ss} - L_{sm})i_Q + \left(\frac{3}{2}L_{srm}\right) i_q \tag{10.264}$$

Hence, substituting for λ_D and λ_Q from Eqs. (10.263) and (10.264) gives

$$T_{dev} = \frac{3}{2}\left(\frac{p}{2}\right)\left(\frac{3}{2}L_{srm}\right)(i_Q i_d - i_D i_q) \tag{10.265}$$

or

$$T_{dev} = (9/4)pL_{srm}(i_Q i_d - i_D i_q) \tag{10.266}$$

Here, one can write the equation of rotational motion of an induction motor with a mechanical load torque, T_L, as

$$T_{dev} - T_L = J(d\dot{\sigma_m}/dt) \tag{10.267}$$

where J is the total polar moment of inertia of all the rotating masses, the motor's rotor, mechanical coupling, and load moments of inertia, and $\dot{\sigma_m}$ is the mechanical angular speed of the motor shaft (rotor).

The starting transient of an induction motor-mechanical load system is accomplished here by simultaneously, numerically solving the electrical state-space motor system in Eq. (10.144) or Eq. (10.146), and the mechanical motor-load rotational motion model in Eq. (10.267). Such a starting transient torque and current profiles are given in Figures 10.8 and 10.9, respectively, for the example, 100 hp motor-load system given in Chapter 6, Section 6.6.

In order to obtain an expression for T_{dev} in the *ABC–abc* frame of reference for Case 10.4, we apply the definition of the transformation, $\underline{T}_s$, to the torque expression in Eq. (10.262), where we have

$$\begin{bmatrix} \lambda_D \\ \lambda_Q \\ \lambda_0 \end{bmatrix} = \underline{T}_s \cdot \begin{bmatrix} \lambda_A \\ \lambda_B \\ \lambda_C \end{bmatrix} \tag{10.268}$$

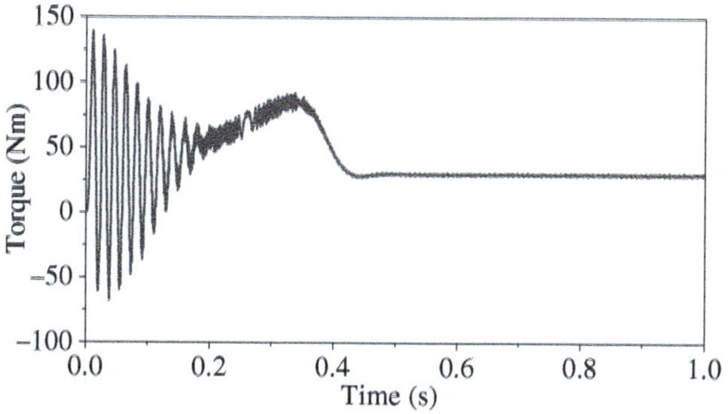

Figure 10.8 The torque profile of the starting transient of the 100 hp motor-load system of Section 6.6.

Figure 10.9 The starting transient of the phase (*A*) current of the 100 hp motor-load system of Section 6.6.

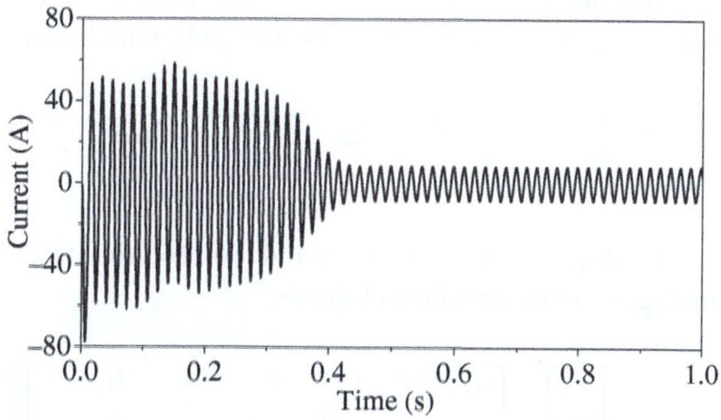

and

$$\begin{bmatrix} i_D \\ i_Q \\ i_0 \end{bmatrix} = \underline{T}_s \cdot \begin{bmatrix} i_A \\ i_B \\ i_C \end{bmatrix} \tag{10.269}$$

where

$$\underline{T}_s = (2/3)\begin{bmatrix} \cos\sigma & \cos(\sigma - 2\pi/3) & \cos(\sigma - 4\pi/3) \\ -\sin\sigma & -\sin(\sigma - 2\pi/3) & -\sin(\sigma - 4\pi/3) \\ (1/2) & (1/2) & (1/2) \end{bmatrix} \tag{10.270}$$

and

$$T_{dev} = (3/2)(p/2)(\lambda_D i_Q - \lambda_Q i_D) \tag{10.271}$$

where p is the number of poles.

Substituting from Eq. (10.270) for $\underline{T}_s$ into Eqs. (10.268) and (10.269), we obtain the following for λ_D, λ_Q, i_D, and i_Q:

$$\lambda_D = (2/3)[\lambda_A \cos\sigma + \lambda_B \cos(\sigma - 2\pi/3) + \lambda_C \cos(\sigma - 4\pi/3)] \tag{10.272}$$

$$\lambda_Q = (2/3)[-\lambda_A \sin\sigma - \lambda_B \sin(\sigma - 2\pi/3) - \lambda_C \sin(\sigma - 4\pi/3)] \quad (10.273)$$

$$i_D = (2/3)[i_A \cos\sigma + i_B \cos(\sigma - 2\pi/3) + i_C \cos(\sigma - 4\pi/3)] \quad (10.274)$$

$$i_Q = (2/3)[-i_A \sin\sigma - i_B \sin(\sigma - 2\pi/3) - i_C \sin(\sigma - 4\pi/3)] \quad (10.275)$$

Substituting for λ_D, λ_Q, i_D, and i_Q from Eqs. (10.272)–(10.275), respectively, into the developed torque equation (10.271), and using the appropriate trigonometric identities and algebraic manipulations, the reader should be able to develop, T_{dev}, Eq. (10.271) in the following form:

$$T_{dev} = (p/2\sqrt{3})[(i_A(\lambda_C - \lambda_B) - i_B(\lambda_A - \lambda_C) - i_C(\lambda_B - \lambda_A)] \quad (10.276)$$

Case 10.5 ***A stationary stator frame of reference at a rotor speed, $\dot{\sigma} = (d\sigma/dt)$ in e.rad/s.*** The torque expression in the stationary stator reference frame can be developed from Eq. (10.243), which is written as follows:

$$T_{dev} = \frac{P_{em}}{\dot{\sigma}_m} = -\frac{K_s}{\dot{\sigma}_m} \cdot \underline{I}^t_{DQ0} \cdot \underline{G}_s \cdot (\dot{\underline{T}}_s \cdot \underline{T}_s^{-1}) \cdot \underline{\Lambda}_{DQ0} - \frac{K_r}{\dot{\sigma}_m} \cdot \underline{I}^t_{dq0} \cdot \underline{G}_r \cdot (\dot{\underline{T}}_r \cdot \underline{T}_r^{-1}) \cdot \underline{\Lambda}_{dq0} \quad (10.277)$$

From Eqs. (10.164), (10.165), and (10.166), the stator flux linkages and rotor flux linkages can be written as follows:

$$\underline{\Lambda}_{DQO} = \begin{bmatrix} \lambda_D \\ \lambda_Q \\ \lambda_0 \end{bmatrix} = \begin{bmatrix} (L_{ss} - L_{sm}) & 0 & 0 \\ 0 & (L_{ss} - L_{sm}) & 0 \\ 0 & 0 & (L_{ss} + 2L_{sm}) \end{bmatrix} \begin{bmatrix} i_D \\ i_Q \\ i_0 \end{bmatrix} + \begin{bmatrix} \frac{3}{2}L_{srm} & 0 & 0 \\ 0 & \frac{3}{2}L_{srm} & 0 \\ 0 & 0 & 0 \end{bmatrix} \begin{bmatrix} i_d \\ i_q \\ i_0 \end{bmatrix} \quad (10.278)$$

and

$$\underline{\Lambda}_{dq0} = \begin{bmatrix} \lambda_d \\ \lambda_q \\ \lambda_0 \end{bmatrix} = \begin{bmatrix} \frac{3}{2}L_{srm} & 0 & 0 \\ 0 & \frac{3}{2}L_{srm} & 0 \\ 0 & 0 & 0 \end{bmatrix} \begin{bmatrix} i_D \\ i_Q \\ i_0 \end{bmatrix} + \begin{bmatrix} (L_{rr} - L_{rm}) & 0 & 0 \\ 0 & (L_{rr} - L_{rm}) & 0 \\ 0 & 0 & (L_{rr} + 2L_{rm}) \end{bmatrix} \begin{bmatrix} i_d \\ i_q \\ i_0 \end{bmatrix} \quad (10.279)$$

and from previous developments in Section 10.6, we have the following:

$$K_s = K_r = 3/2 \quad (10.280)$$

$$\underline{G}_s = \underline{G}_r = \begin{bmatrix} 1 & 0 & 0 \\ 0 & 1 & 0 \\ 0 & 0 & 2 \end{bmatrix} \quad (10.281)$$

and

$$\underline{\dot{T}}_s \cdot \underline{T}_s^{-1} = \begin{bmatrix} 0 & 0 & 0 \\ 0 & 0 & 0 \\ 0 & 0 & 0 \end{bmatrix} \tag{10.282}$$

Also,

$$\underline{\dot{T}}_r \cdot \underline{T}_r^{-1} = \frac{2}{3}\dot{\sigma} \begin{bmatrix} -\sin(\sigma) & -\sin(\sigma + 2\pi/3) & -\sin(\sigma + 4\pi/3) \\ \cos(\sigma) & \cos(\sigma + 2\pi/3) & \cos(\sigma + 4\pi/3) \\ 0 & 0 & 0 \end{bmatrix} \cdot \begin{bmatrix} \cos\sigma & -\sin\sigma & 1 \\ \cos(\sigma + 2\pi/3) & -\sin(\sigma + 2\pi/3) & 1 \\ \cos(\sigma + 4\pi/3) & -\sin(\sigma + 4\pi/3) & 1 \end{bmatrix} \tag{10.283}$$

Hence,

$$\underline{\dot{T}}_r \cdot \underline{T}_r^{-1} = \begin{bmatrix} 0 & -\dot{\sigma} & 0 \\ \dot{\sigma} & 0 & 0 \\ 0 & 0 & 0 \end{bmatrix} \tag{10.284}$$

Substituting for $\underline{G}_s$, $\underline{G}_r$, K_s, K_r, $\underline{\dot{T}}_s$, $\underline{T}_s^{-1}$, $\underline{\Lambda}_{DQ0}$, and $\underline{\Lambda}_{dq0}$ into Eq. (10.277) gives

$$T_{dev} = \frac{K_r}{\dot{\sigma_m}} \cdot \underline{I}_{dq0}^t \cdot \left(\underline{\dot{T}}_r \cdot \underline{T}_r^{-1}\right) \cdot \underline{\Lambda}_{dq0} \tag{10.285}$$

That is,

$$T_{dev} = \frac{3/2}{\dot{\sigma_m}} \begin{bmatrix} i_d & i_q & i_0 \end{bmatrix} \begin{bmatrix} 1 & 0 & 0 \\ 0 & 1 & 0 \\ 0 & 0 & 2 \end{bmatrix} \begin{bmatrix} 0 & -\dot{\sigma} & 0 \\ \dot{\sigma} & 0 & 0 \\ 0 & 0 & 0 \end{bmatrix} \begin{bmatrix} \lambda_d \\ \lambda_q \\ \lambda_0 \end{bmatrix} \tag{10.286}$$

Hence,

$$T_{dev} = -\frac{3/2}{\dot{\sigma_m}} \begin{bmatrix} i_d & i_q & i_0 \end{bmatrix} \begin{bmatrix} -\dot{\sigma}\lambda_q \\ \dot{\sigma}\lambda_d \\ 0 \end{bmatrix} \tag{10.287}$$

However, $\dot{\sigma_m} = \dot{\sigma}/(p/2)$. Hence, it follows that

$$T_{dev} = \left(\frac{3}{2}\right)\left(\frac{p}{2}\right)(i_d\lambda_q - i_q\lambda_d) \tag{10.288}$$

In term of currents and inductances, one can write the following:

$$\lambda_d = \frac{3}{2}L_{srm}i_D + (L_{rr} - L_{rm})i_d \tag{10.289}$$

$$\lambda_q = \frac{3}{2}L_{srm}i_Q + (L_{rr} - L_{rm})i_q \tag{10.290}$$

Hence,

$$T_{dev} = \frac{3}{2}\left(\frac{p}{2}\right)\left(\frac{3}{2}L_{srm}\right)(i_Q i_d - i_D i_q) \tag{10.291}$$

Case 10.6 ***The synchronously rotating frame of reference at a rotor speed,*** $\dot{\sigma} = (d\sigma/dt)$ ***in e.rad/s.*** The torque expression in the synchronously rotating reference frame can be developed from Eq. (10.243), which is, again, written as follows:

$$T_{dev} = \frac{P_{em}}{\dot{\sigma}_m} = -\frac{K_s}{\dot{\sigma}_m} \cdot \underline{I}^t_{DQ0} \cdot \underline{G}_s \cdot \left(\dot{\underline{T}}_s \cdot \underline{T}_s^{-1}\right) \cdot \underline{\Lambda}_{DQ0} - \frac{K_r}{\dot{\sigma}_m} \cdot \underline{I}^t_{dq0} \cdot \underline{G}_r \cdot \left(\dot{\underline{T}}_r \cdot \underline{T}_r^{-1}\right) \cdot \underline{\Lambda}_{dq0} \tag{10.292}$$

In this case, the stator flux linkages and rotor flux linkages are the same as Eqs. (10.278) and (10.279) and are rewritten as follows:

$$\underline{\Lambda}_{DQ0} = \begin{bmatrix} \lambda_D \\ \lambda_Q \\ \lambda_0 \end{bmatrix} = \begin{bmatrix} L_{ss} - L_{sm} & 0 & 0 \\ 0 & L_{ss} - L_{sm} & 0 \\ 0 & 0 & L_{ss} + 2L_{sm} \end{bmatrix} \begin{bmatrix} i_D \\ i_Q \\ i_0 \end{bmatrix} + \begin{bmatrix} \frac{3}{2}L_{srm} & 0 & 0 \\ 0 & \frac{3}{2}L_{srm} & 0 \\ 0 & 0 & 0 \end{bmatrix} \begin{bmatrix} i_d \\ i_q \\ i_0 \end{bmatrix} \tag{10.293}$$

$$\underline{\Lambda}_{dq0} = \begin{bmatrix} \lambda_d \\ \lambda_q \\ \lambda_0 \end{bmatrix} = \begin{bmatrix} \frac{3}{2}L_{srm} & 0 & 0 \\ 0 & \frac{3}{2}L_{srm} & 0 \\ 0 & 0 & 0 \end{bmatrix} \begin{bmatrix} i_D \\ i_Q \\ i_0 \end{bmatrix} + \begin{bmatrix} L_{rr} - L_{rm} & 0 & 0 \\ 0 & L_{rr} - L_{rm} & 0 \\ 0 & 0 & L_{rr} + 2L_{rm} \end{bmatrix} \begin{bmatrix} i_d \\ i_q \\ i_0 \end{bmatrix} \tag{10.294}$$

and from previous developments in Section 10.6, we have the following:

$$K_s = K_r = 3/2 \tag{10.295}$$

$$\underline{G}_s = \underline{G}_r = \begin{bmatrix} 1 & 0 & 0 \\ 0 & 1 & 0 \\ 0 & 0 & 2 \end{bmatrix} \tag{10.296}$$

$$\dot{\underline{T}}_s \cdot \underline{T}_s^{-1} = \frac{2}{3}\omega_s \begin{bmatrix} -\sin(\sigma_s) & -\sin(\sigma_s - 2\pi/3) & -\sin(\sigma_s - 4\pi/3) \\ -\cos(\sigma_s) & -\cos(\sigma_s - 2\pi/3) & -\cos(\sigma_s - 4\pi/3) \\ 0 & 0 & 0 \end{bmatrix} \cdot \begin{bmatrix} \cos(\sigma_s) & -\sin\sigma & 1 \\ \cos(\sigma_s - 2\pi/3) & -\sin(\sigma_s + 2\pi/3) & 1 \\ \cos(\sigma_s - 4\pi/3) & -\sin(\sigma_s + 4\pi/3) & 1 \end{bmatrix} \tag{10.297}$$

which yields

$$\dot{\underline{T}}_s \cdot \underline{T}_s^{-1} = \begin{bmatrix} 0 & -\omega_s & 0 \\ \omega_s & 0 & 0 \\ 0 & 0 & 0 \end{bmatrix} \tag{10.298}$$

Also,

$$\underline{\dot{T}}_r \cdot \underline{T}_r^{-1} = \frac{2}{3}(\omega_s - \dot{\sigma}) \begin{bmatrix} -\sin(\sigma_s - \sigma) & -\sin((\sigma_s - \sigma) - 2\pi/3) & -\sin((\sigma_s - \sigma) - 4\pi/3) \\ -\cos(\sigma_s - \sigma) & -\cos((\sigma_s - \sigma) - 2\pi/3) & -\cos((\sigma_s - \sigma) - 4\pi/3) \\ 0 & 0 & 0 \end{bmatrix} \cdot \begin{bmatrix} \cos(\sigma_s - \sigma) & -\sin(\sigma_s - \sigma) & 1 \\ \cos((\sigma_s - \sigma) - 2\pi/3) & -\sin((\sigma_s - \sigma) - 2\pi/3) & 1 \\ \cos((\sigma_s - \sigma) - 4\pi/3) & -\sin((\sigma_s - \sigma) - 4\pi/3) & 1 \end{bmatrix} \tag{10.299}$$

which yields

$$\underline{\dot{T}}_r \cdot \underline{T}_r^{-1} = \begin{bmatrix} 0 & (\omega_s - \dot{\sigma}) & 0 \\ -(\omega_s - \dot{\sigma}) & 0 & 0 \\ 0 & 0 & 0 \end{bmatrix} \tag{10.300}$$

Substituting for $\underline{G}_s$, $\underline{G}_r$, K_s, K_r, $\underline{\dot{T}}_s$, $\underline{T}_s^{-1}$, $\underline{\Lambda}_{DQ0}$, and $\underline{\Lambda}_{dq0}$ into Eq. (10.292) gives

$$T_{dev} = -\frac{K_s}{\dot{\sigma}_m} \cdot \underline{I}_{DQ0}^t \cdot \underline{G}_s \cdot \left(\underline{\dot{T}}_s \cdot \underline{T}_s^{-1}\right) \cdot \underline{\Lambda}_{DQ0} - \frac{K_r}{\dot{\sigma}_m} \cdot \underline{I}_{dq0}^t \cdot \underline{G}_r \cdot \left(\underline{\dot{T}}_r \cdot \underline{T}_r^{-1}\right) \cdot \underline{\Lambda}_{dq0} \tag{10.301}$$

Hence,

$$\begin{aligned} T_{dev} = &-\frac{3/2}{\dot{\sigma}_m} \begin{bmatrix} i_D & i_Q & i_0 \end{bmatrix} \begin{bmatrix} 1 & 0 & 0 \\ 0 & 1 & 0 \\ 0 & 0 & 2 \end{bmatrix} \begin{bmatrix} 0 & \omega_s & 0 \\ -\omega_s & 0 & 0 \\ 0 & 0 & 0 \end{bmatrix} \begin{bmatrix} \lambda_D \\ \lambda_Q \\ \lambda_0 \end{bmatrix} \\ &-\frac{3/2}{\dot{\sigma}_m} \begin{bmatrix} i_d & i_q & i_0 \end{bmatrix} \begin{bmatrix} 1 & 0 & 0 \\ 0 & 1 & 0 \\ 0 & 0 & 2 \end{bmatrix} \begin{bmatrix} 0 & (\omega_s - \dot{\sigma}) & 0 \\ -(\omega_s - \dot{\sigma}) & 0 & 0 \\ 0 & 0 & 0 \end{bmatrix} \begin{bmatrix} \lambda_d \\ \lambda_q \\ \lambda_0 \end{bmatrix} \end{aligned} \tag{10.302}$$

That is,

$$T_{dev} = -\frac{3/2}{\dot{\sigma}_m} \begin{bmatrix} i_d & i_q & i_0 \end{bmatrix} \begin{bmatrix} \omega_s \lambda_Q \\ -\omega_s \lambda_D \\ 0 \end{bmatrix} - \frac{3/2}{\dot{\sigma}_m} \begin{bmatrix} i_d & i_q & i_0 \end{bmatrix} \begin{bmatrix} (\omega_s - \dot{\sigma})\lambda_q \\ -(\omega_s - \dot{\sigma})\lambda_d \\ 0 \end{bmatrix} \tag{10.303}$$

which gives

$$T_{dev} = -\frac{3/2}{\dot{\sigma}_m}[\omega_s(i_D\lambda_Q - i_Q\lambda_D) + (\omega_s - \dot{\sigma})(i_d\lambda_q - i_q\lambda_d)] \tag{10.304}$$

Therefore, we can write the following:

$$T_{dev} = -\frac{3/2}{\dot{\sigma}_m}[\omega_s(i_D\lambda_Q - i_Q\lambda_D + i_d\lambda_q - i_q\lambda_d) - \dot{\sigma}(i_d\lambda_q - i_q\lambda_d)] \tag{10.305}$$

where the $\omega_s(i_D\lambda_Q - i_Q\lambda_D + i_d\lambda_q - i_q\lambda_d)$ term can be further developed as follows:

$$\lambda_Q = (L_{ss} - L_{sm})i_D + \frac{3}{2}L_{srm}i_d \tag{10.306}$$

$$\lambda_Q = (L_{ss} - L_{sm})i_Q + \frac{3}{2}L_{srm}i_d \tag{10.307}$$

$$\lambda_d = \frac{3}{2}L_{srm}i_D + (L_{rr} - L_{rm})i_d \tag{10.308}$$

and

$$\lambda_q = \frac{3}{2}L_{srm}i_Q + (L_{rr} - L_{rm})i_q \tag{10.309}$$

Hence,

$$\begin{aligned}\omega_s(i_D\lambda_Q - i_Q\lambda_D + i_d\lambda_q - i_q\lambda_d) = \omega_s \Big\{ i_D\left[(L_{ss} - L_{sm})i_Q + \frac{3}{2}L_{srm}i_q\right] \\ -i_Q\left[(L_{ss} - L_{sm})i_D + \frac{3}{2}L_{srm}i_d\right] + i_d\left[\frac{3}{2}L_{srm}i_Q + (L_{rr} - L_{rm})i_q\right] \\ -i_q\left[\frac{3}{2}L_{srm}i_D + (L_{rr} - L_{rm})i_d\right]\Big\}\end{aligned} \tag{10.310}$$

yielding

$$\omega_s(i_D\lambda_Q - i_Q\lambda_D + i_d\lambda_q - i_q\lambda_d) = \omega_s\left(\tfrac{3}{2}L_{srm}i_Di_q - \tfrac{3}{2}L_{srm}i_Qi_d + \tfrac{3}{2}L_{srm}i_Qi_d - \tfrac{3}{2}L_{srm}i_Di_q\right) = 0 \tag{10.311}$$

That is,

$$\omega_s(i_D\lambda_Q - i_Q\lambda_D + i_d\lambda_q - i_q\lambda_d) = 0 \tag{10.312}$$

Hence,

$$T_{dev} = -\frac{3/2}{\dot{\sigma}_m}(\dot{\sigma}(i_d\lambda_q - i_q\lambda_d)) \tag{10.313}$$

However,

$$\dot{\sigma}_m = \frac{\dot{\sigma}}{p/2} \tag{10.314}$$

Accordingly, we can write the following:

$$T_{dev} = \frac{3}{2}\left(\frac{p}{2}\right)(i_d\lambda_q - i_q\lambda_d) \tag{10.315}$$

In term of currents and inductances, one can write the following:

$$T_{dev} = \frac{3}{2}\left(\frac{p}{2}\right)\left(i_d\frac{3}{2}L_{srm}i_Q + (L_{rr} - L_{rm})i_q - i_q\left(\frac{3}{2}L_{srm}i_D + (L_{rr} - L_{rm})i_d\right)\right) \tag{10.316}$$

Hence, we have the following:

$$T_{dev} = \frac{3}{2}\left(\frac{p}{2}\right)\left(\frac{3}{2}L_{srm}\right)(i_Q\lambda_d - i_D\lambda_q) \tag{10.317}$$

The torque expression can also be expressed in terms of stator currents and flux linkages. The developments are given as follows:

$$T_{dev} = \frac{3}{2}\left(\frac{P}{2}\right)\left(\frac{3}{2}L_{srm}\right)(i_Q\lambda_d - i_D\lambda_q) \tag{10.318}$$

That is,

$$T_{dev} = \frac{3}{2}\left(\frac{P}{2}\right)\left[\frac{3}{2}L_{srm}i_Q i_d + i_Q(L_{ss} - L_{sm})i_D - \frac{3}{2}L_{srm}i_D i_q - i_Q(L_{ss} - L_{sm})i_D\right] \tag{10.319}$$

Hence,

$$T_{dev} = \frac{3}{2}\left(\frac{P}{2}\right)\left\{i_Q\left[(L_{ss} - L_{sm})i_D + \frac{3}{2}L_{srm}i_d\right] - i_D\left[(L_{ss} - L_{sm})i_Q + \frac{3}{2}L_{srm}i_q\right]\right\} \tag{10.320}$$

which means that Eq. (10.320) can be rewritten as follows:

$$T_{dev} = \frac{3}{2}\left(\frac{P}{2}\right)[i_Q\lambda_D - i_D\lambda_Q] \tag{10.321}$$

We next treat the various inductance parameters in the induction machine's state-space dynamic models.

10.8 On Inductance Parameters in Induction Machine State-Space Models

In previous sections of this chapter, specifically Section 10.3, we treated the airgap in an induction machine as a smooth cylindrical shell, thus neglecting the effects of rotor and stator slotting on flux fringing and the resulting "effective height, h_g" of an actual airgap in such a machine. We therefore introduce these effects.

The presence of slots or ventilation ducts increases the effective reluctance of an airgap by restricting the effective area through which the magnetic flux flows, due to the so-called "fringing effect." Thus, this is equivalent to an airgap with a larger length or an airgap with a smaller area though which flux flows. This effect depends on the ratio of a slot opening or duct opening to the airgap height, "h_g" or "g," see Figures 10.10 and 10.11, This is done through the use of Carter's coefficients, see Figure 10.12.

The constriction on the area of flux flow through the airgap due to slots is expressed through the use of Carter coefficients, using the symbolism of Figures 10.10–10.12 as follows:

$$h_g(\text{effective}) = h_{ge} = g_e = k_g h_g = k_g g \tag{10.322}$$

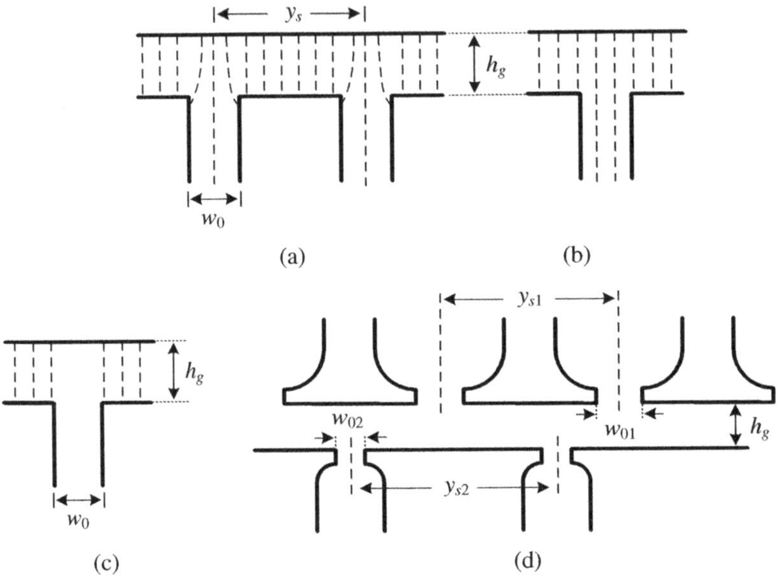

Figure 10.10 Effect of slotting on airgap flux (from Reference [13]).

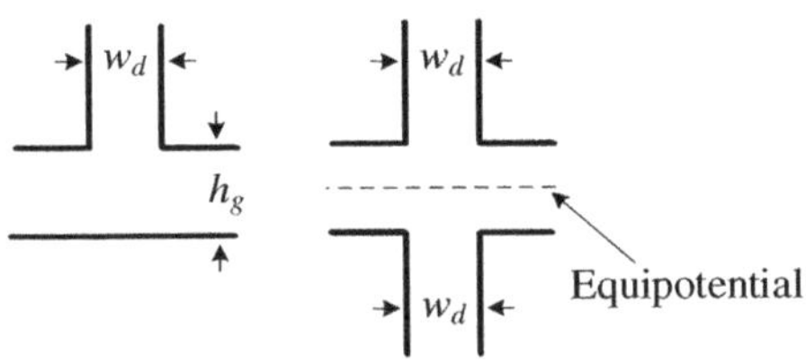

Figure 10.11 Effect of radial ventilation ducts on airgap flux (from Reference [13]).

where

(1) For Case (a) of Figure 10.10, one can write:

$$k_g = y_s/(y_s - k_0 w_0) \tag{10.323}$$

where from the Carter function "*f*" in Figure 10.12, one writes the following:

$$k_0 = f(w_0/h_g) \tag{10.324}$$

(2) For Case (d) of Figure 10.10, one can write

$$k_g = k_{g1} \cdot k_{g2} = \left(\frac{y_{s1}}{y_{s1} - k_{01} w_{01}}\right) \cdot \left(\frac{y_{s2}}{y_{s2} - k_{02} w_{02}}\right) \tag{10.325}$$

where from the Carter function "*f*" in Figure 10.12, one write the following:

$$k_{01} = f(w_{01}/h_g) \text{ and } k_{02} = f(w_{02}/h_g) \tag{10.326}$$

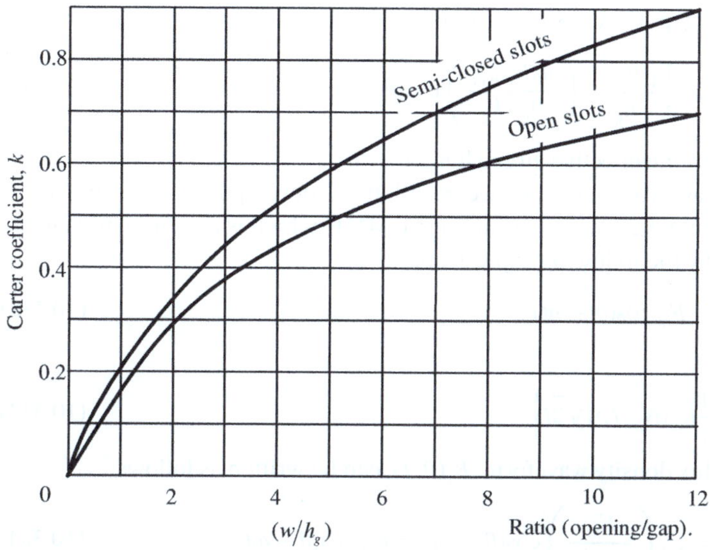

Figure 10.12 Carter coefficients for airgaps (from Reference [13]).

Here again, both k_{01} and k_{02} are obtained from Figure 10.12.

(3) For ventilation type ducts, see Figure 10.11, one can write

$$k_g = \frac{l}{l - k_d(n_d w_d)} \tag{10.327}$$

Here, $k_d = f(w_d/h_g)$ is obtained from the Carter function "f" of Figure 10.12, and l is the effective axial length of the armature conductors. In all the aforementioned, k_g is the Carter airgap coefficient.

For example, in the case of an MMF, $\mathbf{F}_1(\theta)$ is given by

$$\mathbf{F}_1(\theta) = \frac{4}{\pi} k_w \left(\frac{N_{ph}}{p} \right) i \cos(\theta) \tag{10.328}$$

acting on a slotted-type airgap of height, h_g, and a Carter airgap coefficient, k_g, the resulting flux density waveform, $B_1(\theta)$, can be expressed as follows:

$$B_1(\theta) = \mu_0 \left(\frac{\mathbf{F}_1(\theta)}{h_g \cdot k_g} \right) = \left\{ \frac{\mu_0 \left[\frac{4}{\pi} k_w (N_{ph}/p) i \right]}{h_g \cdot k_g} \right\} \cos(\theta) \tag{10.329}$$

Hence, in general for an MMF, $\mathbf{F}_1(\theta)$, given by

$$\mathbf{F}_1(\theta) = F_{1\,\max} \cos(\theta) \tag{10.330}$$

the resulting airgap flux density, $B_1(\theta)$, is

$$B_1(\theta) = \mu_0 \frac{\mathbf{F}_1(\theta)}{h_g \cdot k_g} = \left(\frac{\mu_0 F_{1\max}}{h_g \cdot k_g} \right) \cos(\theta) \tag{10.331}$$

where $(k_g \cdot h_g)$ in the effective airgap height.

In a two-phase AC machine, we recall from developments in Chapter 5, Section 5.5, that one can write the armature MMF, which is a rotating waveform, $\mathbf{F}_1(\theta, t)$, given by the following expression:

$$\mathbf{F}_1(\theta, t) = F_m \cos(\theta - \omega t) \tag{10.332}$$

where

$$F_m = \left[\frac{4}{\pi} k_w (N_{ph}/p) \sqrt{2} I \right] \tag{10.333}$$

Hence, the flux density waveform, $B_1(\theta, t)$, can be written as follows:

$$B_1(\theta, t) = \underbrace{\mu_0 \left(\frac{F_m}{h_g \cdot k_g} \right)}_{B_m} \cos(\theta - \omega t) = B_m \cos(\theta - \omega t) \tag{10.334}$$

where

$$B_m = \mu_0 [F_m/(k_g \cdot h_g)] = \mu_0 \left[\frac{4\sqrt{2}}{\pi} k_{w1} (N_{ph}/p) I \right] \Big/ (k_g \cdot h_g) \tag{10.335}$$

where again, k_g is Carter's airgap coefficient of Eq. (10.323). Now, the airgap flux, Φ, per pole is

$$\Phi = \frac{2}{\pi} B_m \tau_p l = \frac{2}{\pi} B_m \left(\frac{\pi D}{p} \right) l = \frac{2 B_m D l}{p} \tag{10.336}$$

Substituting from Eq. (10.335) into (10.336) gives

$$\Phi = 2\mu_0 \left[\frac{4\sqrt{2}}{\pi} k_{w1} (N_{ph}/p) I \right] \left(\frac{Dl}{k_g h_g p} \right)$$

Hence,

$$\Phi = \frac{8\sqrt{2}}{\pi} \left(\frac{\mu_0 D l}{k_g h_g} \right) \left(\frac{N_1 k_{w1}}{p^2} \right) I \tag{10.337}$$

Therefore, the EMF induced per phase in the stator armature, E_1, due to this flux, can be expressed in magnitude as follows:

$$E_1 = \frac{2\pi f}{\sqrt{2}} (N_1 k_{w1}) \Phi = X_\phi I \tag{10.338}$$

where X_ϕ is the magnetizing reactance. Substituting form Eq. (10.337) into Eq. (10.338) gives the following:

$$E_1 = X_\phi I = \frac{2\pi f}{\sqrt{2}} \cdot \frac{8\sqrt{2}}{\pi}\left(\frac{\mu_0 Dl}{k_g h_g}\right)\left(\frac{N_1 k_{w1}}{p}\right)^2 I_1$$

Hence, for a two-phase AC machine, X_ϕ is given by

$$X_\phi = \frac{E_1}{I} = \frac{16}{\pi} f\left(\frac{\mu_0 \pi Dl}{k_g h_g}\right)\left(\frac{N_1 k_{w1}}{p}\right)^2 \tag{10.339}$$

However,

$$X_\phi = 2\pi f L_\phi \tag{10.340}$$

Hence, the corresponding magnetizing inductance, L_ϕ, is given by

$$L_\phi = \frac{X_\phi}{2\pi f} = \frac{8}{\pi^2}\left(\frac{\mu_0 \pi Dl}{k_g h_g}\right)\left(\frac{N_1 k_{w1}}{p}\right)^2 \tag{10.341}$$

Accordingly, from earlier developments in Chapter 5, in a three-phase AC machine, one can write the rotating MMF waveform as follows:

$$\mathbf{F}_1(\theta, t) = \frac{3}{2} F_m \cos(\theta - \omega t) \tag{10.342}$$

Hence,

$$B_1(\theta, t) = \mu_0 \left(\frac{(3/2)F_m}{k_g h_g}\right)\cos(\theta - \omega t) \tag{10.343}$$

Accordingly,

$$\Phi = \frac{12\sqrt{2}}{\pi}\left(\frac{\mu_0 Dl}{k_g h_g}\right)\left(\frac{N_1 k_{w1}}{p}\right) I \tag{10.344}$$

Consequently,

$$X_\phi = \frac{24}{\pi} f\left(\frac{\mu_0 \pi Dl}{k_g h_g}\right)\left(\frac{N_1 k_{w1}}{p}\right)^2 \tag{10.345}$$

and

$$L_\phi = \frac{X_\phi}{2\pi f} = \frac{12}{\pi^2}\left(\frac{\mu_0 \pi Dl}{k_g h_g}\right)\left(\frac{N_1 k_{w1}}{p}\right)^2 \tag{10.346}$$

Now that we have detailed the derivation of the magnetizing inductance/reactance of poly-phase induction machines, L_ϕ/X_ϕ, we detail next the derivation of the leakage inductance/reactance, L_l/X_l, of such machines.

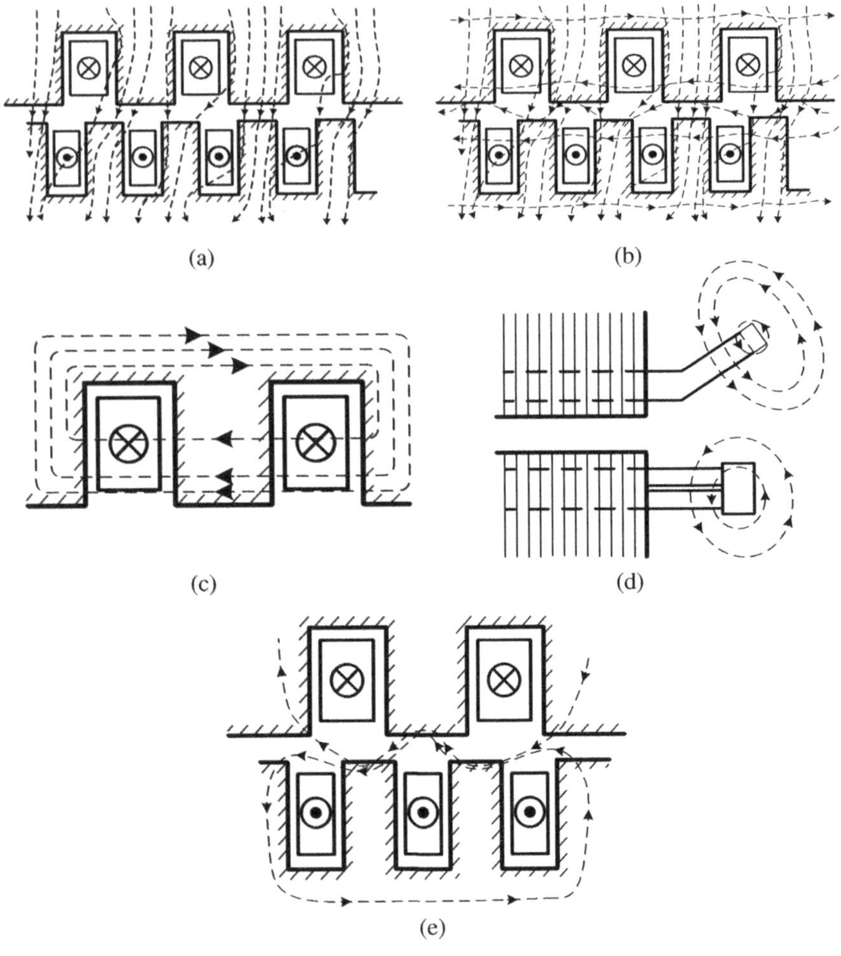

Figure 10.13 Leakage flux patterns (from Reference [13]). (a) Total flux. (b) Working and leakage fluxes. (c) Slot leakage (phase belt). (d) Overhang leakage. (e) Zigzag leakage.

Accordingly, samples of leakage flux patterns are shown in Figure 10.13. The most important of those leakage components and most amenable to closed-form expression-type calculation is the slot-leakage flux. A key component in the calculation of this slot-leakage flux is a quantity called the slot permeance, which consists of a number of parallel paths. Each of these paths has a permeance per meter of axial machine length or "useful conductor length," l.

Referring to the different slot shapes in Figure 10.14, the permeance factors for each slot (a) through (c), assuming laminations of very high (near infinite) permeability, are determined as follows:

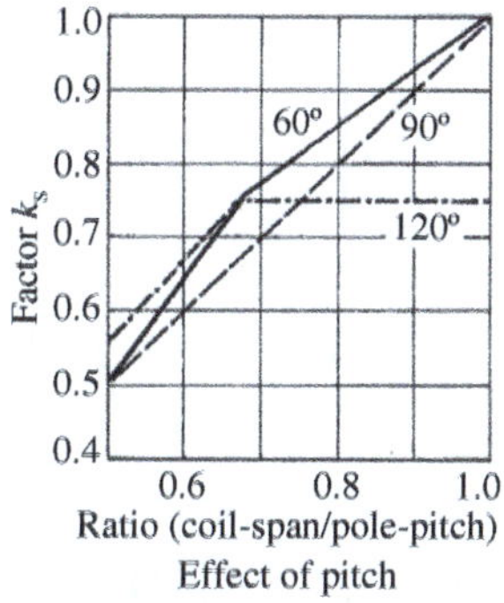

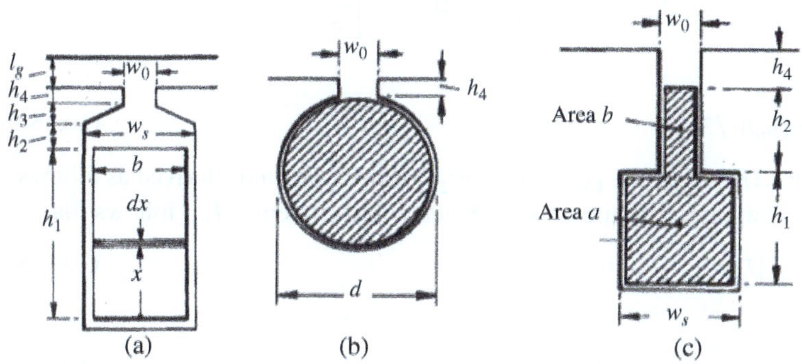

Figure 10.14 Various stator slot geometries and the related permeance calculation (a) rectangular slot (b) round slot and (c) deep rectangular slot.

Permeance calculation for Figure 10.14 (a):

For $0 \le x \le h_1$, the permeance, $\mathbf{P}_1$, for this portion is derived as follows:

At height, x, applying Ampere's law, the enclosed current $=(x/h_1)I$, it follows that

$$H(x) = \left[\left(\frac{x}{h_1}\right) I \Big/ w_s\right] \tag{10.347}$$

Hence,

$$B(x) = \mu_0 H(x) = \mu_0 \left[\left(\frac{x}{h_1}\right) I \Big/ w_s\right] \tag{10.348}$$

and the differential flux is

$$d\Phi(x) = B(x)dx = \mu_0 \left[\left(\frac{x}{h_1}\right) I \Big/ w_s\right] dx \tag{10.349}$$

Accordingly, the differential flux linkage is

$$d\Lambda(x) = \left\{\left(\frac{x}{h_1}\right) I \Big/ I\right\} d\Phi(x) = \mu_0 \left[\left(\frac{x}{h_1}\right)^2 I \Big/ w_s\right] dx \tag{10.350}$$

and, hence, the differential permeance is

$$d\mathbf{P}(x) = d\Lambda(x)/I = \mu_0 \left[\left(\frac{x}{h_1} \right)^2 \Big/ w_s \right] dx \tag{10.351}$$

Therefore, $\mathbf{P}_1$ becomes

$$\mathbf{P}_1 = \mathbf{P}(0 \le x \le h_1) = \int_0^{h_1} d\mathbf{P}(x) = \int_0^{h_1} \mu_0 \left[\frac{x^2}{h_1{}^2} \Big/ w_s \right] dx \tag{10.352}$$

That is,

$$\mathbf{P}_1 = \mathbf{P}(0 \le x \le h_1) = \frac{\mu_0}{\omega_s h_1{}^2} \int_0^{h_1} x^2 dx = \mu_0 \left[\frac{x^3}{3} \right]_0^{h_1} \Big/ w_s h_1{}^2 \tag{10.353}$$

Hence,

$$\mathbf{P}_1 = [\mu_0 h_1 / 3 w_s] \tag{10.354}$$

For $h_1 \le x \le (h_1 + h_2)$, the permeance, $\mathbf{P}_2$, for this portion is derived as follows: At height, x, applying Ampere's law, the enclosed current $= I$, it follows that

$$H(x) = [I/w_s] \tag{10.355}$$

Hence,

$$B(x) = \mu_0 H(x) = \mu_0 [I/w_s] \tag{10.356}$$

and the differential flux is

$$d\Phi(x) = B(x) dx = \mu_0 [I/w_s] dx \tag{10.357}$$

Accordingly, the differential flux linkage is

$$d\Lambda(x) = \{I/I\} d\Phi(x) = \mu_0 [I/w_s] dx \tag{10.358}$$

and, hence, the differential permeance is

$$d\mathbf{P}(x) = d\Lambda(x)/I = \mu_0 [1/w_s] dx \tag{10.359}$$

Therefore, $\mathbf{P}_2$ becomes

$$\mathbf{P}_2 = \mathbf{P}(h_2 \le x \le h_1 + h_2) = \int_{h_1}^{h_1+h_2} d\mathbf{P}(x) = \int_{h_1}^{h_1+h_2} [\mu_0/w_s] dx \tag{10.360}$$

That is,

$$\mathbf{P}_2 = \frac{\mu_0}{w_s} [x]_{h_1}^{h_1+h_2} = \frac{\mu_0 h_2}{w_s} \tag{10.361}$$

Hence,

$$\mathbf{P}_2 = [\mu_0 h_2 / w_s] \tag{10.362}$$

For $(h_1 + h_2) \le x \le (h_1 + h_2 + h_3)$, the permeance, $\mathbf{P}_3$, for this portion is derived as follows:

At height, x, applying Ampere's law, the enclosed current $= I$, and the average width is $(w_s + w_0)/2$, hence, it follows that

$$H(x) \simeq I/[(w_s + w_0)/2] = 2I/(w_s + w_0) \tag{10.363}$$

Hence,

$$B(x) = \mu_0 H(x) \cong \mu_0[2I/(w_s + w_0)] \tag{10.364}$$

and the differential flux is

$$d\Phi(x) = B(x)dx \cong \mu_0[2I/(w_s + w_0)]dx \tag{10.365}$$

Accordingly, the differential flux linkage is

$$d\Lambda(x) = \{I/I\}d\Phi(x) \cong \mu_0[2I/(w_s + w_0)]dx \tag{10.366}$$

and, hence, the differential permeance is

$$d\mathbf{P}(x) = d\Lambda(x)/I \cong \mu_0[2/(w_s + w_0)]dx \tag{10.367}$$

Therefore, $\mathbf{P}_3$ becomes

$$\mathbf{P}_3 = [(h_1 + h_2) \le x \le (h_1 + h_2 + h_3)] = \int_{(h_1+h_2)}^{(h_1+h_2+h_3)} d\mathbf{P}(x) \tag{10.368}$$

That is,

$$\mathbf{P}_3 \cong \int_{(h_1+h_2)}^{(h_1+h_2+h_3)} \mu_0[2/(w_s + w_0)]dx = \frac{2\mu_0 h_3}{(w_s + w_0)} \tag{10.369}$$

Hence,

$$\mathbf{P}_3 \cong [2\mu_0 h_3/(w_s + w_0)] \tag{10.370}$$

or, more accurately,

$$\mathbf{P}_3 = \mu_0[\{h_3/(w_s - w_0)\} \ln(w_s/w_0)] \tag{10.371}$$

For $(h_1 + h_2 + h_3) \le x \le (h_1 + h_2 + h_3 + h_4)$, the permeance, $\mathbf{P}_4$, for this portion is derived as follows:

This portion of the slot is similar to that of, $\mathbf{P}_2$, and hence,

$$\mathbf{P}_4 = [\mu_0 h_4/w_0] \tag{10.372}$$

Hence, the total permeance factor of the slot, $\mathbf{P}_s$, is the sum of all the permeances, $\mathbf{P}_1$ through $\mathbf{P}_4$, in parallel. Hence,

$$\mathbf{P}_s = \mathbf{P}_1 + \mathbf{P}_2 + \mathbf{P}_3 + \mathbf{P}_4 \tag{10.373}$$

That is,

$$\mathbf{P}_s = \mu_0 \left[\frac{h_1}{3w_s} + \frac{h_2}{w_s} + \frac{2h_3}{w_s + w_0} + \frac{h_4}{w_0} \right] \tag{10.374}$$

Accordingly, the permeance of the number of slots per pole per phase (phase-belt), q, is

$$\mathbf{P}_{sq} = \mathbf{P}_s/q = \frac{\mu_0}{q} \left[\frac{h_1}{3w_s} + \frac{h_2}{w_s} + \frac{2h_3}{w_s + w_0} + \frac{h_4}{w_0} \right] \tag{10.375}$$

Hence, the total permeance per pole, per phase for a core with an effective conductor length, l, is

$$\mathbf{P}_{phase/pole} = l\mathbf{P}_{sq} = \frac{\mu_0 l}{q} \left[\frac{h_1}{3w_s} + \frac{h_2}{w_s} + \frac{2h_3}{w_s + w_0} + \frac{h_4}{w_0} \right] \tag{10.376}$$

Hence,

$$\mathbf{P}_{phase/pair\ of\ poles} = 2l\mathbf{P}_{sq} = \frac{2\mu_0 l}{q} \left[\frac{h_1}{3w_s} + \frac{h_2}{w_s} + \frac{2h_3}{w_s + w_0} + \frac{h_4}{w_0} \right] \tag{10.377}$$

The number of turns per phase, per pole pair $= N_{ph}/(p/2)$.
Hence, the leakage inductance per phase per pair of poles, $L_{lph/pole\ pair}$, is

$$L_{lph/pole\ pair} = \left(\frac{N_{ph}}{p/2} \right)^2 \mathbf{P}_{phase/pole\ of\ poles} = \frac{4N_{ph}^{\ 2}}{p^2}(2l\mathbf{P}_{sq}) = \frac{4N_{ph}^{\ 2}}{p^2}(2l\mathbf{P}_s/q) \tag{10.378}$$

Hence, the leakage inductance per phase, L_l, if all the turns are connected in series in each phase is

$$L_l = \frac{p}{2} \cdot L_{lph/pole\ pair} = \frac{p}{2} \cdot \frac{4N_{ph}^{\ 2}}{p^2}(2l\mathbf{P}_s/q) \tag{10.379}$$

$$L_l = \frac{4N_{ph}^{\ 2}}{pq}(l\mathbf{P}_s)k_s \tag{10.380}$$

where k_s is a factor that depends on the phase-belt spread, 60°e or 120°e, see Figure 10.15 for a 120°e double-layer phase belt layout.

Hence, the leakage reactance, $X_l = \omega L_l = 2\pi f L_l$, can be written as

$$X_l = \left[4\omega N_{ph}^{\ 2} l/(pq)\right] k_s \mathbf{P}_s \tag{10.381}$$

or

$$X_l = \left[8\pi f N_{ph}^{\ 2} l/(pq)\right] k_s \mathbf{P}_s \tag{10.382}$$

For slot configuration (a) of Figure 10.14, the leakage reactance per phase, X_l, is

$$X_l \cong \left[8\pi f N_{ph}^{\ 2} l/(pq)\right] k_s \mu_0 \left[\frac{h_1}{3w_s} + \frac{h_2}{w_s} + \frac{2h_3}{w_s + w_0} + \frac{h_4}{w_0} \right] \tag{10.383}$$

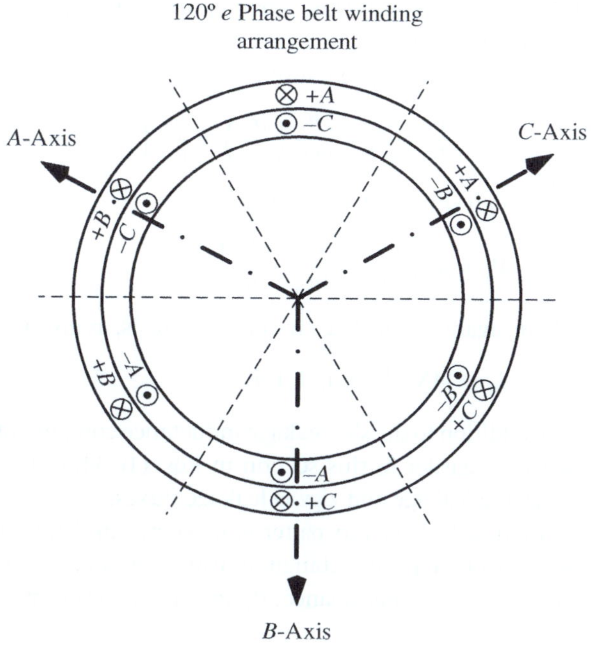

Figure 10.15 The layout for a two-pole, 120°*e* phase-belt configuration in a double-layer three-phase winding.

or, more accurately,

$$X_l = \left[8\pi f N_{ph}{}^2 l/(pq)\right] k_s \mu_0 \left[\frac{h_1}{3w_s} + \frac{h_2}{w_s} + \left(\frac{h_3}{w_s - w_0}\right) \ln\left(\frac{w_s}{w_0}\right) + \frac{h_4}{w_0}\right] \tag{10.384}$$

In summary, the slot permeances, $\mathbf{P}_s$, for the three slot shapes (a), (b), and (c) given in Figure 10.14 are given as follows:

Permeance calculation for the rectangular slot shape shown in Figure 10.14 (a):

$$\mathbf{P}_s \cong \mu_0 \left[\frac{h_1}{3w_s} + \frac{h_2}{w_s} + \frac{2h_3}{w_s + w_0} + \frac{h_4}{w_0}\right] \tag{10.385}$$

Permeance calculation for the round slot shape shown in Figure 10.14 (b):

$$\mathbf{P}_s \cong \mu_0[(0.623 + h_4/w_0)] \tag{10.386}$$

or

$$\mathbf{P}_s \cong \mu_0[(0.66 + h_4/w_0)] \tag{10.387}$$

Permeance calculation for the deep rectangular slot shape shown in Figure 10.14 (c):

For low slip (high speed normal operating/running condition)

$$\mathbf{P}_s = \left\{ \frac{\mu_0}{(a+b)^2} \left[\frac{h_1}{w_s} \left(\frac{a^2}{3} \right) + \frac{h_2}{w_0} \left(a^2 + ab + \frac{b^2}{3} \right) \right] + \frac{\mu_0 h_4}{w_0} \right\} \tag{10.388}$$

For high slip (starting conditions)

$$\mathbf{P}_s = \mu_0 \left[\frac{h_1}{2w_s} + \frac{h_2}{3w_0} + \frac{h_4}{w_0} \right] \tag{10.389}$$

where, again, the leakage inductance, L_l, is, as given earlier by

$$L_l = \left[4N_{ph}{}^2/(pq) \right] (k_s \mathbf{P}_s l) \tag{10.390}$$

In addition to the slot leakage inductance component of the leakage inductance, analyzed earlier in this section in Eqs. (10.347)–(10.390), there are end-turn or overhang leakage and zigzag leakage fluxes.

All these leakage flux patterns are shown in Figure 10.13, and a typical slot leakage flux of a typical rectangular armature slot is shown in detail in Figure 10.16. The slot leakage permeance, $\mathbf{P}_s$, for this slot is given in Eqs. (10.374) and (10.385)

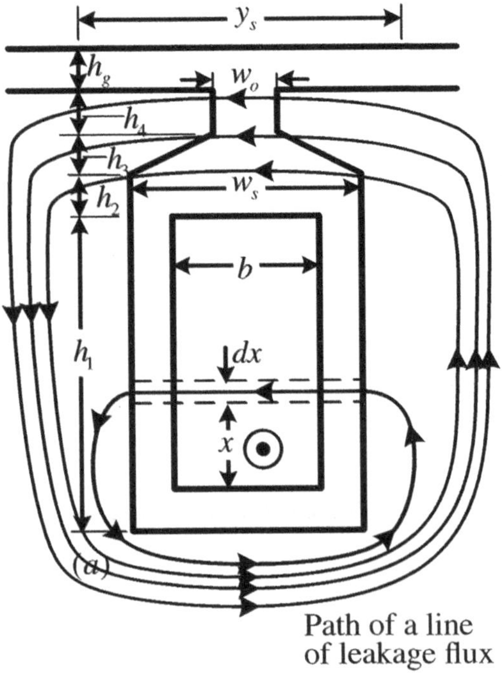

Figure 10.16 Slot leakage flux pattern.

and is repeated here for convenience:

$$\mathbf{P}_s \cong \mu_0 \left[\frac{h_1}{3w_s} + \frac{h_2}{w_s} + \frac{2h_3}{w_s + w_0} + \frac{h_4}{w_0} \right] \tag{10.391}$$

and the leakage inductance, L_l, and reactance, X_l, per phase, assuming all armature phase-to-neutral turns are connected in series, are as follows, see Eqs. (10.380)–(10.390):

$$L_l = \frac{4N_{ph}{}^2 l}{pq} \mu_0 \left[\frac{h_1}{3w_s} + \frac{h_2}{w_s} + \frac{2h_3}{w_s + w_0} + \frac{h_4}{w_0} \right] \tag{10.392}$$

and

$$X_l = \frac{8\pi f N_{ph}{}^2 l}{pq} \mu_0 \left[\frac{h_1}{3w_s} + \frac{h_2}{w_s} + \frac{2h_3}{w_s + w_0} + \frac{h_4}{w_0} \right] \tag{10.393}$$

Furthermore, according to empirical experience, see Reference [13], one can write the following:

$$L_{lzig-zag} = (5/6)L_\phi \left\{ \begin{matrix} [1/(\text{Number of stator slots per-pole})^2] \\ +[1/(\text{Number of rotor slots per-pole})^2] \end{matrix} \right\} \tag{10.394}$$

and

$$X_{lzig-zag} = (5/6)x_\phi \left\{ \begin{matrix} [1/(\text{Number of stator slots per-pole})^2] \\ +[1/(\text{Number of rotor slots per-pole})^2] \end{matrix} \right\} \tag{10.395}$$

Hence, from Eqs. (10.392) and (10.393), it follows that L_l and X_l are proportional to slot depths, h_1 through h_4, and, L_l and X_l are proportional to the effective armature length, l, and, L_l and X_l are proportional to the square of the number of turns per phase, N_{ph}, and also, L_l and X_l are inversely proportional to the number of poles, p, and the number of slots per pole per phase, q. Furthermore, the airgap height, h_g or g, directly impacts the zigzag portion of the leakage. The larger the airgap, the lesser the zigzag leakage, see Figure 10.13.

In addition, winding layouts affect the inductances (leakage and magnetizing) through the phase belts, such as slots per pole per phase, q, phase belt width, 120°e versus 60°e, and winding pitch, distribution and skew factors, all included in $k_w = k_d k_p k_s$.

For the end-turns (or overhang), see Figure 10.17, the coil/turn pitch, τ_c, can be estimated as

$$\tau_c(\text{at mid-slot}) = \left(\frac{\pi D_{\text{mid-slot}}}{p} \right)$$

and its permeance, $\mathbf{P}_{\text{end turns}}$, as

$$\mathbf{P}_{end\ turns} \simeq \frac{2\tau_c}{2l}(\mathbf{P}_s) = \frac{\tau_c}{l}(\mathbf{P}_s) \tag{10.396}$$

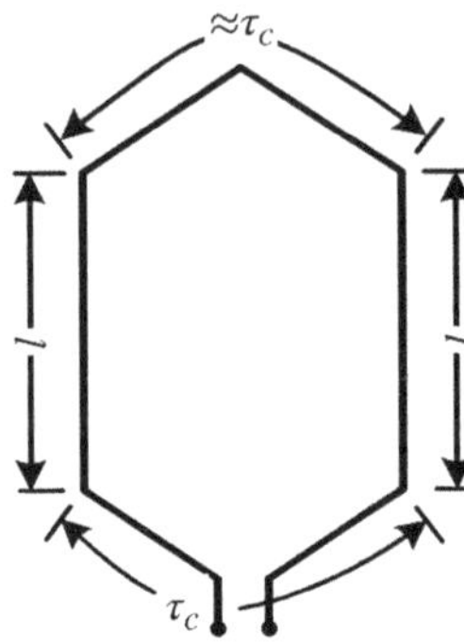

Figure 10.17 Coil and end-leads/turns geometry.

Hence, including end-turn leakage

$$L_{l_{\text{slot and end}}} = \frac{4(N_{ph})^2}{mpq}\left[\mathbf{P}_s + \left(\frac{\tau_c}{l}\right)\mathbf{P}_s\right]k_w \tag{10.397}$$

or

$$L_{l_{\text{slot and end}}} = \frac{4(N_{ph})^2}{mpq}\left[1 + \left(\frac{\tau_c}{l}\right)\right]\mathbf{P}_s k_w \tag{10.398}$$

where N_{ph} is the effective number of turns per phase, k_w is the winding factor $= k_d k_p k_s$, m is the number of parallel paths per-phase, and again

$$L_{lzig-zag} = (5/6)L_\phi \begin{Bmatrix} [1/(\text{Number of stator slots per-pole})^2] \\ +[1/(\text{Number of rotor slots per-pole})^2] \end{Bmatrix} \tag{10.399}$$

or

$$L_{lzig-zag} = (5/6)L_m \begin{Bmatrix} [1/(\text{Number of stator slots per-pole})^2] \\ +[1/(\text{Number of rotor slots per-pole})^2] \end{Bmatrix} \tag{10.400}$$

Thus, the total leakage inductance per phase is

$$L_l = \frac{4(N_{ph})^2}{mpq}\left[1 + \left(\frac{\tau_c}{l}\right)\right]\mathbf{P}_s k_w + (5/6)L_m \begin{Bmatrix} [1/(\text{Number of stator slots per-pole})^2] \\ +[1/(\text{Number of rotor slots per-pole})^2] \end{Bmatrix} \tag{10.401}$$

Once again, the magnetizing reactance, X_ϕ, and inductance, L_ϕ, for induction machines, from Eqs. (10.345) and (10.346), are as follows:

$$X_m = X_\phi = \frac{24}{\pi}f\left(\frac{\mu_0 \pi D l}{k_g l_g}\right)\left(\frac{N_1 k_w}{mp}\right)^2 \tag{10.402}$$

$$L_m = L_\phi = \frac{12}{\pi^2}\left(\frac{\mu_0 \pi D l}{k_g l_g}\right)\left(\frac{N_1 k_w}{mp}\right)^2 \tag{10.403}$$

Hence, X_ϕ and L_ϕ are linearly proportional to mid-gap diameter, D, and effective armature length, l, and are proportional to the square of the effective stator turns per phase, $(N_1 k_{w1})^2$, and are inversely proportional to the effective airgap height, $k_g h_g$, and are inversely proportional to the square of the number of poles, p^2, and are indirectly affected by the slot shape (slot opening or w_0/y_s) through Carter's airgap coefficient, k_g, see Figure 10.16.

At this stage, the reader is urged to recall from Section 10.3, Eqs. (10.48)–(10.59), that all the inductance parameters of the *ABC–abc* state-space model of the induction machine were defined in terms of the magnetizing inductance, L_ϕ, and the leakage inductance, L_l, where L_{sl} was the leakage inductance for the stator winding, and L_{rl} was the leakage inductance of the rotor winding, per phase. In Section 10.9, we discuss some design and parameter computation aspects for induction machines.

10.9 On Design and Parameter Computation Aspects in Induction Machines

In this section, we make use of earlier developments and formulations presented in this chapter as well as material presented in Chapters 5 and 6 and apply these formulations and other development to an example 1.0 hp, 3-phase, Y-connected, 200 V (line-to-line), 2-pole, 60 Hz, squirrel-cage induction motor, which runs at a rated speed of 3450 rad/min.

This motor has a *T*-equivalent circuit whose parameters per phase, referred to the stator side, are as follows, see References [1–11]:

$R_1 = 3.5\,\Omega$, $jX_1 = j3.016\,\Omega$, $R_2 = 2.46\,\Omega$, $jX_2 = j3.016\,\Omega$, and $jX_m (= jX_\phi) = j95.004\,\Omega$.

This motor's design is shown in cross-section in Figure 10.18. As can be seen in the figure, the 3-phase stator armature is housed in 24 armature slots which are labeled by $A, -A, B, -B, C$, and $-C$. Meanwhile, the rotor's squirrel-cage consists of 34 bars. The stator's armature consists of 4 coils per phase, while each coil contains 44 turns. Each phase consists of two parallel paths. That is, $N_1 = 176$ turns/phase and $m = 2$. The stator conductors are of the No. 23 AWG wire type. The stator and rotor slot geometries are detailed in Figures 10.19 and 10.20, in which the geometric dimensions are given in inches (1.0 in. = 2.45 cm = 2.54×10^{-2} m).

The rotor's outer diameter is 3.093 in., and the radial airgap height is 0.0064 in., while the stator core's outer diameter is 5.625 in., and the stator core (stack) axial length is 2.5 in. That is, the useful stator conductor length is $l = 2.5$ in.

In this section, we will perform the following design analysis tasks:

1. Deduce the Thevenin equivalent circuit of this induction motor per phase, with all its parameters, V_{1a}, R_{e1}, X_{e1}, etc.
2. At rated voltage and rated speed, obtain the following:
   2.1. The rotor current, $\bar{I}_2$, referenced to the stator and the rotor current, $\bar{I}_R$, in conventional consumer notation ($\bar{I}_R = -\bar{I}_2$).

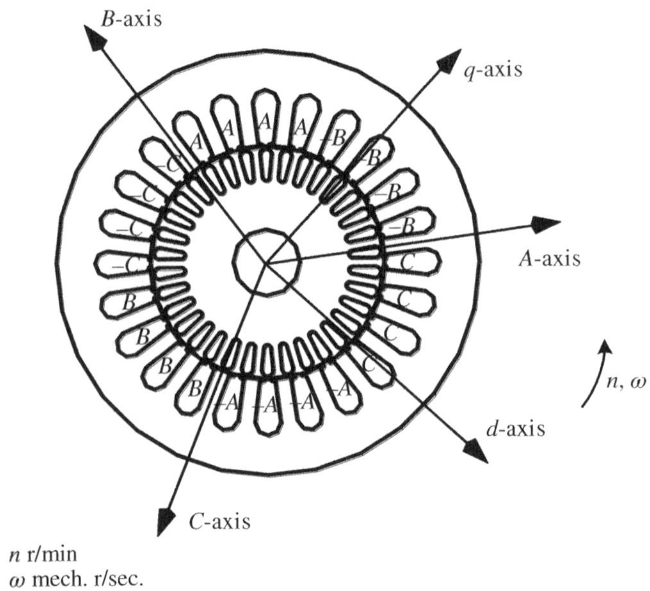

Figure 10.18 Magnetic circuit, winding layout, and axis labeling.

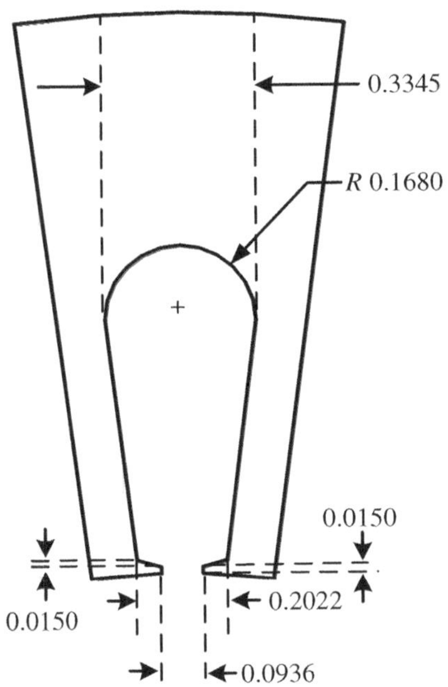

Figure 10.19 Stator slot dimensions in inches.

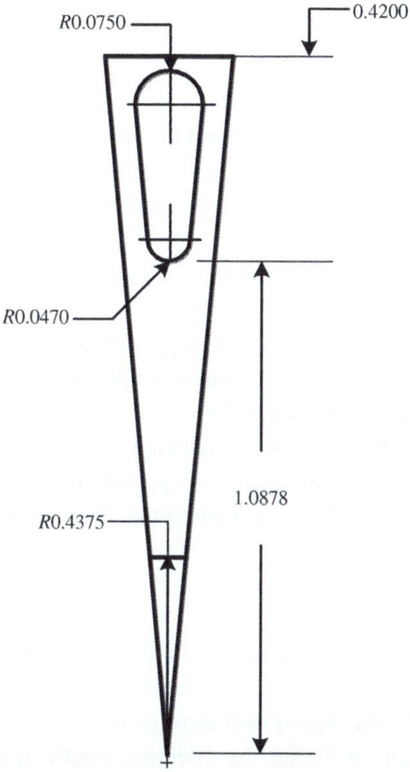

Figure 10.20 Rotor slot dimension in inches.

2.2. The developed electromagnetic (electromechanical) power in Watts and the developed (gross mechanical) torque in Newton-meters.
2.3. The resultant magnetizing (airgap) EMF, $\overline{E}_1$, referred to the stator, and the corresponding magnetizing current, $\overline{I}_m$.
2.4. The stator current, $\overline{I}_S = \overline{I}_1$, the input power, the terminal power factor angle, and power factor.
2.5. Neatly sketch the complete phasor diagram showing on it the accurate orientation of the following phasor and space-vector quantities:
  a. The terminal voltage, $\overline{V}_1$, as the reference.
  b. The magnetizing EMF, $\overline{E}_m = \overline{E}_1$.
  c. The stator current, $\overline{I}_S = \overline{I}_1$, and the stator MMF, $\overline{F}_S$.
  d. The magnetizing current, $\overline{I}_m$, and the resultant MMF, $\overline{F}_{SR}$, as well as the resultant flux, $\overline{\Phi}_{SR}$.
  e. The rotor current, $\overline{I}_R = -\overline{I}_2$, and the rotor MMF, $\overline{F}_R$.
2.6. Find and designate on your phasor diagram of (2.5) the torque angles, δ_r, δ_S and δ_{SR}, in electrical degrees.

3. In order to analyze and synthesize this motor's design, perform the following tasks:
   3.1. Find the resultant airgap flux per pole, Φ_{SR}, in Webers from the knowledge of the fact that $E_m = E_1 = 4.44f(N_1k_{W1}/m)\Phi_{SR}$ and find the resultant maximum airgap flux density, $B_{SR\max}$, in Tesla.
   3.2. Again, find this resultant maximum airgap flux density, $B_{SR\max}$, from the knowledge of the fact that $B_{SR\max} = (\mu_0 F_{SR}/g)$, where $F_{SR} = (3/2)[(4/\pi)(N_1k_{W1}/p)\sqrt{2}(I_m/m)]$, and compare your answers with that obtained in (3.1).
   3.3. Obtain the magnetizing reactance, X_m, from the knowledge of the fact that $X_m = 24f(\mu_0 Dl/g)(N_1k_{W1}/mp)^2$ and compare this value with that given to you in the original data of the statement of the problem.
   3.4. Calculate the developed (gross mechanical or electromagnetic) torque in Newton-meters and developed (gross mechanical) power in Watts from the knowledge of the following three developments detailed in Chapters 5 and 6:
      a. $T_{em} = (\pi/2)(p/2)^2\Phi_{SR}F_S \sin\delta_S$
      b. $T_{em} = (p/4)(\mu_0 Dl/g)F_S F_{Sr} \sin\delta_S$
      c. $T_{em} = \sqrt{2}(k_W B_{SR\max} A \sin\delta_S)(\pi D^2 l/4)$, where the ampere loading $A = 6N_1I_S/m(\pi D)$, amperes/m at the mid-airgap circumference.
   3.5. From the knowledge of the fact that the stator armature turns are made of No. 23 AWG type of wire and the fact that the effective length of a stator conductor, $l = 2.5$ inches, and that the stator winding resistance, $R_1 = 3.5\ \Omega$/phase, find the average length of a stator armature winding turn, and the length of the wires in the two end turn regions of a single turn. Draw an approximate sketch of the dimensions of this average stator armature turn.
4. Compare your answers obtained from the Thevenin equivalent circuit calculations in (2.0) with those obtained from motor dimensions-design type calculations in (3.0) for the following quantities and give comments whenever possible
   a) The magnetizing reactance, X_m.
   b) The developed (gross mechanical or electromagnetic) torque and developed power obtained in (2.2) and (3.4).
5. Repeat (2) through (4) at maximum torque (stall torque) condition, that is, at $s = s_{\max} = \left(R_2/\sqrt{R_{e1}^2 + (X_{e1} + X_2)^2}\right)$
6. Repeat (2) through (4) for the starting condition, that is, at $s = 1.0$.
7. Sketch the torque–speed characteristic of the motor obtained from the Thevenin equivalent circuit calculations and from the motor dimensions/design-type calculations and comment.

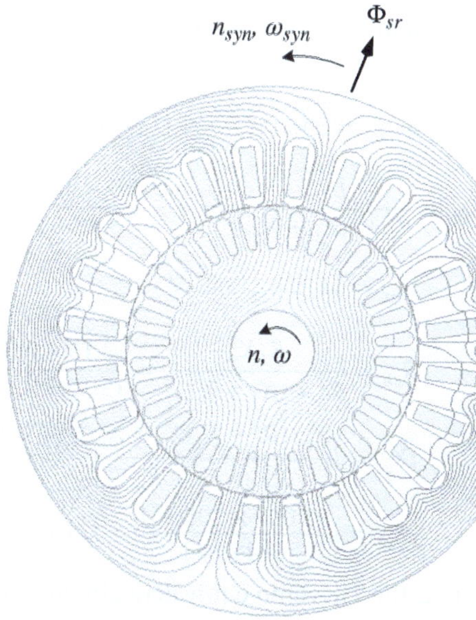

Figure 10.21 Rated load case FE-computed flux plot at $\theta = 0°$, at $n = 3450$ rev/min, $n_{syn} = 3600$ rev/min, and $\omega_{syn} = 115\pi$ mech. rad/s.

Given in Figure 10.21 is the FE-computed flux plot under the rated load condition, and a rated speed of 3450 rad/min. Notice the flux-lines crowding along an axis in quadrature with the resultant flux, Φ_{SR}, axis, thus leading to saturation-induced saliency-effects that are well elaborated in Refs. [4–15]. Now, for this Y-connected 2-pole squirrel-cage induction motor, we have line-to-line voltage 200 V, at 60 Hz, with a rated speed of 3450 rad/min, which has the following equivalent circuit:

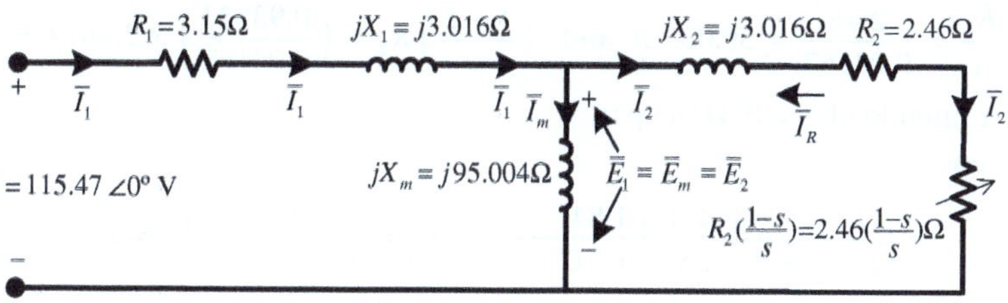

$$\overline{V}_1 = \frac{200}{\sqrt{3}}\angle 0° = 115.47\angle 0° \text{ V}$$

$$n_{syn} = \frac{120f}{p} = \frac{120 \times 60}{2} = 3600 \text{ rev/ min}$$

$$\omega_{syn} = \left(\frac{2\pi}{60} n_{syn}\right) = \frac{2\pi}{60} \times 3600 = 377 \text{ mech.rad/s}$$

(1) At rated load $s = \dfrac{3600 - 3450}{3600} = \dfrac{150}{3600} = 0.04167$

Hence,

$$\omega_{rotor} = 3450 \times \frac{2\pi}{60} = (1 - s)\omega_{syn} = 361.283 \text{ mech.rad/s}$$

$$\overline{V}_{1a} = \frac{jX_m}{R_1 + j(X_1 + X_m)}\overline{V}_1 = \frac{j95.004 \times 115.47\angle 0^\circ}{3.15 + j98.02} = \frac{95.004\angle 90^\circ \times 115.47\angle 0^\circ}{98.071\angle 88^\circ}$$

$$\overline{V}_{1a} = 111.86\angle 1.84^\circ \text{ V}$$

$$R_{e1} = \frac{R_1 X_m^2}{R_1^2 + (X_1 + X_m)^2} = \frac{3.15 \times (95.004)^2}{(3.15)^2 \times (98.02)^2} = 2.956\ \Omega$$

$$jX_{e1} = j\frac{X_m\left[R_1^2 + X_1(X_1 + X_m)\right]}{R_1^2 + (X_1 + X_m)^2} = j\frac{95.004[(3.15)^2 + 3.016(98.02)]}{(3.15)^2 + (98.02)^2}$$

$$jX_{e1} = j\frac{95.004 \times 305.551}{9617.843} = j3.018\ \Omega$$

Hence, the Thevenin equivalent circuit is as follows:

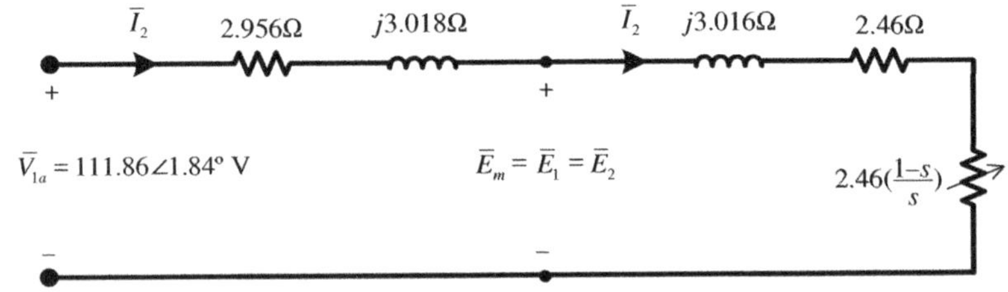

$$\frac{R_2}{s} = \frac{2.46}{0.04167} = 59.035\ \Omega \text{ and } \left(\frac{1-5}{s}\right)R_2 = \left(\frac{0.93833}{0.04167}\right) \times 2.46 = 56.575\ \Omega$$

(2) At rated load, $s = 0.04167$ p.u.

Hence,

$$\overline{I}_2 = \frac{111.86\angle 1.84^\circ}{(2.956 + 59.035) + j(3.018 + 3.016)} = \frac{111.86\angle 1.84^\circ}{62.884\angle 5.56^\circ} = 1.796\angle -3.72^\circ$$

and

$$\overline{E}_m = (59.035 + j3.016) \times 1.796\angle -3.72^\circ = 59.112\angle 2.92^\circ \times 1.796\angle -3.72^\circ$$

$$\overline{E}_m = 106.165\angle -0.8^\circ \text{ V}$$

Therefore,

$$\overline{I}_m = \frac{\overline{E}_m}{jX_m} = \frac{106.165\angle -0.8^\circ}{95.004\angle 90^\circ} = 1.117\angle -90.8^\circ \text{ A}$$

and

$$\overline{I}_1 = \overline{I}_2 + \overline{I}_m = 1.796\angle -3.72^\circ + 1.117\angle -90.8^\circ = 1.776 - j1.234$$

$$\overline{I}_1 = 2.163\angle -34.79^\circ \text{ A}$$

$$\overline{I}_R = -\overline{I}_2 = 1.796\angle -183.72^\circ \text{ A}, \quad \overline{I}_S = \overline{I}_1 = 2.163\angle -34.79^\circ \text{ A}$$

Accordingly,

$$T_{dev} = T_{em} = \frac{3I_2^2[(1-s)R_2/s]}{\omega_{rotor}} = \frac{3I_2^2 R_2/s}{\omega_{syn}} = \frac{3 \times (1.796)^2 \times 59.035}{377} = 1.515 \text{ Nm}$$

The time phasor-space-vector diagram, at $t = 0$, is as shown in the following:

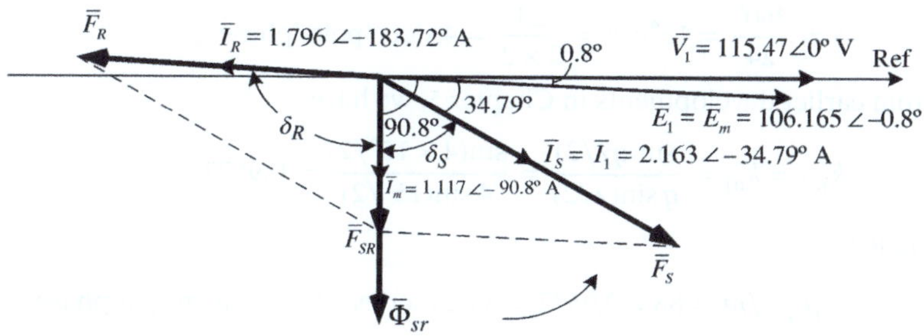

from which

$$\delta_s = 90.8^\circ - 34.79^\circ = 56.01^\circ$$

and

$$\delta_R = 183.72 - 90.8^\circ = 92.92^\circ, \delta_{SR} = \delta_S + \delta_R = 148.93^\circ$$

(1) Design analysis and synthesis

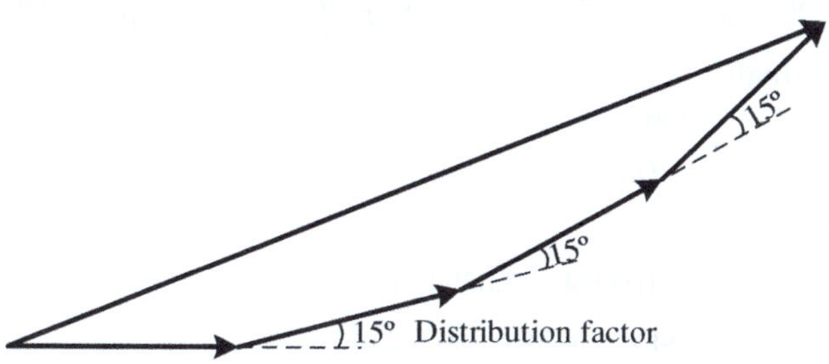

$$\overline{E}_1 = \overline{E}_m = 106.165\angle -0.8^\circ \text{ V}$$

Hence,

$$E_m = 106.165\text{ V}$$

However,

$$\overline{E}_1 = \overline{E}_m = 4.44(N_1 k_{w1}/m) f \Phi_{SR}$$

$$N_1 = 176\text{ Turns}, m = 2\text{ paths}$$

Hence,

$$N_1/m = 88\text{ series turns/phase}$$

Now, k_{w1}

$$\gamma = \frac{360^\circ}{24} = 15^\circ\text{e}, q = \frac{24}{2 \times 3} = 4\text{ slots/pole/phase}$$

from earlier developments in Chapter 5, we have

$$k_{W1} = k_{d1} = \frac{\sin(q\gamma/2)}{q\sin(\gamma/2)} = \frac{\sin(4 \times 15^\circ/2)}{4\sin(15^\circ/2)} = 0.9577$$

Hence,

$$N_1 k_{W1}/m = 88 \times 0.9577 = 84.28\text{ series effective turns per phase}$$

and

$$E_1 = E_m = 106.165 = 4.44 \times 60 \times 84.28 \times \Phi_{SR}$$

Hence,

$$\Phi_{SR} = \frac{106.165}{4.44 \times 60 \times 84.24} = 47.285 \times 10^{-4}\text{ Wb}$$

and since,

$$\Phi_{SR} = \frac{2}{\pi} \cdot \tau_p \cdot l \cdot B_{SR\max} = \frac{2}{\pi}\left(\frac{\pi D}{p}\right) l B_{SR\max} = \frac{2Dl}{p} B_{SR\max}$$

It follows that

$$B_{SR\max} = \frac{p\Phi_{SR}}{2Dl}$$

However,

$$\begin{aligned} D &= 3.093 + 0.0064 = 3.0394\text{ in.} \\ &= 3.0394 \times 0.0254 = 0.0772\text{ m} \\ l &= 2.5 \times 0.0254 = 0.0635\text{ m} \\ g &= 0.0064 \times 0.0254 = 0.000163\text{ m} \end{aligned}$$

Hence,

$$B_{SR\max} = \frac{2 \times 47.285 \times 10^{-4}}{2 \times 0.0772 \times 0.0635} = 0.9646 \text{ Tesla}$$

Meanwhile,

$$F_{SR} = \frac{3}{2}\left[\frac{4}{\pi}\left(\frac{N_1 k_{W1}}{p}\right)\frac{\sqrt{2}I_m}{m}\right] = \frac{3}{2}\left[\frac{4}{\pi}\left(\frac{N_1 k_{W1}}{m}/p\right)\sqrt{2}I_m\right]$$

Hence,

$$F_{SR} = \frac{3}{2}\left[\frac{4}{\pi} \times \frac{84.28}{2} \times \sqrt{2} \times 1.117\right] = 127.135 \text{ AT/pole}$$

However,

$$B_{SR\max} = \mu_0(F_{SR}/g)$$

$$B_{SR\max} = \frac{4\pi \times 10^{-7} \times 127.135}{0.000163} = 0.9801 \text{ Tesla}$$

Here, the magnetizing reactance can be calculated as follows:

$$X_\phi = X_m = 24f\left(\frac{\mu_0 Dl}{g}\right)\left(\frac{N_1 k_{W1}}{mp}\right)^2$$
$$= 24 \times 60 \times \left(\frac{4\pi \times 10^{-7} \times 0.0772 \times 0.0635}{0.000163}\right)\left(\frac{84.28}{2}\right)^2$$

Hence,

$$X_\phi = X_m = 96.64\,\Omega$$

Also, the developed torque can be calculated as follows:

$$T_{em} = T_{dev} = \sqrt{2}k_{w1}(B_{SR\max} \cdot A)\sin\delta_s\left(\frac{\pi D^2 l}{4}\right)$$
$$k_{W1} = 0.9577, B_{SR\max} = 0.9801 \text{ Tesla}$$

Here,

$$A = \frac{6N_1(I_1/m)}{\pi D} = \frac{6 \times 176 \times 2.163/2}{\pi \times 0.772} = 4708.94 \text{ A/m}$$

and

$$\sin\delta_s = \text{Sin}(56.01°) = 0.8291$$

and, in this case, we have

$$\frac{\pi D^2 l}{4} = \frac{\pi \times (0.0772)^2 \times 0.0635}{4} = 2.9723 \times 10^{-4}\text{ m}^3$$

Therefore, one can compute the developed torque as follows:

$$T_{em} = T_{dev} = \sqrt{2} \times 0.9577 \times 0.9801 \times 4708.94 \times 0.8291 \times 2.9723 \times 10^{-4}$$
$$= 1.5404 \text{ Nm}$$

(3) For comparison:

From motor data	From design synthesis
$X_m = 95.004\ \Omega$	$X_m = 96.64\ \Omega$
$T_{dev} = 1.515$ Nm	$T_{dev} = 1.5404$ Nm
$B_{SR\max} = 0.9646$ Tesla	$B_{SR\max} = 0.9801$ Tesla

The table shows close agreement/correlation between the motor's design synthesis data and the manufacturer's data, thus confirming the validity of the formulations provided in Chapters 5, 6, and 10. Next, we provide a sample approximate computation of the slot leakage reactance for this motor. Insofar as the stator armature leakage reactance, $X_2 = X_l$, the stator slot geometry for this machine is shown as follows:

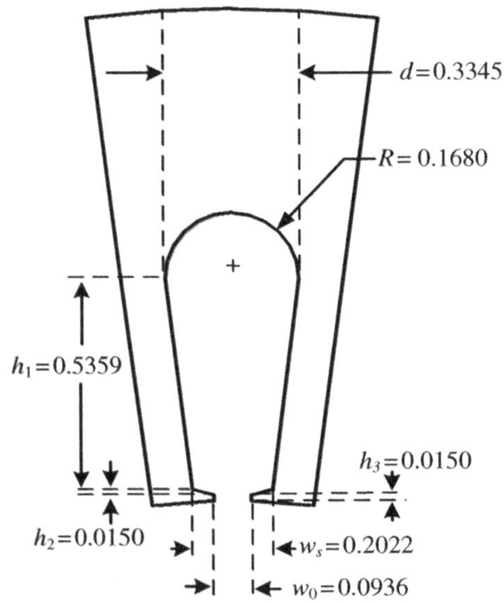

Stator armature slot

The slot permeance, $\mathbf{P}_s$, is given here by

$$\mathbf{P}_s = \mu_0\left[\frac{1}{2}(0.623) + \frac{2h_1}{d+w_s} + \frac{2h_2}{w_s+w_0} + \frac{h_3}{w_0}\right]$$

Hence,

$$X_l = \frac{8\pi f(N_{ph})^2 l}{mpq}k_s\mathbf{P}_s = \frac{8\pi f(N_{ph})^2 l}{mpq}\mu_0\left[\frac{1}{2}\left(0.623 + \frac{h_3}{w_0}\right) + \frac{2h_1}{d+w_s} + \frac{2h_2}{w_s+w_0} + \frac{h_3}{w_0}\right]$$

Here, (coil span/pole pitch) = 1.0, hence, $k_s = 1.0$, see Figure 10.14,

$$\mathbf{P}_s = 4\pi \times 10^{-7}\left[\frac{1}{2}(0.623) + \frac{2 \times 0.5359}{0.3345 + 0.2022} + \frac{2 \times 0.015}{0.2022 + 0.0936} + \frac{0.015}{0.0936}\right]$$

$$\mathbf{P}_s = 4\pi \times 10^{-7}[0.3115 + 1.9970 + 0.1014 + 0.1603] = 3.2298 \times 10^{-6}\text{H/m}$$

Also,

$$f = 60\text{ Hz}, N_{ph} = N_1 = 176\text{ turns}, l = 0.0635\text{ m},$$

$$m = 2\text{ paths}, p = 2\text{ poles}, q = 4\text{ slots/pole/phase}$$

Hence,

$$X_2 = X_l = \frac{8 \times \pi \times 60 \times (176)^2 \times 0.0635}{2 \times 2 \times 4} \times 3.2298 \times 10^{-6}\ \Omega$$

$$= 30\pi \times (176)^2 \times 0.0635 \times 3.2298 \times 10^{-6}\ \Omega$$

$$X_2 = X_l = 0.599\ \Omega$$

$$X_2(\text{due to slot leakage}) = 0.599\ \Omega,$$

which should be compared with $X_2 = 3.016\ \Omega$ as suggested by the manufacturer's data. The difference is in other leakage flux components, such as end-turn, leads connections, and zigzag leakage components.

10.10 Direct Torque Control of Induction Motors

Unlike the field oriented control (FOC) method introduced in Section 6.8 of Chapter 6, where the control is based on imitating the electromagnetic behavior of DC motors, another more straightforward vector control method named direct torque control (DTC) will be introduced in this section for induction motors. The DTC method directly controls the torque and stator flux by controlling the stator voltages of the motors through varying switching states of ASDs, without requiring any transformations of the synchronous reference frame or current controls. This control method has fast dynamic response and simple implementation, which has been widely applied in industrial motor-drive systems.

DTC was invented by Dr. Manfred Depenbrock in Germany in 1984 and was named as Direct Self-Control in his patent [16], while the Japanese scholars Dr. Isao Takahashi and Mr. Toshihiko Noguchi proposed a similar control technique termed as Direct Torque Control and published this work in a journal paper in 1984 [17]. Thus, the invention of DTC is usually credited to all these three researchers. The operating principle of DTC can be explained as follows:

First of all, for induction machines, the stator flux space vector, $\vec{\lambda}_s$, in the stationary reference frame can be expressed as

$$\vec{\lambda}_s = \lambda_{ds} + j\lambda_{qs} = \int (v_{ds} - R_s i_{ds})dt + j\int (v_{qs} - R_s i_{qs})dt \tag{10.404}$$

where λ_{ds} and λ_{qs} refer to the stator flux space-vector components in the direct and quadrature axes, respectively, v_{ds} and v_{qs} refer to the stator voltage space-vector components in the direct and quadrature axes, respectively, i_{ds} and i_{qs} refer to the stator current space-vector components in the direct and quadrature axes, respectively, R_s denotes the stator resistance. Equation (10.404) illustrates that the stator flux space vector can be calculated online based on the sensed stator voltages and currents, if the stator resistance of the induction motor is given (known).

Also, the magnitude and phase angle of the stator flux space vector can be expressed as follows:

$$\begin{cases} \lambda_s = \sqrt{\lambda_{ds}^{\ 2} + \lambda_{qs}^{\ 2}} \\ \theta_s = tan^{-1}(\lambda_{qs}/\lambda_{ds}) \end{cases} \tag{10.405}$$

Accordingly, the developed electromagnetic torque can be calculated based on the stator flux space-vector components, which can be expressed as follows:

$$T_e = 3P(i_{qs}\lambda_{ds} - i_{ds}\lambda_{qs}) \tag{10.406}$$

where P is the number of poles in the motor. Combining (10.404) and (10.406) together, one can conclude that the developed torque can be calculated online based on the sensed stator voltage and current components. This also shows that the torque can be controlled by regulating the stator flux space vector, or in essence the stator voltage. No other motor parameters are required except the stator resistance, R_s, which is different from the FOC method introduced in Section 6.8 of Chapter 6 that generally requires more motor parameters. Furthermore, the stator voltage can be regulated with a voltage space-vector selection strategy using a switching table in the associated ASDs.

The functional block diagram of illustrating the main procedures of implementing a DTC method is shown in Figure 10.22. As can be seen, the measured line voltage and current signals are provided as the input for the torque and stator flux space vector estimators (sometimes also called observers), and the output of such estimators is used in comparison with the reference values of the torque and stator flux space vector. The comparison errors of the torque and stator flux space vector serve as the inputs for their respective hysteresis controllers, and the control outputs determine the voltage space-vector patterns from the switching look-up table embedded in a microcontroller to regulate the output voltages of the associated ASD. In practice, the torque reference value (T_e^*) is generally derived from an outer speed controller, as shown in Figure 10.22.

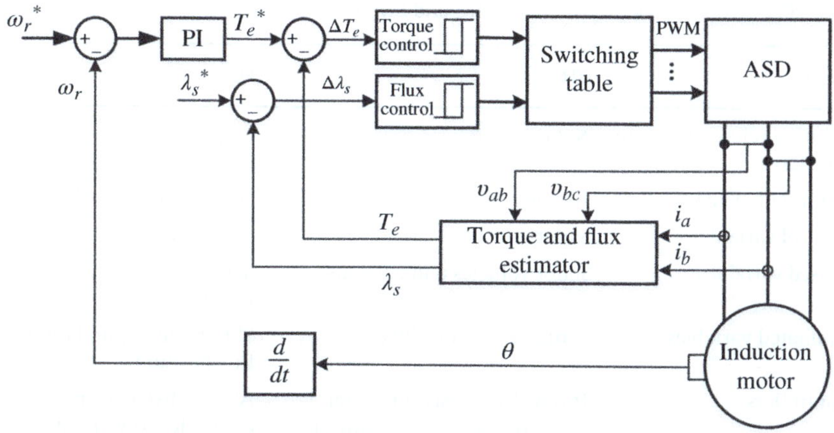

Figure 10.22 Functional block diagram of direct torque control of an induction motor.

The DTC method has much faster dynamic response compared to the FOC approach, since the torque and stator flux space vector are directly controlled. Additionally, this method does not require any reference frame transformation and many motor parameters associated with the FOC approach. However, it should be noted that the switching frequency is not constant in the DTC method and it changes with respect to the motor speed and the hysteresis bands, which may pose certain challenges regarding the torque ripple minimization, harmonic filter design, and semiconductor loss prediction of the inverters of such ASDs. There are research papers published in the literature to achieve approximately constant average switching frequency in the DTC method by controlling the width of the tolerance bands or other approaches, see [18–20]. However, these aspects will not be detailed in this textbook. A comparison table showing the difference between the conventional FOC and the DTC methods is provided in Table 10.1.

Homework Problems

10.1 Consider the 100-hp, 3-phase, 4-pole, 460-V, 60-Hz, Y-connected, wound rotor induction motor, and its mechanical load with all their parameters and particulars, as given earlier in Project 10.1 and Chapter 2 of this textbook, to be the motor-mechanical load case-study system under investigation in this project. Refer to the data of the motor-mechanical load system given in Chapter 2, specifically, all the T-equivalent circuit parameter

Table 10.1 Comparison between the DTC and FOC methods.

Factor	DTC	FOC
Rotating reference frame transform	Not required	Required
Control objectives	Torque and stator flux	Torque and rotor flux
Control variables	Stator voltages	Stator currents
Sensed variables	Stator voltages and currents	Stator currents and rotor position
Estimated variables	Torque and stator flux	Slip frequency and rotor flux position
Controllers	Torque hysteresis controller Stator flux hysteresis controller	At least two current controllers (i.e., i_d and i_q)
Sensitivity to motor parameter variations	Medium	Very sensitive
Dynamic response	Very fast	Medium
PWM strategy	Hysteresis bands	Carrier PWM, SVM, or hysteresis bands
Switching frequency	Variable	Fixed for carrier PWM and SVM

data, the mechanical load-motor inertia, load torque-speed characteristics, pulley ratio, (r_L/r_m), etc. Specifically, recall that the gross mechanical load torque, T'_m, in N_m, versus mechanical speed, ω_{rm}, in mechanical radians per second was given by

$$T'_m = \left[0.007818\,\omega^2{}_{rm} + 0.06299\,\omega_{rm}\right]\ \text{Nm}$$

Using notational symbolism for the induction motor stator and rotor phase circuit parameters given below:

$r_s = 0.042320\,\Omega$, $r_r = 0.0067712\,\Omega$,
$L_{1\ell} = 0.505157$ mH, $L_{2\ell} = 0.080825$ mH,
$L_\phi = 19.645030$ mH,
$L_{ss} = 13.601844$ mH, $L_{sm} = -6.548343$ mH,
$L_{rr} = 2.176295$ mH, $L_{rm} = -1.047735$ mH,
$L_{srm} = 5.2386747$ mH.

(a) Check whether the following inductance parameters of Eq. (10.8) are reasonably accurate:
$r_A = r_B = r_C = 0.042320\,\Omega$,
$r_a = r_b = r_c = 0.0067712\,\Omega$,

$L_{AA} = L_{BB} = L_{CC} = 13.601844\,\text{mH}$,
$L_{AB} = L_{BA} = L_{BC} = L_{CB} = L_{CA} = L_{AC} = -6.548343\,\text{mH}$
$L_{aa} = L_{bb} = L_{cc} = 2.176295\,\text{mH}$,
$L_{ab} = L_{ba} = L_{bc} = L_{cb} = L_{ca} = L_{ac} = -1.047735\,\text{mH}$,
$L_{Aa} = L_{aA} = L_{Bb} = L_{bB} = L_{Cc} = L_{cC} = 5.2386747\cos(\sigma)\,\text{mH}$,
$L_{Ab} = L_{bA} = L_{Bc} = L_{cB} = L_{Ca} = L_{aC} = 5.2386747\cos(\sigma - 4\pi/3)\,\text{mH}$,
and $L_{Ac} = L_{cA} = L_{Ba} = L_{aB} = L_{Cb} = L_{bC} = 5.2386747\cos(\sigma - 2\pi/3)\,\text{mH}$,

(b) Using MATLAB or other software of your own choosing and the state model formulation of *either form* of Eq. (10.8), simulate a line starting transient of the motor-mechanical load system of Project 10.1 for the case of wound-rotor induction motor, with identical conditions to those you previously simulated in item (a) of Project 10.1. That is, only replacing the T-equivalent circuit model with the state-space model of Eq. (10.8). Here, you can assume the following for the stator phase, A, B, and C, line-to-neutral voltages:

$$V_A = \frac{460\sqrt{2}}{\sqrt{3}}\cos(377t) = 375.59\cos(377t)\ \text{V},$$
$$V_B = 375.59\cos(377\text{t} - 2\pi/3)\ \text{V},$$
and $V_C = 275.59\cos(377t - 4\pi/3)$ V.

Notice, your integration time-step length should be allowing for at least a minimum of (200) or more integration time steps per ac cycle. Use the same time-step for both the electrical and mechanical system dynamics at least as a first attempt in your simulations.

(c) Perform a thorough comparison between these results for starting time, speed, developed torque, efficiency, cumulative energy losses, and those you obtained earlier in item (a) of Project 10.1. (Hint: Avoid an apples and oranges type of comparison, that is to be consistent.)

(d) Plot the speed, developed torque, gross mechanical load torque, stator and rotor phase current transients versus time, and comment on the steady-state developed torque ripple (largest rises and dips above and below the average developed torque magnitude in per unit or percent of the average torque, respectively).

10.2 Consider the 100-hp, 3-phase, 6-pole, 460-volts, 60-Hz, Y-connected, induction motor mentioned earlier. (a) Repeat the work in items (b), (c), and (d) of Project 10.1, using state-space model formulations in the $DQ0$–$dq0$ frame of reference. (b) Compare, on a one-to-one time-domain profile by time-domain profile basis, the results you obtained here with those obtained in your work of Project (10.1) and comment on the similarities or differences and the reasons why.

10.3 Consider a 5-hp, 3-phase, 6-pole, 460-V, 60-Hz, Y-connected, squirrel-cage, 1165 rad/min, induction motor. The motor and its mechanical load have a total polar moment of inertia, $J_{\text{total}} = 0.14\,\text{kg m}^2$. The motor has a T-equivalent circuit whose parameters referred to the stator side are as follows:
$r_1 = r_s = 1.342\,\Omega$, $r_2 = r'_r = 1.773\,\Omega$,
$L_{1\ell} = 0.0109\,\text{H}$, $L'_{2\ell} = 0.0109\,\text{H}$,
and $L_\phi = 0.262\,\text{H}$,
Consider for the purposes of modeling that this machine is equivalent to a wound-rotor motor with a stator to rotor turns ratio, $N_1/N_2 = N_s/N_r = 1.0$. The motor was directly coupled to the load. That is, no gear ratio or pulley ratio is involved. Hence, the mechanical angular speeds of the motor and load are equal. Thus, the gross mechanical torque including the load and motor rotational losses, T'_m, as a function of the mechanical angular speed, ω_{rm}, in rad/s, is given by

$$T'_m = 0.2459\omega_{rm} \quad \text{Nm}$$

(a) Using notational symbolism for the induction motor stator and rotor phase circuit parameters as given in Eq. (10.8), obtain the following inductive parameters:

$$L_{ss}, L_{sm}, L_{srm}, L_{rr}, \text{and } L_{rm}.$$

(Verify that $L_{ss} = 0.185\,\text{H}$, $L_{sm} = -0.087\,\text{H}$, $L_{srm} = 0.1746\,\text{H}$, $L_{rr} = 0.185\,\text{H}$, and $L_{rm} = -0.087\,\text{H}$.)

(b) Obtain the parameter for the *ABC–abc* state-space inductance and resistance matrices. That is, find
$L_{AA} = L_{BB} = L_{CC}$, $L_{AB} = L_{BA} = L_{BC} = L_{CB} = L_{CA} = L_{AC}$,
$L_{aa} = L_{bb} = L_{cc}$, $L_{ab} = L_{ba} = L_{bc} = L_{cb} = L_{ca} = L_{ac}$,
$L_{Aa} = L_{aA} = L_{Bb} = L_{bB} = L_{Cc} = L_{cC}$,
$L_{Ab} = L_{bA} = L_{Bc} = L_{cB} = L_{Ca} = L_{aC}$,
and $L_{Ac} = L_{cA} = L_{Ba} = L_{aB} = L_{Cb} = L_{bC}$, in Henries.

(c) Using MATLAB or other software of your own choosing, and the state model formulation of *either form* of Eq. (10.8), simulate a line starting transient of the motor-mechanical load system. Here, you can assume the following for the stator phase, A, B, and C, line-to-neutral voltages:
$V_A = 375.59\cos(377t)\,\text{V}$,
$V_B = 375.59\cos(377t - 2\pi/3)\,\text{V}$,
$V_C = 375.59\cos(377t - 4\pi/3)\,\text{V}$,

(d) Obtain the starting time-domain profiles of the motor A, B, and C phase currents, speed, developed torque, and efficiency and plot your results.

(e) How long does the motor-load system take to reach steady-state speed? Are there any torque pulsations? And why?

(f) Repeat the work in (c) through (e) if the cosinusoidal line-to-neutral voltages are replaced by six-step inverter output voltages with a fundamental peak line-to-neutral voltage of 375.59 V (you may need to consult a textbook on power electronics for the V_A, V_B, and V_C, waveforms' shape to program the needed forcing function vector for your simulations).

(g) Compare your results with these given to you from time-stepping finite element (TSFE) starting transient simulations and comment particularly on the ripple content in the steady-state developed torque time-domain profile, with appropriate physical explanation for any differences that you observe.

10.4 Consider the 5-hp, 3-phase, 6-pole, 460-V, 60-Hz, Y-connected, 1165 rad/min motor of Project 10.4. (a) Repeat the work in items (c), (d), (e), (f), and (g) of Project 10.4, using state-space model formulations in the *DQ*0 − *dq*0 frame of reference. (b) Compare, on a one-to-one time-domain profile by time-domain profile basis, the results you obtained here with those obtained in your work of Project 10.4, and comment on the similarities or differences and the reasons why. This includes the TSFE starting transient simulations given to you in class.

References

1 Nehl, T.W., Fouad, F.A., and Demerdash, N.A. (1982). Determination of saturated values of rotating machinery incremental and apparent inductances by an energy perturbation method. *IEEE Transactions on Power Apparatus and Systems* PAS-101: 4441–4451.

2 Demerdash, N.A., Fouad, F.A., and Nehl, T.W. Determination of winding inductances in ferrite type permanent magnet electric machinery by finite elements. *IEEE Transactions on Magnetics* MAG-18: 1052–1054.

3 Demerdash, N.A., Hijazi, T.M., and Arkadan, A.A. (1988). Computation of winding inductances of permanent magnet brushless DC motors with damper windings by energy perturbation. *IEEE Transactions on Energy Conversion* EC-3 (3): 705–713.

4 Demerdash, N.A. and Nehl, T.W. (1999). Electric machinery parameters and torques by current and energy perturbations from field computations – Part I: theory and formulation. *IEEE Transactions on Energy Conversion* 14 (4): 1507–1513.

5 Demerdash, N.A. and Nehl, T.W. (1999). Electric machinery parameters and torques by current and energy perturbations from field computations – Part II: applications and results. *IEEE Transactions on Energy Conversion* 14 (4): 1514–1522.

6 Demerdash, N.A., Bangura, J.F., and Arkadan, A.A. (1999). A time-stepping coupled finite element – state space model for induction motor drives – Part I: model formulation and machine parameter computation. *IEEE Transactions on Energy Conversion* 14 (4): 1465–1471.

7 Bangura, F.J., Isaac, F.N., Demerdash, N.A., and Arkadan, A.A. (1999). A time-stepping coupled finite element – state space model for induction motor drives – Part II: machine performance computation and verifications. *IEEE Transactions on Energy Conversion* 14 (4): 1472–1478.

8 Demerdash, N.A. and Bangura, J.F. (1999). Characterization of induction motors in adjustable-speed drives using a time-stepping coupled finite-element state space method including experimental validation. *IEEE Transactions on Industry Applications* 35 (4): 790–802.

9 Bangura, F.J. and Demerdash, N.A. (1999). Simulation of inverter-fed induction motor drives with pulse-width modulation by a time-stepping coupled finite element- flux linkage-based state space model. *IEEE Transactions on Energy Conversion* 14 (3): 518–525.

10 Bangura, F.J. and Demerdash, N.A. (1999). Performance and torque-ripple characterization in induction motor in adjustable-speed drives using a time-stepping coupled finite-element state space techniques. *IEEE Transactions on Industry Applications* 35 (5): 982–990.

11 Bangura, F.J. and Demerdash, N.A. (1999). Diagnosis and characterization of effects of broken bars and connectors in squirrel-cage induction motors by a time-stepping coupled finite element-state space modeling approach. *IEEE Transactions on Energy Conversion* 14 (4): 1167–1176.

12 Solveson, M., Mirafzal, B., and Demerdash, N.A.O. (2006). Soft started induction motor modeling and heating issues for different starting profiles using a flux linkage ABC-frame of reference. *IEEE Transactions on Industry Applications* 42 (4): 973–982.

13 Say, M.G. (1976). *Alternating Current Machines*. Pitman.

14 Demerdash, N.A. and Baldassari, P. (1992). A combined finite element-state space modeling environment for induction motors in the ABC frame of reference: the no-load condition. *IEEE Transactions on Energy Conversion* 7 (4): 698–709.

15 Baldassari, P. and Demerdash, N.A. (1992). A combined finite element-state space modeling environment for induction motors in the ABC frame of reference: the blocked-rotor and sinusoidally energized load conditions. *IEEE Transactions on Energy Conversion* 7 (4): 710–720.

16 Depenbrock, M. (1984). Direct self-control of the flux and rotary moment of a rotary-field machine. Patent number 4,678,248, filed on Oct. 20, 1984.

17 Noguchi, T. and Takahashi, I. (1984). Quick torque response control of an induction motor based on a new concept. *IEEJ Technical Meetings on Rotating Machine RM84-76* 61–70.

18 Kang, J.-K. and Sul, S.-K. (1999). New direct torque control of induction motor for minimum torque ripple and constant switching frequency. *IEEE Transactions on Industry Applications* 35 (5): 1076–1082.

19 Lai, Y.-S. and Chen, J.-H. (2001). A new approach to direct torque control of induction motor drives for constant inverter switching frequency and torque ripple reduction. *IEEE Transactions on Energy Conversion* 16 (3): 220–227.

20 Toh, C.L., Idris, N.R.N., and Yatim, A.H.M. (2005). Constant and high switching frequency torque controller for DTC drives. *IEEE Power Electronics Letters* 3 (2): 76–80.

11

Single-Phase Induction Motors and Other Special Motors

In this chapter, the operating principles and main characteristics of single-phase induction motors will be introduced, including split-phase induction motors, capacitor-start induction motors, and capacitor-start capacitor-run induction motors. Afterwards, other special motors such as switched reluctance motors are also discussed.

11.1 Single-Phase Induction Motors

Single-phase induction motors, which only require a single-phase AC supply to energize the stator windings, have been broadly used in residential and commercial electrical appliances and gadgets such as refrigerators, washing machines, ceiling fans, and many other applications. The electrical and mechanical structure of a single-phase induction motor is very similar to that of a three-phase induction motor, having windings on the stator and squirrel cage on the rotor. Compared to the general three-phase induction motors, single-phase induction motors are simpler in construction, more reliable, and more economical. However, single-phase induction motors do not inherently have a starting torque capability, since there is only one phase on the stator winding and the resulting magnetic field does not rotate. Based on various techniques to start such motors, there are mainly three types of single-phase induction machines used broadly in residential, commercial, and industrial applications, namely split-phase induction motors, capacitor-start induction motors, and capacitor-start capacitor-run induction motors, which will be elaborated in Sections 11.1.1–11.1.3.

11.1.1 Split-Phase Induction Motors

As a single-phase induction motor, a split-phase induction motor has two stator windings, including a main winding and an axillary winding, the magnetic

Electric Machinery and Drives: An Electromagnetics Perspective, First Edition.
Nabeel A. O. Demerdash, JiangBiao He, and Hao Chen.

Companion website: www.wiley.com/go/demerdash/electricmachineryanddrives

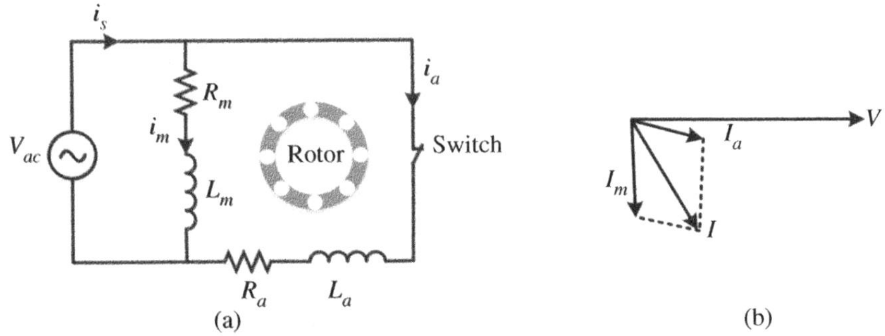

Figure 11.1 (a) A split-phase induction motor. (b) The phasor diagram of the voltage and current. (Here, the subscripts "m" and "a" refer to the main winding and auxiliary winding, respectively.)

axes of which are spatially displaced by 90°*e* apart from each other, as shown schematically in Figure 11.1a. The auxiliary winding is designed to have a higher resistance/reactance (i.e., R/X) ratio than the main winding. Specifically, the main winding should have low resistance and high inductive reactance, while the auxiliary winding should have high resistance but low inductive reactance. This is to ensure that the current flowing in the main winding lags behind the current in the auxiliary winding, although the two winding currents are unequal in magnitude (i.e., thus resulting in a non-uniform rotating magnetic field). Upon starting the motor, both the two windings are electrically connected in parallel, enabling a starting torque to rotate the motor (e.g., the starting torque is usually 1.5–2 times the rated torque). Once the motor operating speed approaches the desired value such as 70–80% of the synchronous speed, the auxiliary winding will be electrically disconnected from the motor circuit by turning off the auxiliary switch.

As is shown in Figure 11.1b, the current in the main winding (i_m) lags behind the supply AC voltage by nearly 90°*e*, whereas the current in the auxiliary winding (i_{aux}) is almost in phase with the supply voltage due to the high resistance/reactance ratio. The phase difference between i_m and i_{aux} is generally lower than 90°*e* but should be sufficient to produce a rotating magnetic field component and hence start the motor. This is the reason that split-phase induction motors are sometimes called resistor-start induction motors, to differentiate the starting technique from capacitor-start induction motors which will be introduced in Section 11.1.2.

For low-power applications (e.g., below 100 W), a relay can be used to connect/disconnect an auxiliary winding. A centrifugal switch may need to be used for high-power applications (e.g., above 100 W). Split-phase induction motors generally have a relatively low starting current and moderate starting torque. They are widely used for applications which do not need high starting torque, such as air conditioning, fans, washing machines, grinders, blowers, and centrifugal pumps. Most such split-phase induction motors are sized in the fractional-horsepower range and are low cost.

11.1.2 Capacitor-Start Induction Motors

A capacitor-start induction motor is another type of single-phase induction motor that can provide higher starting torque by producing a larger phase difference between the currents in the main winding and the auxiliary winding. In such motors, a capacitor is connected in series with the auxiliary windings, as shown schematically in Figure 11.2.

Whenever the motor starts, the centrifugal switch will be turned on, and the two windings the magnetic axes of which are physically arranged 90 electrical degrees from each other, will produce a rotating magnetic field. In other words, a single-phase supply current is split into two phases, and the related two magnetomotive forces (MMFs) are equal in magnitude but $90^{\circ}e$ in phase difference, which actually acts in a similar fashion as in a two-phase induction motor. By properly sizing the capacitor, the starting torque can be multiple times (e.g., 3–5 times) higher than the rated torque of the motor. Compared to a split-phase induction motor under the same operating conditions, a capacitor-start induction motor counterpart has a relatively higher cost, but the starting torque is much higher. Such capacitor-type induction motors are suitable for the low-power applications with high-inertial loads, where frequent starting is required, such as conveyors, machine tools, compressors, and many other similar applications.

11.1.3 Capacitor-Start Capacitor-Run Induction Motors

As introduced in Section 11.1.2, the starting capacitor plays an important role in improving the torque-speed characteristic of a capacitor-start single-phase

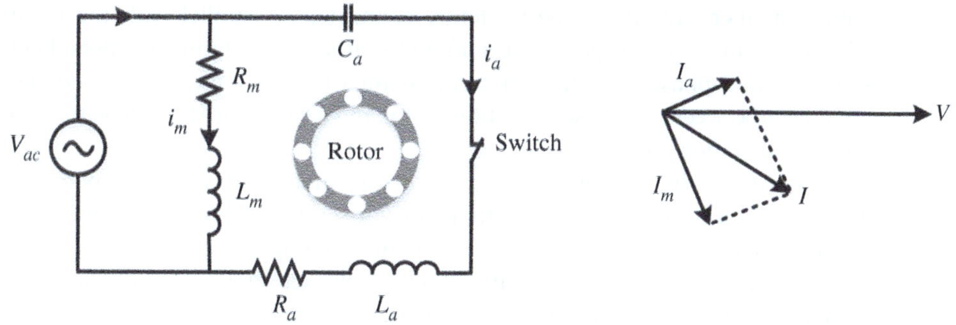

Figure 11.2 A capacitor-start induction motor and the phasor diagram of voltage and currents.

induction motor. In certain applications, the starting capacitor in the auxiliary winding is sometimes left permanently connected in the motor circuit. If the capacitor's value is chosen correctly, such a motor will have a near-uniform circular rotating magnetic field at some specific load, and it will behave just as in a three-phase induction motor at that point. Such a design is called a permanent split-capacitor motor. Such permanent split capacitor motors are simpler and more economic in terms of hardware implementation compared to the capacitor start motors introduced in Section 11.1.2. The reason is that, the auxiliary starting switch and the related control circuit can be eliminated, leading to lower cost and simpler hardware structure. Under rated loading conditions, the permanent split capacitor motors may have higher efficiency, higher power factor, and lower torque ripple, in comparison to the conventional single-phase induction motors.

Nevertheless, such permanent split capacitor motors may have a lower starting torque compared to the the capacitor-start motors, due to the fact that the starting capacitor needs to be sized to balance the currents in the main and axillary windings at normal load conditions. Considering that the starting current typically is much larger than the normal load current, a capacitor that balances the phases under normal loads leaves them very unbalanced under starting conditions [1].

To concurrently have large starting torque and desired running conditions, two capacitors can be used to constitute two auxiliary winding paths in parallel to the main winding, as shown in Figure 11.3. Accordingly, such a motor with two capacitors are named as capacitor-start capacitor-run induction motor. Specifically, as shown in Figure 11.3, the starting capacitor, C_{a1}, is generally sized large enough (e.g., over 5 times larger than C_{a2}), to guarantee the currents in the main winding and auxiliary windings are approximately balanced, resulting in high starting torque. Once the motor operates close to the rated speed, the auxiliary winding consisting of C_{a1} will be disconnected from the motor circuit by turning off the auxiliary switch. Consequently, only the 2nd auxiliary winding with a smaller capacitor is permanently connected in parallel with the main winding, to maintain proper phase shift and synthesize a rotating magnetic field similar to a 3-phase induction motor. A comparison of motor cross sectional topologies between a 3-phase induction motor and a capacitor-start capacitor-run motor is also shown in Figure 11.3.

In industries, the capacitor-start capacitor-run motors are mainly applied for driving loads with high inertia and frequent starts, while requires high efficiency and high start torque. Such applications include pumps, compressors, refrigerators, washing machines, and many others. However, one potential drawback with such machines is that, the capacitors have low reliability, especially under harsh operating conditions, so regular inspection and maintenance would be necessary.

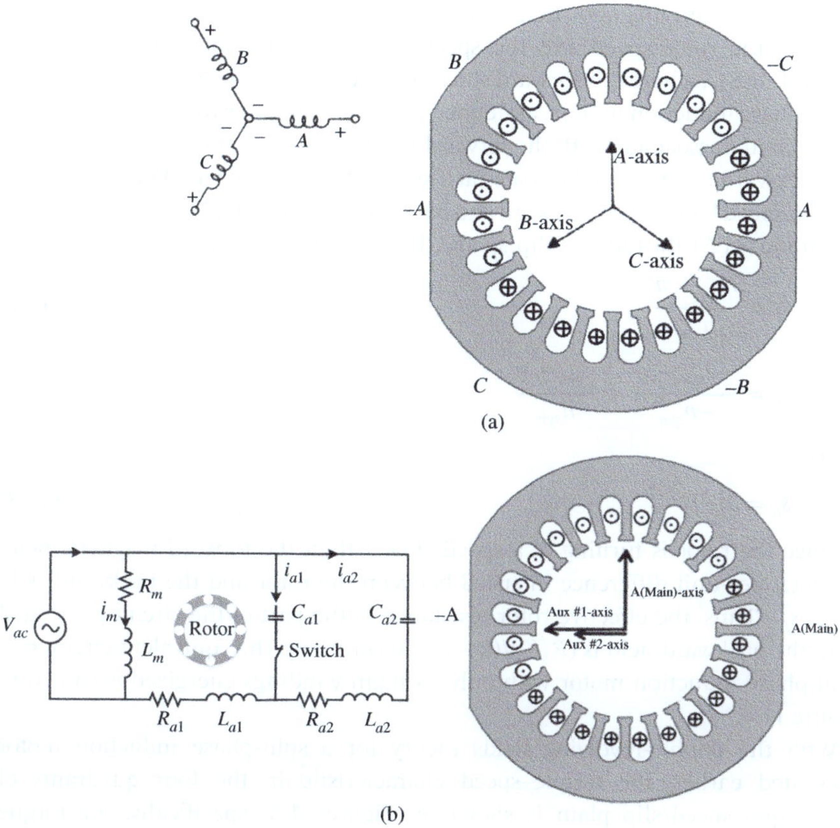

Figure 11.3 Comparison of motor cross-sectional topologies between a three-phase induction motor and a capacitor-start capacitor-run induction motor. (a) Three-phase induction motor. (b) Circuit diagram and cross section topology of the capacitor-start capacitor-run induction motor.

11.2 Torque–Speed Characteristics of Single-Phase Induction Motors

According to the double-revolving field theory of single-phase induction motors, a stationary pulsating magnetic field can be resolved, see earlier developments in Chapter 5, into two rotating magnetic fields (MMFs), namely a forward rotating magnetic field (MMF) and a backward rotating magnetic field (MMF). Both the magnetic fields are of equal magnitude but rotating in opposite directions. The motor responds to each magnetic field separately and the net torque produced in the motor is equal to the sum of the torques due to each of the two magnetic fields.

For the forward magnetic field rotating at a speed, n_{syn}, the per-unit difference between the rotor speed and the speed of the forward magnetic field is the forward slip, s_f, where this forward slip is defined in the same manner as it was for three-phase induction motors, as defined in (11.1). The rotor resistance in the part of the circuit associated with the forward magnetic field is thus $(R_2/2)/(s_f)$. Since the backward magnetic field rotates at a speed of $(-n_{syn})$, the total per-unit difference in speed between the forward and backward rotating fields is 2, as deduced in (11.2) and (11.3), the forward slip, s_f, and the backward slip, s_b, are given as follows:

$$s_f = \frac{n_{syn} - n}{n_{syn}} \tag{11.1}$$

$$s_b = \frac{-n_{syn} - n}{-n_{syn}} = \frac{n_{syn} + n}{n_{syn}} \tag{11.2}$$

Thus,

$$s_b = 2 - s_f \tag{11.3}$$

Since the rotor is turning at a speed slower than the forward magnetic field, the total per-unit difference in speed between the rotor and the backward field is $(2-s_f)$. Thus, the effective rotor resistance in the part of the circuit associated with the backward field is $(R_2/2)/(2-s_f)$. Accordingly, the equivalent circuit of a split-phase induction motor with only its main windings energized is shown in Figure 11.4.

With the double-revolving fields theory for a split-phase induction motor presented earlier, the torque–speed characteristic in the four quadrants of the torque–speed/slip plain is shown in Figure 11.5. Specifically, the torque component of the single-phase motor consists of the torques produced by the

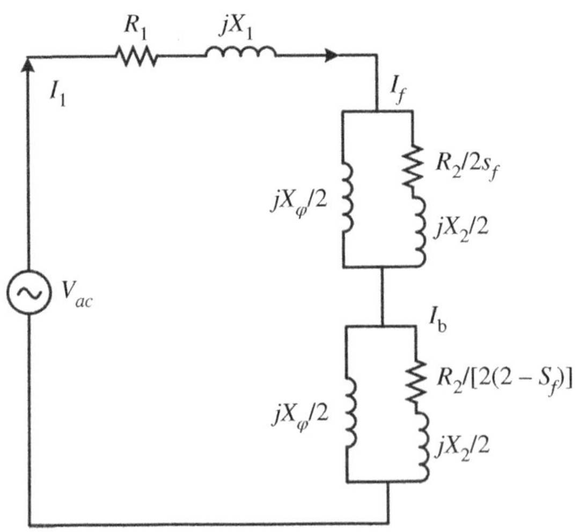

Figure 11.4 The equivalent circuit of a split-phase induction motor operating at a speed other than synchronous with only its main winding energized.

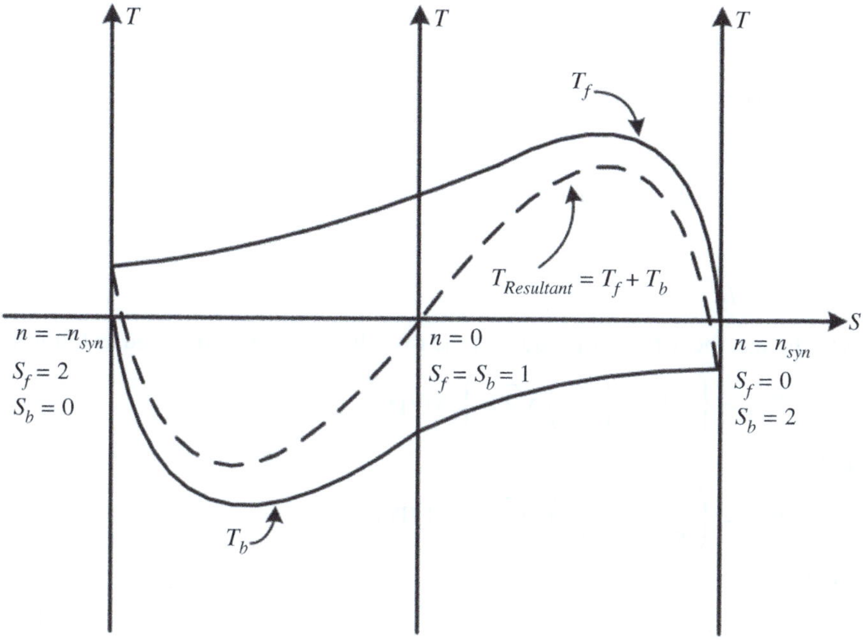

Figure 11.5 Torque–speed characteristics of a split-phase induction motor.

forward-rotating magnetic field component and the backward-rotating magnetic field component. When the rotor speed is zero (i.e., stall condition), the two slips become: $s_f = s_b = 1$; when the rotor speed is equal to the forward magnetic field speed (i.e., n_{syn}), the two slips becom $s_f = 0$ and $s_b = 2$. Likewise, when the rotor speed is equal to the backward field speed (i.e., $-n_{syn}$), the two slips become $s_f = 2$ and $s_b = 0$.

11.3 Field Analysis of Split-Phase and Capacitor-Start Induction Motors

In this section, field analysis of split-phase induction motors and capacitor-start induction motors will be discussed based on a rotating (traveling) MMF wave point of view and an elliptic field point of view.

11.3.1 Rotating Wave Point of View

This is a case which is very similar to the balanced two-phase case except that the two windings (A) and (B) normally represent a main and an auxiliary set of phase windings, in which the numbers of turns per phase are unequal (i.e., $N_a \neq N_b$), and the current-carrying capacities of phases (A) and (B) are also unequal. Thus, the root mean square (RMS) values of the currents in the two phases, $I_a \neq I_b$. Hence, the currents in the two phases can be expressed as follows:

$$i_a(t) = \sqrt{2}\, I_a \cos(\omega t) \tag{11.4}$$

and

$$i_b(t) = \sqrt{2}\, I_b \cos\left(\omega t - \frac{\pi}{2}\right) \tag{11.5}$$

Again,

$$I_a \neq I_b \tag{11.6}$$

And, hence, the MMF amplitudes of the two phases can be expressed as follows:

$$F_{ma} = \left\{ \frac{4}{\pi} K_{wa1} \left(\frac{N_a}{p}\right) \sqrt{2}\, I_a \right\} \text{AT/pole} \tag{11.7}$$

and

$$F_{mb} = \left\{ \frac{4}{\pi} K_{wb1} \left(\frac{N_b}{p}\right) \sqrt{2}\, I_b \right\} \text{AT/Pole} \tag{11.8}$$

where

$$F_{ma} \neq F_{mb} \tag{11.9}$$

Hence, the general form for the two MMFs produced by the two phases (A) and (B) yields fundamental components of MMFs, $F_{a1}(\theta, t)$ and $F_{b1}(\theta, t)$, which can be expressed as follows:

$$F_{a1}(\theta, t) = F_{ma} \cos(\theta) \cos(\omega t) \tag{11.10}$$

and

$$F_{b1}(\theta, t) = F_{mb} \cos\left(\theta - \frac{\pi}{2}\right) \cos\left(\omega t - \frac{\pi}{2}\right) \tag{11.11}$$

Therefore,

$$F_{a1}(\theta, t) = \frac{F_{ma}}{2} \cos(\theta - \omega t) + \frac{F_{ma}}{2} \cos(\theta + \omega t) \tag{11.12}$$

and

$$\begin{aligned} F_{b1}(\theta, t) &= \frac{F_{mb}}{2} \cos(\theta - \omega t) + \frac{F_{mb}}{2} \cos(\theta + \omega t - \pi) \\ &= \frac{F_{mb}}{2} \cos(\theta - \omega t) - \frac{F_{mb}}{2} \cos(\theta + \omega t) \end{aligned} \tag{11.13}$$

Hence, the total (i.e., resultant) MMF has a fundamental component, $F_1(\theta, t)$, as follows:

$$F_1(\theta, t) = F_{a1}(\theta, t) + F_{b1}(\theta, t) \tag{11.14}$$

$$F_1(\theta, t) = \frac{1}{2}(F_{ma} + F_{mb}) \cos(\theta - \omega t) + \frac{1}{2}(F_{ma} - F_{mb}) \cos(\theta + \omega t) \tag{11.15}$$

Equation (11.15) constitutes a forward rotating/traveling MMF wave with an amplitude of $\frac{1}{2}(F_{ma}+F_{mb})$ and a reverse (i.e., backward) rotating/traveling MMF wave with an amplitude of $\frac{1}{2}(F_{ma}-F_{mb})$. Thus, there could be forward- and reverse-oriented torque components produced. If phase (B) is disconnected, Eq. (11.15) reduces to that of a single-phase case.

11.3.2 Elliptic Field Point of View

Now, at any instant of time, t, and phase angle, (θ), the locus of the tip of the space vector representing the MMF, $F_{a1}(\theta, t)$, where

$$F_1(0, t) = \frac{1}{2}(F_{ma}+F_{mb})\cos(\theta-\omega t)+\frac{1}{2}(F_{ma}-F_{mb})\cos(\theta+\omega t) \tag{11.16}$$

can be developed in the x–y plane as follows:

Along the x-axis, one can write, ($\theta = 0$), hence,

$$x = F_1(0, t) = \frac{1}{2}(F_{ma}+F_{mb})\cos(\omega t)+\frac{1}{2}(F_{ma}-F_{mb})\cos(\omega t) \tag{11.17}$$

thus,

$$x = F_{ma}\cos(\omega t) \tag{11.18}$$

Accordingly,

$$\frac{x}{F_{ma}} = \cos(\omega t) \tag{11.19}$$

Along the y-axis, one can write $\left(\theta = \frac{\pi}{2}\right)$, hence,

$$y = F_1\left(\frac{\pi}{2}, t\right) = \frac{1}{2}(F_{ma}+F_{mb})\cos\left(\frac{\pi}{2}-\omega t\right)+\frac{1}{2}(F_{ma}-F_{mb})\cos\left(\frac{\pi}{2}+\omega t\right) \tag{11.20}$$

$$y = \frac{1}{2}(F_{ma}+F_{mb})\sin(\omega t)-\frac{1}{2}(F_{ma}-F_{mb})\sin(\omega t) \tag{11.21}$$

thus,

$$y = F_{mb}\sin(\omega t) \tag{11.22}$$

Therefore,

$$\frac{y}{F_{mb}} = \sin(\omega t) \tag{11.23}$$

Accordingly, one can write the following expression:

$$\left(\frac{x}{F_{ma}}\right)^2+\left(\frac{y}{F_{mb}}\right)^2 = \cos(\omega t)^2+\sin(\omega t)^2 = 1 \tag{11.24}$$

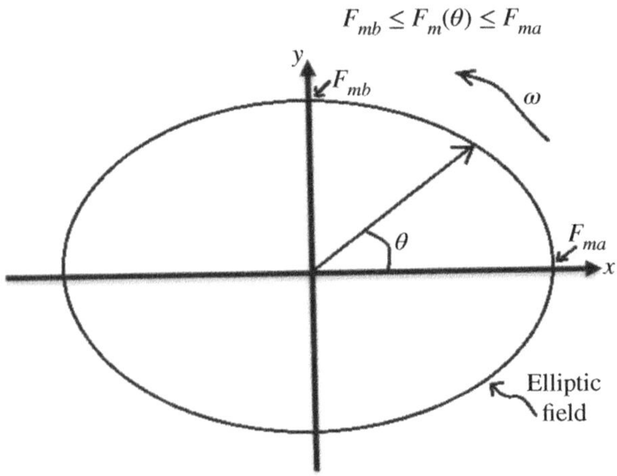

Figure 11.6 The elliptic field represents the most general case of magnetic field (MMF waveform) which can exist in an electric machine.

As is well known, the equation $\left(\frac{x}{F_{ma}}\right)^2 + \left(\frac{y}{F_{mb}}\right)^2 = 1$ refers to the locus of the tip of the MMF space vector, which represents an ellipse shown as in Figure 11.6. The resultant field (MMF) space vector changes both its magnitude and spatial position (phase) with time in a continuous fashion such that its endpoint (tip locus) describes an ellipse as shown in Figure 11.6.

11.4 Power and Torque for Split-Phase Induction Motors

Similar to the procedures for three-phase induction motors, the torque and power for split-phase induction motors can be calculated by analyzing the T-equivalent circuits shown in Figure 11.7 for standstill and rotating conditions. First of all, for single-phase induction motors, the MMF consists of the forward MMF F_f and the backward MMF F_b, which can be expressed as follows:

$$F(\theta, t) = \overline{F_f} + \overline{F_b} \tag{11.25}$$

$$F(\theta, t) = \frac{F_m}{2}\cos(\theta - \omega t + \emptyset) + \frac{F_m}{2}\cos(\theta + \omega t - \emptyset) \tag{11.26}$$

Here,

$$\overline{E}_1 \propto (\overline{F_f} + \overline{F_b}) \tag{11.27}$$

Hence,

$$\overline{E}_1 = \overline{E}_f + \overline{E}_b = \frac{1}{2} jX_m \overline{I}_{mf} + \frac{j}{2} X_m \overline{I}_{mb} \tag{11.28}$$

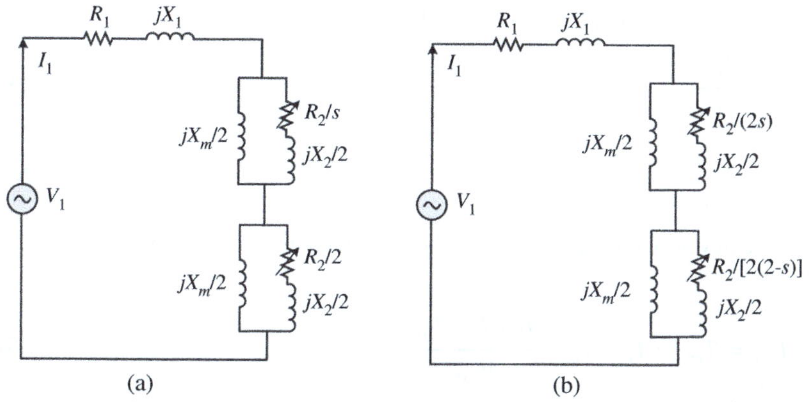

Figure 11.7 Equivalent circuit of a single-phase induction motor at various conditions. (a) At standstill condition ($\omega_r = 0$) and (b) at running load condition ($\omega_r \neq 0$).

at a standstill condition ($\omega_r = 0$), the related T-equivalent circuit is shown in Figure 11.7a.

However, at running load condition ($\omega_r \neq 0$), there exists a forward slip, s_f, and a backward slip, s_b, where

$$s_f = \frac{\omega - \omega_r}{\omega} = s \tag{11.29}$$

and

$$s_b = \frac{-\omega - \omega_r}{-\omega} = \frac{\omega + \omega_r}{\omega} = 2 - s \tag{11.30}$$

Hence, at running/rotating load condition ($\omega_r \neq 0$), the associated T-equivalent circuit is shown in Figure 11.7b.

According to the equivalent circuit, the impedances for the forward magnetic field component and backward field component can be written as follows:

$$\overline{Z}_f = R_f + jX_f = \left(\frac{\frac{R_2}{2}}{s} + j\frac{X_2}{2}\right) // \left(j\frac{X_m}{2}\right)$$
$$= \left(\frac{\frac{R_2}{2}}{s} + j\frac{X_2}{2}\right) * \left(j\frac{X_m}{2}\right) \Big/ \left[\left(\frac{\frac{R_2}{2}}{s} + j\frac{X_2}{2}\right) + \left(j\frac{X_m}{2}\right)\right] \tag{11.31}$$

$$\overline{Z}_b = R_b + jX_b = \left(\frac{\frac{R_2}{2}}{(2-s)} + j\frac{X_2}{2}\right) // \left(j\frac{X_m}{2}\right)$$
$$= \left(\frac{\frac{R_2}{2}}{(2-s)} + j\frac{X_2}{2}\right) * \left(j\frac{X_m}{2}\right) \Big/ \left[\left(\frac{\frac{R_2}{2}}{(2-s)} + j\frac{X_2}{2}\right) + \left(j\frac{X_m}{2}\right)\right] \tag{11.32}$$

Hence, the torque and power can be expressed as follows:

$$T_f = \frac{I_1^2 R_f}{\omega_{syn}} \tag{11.33}$$

$$T_b = \frac{I_1^2 R_b}{\omega_{syn}} \tag{11.34}$$

Thus, the developed torque, T_{dev}, can be expressed as the difference between T_f and T_b:

$$T_{dev} = T_{em} = T_f - T_b = \frac{I_1^2 (R_f - R_b)}{\omega_{syn}} \tag{11.35}$$

Accordingly, the developed power of the single-phase motor can be written as follows:

$$P_{dev} = P_{em} = \omega_r T_{dev} = \omega_r T_{em} = (1 - s)\, \omega_{syn}\, T_{dev} \tag{11.36}$$

From Eqs. (11.35) and (11.36), one can obtain the following:

$$P_{dev} = P_{em} = (1 - s)\, I_1^2 (R_f - R_b) \tag{11.37}$$

Also, the rotor ohmic losses can be expressed as follows:

$$P_{rotor\ ohmic} = s I_1^2 (R_f - R_b) \tag{11.38}$$

The stator ohmic losses can be written as follows:

$$P_{stator\ ohmic} = I_1^2 R_1 \tag{11.39}$$

Thus, the output power, P_{out}, of the motor can be obtained by subtracting the rotational losses from the developed power:

$$P_{out} = P_{dev} - P_{rotational} \tag{11.40}$$

Accordingly, the output torque of the motor can be expressed as follows:

$$T_{out} = \frac{P_{out}}{\omega_r} = \frac{P_{dev} - P_{rotational}}{\omega_r} \tag{11.41}$$

By substituting P_{dev}, from Eq. (11.37), we can further express the output torque as follows:

$$T_{out} = \frac{(1 - s)\, I_1^2 (R_f - R_b) - P_{rotational}}{(1 - s)\omega_{syn}} \tag{11.42}$$

Also, the motor input power, P_{in}, can be obtained as follows:

$$P_{in} = Re\left(\overline{V}_1\, \overline{I}_1^{\,*}\right) \tag{11.43}$$

Thus, the efficiency of the motor can be obtained as follows:

$$\eta = \frac{P_{in}}{P_{out}} \tag{11.44}$$

Now, we proceed to a numerical example of the analysis of such a motor.

Example 11.1

The following are the equivalent-circuit parameters of a 0.25-hp, 60-Hz, 115-V, 4-pole capacitor-start motor, $R_1 = 2.15\ \Omega, jX_1 = j\,3.01\ \Omega, R_2 = 4.45\ \Omega, jX_2 = j\,2.35\ \Omega$, $jX_m = j\,70.5\ \Omega$, and the rotation losses (i.e., core and windage) are 40 W. Calculate for a slip, $s = 0.05\ p.\ u$, the (a) current $\overline{I}_1$, (b) power factor, (c) output power and torque, as well as efficiency.

Solution

Referring to the T-equivalent circuit of the single-phase induction motor, Figure 11.7, we have

(a) $\overline{V}_1 = 115\angle 0^\circ, \overline{I}_1 = 115\angle 0^\circ$

$$\overline{Z}_f = R_f + j\,X_f = \left(\frac{\frac{R_2}{2}}{s} + j\frac{X_2}{2}\right) \parallel \left(j\frac{X_m}{2}\right) = \frac{(j\,35.25)\,(44.5 + j\,1.175)}{44.5 + j\,36.43}$$

$$\overline{Z}_f = R_f + j\,X_f = 27.2\angle 52.1^\circ\ \Omega = 16.72 + j\,21.45\ \Omega$$

$$\overline{Z}_b = R_b + j\,X_b = \left(\frac{\frac{R_2}{2}}{(2-s)} + j\frac{X_2}{2}\right) \parallel \left(j\frac{X_m}{2}\right) = \frac{(j\,35.25)\,(1.14 + j\,1.175)}{1.14 + j\,36.43}$$

$$\overline{Z}_b = R_b + jX_b = 1.58\angle 47.6^\circ\ \Omega = 1.06 + j\,1.17\ \Omega$$

Hence,

$$R_1 + j\,X_1 + \overline{Z}_f + \overline{Z}_b = (2.15 + 16.72 + 1.06) + j\,(3.01 + 21.45 + 1.17)$$

$$R_1 + j\,X_1 + \overline{Z}_f + \overline{Z}_b = 19.93 + j\,25.63 = 32.5\angle - 52.1^\circ\ \Omega$$

Thus, the current $\overline{I}_1$ can be calculated as follows:

$$\overline{I}_1 = \frac{\overline{V}_1}{(R_1 + jX_1) + \overline{Z}_f + \overline{Z}_b} = \frac{115\angle 0^\circ}{32.5\angle 52.1^\circ} = 3.54\angle - 52.1^\circ\ A$$

(b) Accordingly, the power factor can be calculated as: $PF = \cos\,(52.1^\circ) = 0.615$ *lagging*.

(c) The developed torque of the motor can be computed as follows:

$$P_{dev} = P_{em} = (1-s)\,I_1^{\,2}(R_f - R_b) = 0.95\,(3.54)^2\,(16.72 - 1.06)$$

$$P_{dev} = 186\ \text{W}$$

$$P_{out} = P_{dev} - P_{rotational} = 186 - 40 = 146\text{ W}$$

$$P_{out} = 146\text{ W} = 0.196\text{ hp}$$

$$T_{out} = \frac{P_{out}}{(1-s)\omega_{syn}} = \frac{146}{(0.95)\left(\frac{2\pi}{60} \times 1800\right)} = 0.816\text{ Nm}$$

$$P_{in} = Re\left(\overline{V}_1 \overline{I}_1^{\,*}\right) = Re(115\angle 0^{\circ} \times (3.54\angle -52.1^{\circ})^{*}) = 250\text{ W}$$

$$\eta = \frac{P_{out}}{P_{in}} = \frac{146}{250} = 0.584 = 58.4\%$$

11.5 Power and Torque for Capacitor-Start Induction Motors

In order to compute the power and torque for capacitor-start induction motors, let us examine $\overline{E}_{1M}$ and $\overline{E}_{1A}$, in the main and auxiliary windings, which are the result of the forward and backward rotating MMF components of the main and the auxiliary windings, $\overline{F}_{fM}$, $\overline{F}_{bM}$, $\overline{F}_{fA}$, and $\overline{F}_{bA}$, respectively.

(1) First, let us start by assuming that only the auxiliary winding is energized:
- In this case, $\overline{F}_{fM} = \overline{F}_{bM} = 0$ and $\overline{F}_{fA} \neq 0$ and $\overline{F}_{bA} \neq 0$. Hence, one can write the following:

$$\overline{E}_{1A} = \overline{E}_{fA} + \overline{E}_{bA} = \overline{Z}_{fA}\overline{I}_A + \overline{Z}_{bA}\overline{I}_A \tag{11.45}$$

Meanwhile, with $(N_A/N_M) = a$

$$\overline{E}_{1M} = -j\frac{\overline{E}_{fA}}{a} + j\frac{\overline{E}_{bA}}{a} = -j\frac{\overline{Z}_{fA}}{a}\overline{I}_A + j\frac{\overline{Z}_{bA}}{a}\overline{I}_A \tag{11.46}$$

See windings *A* and *M* orientation with respect to the forward and backward rotating fields (leading/lagging).

(2) Next, let us assume that only the main winding is energized:
- In this case; $\overline{F}_{fM} \neq 0$, $\overline{F}_{bM} \neq 0$, and $\overline{F}_{fA} = \overline{F}_{bA} = 0$.

Hence, one can write the following:

$$\overline{E}_{1M} = \overline{E}_{fM} + \overline{E}_{bM} = \overline{Z}_{fM}\overline{I}_M + \overline{Z}_{bM}\overline{I}_M \tag{11.47}$$

Meanwhile, examining the *A* and *M* winding orientations with respect to the forward and backward rotating fields (leading/lagging), one can write the following:

$$\overline{E}_{1A} = ja\,\overline{E}_{fM} - ja\,\overline{E}_{bM} = ja\,\overline{Z}_{fM}\overline{I}_M - ja\overline{Z}_{bM}\overline{I}_M \tag{11.48}$$

(3) Next, when both A and M windings are energized, one can write the following:

$$\overline{E}_{1M} = -j\frac{\overline{E}_{fA}}{a} + j\frac{\overline{E}_{bA}}{a} + \overline{E}_{fM} + \overline{E}_{bM} \tag{11.49}$$

and

$$\overline{E}_{1A} = \overline{E}_{fA} + \overline{E}_{bA} + ja\,\overline{E}_{fM} - ja\overline{E}_{bM} \tag{11.50}$$

Hence, the overall model and corresponding equivalent circuits are as follows:

$$\overline{V}_M = (R_{1m} + jX_{1m})\overline{I}_M + \overline{E}_{fM} + \overline{E}_{bM} - j\frac{\overline{E}_{fA}}{a} + j\frac{\overline{E}_{bA}}{a} \tag{11.51}$$

$$\overline{V}_A = (R_{1A} + jX_{1A} + R_c + jX_c)\overline{I}_A + \overline{E}_{fA} + \overline{E}_{bA} + ja\overline{E}_{fM} - ja\overline{E}_{bM} \tag{11.52}$$

However, recall that

$$\overline{E}_{fM} = \overline{Z}_{fM}\overline{I}_M, \overline{E}_{bM} = \overline{Z}_{bM}\overline{I}_M, \overline{E}_{fA} = \overline{Z}_{fA}\overline{I}_A \text{ and } \overline{Z}_{bA} = \overline{Z}_{bA}\overline{I}_A$$

where

$$\overline{Z}_{fM} = \overline{Z}_f, \overline{Z}_{fA} = a^2\overline{Z}_f \text{ and } \overline{Z}_{bM} = \overline{Z}_b, \overline{Z}_{bA} = a^2\overline{Z}_b$$

Hence, one can write the following:

$$\overline{V}_M = (R_{1m} + jX_{1m})\overline{I}_M + \overline{Z}_f\overline{I}_M + \overline{Z}_b\overline{I}_M - j\frac{a^2\overline{Z}_f}{a}\overline{I}_A + j\frac{a^2\overline{Z}_b}{a}\overline{I}_A \tag{11.53}$$

$$\overline{V}_M = (R_{1m} + jX_{1m})\overline{I}_M + \overline{Z}_f\overline{I}_M + \overline{Z}_b\overline{I}_M - ja\,\overline{Z}_f\overline{I}_A + ja\,\overline{Z}_b\overline{I}_A \tag{11.54}$$

and

$$\overline{V}_A = (R_{1A} + jX_{1A} + R_c + jX_c)\overline{I}_A + a^2\overline{Z}_f\overline{I}_A + a^2\overline{Z}_b\overline{I}_A + ja\,\overline{Z}_f\overline{I}_M - ja\,\overline{Z}_b\overline{I}_M \tag{11.55}$$

Thus,

$$\overline{V}_A = (R_{1A} + jX_{1A} + R_c + jX_c)\overline{I}_A + a^2\overline{Z}_f\overline{I}_A + a^2\overline{Z}_b\overline{I}_A + ja\,\overline{Z}_f\overline{I}_M - ja\,\overline{Z}_b\overline{I}_M \tag{11.56}$$

Hence, the equivalent circuit can be configured, as shown in Figure 11.8.

The electromagnetic airgap power, P_g, becomes

$$P_g = P_{gf} - P_{gb} = Re\left[(\overline{E}_f - \overline{E}_b)\overline{I}_M^{\,*} + ja\,(\overline{E}_f + \overline{E}_b)\overline{I}_A^{\,*}\right] \tag{11.57}$$

and the developed electromagnetic torque becomes

$$T_{dev} = T_{em} = \frac{P_g}{\omega_{syn}} = \frac{P_{gf} - P_{gb}}{\omega_{syn}} \tag{11.58}$$

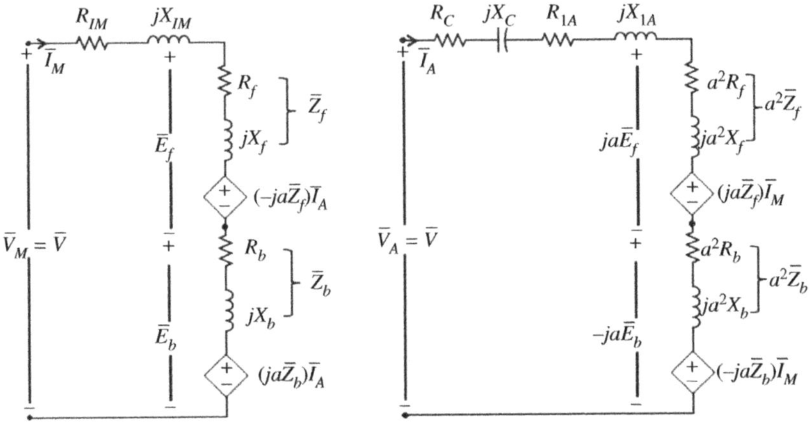

Figure 11.8 Equivalent circuit of a capacitor-start induction motor. (a) Only the main winding is energized and (b) only the auxiliary winding is energized.

By substituting Eq. (11.57) into (11.58), one can further express the developed torque as follows:

$$T_{dev} = T_{em} = \frac{Re\left[(\overline{E}_f - \overline{E}_b)\overline{I}_M^{\,*} + ja\,(\overline{E}_f + \overline{E}_b)\overline{I}_A^{\,*}\right]}{\omega_{syn}} \tag{11.59}$$

Thus, the equivalent circuit equations in Matrix form with $\overline{V}_M = \overline{V}_A = \overline{V}$ become as follows:

$$\begin{bmatrix} (R_{1m} + jX_{1m} + \overline{Z}_f + \overline{Z}_b) & ja\,(\overline{Z}_b - \overline{Z}_f) \\ ja\,(\overline{Z}_f - \overline{Z}_b) & (R_{1A} + jX_{1A} + R_c + jX_c + a^2(\overline{Z}_f + \overline{Z}_b)) \end{bmatrix} \begin{bmatrix} \overline{I}_M \\ \overline{I}_A \end{bmatrix} = \begin{bmatrix} \overline{V} \\ \overline{V} \end{bmatrix} \tag{11.60}$$

Now, we are ready for another numerical example as follows:

Example 11.2
The following are the equivalent-circuit parameters for the main and auxiliary windings of a 0.25-hp, 115-V, 60-Hz, 4-pole, capacitor motor of the permanent capacitor split-phase type:

$$a = \frac{N_A}{N_M} = 1.18, R_{1m} = 2.20\ \Omega, jX_{1m} = j\,3.05\ \Omega, jX_m = jX_\emptyset = j\,73.0\ \Omega$$

$$R_{1A} = 7.80\ \Omega, jX_{1A} = j\,3.52\ \Omega, R_2 = 4.5\ \Omega, jX_2 = j\,2.32\ \Omega$$

$$R_c = 2.00\ \Omega, jX_c = -j180\ \Omega$$

The rotational losses (i.e., friction, windage, and core losses) are 38 W.

At a slip, $s = 0.06$ p.u., calculate the following:

(a) The currents in the main and auxiliary windings.
(b) The line current.
(c) The motor power factor.
(d) The voltage across the capacitor.
(e) The developed torque.
(f) The mechanical output power.

Solution

Here, for this motor, the voltage across the main winding equals to the voltage across the auxiliary winding:

$$\overline{V}_M = \overline{V}_A = 115\angle 0^\circ \text{ V}$$

$$R_{1m} + jX_{1m} = 2.20 + j\,3.05\ \Omega$$

$$\overline{Z}_f = \frac{j\left(\frac{X_\phi}{2}\right)\left[\frac{R_2}{s} + j\frac{X_2}{2}\right]}{\left(\frac{R_2}{s}\right) + j\left(\frac{X_\phi}{2} + \frac{X_2}{2}\right)} = \frac{j\left(\frac{73}{2}\right)\left[\frac{4.5}{0.06} + j2.32\right]}{\left(\frac{4.5}{0.06}\right) + j(73 + 2.32)}$$

$$= 25.7\angle 46.6^\circ\ \Omega = 17.67 + j18.7\ \Omega$$

$$\overline{Z}_b = \frac{j\left(\frac{X_\phi}{2}\right)\left[\frac{R_2}{2(2-s)} + j\frac{X_2}{2}\right]}{\left(\frac{R_2}{2(2-s)}\right) + j\left(\frac{X_\phi}{2} + \frac{X_2}{2}\right)} = \frac{j(36.5)[2.32 + j2.32]}{(2.32) + j(75.32)}$$

$$= 1.59\angle 46.8^\circ\ \Omega = 1.09 + j\,1.16\ \Omega$$

Thus,

$$R_{1m} + jX_{1m} + \overline{Z}_f + \overline{Z}_b = 31.1\angle 47.55^\circ\ \Omega = 20.96 + j22.91\Omega$$

$$ja(\overline{Z}_f - \overline{Z}_b) = j1.18(16.58 + j17.54)$$

$$= -20.67 + j19.55\ \Omega$$

$$= -28.5\angle -43.4^\circ\ \Omega$$

$$(R_{1A} + jX_{1A} + R_c + jX_c + a^2(\overline{Z}_f + \overline{Z}_b))$$

$$= 7.80 + j3.52 + 2.00 - j180 + (1.18)^2(18.76 + j19.86)$$

$$= 35.95 - j148.8 = 153.0\angle -76.4^\circ\ \Omega$$

$$\begin{bmatrix} 31.1\angle 47.55^\circ & 28.5\angle -43.4^\circ \\ -28.5\angle -43.4^\circ & 153.0\angle -76.4^\circ \end{bmatrix}\begin{bmatrix} \overline{I}_M \\ \overline{I}_A \end{bmatrix} = \begin{bmatrix} 115\angle 0^\circ \\ 115\angle 0^\circ \end{bmatrix}$$

By solving this matrix, the currents in the main winding and the auxiliary winding can be calculated as follows:

(a) $$\bar{I}_M = 1.95 - j2.085 = 2.86\angle - 47^\circ\ A$$

$$\bar{I}_A = 0.695 + j0.612 = 0.92\angle 40.9^\circ\ A$$

(b) $$\bar{I} = \bar{I}_M + \bar{I}_A = 2.65 - j1.47 = 3.03\angle - 29.1^\circ\ A$$

(c) $$PF = \cos(29.1^\circ) = 0.875\ lagging$$

However, one can write the following:

$$P_{gf} = Re\left\{\bar{E}_f\bar{I}_M^{\ *} + ja\bar{E}_f\bar{I}_A^{\ *}\right\}$$

$$P_{gb} = Re\left\{\bar{E}_b\bar{I}_M^{\ *} - ja\bar{E}_b\bar{I}_A^{\ *}\right\}$$

$$P_g = P_{gf} - P_{gb}$$

$$P_{dev} = (1 - s)P_g$$

$$T_{dev} = \frac{P_{dev}}{\omega_{rotor}} = \frac{(1 - s)P_g}{(1 - s)\omega_{syn}} = \frac{P_g}{\omega_{syn}}$$

$$P_{out} = P_{dev} - P_{rotational}$$

Here,

$$\bar{E}_f = (\bar{Z}_f\bar{I}_M - ja\,\bar{Z}_f\bar{I}_A)$$

$$ja\,\bar{E}_f = (a^2\,\bar{Z}_f\bar{I}_A + ja\,\bar{Z}_f\bar{I}_M)$$

$$\bar{E}_b = (\bar{Z}_b\bar{I}_M + ja\,\bar{Z}_b\bar{I}_A)$$

$$-ja\,\bar{E}_b = (a^2\,\bar{Z}_b\bar{I}_A - ja\,\bar{Z}_b\bar{I}_M)$$

(d) The voltage across the capacitor $= \|(2.00 - j180) \times 0.92\angle 40.9^\circ\|$

$= 165.61$ V

(e) $$P_g = Re\left[\bar{Z}_f\bar{I}_M - ja\,\bar{Z}_f\bar{I}_A)\bar{I}_M^{\ *} + ja\,(\bar{Z}_f\bar{I}_M - ja\,\bar{Z}_f\bar{I}_A)\bar{I}_A^{\ *}\right]$$

$$P_g = Re\left[\bar{Z}_f I_M^{\ 2} + ja\,\bar{Z}_f\left(\bar{I}_M\bar{I}_A^{\ *} - \bar{I}_M^{\ *}\bar{I}_A\right) + a^2\,\bar{Z}_f I_A^{\ 2}\right]$$

$$P_g = 271\ W, T_{dev} = \frac{P_g}{\omega_{syn}}$$

(f) The developed mechanical power, P_{dev},

$$P_{dev} = (1 - s)P_g = (1 - 0.06) \times 271 = 254.4\ \text{W}$$

$$P_{mech\ out} = P_{dev} - P_{rotational} = 254.4 - 38 = 216.5\ \text{W}$$

11.6 Switched Reluctance Motors

Differing from the operating principle of induction motors or PMSMs, where torque is produced by the interaction of two magnetic fields (one by the stator and another by the rotor), a switched reluctance motor (SRM) produces torque based on variable reluctance in the air gap between the rotor and the stator. Specifically, when an SRM stator winding is energized, producing a single magnetic field, reluctance torque is developed by the tendency of the rotor to move to its minimum reluctance position, which is to align with a stator tooth to minimize the airgap in between. The formation of the rotating magnetic field can be achieved by switching the power electronic circuits in a predetermined continuous sequence. By sequentially switching phase currents into the stator windings at a commanded frequency, the rotor will be rotated at a desired speed.

The construction of a four-phase switched reluctance motor is shown in Figure 11.9. This motor includes eight stator poles and six rotor poles. The construction of the stator can be achieved by using silicon steel laminated inside the projected poles, wound by copper windings. Traditional SRMs have an even number of poles within the stator which have field coils. When the poles are opposite, then the field coils will be connected in series. Thus, their MMFs are additive and the associated coils are called phase windings. A set of coils can comprise one phase winding of the SRM. Each winding can be connected to the

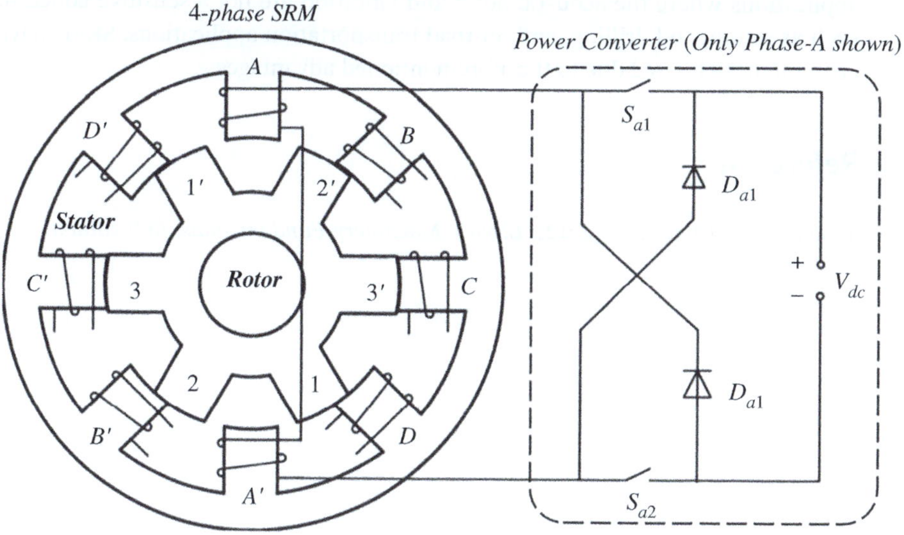

Figure 11.9 Topology of a four-phase-switched reluctance motor with eight poles in the stator and six poles in the rotor.

motor terminal and these are properly connected toward the opposite terminals of a power electronic circuit. The construction of the rotor can be implemented with silicon steel stampings through externally projected poles. The rotor of an SRM generally has an even number of poles, requiring no winding coils, permanent magnets, commutators, or brushes. In most of the existing motors, the rotor poles are based on the number of stator poles. The shaft of the rotor generally holds a position sensor and provides the rotor position signals, so the operation of various semiconductor switches in the power electronic circuitry is mainly controlled through the rotor position signals. In SRMs, both the stator and rotor include projected poles that are constructed with a soft iron as well as silicon steel stampings which are used for reducing eddy current and hysteresis losses.

SRMs have multiple advantages, including rigid structure, low production cost, the absence of permanent magnets, excellent power-speed characteristics, and high reliability and robustness. Also, there are no DC-bus shoot-through faults due to the presence of the phase winding interconnected between the upper switch and the lower switch (see Figure 11.9 as an example). However, the vibration and acoustic noise is generally larger than induction motors and permanent magnet synchronous motors, due to the doubly salient structure. Another factor constraining the wide adoption of SRMs by industries is that, the traditional 3-phase half-bridge circuits that have been intensively used for induction motors and PMSMs unfortunately do not work for SRMs, which requires different circuit topologies and perhaps customized semiconductor packaging. For industrial applications where the acoustic noise and vibration are not a sensitive concern, such as mining, oil drilling, and off-road transportation applications, SRMs have been intensively used due to the aforementioned advantages.

Reference

1 Stephen, J. Chapman. (2012). *Electric Machinery Fundamentals, fifth edition.* McGraw Hill.

12

Emerging Applications, Technical Trends, and Challenges

Although the conventional technologies of electric machinery and power electronic drives have been well developed, the emerging applications with various operating conditions pose new technical challenges. Thus, it is necessary to understand these new applications, technical trends, and challenges, and, most importantly, further improve the performance of electric machines and drives based on the new applications' requirements. In this chapter, representative emerging applications, future technical trends, and challenges for electric machinery and drives will be briefly reviewed and summarized.

12.1 Emerging Applications

12.1.1 Transportation Electrifications

Transportation electrifications have received intensive attention in the recent decade, mainly due to the benefits of much higher energy efficiency, dramatically reduced or zero emission of greenhouse gas (GHG), and lower operation cost and acoustic noise, compared to the conventional vehicles based on internal combustion engines powered by fossil fuels. At present, many countries have set goals to achieve carbon-neutral sustainable economy in the coming decades, while the transportation sector accounts for a large percentage in GHG emissions. For instance, GHG emissions from transportation account for about 27% of total US GHG emissions, making it the largest contributor of US GHG emissions [1]. Thus, electrifying transportation equipment such as ground vehicles, airplanes, and shipboard propulsion is critical to reach the GHG reduction goals.

Driven by the ground transportation market, electric vehicles (EVs) have been well developed and commercialized by many automotive manufacturers in recent years, while increasing effort is still on-going to develop and

Electric Machinery and Drives: An Electromagnetics Perspective, First Edition.
Nabeel A. O. Demerdash, JiangBiao He, and Hao Chen.

Companion website: www.wiley.com/go/demerdash/electricmachineryanddrives

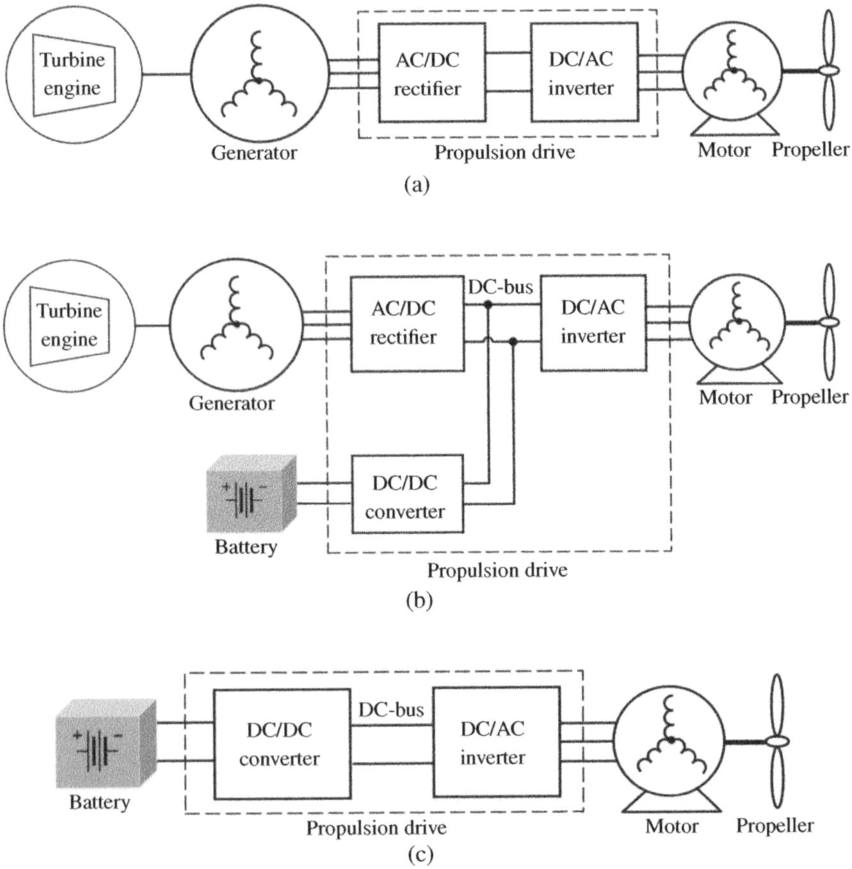

Figure 12.1 Power topologies for various transportation propulsion structures: (a) turbo-electric propulsion, (b) hybrid-electric propulsion, and (c) all-electric propulsion.

commercialize hybrid-electric or all-electric airplanes and ships, in which the operating environment, power capacity, and performance requirements are more challenging. Electric machinery and drives, being the core components of the propulsion drivetrain systems in the electrified transportation equipment, have a great impact on the performance of the overall propulsion systems. Figure 12.1a–c shows the basic power structures of (a) a turbo-electric propulsion system, (b) hybrid-electric propulsion system, and (c) all-electric propulsion system, for transportation electrifications, respectively. The eventual trend is to shift to all-electric propulsion and achieve zero carbon emission. However, currently, the limited energy density with battery energy storage is a technical bottleneck to develop large airplanes or ships, which is why turbo-electric and hybrid-electric propulsion are adopted for large transportation systems (e.g., above 5 MW systems in airplanes or ships).

In transportation electrification applications, the common performance goals for electric machinery and drives generally include high efficiency, high-power density, high reliability, and low cost. For fixed stored energy in the transportation vehicles, energy efficiency and power density mainly determine the driving range and payload capacity, while reliability and cost are also critical for safety and commercialization, respectively. How to concurrently meet these four performance goals requires thorough trade-off studies in multiphysics domains such as electrical, magnetic, thermal, dielectric insulation, and mechanical designs. For instance, to increase the power density of electric machinery, the fundamental operating frequency of the voltages outputting from the propulsion drives is preferred to be much higher than the utility grid frequency. In conventional aircraft power systems, the fundamental operating frequency for propulsion motors is 400 Hz, and, recently, there is a trend to increase the operating frequency to above 1 kHz to achieve higher power density of associated motors [2]. Accordingly, such high-speed motors require high switching frequencies of the associated power electronic drives, which becomes challenging for high-power drives. This is due to the increased switching losses as a result of such high switching frequencies, and, in particular, if there are strict requirements on efficiencies and reliability. Meanwhile, to achieve fast charging of the batteries and reduce the on-board cable weight and ohmic losses, higher DC voltages such as 800 V for EVs and medium voltage (e.g., multiple kilovolts) are recommended for electric airplanes and ships. Thus, using the emerging ultra-efficient wide bandgap semiconductor devices such as high-voltage SiC MOSFET modules and advanced multilevel power converter topologies might be a promising solution.

12.1.2 Renewable Power Generation

Electric machinery and drives also play critical roles in the drivetrain systems of renewable energy generation applications, specifically in onshore and offshore wind turbine systems, ocean wave energy generations, and hydropower generations. Figure 12.2 shows the roles of an electric generator and power electronic drive in a drivetrain system. Typically, a prime mover is usually mechanically connected to the generator with or without a gearbox, and the generated power with unregulated variable frequency will be rectified with an AC/DC converter, which is then inverted into AC power with a fixed grid frequency by a DC/AC inverter, before the AC power is integrated into the power grid through a step-up power transformer.

Similar to the aforementioned transportation applications, high efficiency, high reliability, and low cost are the key performance goals which should be

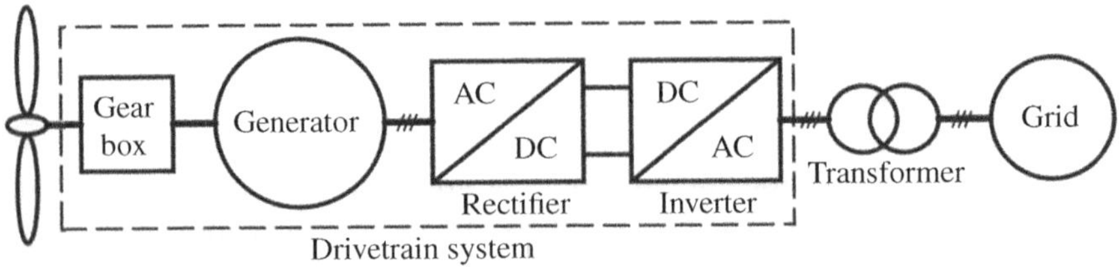

Figure 12.2 Generators and power converters in the drivetrain of renewable power generation systems (wind turbine, wave energy converter, or hydropower).

considered in the design stages of the drivetrains for these renewable power generation systems. However, the requirements on high-power density of drivetrains may not be so sensitive in these applications due to the outdoor nature of such installations, where physical dimensions and weight of the drivetrains are not as strict as in transportation vehicles. Generators used for wind turbine power generation mainly include permanent magnet synchronous machines and doubly-fed induction generators [3]. In ocean wave energy converter systems, linear permanent magnet or induction generators are generally used to convert the mechanical movements into electric energy [4]. For hydropower generation systems, wound-field synchronous generators have been dominating the market, mainly due to the fact that the stator magnetomotive force (MMF) and rotor MMF rotate at the synchronous speed corresponding to the given rated frequency, as well as the convenience of reactive power control, as explained in Chapter 7.

12.1.3 Robotics

Robotics has been a fast-growing area in recent decades, since robots are increasingly needed in many applications, including industrial automation, underwater, nuclear plants, national defense, space explorations, healthcare, entertainment, and agriculture. Most commercial robot arms use brushed or brushless DC motors, stepper motors, or servo motors, depending on the specific applications of the robots. The design consideration factors mainly consist of the following four aspects: mission profiles (rotational, linear, orthogonal, twisting, and revolving) of the robotic joints, torque–speed characteristics, motion precision requirements, and compliance standards of acoustic noise. For instance, in healthcare applications, servo motors might be preferred due to the precise position and speed control based on closed-loop feedback controls. For massive robots used in large warehouses to transport materials, a brushed or brushless DC motors might be preferred due to the lower cost and tolerance to acoustic noise. Regarding the robots used for space explorations, reliability of the motors and drives becomes much more important, due to the maintenance inconvenience and harsh operating environment (e.g., extreme temperatures and air pressure, and cosmic radiations).

12.1.4 Drones

Drones, also named as unmanned aerial vehicles (UAV), have been increasingly utilized in a broad range of applications, including surveillance, national defense, agriculture, forestry, delivery services, disaster rescue, oil and gas industries, and many others. As the core component in a drone system, propulsion motors is experiencing a dramatic market demand. According to [X], the drone motor market is projected to grow at a remarkable compound annual growth rate (CAGR) of 11.6%, ultimately reaching USD 15 billion by 2033. For drone motors, long flight range, high payload capacity, high reliability and resilience under various weather conditions, and low cost are the common performance demands in most applications. Accordingly, the technical requirements for drone motors include: high speed, high efficiency, compact size, light weight, fault tolerance, and low cost. At present, most of the drone motors in the market are brushless DC (BLDC) motors, due to their low cost, high efficiency, and simple control. However, for high-performance drones, PMSMs have received more attention due to higher efficiency, higher torque density, and lower acoustic noise, although the control might be more complicated and the hardware cost might be relatively higher compared to BLDC counterparts.

12.2 Future Technical Trends

The aforementioned emerging representative applications define future technical trends for electric machinery and drives, which will be briefly summarized as follows:

12.2.1 Advanced Materials

To increase the efficiency, power density, and reliability of electric machines, it is imperative to develop advanced materials for winding/coil conductors, magnetic core materials, and dielectric insulation materials, which will be elaborated in the following sections.

(1) *Advanced conductor materials with low resistance*: conductors are used extensively for machine armature or field windings, and the resistance and current-carrying capability of these conductors significantly affect the efficiency and power density of the machines. Currently, copper conductors have

been used for the windings/coils in almost every type of electric machines. However, the increase of resistance and decrease of current-carrying capability at high temperature for copper conductors require advanced conductor material, especially for high-power electric machines. Significant research activities are reported in the literature about developing carbon nanotubes (CNTs) conductors, or CNT-copper composite conductors, which are reported to have better thermal conductivity and structural stability, compared to copper counterparts [5]. Additionally, CNT conductors have much lower skin effect losses compared to copper counterparts [6], making them very attractive for high-speed machine designs. On the other hand, superconducting conductors, having zero resistance below critical temperature, have also received much attention over the past decades to develop compact high-efficiency electric machines. A number of megawatt-scale electric machines have been built and demonstrated based on high-temperature superconducting (HTS) windings where the conductors were operated at temperatures above 77 K [7]. However, due to the requirements of cryogenic and refrigerating systems, superconducting machines are only beneficial for high-power systems in terms of power density and cost, such as these above 8 MW [8].

(2) *Advanced magnetic materials*: at present, the common magnetic core materials used in electric machines are the class of ferromagnetic materials such as Iron (Fe), Nickel (Ni), and Cobalt (Co). Ferromagnetic materials are affordable in price, having good temperature characteristics, and are corrosion-resistant. However, ferromagnetic materials such as silicon steel laminations possess significant hysteresis and eddy current losses at high frequencies. Thus, developing low-loss and high-strength magnetic materials is indispensable to improve the performance of electric machines. For instance, soft magnetic composites (SMCs) have received increasing attention due to the characteristics of easily magnetizing and demagnetizing, high values of saturation magnetization, low coercivity and hysteresis losses, and high permeabilities [9].

On the other hand, there is a strong necessity to reduce the content of heavy rare earth material for permanent magnets, due to the increasing price and unsustainability of heavy rare earth materials, such as Dysprosium (Dy) and Terbium (Tb). Considerable effort has been reported in the literature to develop Dy-free magnets and higher-coercivity Alnico magnets, but the susceptibility to demagnetization and scalability for massive manufacturing still present concerns [9].

(3) *Advanced insulation material with great thermal conduction*: as the switching speed of modern semiconductor devices, such as SiC or GaN transistors, becomes faster and faster, the high switching frequencies and high dv/dt

cause substantial dielectric stress on winding insulation in electric machines. Particularly, developing advanced insulation material is of high necessity and significance in medium-voltage and high-voltage power conversion systems, which have become critical in the emerging renewable power generations, electric ship and airplane transportation vehicles, and others. The higher voltage stress, combined with high switching frequencies and high dv/dt, results in very challenging insulation stress to address for both electric machines and power electronics. In special applications, such as high-altitude aviation or spacecraft, the low air pressure incurs higher electric stress on motor winding insulation, which deserves extra attention during the design stage.

Common insulation materials used in conventional electric machines are composed of polymeric coatings, polymeric films, and special papers, such as polyimide (Kapton) and polyaramid (Nomex), which provide excellent electric insulation but poor thermal conduction. Such a drawback limits heat transfer and thus affects energy efficiency of such electric machines. In addition, they can easily degrade under electric discharge [10]. Thus, developing advanced insulation material with high thermal conductivity, good processability, and manufacturability will be of paramount importance to ensure high-power densities and high reliability of electric machinery and drives.

12.2.2 Advanced Manufacturing

In conventional manufacturing of electric machinery and drives, all the components are fabricated through a process of material removal, such as milling, based on computer-aided-design (CAD) drawings. Significant material waste, complicated production processes, and long lead times are the main drawbacks of such conventional manufacturing processes. Additive manufacturing, as a relatively new manufacturing methodology, also known as 3D printing or rapid prototyping, uses CAD software or 3D object scanners to direct hardware to deposit material, layer upon layer, in precise geometric shapes. Additive manufacturing opens up new space for the production of electric machinery and drives due to faster prototyping, reduced lead times, and shorter supply chains. It enables a tool-free production of components and entire assemblies directly from a CAD file. Presently, additive manufacturing technologies have not been broadly utilized in industrial production yet due to certain limitations (e.g., size limitations, post-processing, and high production cost), but these innovations are gaining more and more popularity as the technologies are becoming more mature. A comprehensive review of additive manufacturing of coils, magnetic core packs, permanent magnets, and mechanical and thermal management components is provided in Reference [11].

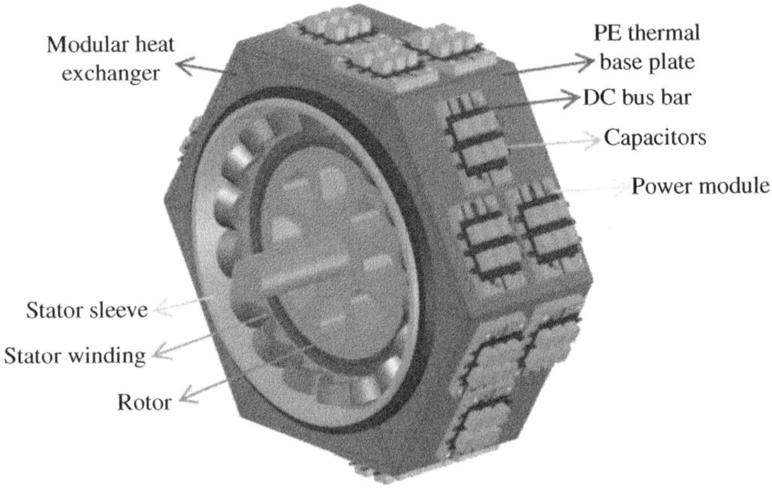

Figure 12.3 An integrated six-phase motor-drive concept. Source: Adapted from Jahns and Sarlioglu [12].

12.2.3 Integrated Motor-Drive Systems

Driven by the requirements of improving power densities and reducing the cost of electric machine-drive systems, integrated motor-drive (IMD) technologies have become very attractive in recent years. IMD is a structural integration of an electric machine with a drive as a single unit, so the machine and drive can share the mechanical frame and thermal management, and no power cables are needed to interconnect the machine and drive. A six-phase IMD concept is illustrated in Figure 12.3, according to Jahns and Sarlioglu [12]; IMD technology improves the systematic power density with 10–20% less volume and reduces 30–40% of the overall system costs of installation and manufacturing. The elimination of power cables between the machine and drive also helps reduce EMI and reflected surge voltage on the machine stator winding terminals. The advent of wide bandgap (WBG) semiconductor switches such as SiC and GaN transistors, in recent years, has facilitated the development of IMD due to smaller sizes of such WBG switches, in addition to higher operating temperature capability.

12.3 Technical Challenges

12.3.1 Extreme Operating Environmental Conditions

Designing electric machinery and power electronic drives for extreme operating environments has been a challenge. These extreme environmental conditions include extremely high or low temperatures, extremes of air pressure, humidity, and cosmic radiation to electronics at high altitudes. The related practical applications include designing machinery and drives for land rovers and robots used in space exploration, machines in special transportation vehicular systems such as generators located adjacent to aircraft main propulsion turbines, machines, and drives at undersea operating conditions, and many others. For instance, technologies for designing motors and drives for land rovers or robots in Venus space exploration are still not mature. The high surface temperature on Venus is about 900 °F (i.e., 475 °C), which poses dramatic challenges during the development of electric machinery and drive components, including windings/coils, magnets, bearings, insulation materials, semiconductor switches, sensors, and microprocessors. Particularly, designs of the following power components at high temperature deserve more effort:

- Almost all the commercial permanent magnet materials have maximum operating temperatures below 250 °C. Above this temperature, the magnets may be subject to irreversible demagnetization or damaging effects, and the magnetic property will degrade significantly.
- Copper winding/coil resistance will increase dramatically at high temperature.
- Machine winding insulation typically has a maximum operating temperature only below 180 °C.
- Conventional bearings made of steel and using organic lubricants, usually rated up to 82 °C (i.e., 180 °F), will need to be replaced by high-temperature-resistant elements.
- Most commercial semiconductor switches, including SiC and GaN transistors, have a maximum operating temperature no more than 175 °C.

Thus, to develop electric machines and drives for such extreme operating conditions, advanced materials, packaging, thermal management, and multiphysics synergistic designs are all required to guarantee the functionality and reliability of such systems.

12.3.2 Systematic Multiphysics Synergistic Designs

To achieve high efficiency, high-power density, and high reliability, multiphysics synergistic designs that concurrently consider the performance in electrical, magnetic, mechanical, thermal, dielectric, and material domains are essential. Considering only electrical, magnetic, and mechanical domains is challenging to achieve desired overall performance. For instance, when designing a high-speed electric machine for medium-voltage high-power propulsion applications, output torque and speed are mainly determined by the design parameters in the electrical, magnetic, and mechanical domains. However, thermal management and dielectric insulation design of the stators, rotors, and bearings also influence the power density and reliability of such motors. Additionally, systematic synergistic designs and analysis of both machines and drives in combination with the typical mission profiles are critical to achieve the optimal systematic performance.

12.3.3 Electromagnetic Interference

The emerging WBG semiconductor switches have been increasingly utilized in power converters over the past few years, in order to enhance efficiency and power density of adjustable speed drives. However, the fast-switching speed of the WBG switches induces notable high-frequency (MHz scale) reflected voltage stress on the stator windings, especially for machine-drive systems involving long cable connections, as shown in Figure 12.4. Figure 12.5 illustrates the relationship between the reflected voltage across the motor terminals and various semiconductors switching rise time and cable length [13]. As can be seen, the shorter the rise time of the semiconductor switches, or the longer the power cables between the drives and motors, the higher the reflected surge voltage stress across the motor windings. Furthermore, the reflected surge voltages are nonlinearly distributed across the windings, and the first few turns closer to the drive side generally share most of the high-frequency surge voltages and have more severe dielectric insulation stress [14]. If no mitigation solutions are available, such high surge voltages can be as high as twice of the DC-bus voltage of the inverters across the motor stator windings, which may damage the winding insulation and cause inter-turn or inter-coil short-circuit faults. This has become a pressing concern for safety-critical applications such as wind turbine generators, electric ships, and electric aircraft where long cables are used to interconnect the drives and machines, particularly, for the medium-voltage and high-voltage power conversion systems where the dv/dt might be dramatic. Traditional solutions to address such a problem are to design and install a dv/dt filter or sinewave passive filter at the output of the drive, but such a solution will reduce the systematic

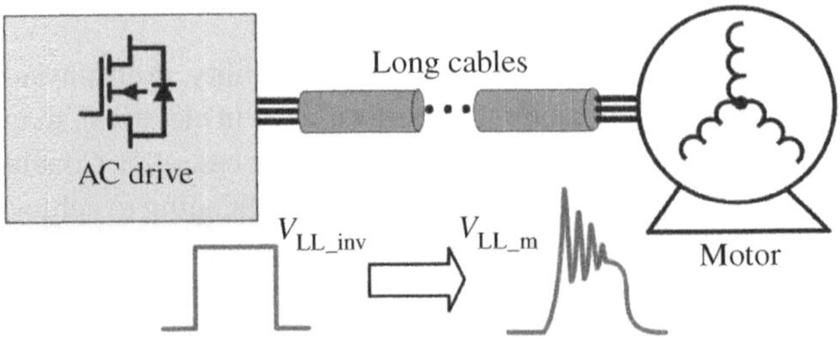

Figure 12.4 Reflected voltage in long-cable-fed motor-drive systems.

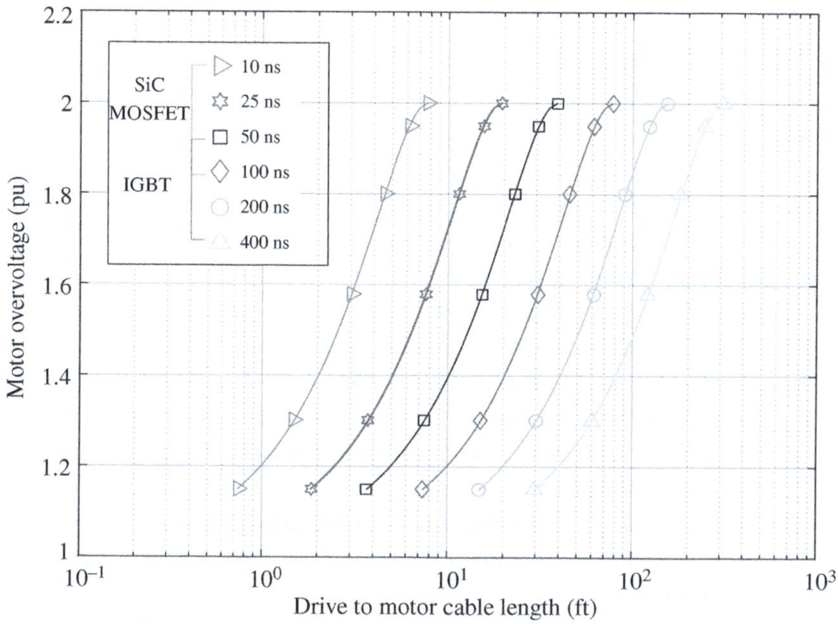

Figure 12.5 The voltage spike versus cable length and semiconductor rise time. Source: Yi et al. [13]/IEEE.

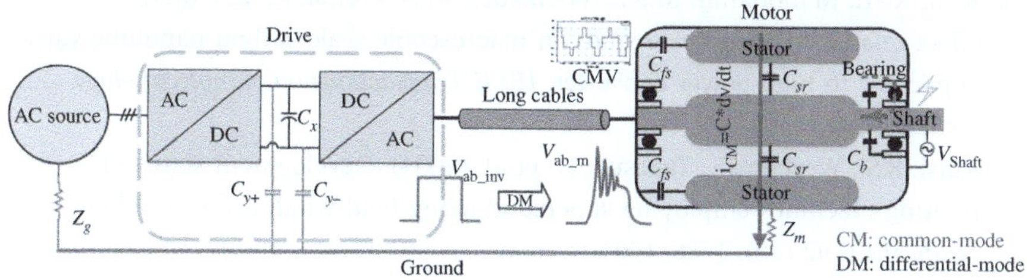

Figure 12.6 Differential-mode and common-mode EMI in a motor-drive system.

power density and efficiency, in addition to increasing the implementation cost due to a number of additional hardware components.

Moreover, the fast-switching of WBG switches in the drives may also induce high-frequency common-mode voltage and parasitic current in the bearings of the machines, resulting in premature bearing aging or failures in the machines, as shown in Figure 12.6. Likewise, the existing solutions to reduce the common-mode voltage and bearing current are to design passive common-mode filters, which will unfortunately decrease the systematic power density, efficiency, and increase the cost. Hence, advanced solutions are needed to mitigate the electromagnetic interference and protect such machine-drive systems from systematic tripping and downtime cost.

References

1 U.S. Environmental Protection Agency. (2022) Carbon pollution from transportation. https://www.epa.gov/transportation-air-pollution-and-climate-change (accessed 3 July 2022).

2 Benzaquen, J., He, J., and Mirafzal, B. (2021). Toward more electric powertrains in aircraft: technical challenges and advancements. *CES Transactions on Electrical Machines and Systems* 5 (3): 177–193.

3 Polinder, H., Ferreira, J.A., Jensen, B.B. et al. (2013). Trends in wind turbine generator systems. *IEEE Journal of Emerging and Selected Topics in Power Electronics* 1 (3): 174–185.

4 Rahman, A., Farrok, O., Islam, M.R., and Xu, W. (2020). Recent progress in electrical generators for oceanic wave energy conversion. *IEEE Access* 8: 138595–138615.

5 Lekawa-Raus, A., Patmore, J., Kurzepa, L. et al. (2014). Electrical properties of carbon nanotube based fibers and their future use in electrical wiring. *Advanced Functional Materials* 24 (24): 3661–3682.

6 Kim, K.-T., Manoharan, M.S., Abdelkader, M.A.T. et al. (2021). Skin effect-related AC resistance study in macroscopic scale carbon nanotube yarn applicable to high-power converter. *IEEE Transactions on Nanotechnology* 20: 417–424.

7 Kalsi, S.S., Weeber, K., Takesue, H. et al. (2004). Development status of rotating machines employing superconducting field windings. *Proceedings of the IEEE* 92 (10): 1688–1704.

8 Qu, R., Liu, Y., and Wang, J. (2013). Review of superconducting generator topologies for direct-drive wind turbines. *IEEE Transactions on Applied Superconductivity* 23 (3): 5201108.

9 El-Refaie, A. (2019). Role of advanced materials in electrical machines. *CES Transactions on Electrical Machines and Systems* 3 (2): 124–132.

10 Yin, W., Yakimov, A., Comanzo, H. et al. (2019). Highly thermally conductive insulation for high power density electric machines. *AIAA/IEEE Electric Aircraft Technologies Symposium (EATS)* 2019: 1–7.

11 Wrobel, R. and Mecrow, B. (2020). A comprehensive review of additive manufacturing in construction of electrical machines. *IEEE Transactions on Energy Conversion* 35 (2): 1054–1064.

12 Jahns, T.M. and Sarlioglu, B. (2020). The incredible shrinking motor drive: accelerating the transition to integrated motor drives. *IEEE Power Electronics Magazine* 7 (3): 18–27.

13 Yi, P., Murthy, P.K.S., and Wei, L. (2016). Performance evaluation of SiC MOSFETs with long power cable and induction motor. *IEEE Energy Conversion Congress and Exposition (ECCE)* 2016: 1–7.

14 He, J., Li, C., Jassal, A. et al. (2019). Multi-domain design optimization of dv/dt filter for SiC-based three-phase inverters in high-frequency motor-drive applications. *IEEE Transactions on Industry Applications* 55 (5): 5214–5222.

15 Future Market Insights Global and Consulting Pvt. Ltd. (2024). Drone Motors Market is Predicted to Reach USD 15 Billion, November 6.

Index

Electric Machinery and Drives: An Electromagnetics Perspective, First Edition.
Nabeel A. O. Demerdash, JiangBiao He, and Hao Chen.

Companion website: www.wiley.com/go/demerdash/electricmachineryanddrives

l

m

n

o

p

q

r

s

IEEE Press Series on Power and Energy Systems

Series Editor: Ganesh Kumar Venayagamoorthy, Clemson University, Clemson, South Carolina, USA.

The mission of the IEEE Press Series on Power and Energy Systems is to publish leading-edge books that cover a broad spectrum of current and forward-looking technologies in the fast-moving area of power and energy systems including smart grid, renewable energy systems, electric vehi- cles and related areas. Our target audience includes power and energy systems professionals from academia, industry and government who are interested in enhancing their knowledge and per- spectives in their areas of interest.

1. *Electric Power Systems: Design and Analysis, Revised Printing*
   Mohamed E. El-Hawary
2. *Power System Stability*
   Edward W. Kimbark
3. *Analysis of Faulted Power Systems*
   Paul M. Anderson
4. *Inspection of Large Synchronous Machines: Checklists, Failure Identification, and Troubleshooting*
   Isidor Kerszenbaum
5. *Electric Power Applications of Fuzzy Systems*
   Mohamed E. El-Hawary
6. *Power System Protection*
   Paul M. Anderson
7. *Subsynchronous Resonance in Power Systems*
   Paul M. Anderson, B.L. Agrawal, J.E. Van Ness
8. *Understanding Power Quality Problems: Voltage Sags and Interruptions*
   Math H. Bollen
9. *Analysis of Electric Machinery*
   Paul C. Krause, Oleg Wasynczuk, and S.D. Sudhoff
10. *Power System Control and Stability, Revised Printing*
    Paul M. Anderson, A.A. Fouad
11. *Principles of Electric Machines with Power Electronic Applications, Second Edition*
    Mohamed E. El-Hawary

12. *Pulse Width Modulation for Power Converters: Principles and Practice*
    D. Grahame Holmes and Thomas Lipo
13. *Analysis of Electric Machinery and Drive Systems, Second Edition*
    Paul C. Krause, Oleg Wasynczuk, and S.D. Sudhoff
14. *Risk Assessment for Power Systems: Models, Methods, and Applications*
    Wenyuan Li
15. *Optimization Principles: Practical Applications to the Operations of Markets of the Electric Power Industry*
    Narayan S. Rau
16. *Electric Economics: Regulation and Deregulation*
    Geoffrey Rothwell and Tomas Gomez
17. *Electric Power Systems: Analysis and Control*
    Fabio Saccomanno
18. *Electrical Insulation for Rotating Machines: Design, Evaluation, Aging, Testing, and Repair*
    Greg C. Stone, Edward A. Boulter, Ian Culbert, and Hussein Dhirani
19. *Signal Processing of Power Quality Disturbances*
    Math H. J. Bollen and Irene Y. H. Gu
20. *Instantaneous Power Theory and Applications to Power Conditioning*
    Hirofumi Akagi, Edson H. Watanabe, and Mauricio Aredes
21. *Maintaining Mission Critical Systems in a 24/7 Environment*
    Peter M. Curtis
22. *Elements of Tidal-Electric Engineering*
    Robert H. Clark
23. *Handbook of Large Turbo-Generator Operation and Maintenance, Second Edition*
    Geoff Klempner and Isidor Kerszenbaum
24. *Introduction to Electrical Power Systems*
    Mohamed E. El-Hawary
25. *Modeling and Control of Fuel Cells: Distributed Generation Applications*
    M. Hashem Nehrir and Caisheng Wang
26. *Power Distribution System Reliability: Practical Methods and Applications*
    Ali A. Chowdhury and Don O. Koval
27. *Introduction to FACTS Controllers: Theory, Modeling, and Applications*
    Kalyan K. Sen, Mey Ling Sen
28. *Economic Market Design and Planning for Electric Power Systems*
    James Momoh and Lamine Mili
29. *Operation and Control of Electric Energy Processing Systems*
    James Momoh and Lamine Mili

30. *Restructured Electric Power Systems: Analysis of Electricity Markets with Equilibrium Models*
Xiao-Ping Zhang
31. *An Introduction to Wavelet Modulated Inverters*
S.A. Saleh and M.A. Rahman
32. *Control of Electric Machine Drive Systems*
Seung-Ki Sul
33. *Probabilistic Transmission System Planning*
Wenyuan Li
34. *Electricity Power Generation: The Changing Dimensions*
Digambar M. Tagare
35. *Electric Distribution Systems*
Abdelhay A. Sallam and Om P. Malik
36. *Practical Lighting Design with LEDs*
Ron Lenk and Carol Lenk
37. *High Voltage and Electrical Insulation Engineering*
Ravindra Arora and Wolfgang Mosch
38. *Maintaining Mission Critical Systems in a 24/7 Environment, Second Edition*
Peter Curtis
39. *Power Conversion and Control of Wind Energy Systems*
Bin Wu, Yongqiang Lang, Navid Zargari, and Samir Kouro
40. *Integration of Distributed Generation in the Power System*
Math H. Bollen and Fainan Hassan
41. *Doubly Fed Induction Machine: Modeling and Control for Wind Energy Generation Applications*
Gonzalo Abad, Jesús López, Miguel Rodrigues, Luis Marroyo, and Grzegorz Iwanski
42. *High Voltage Protection for Telecommunications*
Steven W. Blume
43. *Smart Grid: Fundamentals of Design and Analysis*
James Momoh
44. *Electromechanical Motion Devices, Second Edition*
Paul Krause Oleg, Wasynczuk, and Steven Pekarek
45. *Electrical Energy Conversion and Transport: An Interactive Computer-Based Approach, Second Edition*
George G. Karady and Keith E. Holbert
46. *ARC Flash Hazard and Analysis and Mitigation*
J.C. Das
47. *Handbook of Electrical Power System Dynamics: Modeling, Stability, and Control*
Mircea Eremia and Mohammad Shahidehpour

48. *Analysis of Electric Machinery and Drive Systems, Third Edition*
Paul C. Krause, Oleg Wasynczuk, S.D. Sudhoff, and Steven D. Pekarek
49. *Extruded Cables for High-Voltage Direct-Current Transmission: Advances in Research and Development*
Giovanni Mazzanti and Massimo Marzinotto
50. *Power Magnetic Devices: A Multi-Objective Design Approach*
S.D. Sudhoff
51. *Risk Assessment of Power Systems: Models, Methods, and Applications, Second Edition*
Wenyuan Li
52. *Practical Power System Operation*
Ebrahim Vaahedi
53. *The Selection Process of Biomass Materials for the Production of Bio-Fuels and Co-Firing*
Najib Altawell
54. *Electrical Insulation for Rotating Machines: Design, Evaluation, Aging, Testing, and Repair, Second Edition*
Greg C. Stone, Ian Culbert, Edward A. Boulter, and Hussein Dhirani
55. *Principles of Electrical Safety*
Peter E. Sutherland
56. *Advanced Power Electronics Converters: PWM Converters Processing AC Voltages*
Euzeli Cipriano dos Santos Jr. and Edison Roberto Cabral da Silva
57. *Optimization of Power System Operation, Second Edition*
Jizhong Zhu
58. *Power System Harmonics and Passive Filter Designs*
J.C. Das
59. *Digital Control of High-Frequency Switched-Mode Power Converters*
Luca Corradini, Dragan Maksimoviæ, Paolo Mattavelli, and Regan Zane
60. *Industrial Power Distribution, Second Edition*
Ralph E. Fehr, III
61. *HVDC Grids: For Offshore and Supergrid of the Future*
Dirk Van Hertem, Oriol Gomis-Bellmunt, and Jun Liang
62. *Advanced Solutions in Power Systems: HVDC, FACTS, and Artificial Intelligence*
Mircea Eremia, Chen-Ching Liu, and Abdel-Aty Edris
63. *Operation and Maintenance of Large Turbo-Generators*
Geoff Klempner and Isidor Kerszenbaum
64. *Electrical Energy Conversion and Transport: An Interactive Computer-Based Approach*
George G. Karady and Keith E. Holbert

65. *Modeling and High-Performance Control of Electric Machines*
John Chiasson
66. *Rating of Electric Power Cables in Unfavorable Thermal Environment*
George J. Anders
67. *Electric Power System Basics for the Nonelectrical Professional*
Steven W. Blume
68. *Modern Heuristic Optimization Techniques: Theory and Applications to Power Systems*
Kwang Y. Lee and Mohamed A. El-Sharkawi
69. *Real-Time Stability Assessment in Modern Power System Control Centers*
Savu C. Savulescu
70. *Optimization of Power System Operation*
Jizhong Zhu
71. *Insulators for Icing and Polluted Environments*
Masoud Farzaneh and William A. Chisholm
72. *PID and Predictive Control of Electric Devices and Power Converters Using MATLAB®/Simulink®*
Liuping Wang, Shan Chai, Dae Yoo, Lu Gan, and Ki Ng
73. *Power Grid Operation in a Market Environment: Economic Efficiency and Risk Mitigation*
Hong Chen
74. *Electric Power System Basics for Nonelectrical Professional, Second Edition*
Steven W. Blume
75. *Energy Production Systems Engineering Thomas*
Howard Blair
76. *Model Predictive Control of Wind Energy Conversion Systems*
Venkata Yaramasu and Bin Wu
77. *Understanding Symmetrical Components for Power System Modeling*
J.C. Das
78. *High-Power Converters and AC Drives, Second Edition*
Bin Wu and Mehdi Narimani
79. *Current Signature Analysis for Condition Monitoring of Cage Induction Motors: Industrial Application and Case Histories*
William T. Thomson and Ian Culbert
80. *Introduction to Electric Power and Drive Systems*
Paul Krause, Oleg Wasynczuk, Timothy O'Connell, and Maher Hasan
81. *Instantaneous Power Theory and Applications to Power Conditioning, Second Edition*
Hirofumi, Edson Hirokazu Watanabe, and Mauricio Aredes
82. *Practical Lighting Design with LEDs, Second Edition*
Ron Lenk and Carol Lenk

83. *Introduction to AC Machine Design*
Thomas A. Lipo
84. *Advances in Electric Power and Energy Systems: Load and Price Forecasting*
Mohamed E. El-Hawary
85. *Electricity Markets: Theories and Applications*
Jeremy Lin and Jernando H. Magnago
86. *Multiphysics Simulation by Design for Electrical Machines, Power Electronics and Drives*
Marius Rosu, Ping Zhou, Dingsheng Lin, Dan M. Ionel, Mircea Popescu, Frede Blaabjerg, Vandana Rallabandi, and David Staton
87. *Modular Multilevel Converters: Analysis, Control, and Applications*
Sixing Du, Apparao Dekka, Bin Wu, and Navid Zargari
88. *Electrical Railway Transportation Systems*
Morris Brenna, Federica Foiadelli, and Dario Zaninelli
89. *Energy Processing and Smart Grid*
James A. Momoh
90. *Handbook of Large Turbo-Generator Operation and Maintenance, 3rd Edition*
Geoff Klempner and Isidor Kerszenbaum
91. *Advanced Control of Doubly Fed Induction Generator for Wind Power Systems*
Dehong Xu, Dr. Frede Blaabjerg, Wenjie Chen and Nan Zhu
92. *Electric Distribution Systems, 2nd Edition*
Abdelhay A. Sallam and Om P. Malik
93. *Power Electronics in Renewable Energy Systems and Smart Grid: Technology and Applications*
Bimal K. Bose
94. *Distributed Fiber Optic Sensing and Dynamic Rating of Power Cables*
Sudhakar Cherukupalli and George J Anders
95. *Power System and Control and Stability, Third Edition*
Vijay Vittal, James D. McCalley, Paul M. Anderson, and A. A. Fouad
96. *Electromechanical Motion Devices: Rotating Magnetic Field-Based Analysis and Online Animations, Third Edition*
Paul Krause, Oleg Wasynczuk, Steven D. Pekarek, and Timothy O'Connell
97. *Applications of Modern Heuristic Optimization Methods in Power and Energy Systems*
Kwang Y. Lee and Zita A. Vale
98. *Handbook of Large Hydro Generators: Operation and Maintenance*
Glenn Mottershead, Stefano Bomben, Isidor Kerszenbaum, and Geoff Klempner
99. *Advances in Electric Power and Energy: Static State Estimation*
Mohamed E. El-hawary

100. *Arc Flash Hazard Analysis and Mitigation, Second Edition*
J.C. Das
101. *Maintaining Mission Critical Systems in a 24/7 Environment, Third Edition*
Peter M. Curtis
102. *Real-Time Electromagnetic Transient Simulation of AC-DC Networks*
Venkata Dinavahi and Ning Lin
103. *Probabilistic Power System Expansion Planning with Renewable Energy Resources and Energy Storage Systems*
Jaeseok Choi and Kwang Y. Lee
104. *Power Magnetic Devices: A Multi-Objective Design Approach, Second Edition*
Scott D. Sudhoff
105. *Optimal Coordination of Power Protective Devices with Illustrative Examples*
Ali R. Al-Roomi
106. *Resilient Control Architectures and Power Systems*
Craig Rieger, Ronald Boring, Brian Johnson, and Timothy McJunkin
107. *Alternative Liquid Dielectrics for High Voltage Transformer Insulation Systems: Performance Analysis and Applications*
Edited by U. Mohan Rao, I. Fofana, and R. Sarathi
108. *Introduction to the Analysis of Electromechanical Systems*
Paul C. Krause, Oleg Wasynczuk, and Timothy O'Connell
109. *Power Flow Control Solutions for a Modern Grid using SMART Power Flow Controllers*
Kalyan K. Sen and Mey Ling Sen
110. *Power System Protection: Fundamentals and Applications*
John Ciufo and Aaron Cooperberg
111. *Soft-Switching Technology for Three-phase Power Electronics Converters*
Dehong Xu, Rui Li, Ning He, Jinyi Deng, and Yuying Wu
112. *Power System Protection, Second Edition*
Paul M. Anderson, Charles Henville, Rasheek Rifaat, Brian Johnson, and Sakis Meliopoulos
113. *High Voltage and Electrical Insulation Engineering, Second Edition*
Ravindra Arora and Wolfgang Mosch
114. *Modeling and Control of Modern Electrical Energy Systems*
Masoud Karimi-Ghartemani
115. *Control of Power Electronic Converters with Microgrid Applications*
Armdam Ghosh and Firuz Zare
116. *Coordinated Operation and Planning of Modern Heat and Electricity Incorporated Networks*
Mohammadreza Daneshvar, Behnam Mohammadi-Ivatloo, and Kazem Zare

117. *Smart Energy for Transportation and Health in a Smart City*
Chun Sing Lai, Loi Lei Lai, and Qi Hong Lai
118. *Wireless Power Transfer: Principles and Applications*
Zhen Zhang and Hongliang Pang
119. *Intelligent Data Mining and Analysis in Power and Energy Systems: Models and Applications for Smarter Efficient Power Systems*
Zita Vale, Tiago Pinto, Michael Negnevitsky, and Ganesh Kumar Venayagamoorthy
120. *Introduction to Modern Analysis of Electric Machines and Drives*
Paul C. Krause and Thomas C. Krause
121. *Electromagnetic Analysis and Condition Monitoring of Synchronous Generators*
Hossein Ehya and Jawad Faiz
122. *Transportation Electrification: Breakthroughs in Electrified Vehicles, Aircraft, Rolling Stock, and Watercraft*
Ahmed A. Mohamed, Ahmad Arshan Khan, Ahmed T. Elsayed, and Mohamed A. Elshaer
123. *Modular Multilevel Converters: Control, Fault Detection, and Protection*
Fuji Deng, Chengkai Liu, and Zhe Chen
124. *Stability-Constrained Optimization for Modern Power System Operation and Planning*
Yan Xu, Yuan Chi, and Heling Yuan
125. *Interval Methods for Uncertain Power System Analysis*
Alfredo Vaccaro
126. *Practical Partial Discharge Measurement on Electrical Equipment*
Greg C. Stone, Andrea Cavallini, Glenn Behrmann, and Claudio Angelo Serafino
127. *Graph Database and Graph Computing for Power System Analysis*
Renchang Dai and Guangyi Liu
128. *Microgrids: Theory and Practice*
Edited by Peng Zhang
129. *Electric Machinery and Drives: An Electromagnetics Perspective*
Nabeel A. O. Demerdash, JiangBiao He, and Hao Chen

Printed and bound by CPI Group (UK) Ltd, Croydon, CR0 4YY
17/09/2026
14964255-0006